Linear Models and the Relevant Distributions and Matrix Algebra

Linear Models and the Relevant Distributions and Matrix Algebra: A Unified Approach, Volume 2 covers several important topics that were not included in the first volume. The second volume complements the first, providing detailed solutions to the exercises in both volumes, thereby greatly enhancing its appeal for use in advanced statistics programs. This volume can serve as a valuable reference. It can also serve as a resource in a mathematical statistics course for use in illustrating various theoretical concepts in the context of a relatively complex setting of great practical importance. Together with the first volume, this volume provides a largely self-contained treatment of an important area of statistics and should prove highly useful to graduate students and others.

Key Features:

- Includes solutions to the exercises from both the first and second volumes
- Includes coverage of several topics not covered in the first volume
- Highly valuable as a reference book for graduate students and researchers

David A. Harville served for 10 years as a mathematical statistician in the Applied Mathematics Research Laboratory of the Aerospace Research Laboratories (at Wright-Patterson AFB, Ohio), for 20 years as a full professor in Iowa State University's Department of Statistics (where he now has emeritus status), and 7 years as a research staff member of the Mathematical Sciences Department of IBM's T.J. Watson Research Center. He has extensive experience in the area of linear statistical models, having taught (on numerous occasions) M.S. and Ph.D. level courses on that subject, having been the thesis advisor of 10 Ph.D. graduates, and having authored (or co-authored) 3 books and more than 80 research articles. His work has been recognized through his election as a Fellow of the American Statistical Association and of the Institute of Mathematical Statistics and as a member of the International Statistical Institute.

Linear Models and the Relevant Distributions and Matrix Algebra

A Unified Approach

Volume 2

David A. Harville

CRC Press is an imprint of the
Taylor & Francis Group, an **informa** business
A CHAPMAN & HALL BOOK

Designed cover image: © David A. Harville

First edition published 2024
by CRC Press
6000 Broken Sound Parkway NW, Suite 300, Boca Raton, FL 33487-2742

and by CRC Press
4 Park Square, Milton Park, Abingdon, Oxon, OX14 4RN

CRC Press is an imprint of Taylor & Francis Group, LLC

ISBN: 978-1-032-47123-5 (hbk)
ISBN: 978-1-032-47165-5 (pbk)
ISBN: 978-1-003-38487-8 (ebk)

DOI: 10.1201/9781003384878

Typeset in Nimbus font
by KnowledgeWorks Global Ltd.

Publisher's note: This book has been prepared from camera-ready copy provided by the authors.

Contents

Preface (to Volume 2)

The present volume serves two purposes. First, it provides coverage of some important topics that (for lack of space) were not covered in the first volume. And second, it provides solutions to the exercises, both those from the first volume and those from the present volume—for the convenience of the reader, a format is adopted in which each solution is coupled with a restatement of the exercise.

The approach taken in the present volune is consistent with that taken in the first volume; the first volume has a number of distinctive features including the following.

- As suggested by its title (which is the same as the title of the present volume), the first volume includes coverage of the relevant distributions and matrix algebra; the coverage is more-or-less self-contained and some parts of it are interspersed or integrated with the coverage of the results on linear models. A great deal of the coverage is of a kind that could be characterized as unified.
- The coverage of results on linear models expands on the traditional by including coverage of such topics as extensions to spherically symmetric distributions, false discovery rates, and Monte Carlo determination of percentage points.
- An attempt is made to bridge the gap between the kinds of theoretical developments that may be presented in a course on linear models and those that may be presented in a course on mathematical statistics.
- Proofs are included for almost "everything."
- An attempt is made to keep the presentation at a level (and to include enough details) that the results are accessible to a "relatively broad audience."
- Real data are included as examples and are used for purposes of illustration.
- The volume includes a large number of exercises.
- To the greatest extent "practical," an attempt is made to achieve a presentation that allows the various parts of the volume to be accessed independently of each other.
- The volume includes a considerable amount of material pertaining to predictive inference.

The downside of a volume with such features is that the length of the volume becomes a limiting factor on the number of topics that can be covered. In the case of the first volume of *Linear Models and the Relevant Distributions and Matrix Algebra*, the inclusion of the various features was achieved by excluding from the coverage some mainstream topics that would typically be among the topics covered in a course on linear models. Several of those topics are covered in Part I of Volume 2 including the topic of constrained linear models and various related topics such as the analysis of variance, regression splines, and reparameterization. Others represent potential topics for future editions of Volume 2. Among the potential future topics would be the topic of hierarchical linear models and their various uses including their use (via the introduction of a latent variable) as a basis for making inferences from ordinal data.

The presentation of material in Volume 2 is premised on an assumption that a reader of this volume has access to the first volume (and to the results and the list of references found therein).

In fact, the new material that makes up Part I has been regarded (for purposes of numbering and formatting) as though it forms an additional (eighth) chapter of the first volume. Accordingly, the new material is divided into numbered sections, and (in some cases) the sections into lettered subsections. Sections are identified by two numbers (chapter and section within chapter) separated by a decimal point—thus, the fifth section of the new material (Chapter 8) is referred to as Section 8.5. Within a section, a subsection is referred to by letter alone. A subsection in one of the seven chapters of the first volume or in a different section of "Chapter 8" is referred to by referring to the section and by appending a letter to the section number—for example, in Section 8.7, Subsection f of Section 8.1 is referred to as Section 8.1f. As in the first volume, an exercise in an earlier chapter is referred to by the number obtained by inserting the chapter number (and a decimal point) in front of the exercise number.

Some of the subsections are divided into parts. Each such subsection includes two or more parts that begin with a heading and may or may not include an introductory part (with no heading). On the relatively small number of occasions on which reference is made to one or more of the individual parts, the parts that begin with headings are identified as though they had been numbered $1, 2, \ldots$ in order of appearance.

Some of the displayed "equations" are numbered. An equation number consists of two parts (corresponding to section within chapter and equation within section) separated by a decimal point (and is enclosed in parentheses). As in the first volume, an equation in an earlier chapter is referred to by the "number" obtained by starting with the chapter number and appending a decimal point and the equation number— accordingly, in Chapter 8, result (5.11) of Chapter 3 is referred to as result (3.5.11). For purposes of numbering (and referring to) the equations in the exercises, the exercises in a chapter are regarded as forming Section E of that chapter; and similarly (in the case of Part II of the present volume) the solutions to the exercises are regarded as forming Section S.

David A. Harville
harville@iastate.edu
March 23, 2023

Part I

Additional Topics

8

Constrained Linear Models and Related Topics

Suppose that $\mathbf{y} = (y_1, y_2, \ldots, y_N)'$ is an N-dimensional observable random column vector that follows a G–M, Aitken, or general linear model, in which case

$$\mathbf{y} = \mathbf{X}\boldsymbol{\beta} + \mathbf{e},$$

where $\boldsymbol{\beta}$ is a $P{\times}1$ vector of unconstrained parameters. Now, consider a variation on the G–M, Aitken, or general linear model in which $\boldsymbol{\beta}$ is subject to the constraint

$$\mathbf{A}\boldsymbol{\beta} = \mathbf{d},$$

where (for some nonnegative integer Q) $\mathbf{A}$ is a $Q \times P$ matrix and $\mathbf{d}$ a $Q \times 1$ vector of known constants. And assume that $\mathbf{d} \in \mathcal{C}(\mathbf{A})$ (since otherwise the set of $\boldsymbol{\beta}$-values that satisfy the constraint would be the empty set). Further, depending on whether the model equation pertains to the G–M model, the Aitken model, or the general linear model, let us refer to the model with model equation $\mathbf{y} = \mathbf{X}\boldsymbol{\beta} + \mathbf{e}$ and constraint $\mathbf{A}\boldsymbol{\beta} = \mathbf{d}$ as the constrained G–M model, the constrained Aitken model, or the constrained general linear model. Here, it is implicitly assumed that (as in the unconstrained case) the value(s) of the parameter(s) (σ or the elements of $\boldsymbol{\theta}$) that determine the value of $\text{var}(\mathbf{y})$ are unrelated to the value of $\boldsymbol{\beta}$, so that in the case of the constrained G–M or constrained Aitken model the parameter space is

$$\{\boldsymbol{\beta}, \sigma : \mathbf{A}\boldsymbol{\beta} = \mathbf{d}, \ \sigma > 0\}$$

and in the case of the constrained general linear model the parameter space is

$$\{\boldsymbol{\beta}, \boldsymbol{\theta} : \mathbf{A}\boldsymbol{\beta} = \mathbf{d}, \ \boldsymbol{\theta} \in \boldsymbol{\Omega}\}.$$

The unconstrained (G–M, Aitken, or general linear) model can be regarded as a special case of the corresponding constrained model. Specifically, it can be regarded as the "degenerate" special case where $Q = 0$, that is, the special case where $\mathbf{A}$ has 0 rows and $\mathbf{d}$ has 0 elements. Alternatively, whatever the value of Q, it can be regarded as the special case where $\mathbf{A} = \mathbf{0}$ and $\mathbf{d} = \mathbf{0}$.

Constrained G–M, Aitken, or general linear models can serve as a basis for the analysis of classificatory data and for the design of factorial or fractional factorial experiments (which give rise to classificatory data). Suppose there are K factors with levels $L_1, L_2, \ldots, L_K$, respectively. And suppose that each of these factors is "crossed" with the others, so that all $\prod_{i=1}^{K} L_i$ combinations of levels are possible. Then, among the possible models is a G–M model in which $\boldsymbol{\beta}$ has $\prod_{i=1}^{K} L_i$ elements, each of which represents the expected value of those of the y_i's that pertain to a particular one of the $\prod_{i=1}^{K} L_i$ combinations of levels. As mentioned earlier (in Section 1.3) in the special case of a single factor, a model of this kind is referred to as a cell-means model.

When (in such a setting) $K \geq 2$, there may be one or more factors or combinations of factors whose "effect" on $\text{E}(\mathbf{y})$ is thought not to vary much with the levels of the remaining factors. In such a case, there may be a willingness to assume that the effect of those factors or combinations of factors does not vary at all with the levels of the remaining factors. These kinds of assumptions can be incorporated into the cell-means model by expressing them as a constraint of the form $\mathbf{A}\boldsymbol{\beta} = \mathbf{d}$ (with $\mathbf{d} = \mathbf{0}$).

Results are presented in Section 8.1 that extend to constrained models various of the results (presented in earlier chapters) on unconstrained models. Those results are of a general nature. Coverage of results that are specific to classificatory models and especially to cell-means models is provided by, for example, Hocking (1985).

Results on constrained models can be of interest even in cases where the model is taken to be an unconstrained model. Among results of that kind is a result that leads to an alternative expression for the numerator of the F statistic for testing the null hypothesis $H_0 : \boldsymbol{\Lambda}'\boldsymbol{\beta} = \boldsymbol{\tau}^{(0)}$ versus the alternative hypothesis $H_1 : \boldsymbol{\Lambda}'\boldsymbol{\beta} \neq \boldsymbol{\tau}^{(0)}$. This result is the subject of Section 8.3. And as is to be discussed in Section 8.2, results on constrained models provide a basis for transforming an unconstrained model of less than full rank into a (constrained or unconstrained) model that is of full rank and that can be used interchangeably with the original model.

Corresponding to any constrained (or unconstrained) model is a residual sum of squares and a so-called model sum of squares; these quantities are discussed in Section 8.4. Results on the residual and model sums of squares along with other results pertaining to constrained models provide the theoretical underpinnings for the analysis of variance (in which the "total sum of squares" is subdivided into meaningful components). The analysis of variance is the subject of Section 8.5.

In Section 4.2, a setting was considered where $\mathbf{y} = (y_1, y_2, \ldots, y_N)'$ is an N-dimensional observable random column vector that follows a G–M model, where corresponding to the elements of $\mathbf{y}$ are the values, say $u_1, u_2, \ldots, u_N$, of an explanatory variable u, and where

$$\mathrm{E}(y_i) = \delta(u_i) \qquad (i = 1, 2, \ldots, N)$$

for some polynomial $\delta(\cdot)$ whose coefficients are represented by the elements of $\boldsymbol{\beta}$. Instead of taking $\delta(\cdot)$ to be a polynomial, it could be taken to be some other member of the broader class of functions consisting of piecewise polynomials. There may be choices for $\delta(\cdot)$ in this broader class of functions for which the assumptions inherent in the equalities $\mathrm{E}(y_i) = \delta(u_i)$ $(i = 1, 2, \ldots, N)$ would be much more reasonable than would be the case if the choice were restricted to ordinary polynomials. To insure that the piecewise polynomial is "connected" and is sufficiently "smooth," it may be necessary to impose a parametric constraint of the form $\mathbf{A}\boldsymbol{\beta} = \mathbf{d}$. When $\delta(\cdot)$ is taken to be a piecewise polynomial, the resultant model can (depending on what is assumed about the "join points") be either linear or partially linear. In either case, results on constrained G–M models are highly useful in devising suitable inferential procedures. The use of piecewise polynomials for inferential purposes is the subject of Section 8.6.

The final section of the present chapter (Section 8.7) is devoted to a discussion of reparameterization; reparameterization is an important topic with significant ties to the topic of constrained models. Consider a G–M, Aitken, or general linear model with a model matrix $\mathbf{W}$ (having N rows) that may differ from the model matrix $\mathbf{X}$ of the "original" G–M, Aitken, or general linear model. Under certain conditions (to be discussed in Section 8.7) the two models are said to be reparameterizations of each other and are related in a way that allows them to be used "interchangeably."

8.1 Constrained Models: Basic Definitions and Results

Let us take $\mathbf{y}$ to be an $N \times 1$ observable random vector that follows a G–M, Aitken, or general linear model, so that

$$\mathbf{y} = \mathbf{X}\boldsymbol{\beta} + \mathbf{e}, \tag{1.1}$$

where $\mathrm{E}(\mathbf{e}) = \mathbf{0}$ and where $\mathrm{var}(\mathbf{e}) = \sigma^2\mathbf{I}$ (in the case of the G–M model), $\mathrm{var}(\mathbf{e}) = \sigma^2\mathbf{H}$ (in the case of the Aitken model), and $\mathrm{var}(\mathbf{e}) = \mathbf{V}(\boldsymbol{\theta})$ (in the case of the general linear model). And let us consider how various of the results obtained earlier (in Chapters 5 and 7) in the "unconstrained case," where the parameter space is $\{\boldsymbol{\beta}, \sigma : \boldsymbol{\beta} \in \mathbb{R}^P, \sigma > 0\}$ or $\{\boldsymbol{\beta}, \boldsymbol{\theta} : \boldsymbol{\beta} \in \mathbb{R}^P, \boldsymbol{\theta} \in \boldsymbol{\Omega}\}$, are affected when $\boldsymbol{\beta}$ is subject to the constraint

$$\mathbf{A}\boldsymbol{\beta} = \mathbf{d} \tag{1.2}$$

[where $\mathbf{d} \in \mathcal{C}(\mathbf{A})$] and the parameter space becomes $\{\boldsymbol{\beta}, \sigma \, : \, \mathbf{A}\boldsymbol{\beta} = \mathbf{d}, \, \sigma > 0\}$ (in the case of the G–M or Aitken model) and $\{\boldsymbol{\beta}, \boldsymbol{\theta} \, : \, \mathbf{A}\boldsymbol{\beta} = \mathbf{d}, \, \boldsymbol{\theta} \in \boldsymbol{\Omega}\}$ (in the case of the general linear model).

a. Unbiasedness and estimability

Let $\boldsymbol{\lambda}'\boldsymbol{\beta}$ represent an arbitrary linear combination of the elements of $\boldsymbol{\beta}$. And let $t(\mathbf{y})$ represent a linear estimator of $\boldsymbol{\lambda}'\boldsymbol{\beta}$, in which case $t(\mathbf{y}) = c + \mathbf{a}'\mathbf{y}$ for some constant c and some N-dimensional column vector $\mathbf{a}$ (of constants). Now, supppose that $\boldsymbol{\beta}$ is subject to the restriction (1.2) (but its value is otherwise unknown). Then, by definition, $\boldsymbol{\lambda}'\boldsymbol{\beta}$ is estimated unbiasedly by $c + \mathbf{a}'\mathbf{y}$ if (and only if)

$$\mathrm{E}(c + \mathbf{a}'\mathbf{y}) = \boldsymbol{\lambda}'\boldsymbol{\beta}$$

for every value of $\boldsymbol{\beta}$ for which $\mathbf{A}\boldsymbol{\beta} = \mathbf{d}$.

To insure that $\boldsymbol{\lambda}'\boldsymbol{\beta}$ is estimated unbiasedly by $c + \mathbf{a}'\mathbf{y}$, it is sufficient that

$$\mathbf{a}'\mathbf{X} + \mathbf{r}'\mathbf{A} = \boldsymbol{\lambda}' \quad \text{and} \quad c = \mathbf{r}'\mathbf{d} \tag{1.3}$$

for some (Q-dimensional) column vector $\mathbf{r}$. To see this, suppose that condition (1.3) is satisfied. Then, for every value of $\boldsymbol{\beta}$ for which $\mathbf{A}\boldsymbol{\beta} = \mathbf{d}$, we find that

$$\mathrm{E}(c+\mathbf{a}'\mathbf{y}) = c+\mathbf{a}'\mathbf{X}\boldsymbol{\beta} = \mathbf{r}'\mathbf{d}+\boldsymbol{\lambda}'\boldsymbol{\beta}-\mathbf{r}'\mathbf{A}\boldsymbol{\beta} = \mathbf{r}'\mathbf{d}+\boldsymbol{\lambda}'\boldsymbol{\beta}-\mathbf{r}'\mathbf{d} = \boldsymbol{\lambda}'\boldsymbol{\beta}.$$

Thus, $\boldsymbol{\lambda}'\boldsymbol{\beta}$ is estimated unbiasedly by $c + \mathbf{a}'\mathbf{y}$.

Is condition (1.3) necessary as well as sufficient for $\boldsymbol{\lambda}'\boldsymbol{\beta}$ to be estimated unbiasedly by $c + \mathbf{a}'\mathbf{y}$? The answer is yes, as is evident upon observing that if $\boldsymbol{\lambda}'\boldsymbol{\beta}$ is estimated unbiasedly by $c + \mathbf{a}'\mathbf{y}$, then

$$(\boldsymbol{\lambda}' - \mathbf{a}'\mathbf{X})\boldsymbol{\beta} = c$$

for every value of $\boldsymbol{\beta}$ for which $\mathbf{A}\boldsymbol{\beta} = \mathbf{d}$, implying (in light of Theorem 2.11.7) that

$$(\boldsymbol{\lambda}'-\mathbf{a}'\mathbf{X})[\mathbf{A}^{-}\mathbf{d} + (\mathbf{I}-\mathbf{A}^{-}\mathbf{A})\,\mathbf{s}] = c \tag{1.4}$$

for every $P \times 1$ vector $\mathbf{s}$, so that

$$(\boldsymbol{\lambda}'-\mathbf{a}'\mathbf{X})\mathbf{A}^{-}\mathbf{d} = c \quad \text{and} \quad (\boldsymbol{\lambda}'-\mathbf{a}'\mathbf{X})(\mathbf{I}-\mathbf{A}^{-}\mathbf{A}) = \mathbf{0}$$

[as becomes evident upon applying equality (1.4) with $\mathbf{s} = \mathbf{0}$ and upon recalling Lemma 2.2.2], or equivalently

$$(\boldsymbol{\lambda}'-\mathbf{a}'\mathbf{X})\mathbf{A}^{-}\mathbf{d} = c \quad \text{and} \quad \mathbf{a}'\mathbf{X} + (\boldsymbol{\lambda}'-\mathbf{a}'\mathbf{X})\mathbf{A}^{-}\mathbf{A} = \boldsymbol{\lambda}',$$

and hence that the equalities (1.3) are satisfied for some vector $\mathbf{r}$, namely, $\mathbf{r} = [(\boldsymbol{\lambda}'-\mathbf{a}'\mathbf{X})\mathbf{A}^{-}]'$.

By definition, $\boldsymbol{\lambda}'\boldsymbol{\beta}$ is estimable if there exists a constant c and a column vector $\mathbf{a}$ (of constants) such that $\boldsymbol{\lambda}'\boldsymbol{\beta}$ is estimated unbiasedly by $c + \mathbf{a}'\mathbf{y}$—otherwise (in the absence of any such c and $\mathbf{a}$), $\boldsymbol{\lambda}'\boldsymbol{\beta}$ is said to be nonestimable. Clearly, a necessary condition for $\boldsymbol{\lambda}'\boldsymbol{\beta}$ to be estimable is the existence of column vectors $\mathbf{a}$ and $\mathbf{r}$ such that

$$\mathbf{a}'\mathbf{X} + \mathbf{r}'\mathbf{A} = \boldsymbol{\lambda}'. \tag{1.5}$$

This condition is equivalent to the condition

$$\boldsymbol{\lambda}' \in \mathcal{R}\begin{pmatrix}\mathbf{X}\\ \mathbf{A}\end{pmatrix}. \tag{1.6}$$

And it is sufficient as well as necessary, as is evident upon observing that if $\mathbf{a}$ and $\mathbf{r}$ satisfy equality (1.5), then

$$\mathrm{E}(\mathbf{r}'\mathbf{d} + \mathbf{a}'\mathbf{y}) = \mathbf{r}'\mathbf{d} + \mathrm{E}(\mathbf{a}'\mathbf{y}) = \mathbf{r}'\mathbf{A}\boldsymbol{\beta} + \mathbf{a}'\mathbf{X}\boldsymbol{\beta} = \boldsymbol{\lambda}'\boldsymbol{\beta},$$

and consequently $\boldsymbol{\lambda}'\boldsymbol{\beta}$ is estimated unbiasedly by $\mathbf{r}'\mathbf{d} + \mathbf{a}'\mathbf{y}$.

Note (in light of the results of Section 5.3) that $\boldsymbol{\lambda}' \in \mathcal{R}\begin{pmatrix}\mathbf{X}\\ \mathbf{A}\end{pmatrix}$ if and only if $\boldsymbol{\lambda}'\boldsymbol{\beta}$ is estimable under an unconstrained G–M, Aitken, or general linear model with model matrix $\begin{pmatrix}\mathbf{X}\\ \mathbf{A}\end{pmatrix}$. Thus, $\boldsymbol{\lambda}'\boldsymbol{\beta}$

is estimable under a constrained G–M, Aitken, or general linear model (with model matrix $\mathbf{X}$ and constraint $\mathbf{A}\boldsymbol{\beta} = \mathbf{d}$) if and only if $\boldsymbol{\lambda}'\boldsymbol{\beta}$ is estimable under an unconstrained G–M, Aitken, or general linear model with model matrix $\begin{pmatrix}\mathbf{X}\\ \mathbf{A}\end{pmatrix}$. There is an implication that any result on the estimability or nonestimability of linear combinations (of the elements of $\boldsymbol{\beta}$) obtained under an unconstrained G–M, Aitken, or general linear model (with model equation $\mathbf{y} = \mathbf{X}\boldsymbol{\beta} + \mathbf{e}$) can be readily translated into a result on the estimability or nonestimability of those linear combinations under a constrained G–M, Aitken, or general linear model (with model equation $\mathbf{y} = \mathbf{X}\boldsymbol{\beta} + \mathbf{e}$ and constraint $\mathbf{A}\boldsymbol{\beta} = \mathbf{d}$). It is simply a matter of applying the original (unconstrained) version of the result (in which the model matrix is of dimensions $N \times P$ and is represented by $\mathbf{X}$) with the $(N+Q) \times P$ "augmented" matrix $\begin{pmatrix}\mathbf{X}\\ \mathbf{A}\end{pmatrix}$ in place of $\mathbf{X}$.

In particular, this approach can be used to show that the following properties [which were established earlier (in Section 5.3) for an (unconstrained) G–M, Aitken, or general linear model] extend to a constrained G–M, Aitken, or general linear model (with model equation $\mathbf{y} = \mathbf{X}\boldsymbol{\beta} + \mathbf{e}$ and constraint $\mathbf{A}\boldsymbol{\beta} = \mathbf{d}$):

(1) linear combinations of estimable functions are estimable; and

(2) linear combinations of nonestimable functions are not necessarily nonestimable.

b. Translation equivariance

Let us continue to take $\mathbf{y}$ to be an $N \times 1$ observable random vector that follows a constrained G–M, Aitken, or general linear model (with model equation $\mathbf{y} = \mathbf{X}\boldsymbol{\beta} + \mathbf{e}$ and constraint $\mathbf{A}\boldsymbol{\beta} = \mathbf{d}$). And let us consider further the (point) estimation of a linear combination $\boldsymbol{\lambda}'\boldsymbol{\beta}$ of the elements of $\boldsymbol{\beta}$.

Let $t(\mathbf{y})$ (a function of $\mathbf{y}$) represent an arbitrary estimator of $\boldsymbol{\lambda}'\boldsymbol{\beta}$. There are various criteria that are sometimes imposed on the choice of estimator. These include unbiasedness, which was discussed in Subsection a. Another is translation equivariance. Translation equivariance was discussed previously (in Section 5.2). Let us extend that discussion (which was confined to the special case where $\boldsymbol{\beta}$ is unrestricted) to the present case (where $\boldsymbol{\beta}$ is subject to the constraint $\mathbf{A}\boldsymbol{\beta} = \mathbf{d}$).

Let $\mathbf{k}$ represent a P-dimensional column vector of constants, and consider the transformed N-dimensional observable random column vector $\mathbf{z}$ defined as follows: $\mathbf{z} = \mathbf{y} + \mathbf{X}\mathbf{k}$. Clearly,

$$\mathbf{z} = \mathbf{X}\boldsymbol{\tau} + \mathbf{e}, \tag{1.7}$$

where $\boldsymbol{\tau} = \boldsymbol{\beta} + \mathbf{k}$. And

$$\boldsymbol{\lambda}'\boldsymbol{\tau} = \boldsymbol{\lambda}'\boldsymbol{\beta} + \boldsymbol{\lambda}'\mathbf{k} \quad \text{or, equivalently,} \quad \boldsymbol{\lambda}'\boldsymbol{\beta} = \boldsymbol{\lambda}'\boldsymbol{\tau} - \boldsymbol{\lambda}'\mathbf{k}. \tag{1.8}$$

Moreover, if $\mathbf{k}$ is such that $\mathbf{A}\mathbf{k} = \mathbf{0}$, then

$$\mathbf{A}\boldsymbol{\tau} = \mathbf{d} \quad \Leftrightarrow \quad \mathbf{A}\boldsymbol{\beta} = \mathbf{d},$$

in which case the model for $\mathbf{z}$ with model equation (1.7) (and with $\boldsymbol{\tau}$ assuming the role of $\boldsymbol{\beta}$) is of the same general form as the model for $\mathbf{y}$—if $\mathbf{k}$ is such that $\mathbf{A}\mathbf{k} \neq \mathbf{0}$, then $\mathbf{A}\boldsymbol{\tau} \neq \mathbf{A}\boldsymbol{\beta} = \mathbf{d}$, in which case the parameter space for $\boldsymbol{\tau}$ would differ from that for $\boldsymbol{\beta}$ (in fact, the set $\{\underline{\boldsymbol{\tau}} \in \mathbb{R}^P : \mathbf{A}\underline{\boldsymbol{\tau}} = \mathbf{d}\}$ would have no members in common with the set $\{\boldsymbol{\tau} : \boldsymbol{\tau} = \boldsymbol{\beta}+\mathbf{k},\ \mathbf{A}\boldsymbol{\beta} = \mathbf{d}\}$).

Accordingly, it can be argued that for a procedure to be suitable for estimating $\boldsymbol{\lambda}'\boldsymbol{\beta}$ from $\mathbf{y}$, it should (for the sake of consistency) also be suitable (for every $\mathbf{k}$ such that $\mathbf{A}\mathbf{k} = \mathbf{0}$) for estimating $\boldsymbol{\lambda}'\boldsymbol{\tau}$ from $\mathbf{z}$, implying [in light of relationship (1.8)] that (for every $\mathbf{k}$ such that $\mathbf{A}\mathbf{k} = \mathbf{0}$) the estimator $t(\mathbf{y})$ of $\boldsymbol{\lambda}'\boldsymbol{\beta}$ should satisfy the condition

$$t(\mathbf{y}) + \boldsymbol{\lambda}'\mathbf{k} = t(\mathbf{z})$$

or, equivalently, the condition

$$t(\mathbf{y}) + \boldsymbol{\lambda}'\mathbf{k} = t(\mathbf{y} + \mathbf{X}\mathbf{k}). \tag{1.9}$$

In the case of a linear estimator $c + \mathbf{a}'\mathbf{y}$, condition (1.9) can be restated in the form of the condition

$$c + \mathbf{a}'\mathbf{y} + \boldsymbol{\lambda}'\mathbf{k} = c + \mathbf{a}'(\mathbf{y} + \mathbf{X}\mathbf{k}),$$

which is equivalent to the condition

$$(\boldsymbol{\lambda}' - \mathbf{a}'\mathbf{X})\mathbf{k} = \mathbf{0}. \tag{1.10}$$

The estimator $t(\mathbf{y})$ is said to be translation equivariant if for every $\mathbf{k}$ such that $\mathbf{A}\mathbf{k} = \mathbf{0}$, it satisfies condition (1.9). Thus, a linear estimator $c + \mathbf{a}'\mathbf{y}$ is translation equivariant if for every $\mathbf{k}$ such that $\mathbf{A}\mathbf{k} = \mathbf{0}$, it satisfies condition (1.10). Moreover, for condition (1.10) to be satisfied for every $\mathbf{k}$ such that $\mathbf{A}\mathbf{k} = \mathbf{0}$, it is necessary and sufficient that

$$(\boldsymbol{\lambda}' - \mathbf{a}'\mathbf{X})(\mathbf{I} - \mathbf{A}^{-}\mathbf{A})\mathbf{s} = \mathbf{0} \quad \text{for every } P \times 1 \text{ vector } \mathbf{s}$$

(as is evident from Theorem 2.11.3) or, equivalently, that

$$(\boldsymbol{\lambda}' - \mathbf{a}'\mathbf{X})(\mathbf{I} - \mathbf{A}^{-}\mathbf{A}) = \mathbf{0}. \tag{1.11}$$

And condition (1.11) is satisfied if and only if

$$\boldsymbol{\lambda}' - \mathbf{a}'\mathbf{X} = \mathbf{r}'\mathbf{A} \quad \text{for some } Q \times 1 \text{ vector } \mathbf{r}$$

(as can be readily verified) or, equivalently, if and only if

$$\mathbf{a}'\mathbf{X} + \mathbf{r}'\mathbf{A} = \boldsymbol{\lambda}' \quad \text{for some } Q \times 1 \text{ vector } \mathbf{r}. \tag{1.12}$$

In summary, $c + \mathbf{a}'\mathbf{y}$ is a translation-equivariant estimator of $\boldsymbol{\lambda}'\boldsymbol{\beta}$ if and only if the vector $\mathbf{a}$ satisfies condition (1.11) or, equivalently, condition (1.12). As in the case where $\boldsymbol{\beta}$ is unrestricted, a linear unbiased estimator of $\boldsymbol{\lambda}'\boldsymbol{\beta}$ is translation equivariant, but a linear translation-equivariant estimator is not necessarily unbiased. What is required of the vector $\mathbf{a}$ for the unbiasedness of the linear estimator $c + \mathbf{a}'\mathbf{y}$ is the same as what is required for translation equivariance; however, something is required of the constant c for unbiasedness while nothing is required for translation equivariance.

c. Identifiability

Suppose that $\mathbf{y}$ is an $N \times 1$ observable random vector that follows a G–M, Aitken, or general linear model (with model equation $\mathbf{y} = \mathbf{X}\boldsymbol{\beta} + \mathbf{e}$). In the absence of any constraints on $\boldsymbol{\beta}$, the concept of identifiability and its relationship to the concept of estimability are as discussed previously (in Section 5.3c). Let us extend that discussion to the more general case where $\mathbf{y}$ follows a constrained G–M, Aitken, or general linear model (with model equation $\mathbf{y} = \mathbf{X}\boldsymbol{\beta} + \mathbf{e}$ and constraint $\mathbf{A}\boldsymbol{\beta} = \mathbf{d}$).

As in the unconstrained case, a linear combination $\boldsymbol{\lambda}'\boldsymbol{\beta}$ of the elements of $\boldsymbol{\beta}$ is said to be identifiable if $\boldsymbol{\lambda}'\boldsymbol{\beta}$ has a fixed value for each value of $\mathrm{E}(\mathbf{y})$ (i.e., if the value of $\boldsymbol{\lambda}'\boldsymbol{\beta}$ is determinable from knowledge of the value of $\mathbf{X}\boldsymbol{\beta}$). What is different from the unconstrained case is that in the general case (where $\boldsymbol{\beta}$ is subject to the constraint $\mathbf{A}\boldsymbol{\beta} = \mathbf{d}$), the value of $\mathrm{E}(\mathbf{y})$ may be confined to a proper subset of $\mathcal{C}(\mathbf{X})$. More formally, $\boldsymbol{\lambda}'\boldsymbol{\beta}$ is said to be identifiable if for every pair of P-dimensionable column vectors $\boldsymbol{\beta}_1$ and $\boldsymbol{\beta}_2$ for which $\mathbf{A}\boldsymbol{\beta}_1 = \mathbf{A}\boldsymbol{\beta}_2 = \mathbf{d}$,

$$\mathbf{X}\boldsymbol{\beta}_1 = \mathbf{X}\boldsymbol{\beta}_2 \quad \Rightarrow \quad \boldsymbol{\lambda}'\boldsymbol{\beta}_1 = \boldsymbol{\lambda}'\boldsymbol{\beta}_2.$$

Or, equivalently, $\boldsymbol{\lambda}'\boldsymbol{\beta}$ is identifiable if $\boldsymbol{\lambda}'\boldsymbol{\beta}_1 = \boldsymbol{\lambda}'\boldsymbol{\beta}_2$ for every pair of vectors $\boldsymbol{\beta}_1$ and $\boldsymbol{\beta}_2$ for which $\mathbf{A}\boldsymbol{\beta}_1 = \mathbf{A}\boldsymbol{\beta}_2 = \mathbf{d}$ and $\mathbf{X}\boldsymbol{\beta}_1 = \mathbf{X}\boldsymbol{\beta}_2$.

For $\boldsymbol{\lambda}'\boldsymbol{\beta}$ to be identifiable (when $\boldsymbol{\beta}$ is subject to the constraint $\mathbf{A}\boldsymbol{\beta} = \mathbf{d}$), it is necessary and sufficient that $\boldsymbol{\lambda}'\boldsymbol{\beta}_1^* = \boldsymbol{\lambda}'\boldsymbol{\beta}_2^*$ for every pair of P-dimensional column vectors $\boldsymbol{\beta}_1^*$ and $\boldsymbol{\beta}_2^*$ for which

$$\begin{pmatrix} \mathbf{X} \\ \mathbf{A} \end{pmatrix} \boldsymbol{\beta}_1^* = \begin{pmatrix} \mathbf{X} \\ \mathbf{A} \end{pmatrix} \boldsymbol{\beta}_2^*. \tag{1.13}$$

Let us verify the necessity and sufficiency of this condition, starting with the sufficiency and taking $\mathbf{r}$ to be a $P \times 1$ vector such that $\mathbf{A}\mathbf{r} = \mathbf{d}$ [the existence of which follows from the supposition that $\mathbf{d} \in \mathcal{C}(\mathbf{A})$]. Accordingly, suppose that $\boldsymbol{\lambda}'\boldsymbol{\beta}_1^* = \boldsymbol{\lambda}'\boldsymbol{\beta}_2^*$ for every pair of vectors $\boldsymbol{\beta}_1^*$ and $\boldsymbol{\beta}_2^*$ for which equality (1.13) is satisfied. And let $\boldsymbol{\beta}_1$ and $\boldsymbol{\beta}_2$ represent any pair of vectors for which

$\mathbf{A}\boldsymbol{\beta}_1 = \mathbf{A}\boldsymbol{\beta}_2 = \mathbf{d}$ and $\mathbf{X}\boldsymbol{\beta}_1 = \mathbf{X}\boldsymbol{\beta}_2$, and take $\boldsymbol{\beta}_1^* = \boldsymbol{\beta}_1 - \mathbf{r}$ and $\boldsymbol{\beta}_2^* = \boldsymbol{\beta}_2 - \mathbf{r}$. Then, clearly, $\boldsymbol{\beta}_1^*$ and $\boldsymbol{\beta}_2^*$ satisfy equality (1.13). Thus, to complete the verification of sufficiency, it remains only to observe that

$$\boldsymbol{\lambda}'\boldsymbol{\beta}_1 = \boldsymbol{\lambda}'\boldsymbol{\beta}_1^* + \boldsymbol{\lambda}'\mathbf{r} = \boldsymbol{\lambda}'\boldsymbol{\beta}_2^* + \boldsymbol{\lambda}'\mathbf{r} = \boldsymbol{\lambda}'\boldsymbol{\beta}_2.$$

Turning now to the verification of necessity, suppose that $\boldsymbol{\lambda}'\boldsymbol{\beta}_1 = \boldsymbol{\lambda}'\boldsymbol{\beta}_2$ for every pair of vectors $\boldsymbol{\beta}_1$ and $\boldsymbol{\beta}_2$ for which $\mathbf{A}\boldsymbol{\beta}_1 = \mathbf{A}\boldsymbol{\beta}_2 = \mathbf{d}$ and $\mathbf{X}\boldsymbol{\beta}_1 = \mathbf{X}\boldsymbol{\beta}_2$. Further, let $\boldsymbol{\beta}_1^*$ and $\boldsymbol{\beta}_2^*$ represent any pair of vectors for which equality (1.13) is satisfied, and take

$$\boldsymbol{\beta}_1 = \boldsymbol{\beta}_1^* - \boldsymbol{\beta}_2^* + \mathbf{r} \quad \text{and} \quad \boldsymbol{\beta}_2 = \mathbf{r}.$$

Then, clearly,

$$\mathbf{A}\boldsymbol{\beta}_1 = \mathbf{A}\boldsymbol{\beta}_1^* - \mathbf{A}\boldsymbol{\beta}_2^* + \mathbf{A}\mathbf{r} = \mathbf{A}\mathbf{r} = \mathbf{d},$$
$$\mathbf{A}\boldsymbol{\beta}_2 = \mathbf{A}\mathbf{r} = \mathbf{d},$$

and

$$\mathbf{X}\boldsymbol{\beta}_1 = \mathbf{X}\boldsymbol{\beta}_1^* - \mathbf{X}\boldsymbol{\beta}_2^* + \mathbf{X}\mathbf{r} = \mathbf{X}\mathbf{r} = \mathbf{X}\boldsymbol{\beta}_2.$$

And it follows that

$$\boldsymbol{\lambda}'\boldsymbol{\beta}_1^* = \boldsymbol{\lambda}'(\boldsymbol{\beta}_1 + \boldsymbol{\beta}_2^* - \mathbf{r}) = \boldsymbol{\lambda}'\boldsymbol{\beta}_2^* + \boldsymbol{\lambda}'\boldsymbol{\beta}_1 - \boldsymbol{\lambda}'\boldsymbol{\beta}_2 = \boldsymbol{\lambda}'\boldsymbol{\beta}_2^* + 0 = \boldsymbol{\lambda}'\boldsymbol{\beta}_2^*,$$

which completes the verification of necessity.

It follows almost immediately from what has been established that $\boldsymbol{\lambda}'\boldsymbol{\beta}$ is identifiable under a constrained G–M, Aitken, or general linear model (with model equation $\mathbf{y} = \mathbf{X}\boldsymbol{\beta} + \mathbf{e}$ and constraint $\mathbf{A}\boldsymbol{\beta} = \mathbf{d}$) if and only if $\boldsymbol{\lambda}'\boldsymbol{\beta}$ is identifiable under an unconstrained G–M, Aitken, or general linear model with model matrix $\begin{pmatrix}\mathbf{X}\\ \mathbf{A}\end{pmatrix}$ and hence (in light of the results of Section 5.3c) if and only if $\boldsymbol{\lambda}'\boldsymbol{\beta}$ is estimable under an unconstrained G–M, Aitken, or general linear model with model matrix $\begin{pmatrix}\mathbf{X}\\ \mathbf{A}\end{pmatrix}$. Moreover, it follows from the results of Subsection a that $\boldsymbol{\lambda}'\boldsymbol{\beta}$ is estimable under an unconstrained G–M, Aitken, or general linear model with model matrix $\begin{pmatrix}\mathbf{X}\\ \mathbf{A}\end{pmatrix}$ if and only if $\boldsymbol{\lambda}'\boldsymbol{\beta}$ is estimable under a constrained G–M, Aitken, or general linear model (with model equation $\mathbf{y} = \mathbf{X}\boldsymbol{\beta} + \mathbf{e}$ and constraint $\mathbf{A}\boldsymbol{\beta} = \mathbf{d}$). Thus, $\boldsymbol{\lambda}'\boldsymbol{\beta}$ is identifiable under a constrained G–M, Aitken, or general linear model (with model equation $\mathbf{y} = \mathbf{X}\boldsymbol{\beta} + \mathbf{e}$ and constraint $\mathbf{A}\boldsymbol{\beta} = \mathbf{d}$) if and only if $\boldsymbol{\lambda}'\boldsymbol{\beta}$ is estimable under that model—this result extends the result obtained in Section 5.3c for an unconstrained G–M, Aitken, or general linear model (with model matrix $\mathbf{X}$).

d. Constrained least squares

Least squares estimators and their use in devising procedures for making various sorts of statistical inferences were discussed extensively in Chapters 5 and 7. Throughout that discussion, the underlying model was taken to be a G–M model (or, on occasion, an Aitken or general linear model) with model equation $\mathbf{y} = \mathbf{X}\boldsymbol{\beta} + \mathbf{e}$ and with a parameter space $\{\boldsymbol{\beta}, \sigma : \boldsymbol{\beta} \in \mathbb{R}^P, \sigma > 0\}$ (or $\{\boldsymbol{\beta}, \boldsymbol{\theta} : \boldsymbol{\beta} \in \mathbb{R}^P, \boldsymbol{\theta} \in \boldsymbol{\Omega}\}$) in which $\boldsymbol{\beta}$ is unrestricted.

Now, suppose that $\boldsymbol{\beta}$ is not unrestricted, but rather is subject to the constraint $\mathbf{A}\boldsymbol{\beta} = \mathbf{d}$; and suppose that in implementing the method of least squares, the underlying model is taken to be a constrained G–M, Aitken, or general linear model. Under such circumstances, the method of least squares consists of minimizing the quantity $(\underline{\mathbf{y}} - \mathbf{X}\mathbf{b})'(\underline{\mathbf{y}} - \mathbf{X}\mathbf{b})$ with respect to the P-dimensional vector $\mathbf{b}$ (where $\underline{\mathbf{y}}$ is the observed value of $\mathbf{y}$). Instead of the minimization being unrestricted (as before), it is subject to the restriction $\mathbf{A}\mathbf{b} = \mathbf{d}$. When the minimization is subject to this restriction, let us use the term *constrained* or *restricted least squares* in referring to the method of least squares. And let us consider how and to what extent the results obtained earlier (in Chapters 5 and 7) on the method of least squares can be extended to constrained least squares.

Lagrangian function and the constrained normal equations. Let $q(\mathbf{b}) = (\underline{\mathbf{y}} - \mathbf{X}\mathbf{b})'(\underline{\mathbf{y}} - \mathbf{X}\mathbf{b})$. Then, the least squares minimization problem consists of minimizing $q(\mathbf{b})$ with respect to $\mathbf{b}$ subject to the constraint $\mathbf{Ab} = \mathbf{d}$. And upon letting $\mathbf{r} = (r_1, r_2, \ldots, r_Q)'$ represent an arbitrary Q-dimensional column vector, the Lagrangian function for this problem is expressible as

$$f(\mathbf{b}) = q(\mathbf{b}) - 2\mathbf{r}'(\mathbf{d} - \mathbf{Ab}).$$

Here, the role of the Lagrange multipliers is played by the elements $-2r_1, -2r_2, \ldots, -2r_Q$ of the vector $-2\mathbf{r}$. (Expressing the Lagrangian function in such a way that the vector of Lagrange multipliers is represented by $-2\mathbf{r}$ rather than by $\mathbf{r}$ will be convenient in what follows.)

Upon applying Lagrange's theorem [which is Theorem 10 in Magnus and Neudecker's (1988) Chapter 7], we find that for $q(\mathbf{b})$ to attain its minimum value at a point $\underline{\mathbf{b}}$ (subject to the constraint $\mathbf{Ab} = \mathbf{d}$), it is necessary that there exists a $Q \times 1$ vector $\underline{\mathbf{r}}$ such that for $\mathbf{r} = \underline{\mathbf{r}}$, $\underline{\mathbf{b}}$ is a stationary point of the Lagrangian function f (i.e., such that for $\mathbf{r} = \underline{\mathbf{r}}$, $\partial f / \partial \mathbf{b} = \mathbf{0}$ at $\mathbf{b} = \underline{\mathbf{b}}$). In fact, the theorem indicates that $Q - \text{rank}(\mathbf{A})$ of the elements of the vector $\underline{\mathbf{r}}$ can be taken to be 0. Of course, it is also necessary that $\underline{\mathbf{b}}$ satisfy the constraint (i.e., that $\mathbf{A}\underline{\mathbf{b}} = \mathbf{d}$).

Making use of result (5.4.13) and formula (5.4.7), we find that

$$\frac{\partial f(\mathbf{b})}{\partial \mathbf{b}} = \frac{\partial q(\mathbf{b})}{\partial \mathbf{b}} - 2\frac{\partial (\mathbf{A}'\mathbf{r})'\mathbf{b}}{\partial \mathbf{b}} = 2(\mathbf{X}'\mathbf{X}\mathbf{b} - \mathbf{X}'\underline{\mathbf{y}} + \mathbf{A}'\mathbf{r}).$$

Thus, for $q(\mathbf{b})$ to attain its minimum value at a point $\underline{\mathbf{b}}$ (subject to the constraint $\mathbf{Ab} = \mathbf{d}$), it is necessary that there exists a $Q \times 1$ vector $\underline{\mathbf{r}}$ such that $\mathbf{X}'\mathbf{X}\underline{\mathbf{b}} + \mathbf{A}'\underline{\mathbf{r}} = \mathbf{X}'\underline{\mathbf{y}}$ (and that $\mathbf{A}\underline{\mathbf{b}} = \mathbf{d}$) or, equivalently, it is necessary that there exists a $Q \times 1$ vector $\underline{\mathbf{r}}$ such that $\underline{\mathbf{b}}$ and $\underline{\mathbf{r}}$ are, respectively, the first and second parts of a solution to the linear system

$$\mathbf{X}'\mathbf{X}\mathbf{b} + \mathbf{A}'\mathbf{r} = \mathbf{X}'\underline{\mathbf{y}}, \tag{1.14}$$

$$\mathbf{Ab} = \mathbf{d} \tag{1.15}$$

(in $\mathbf{b}$ and $\mathbf{r}$). Note that the linear system formed by equations (1.14) and (1.15) can be reexpressed as

$$\begin{pmatrix} \mathbf{X}'\mathbf{X} & \mathbf{A}' \\ \mathbf{A} & \mathbf{0} \end{pmatrix} \begin{pmatrix} \mathbf{b} \\ \mathbf{r} \end{pmatrix} = \begin{pmatrix} \mathbf{X}'\underline{\mathbf{y}} \\ \mathbf{d} \end{pmatrix}. \tag{1.16}$$

Equations (1.14) and (1.15) are referred to collectively as the *constrained normal equations*; this term is also used in referring to linear system (1.16).

Consistency of the constrained normal equations. Do the constrained normal equations necessarily have a solution? The answer is yes!

Let us verify the consistency of the constrained normal equations. To do so, it suffices (in light of Theorem 5.3.1) to show that the constrained normal equations are compatible. Accordingly, let $\mathbf{k}$ represent any $P \times 1$ vector and $\boldsymbol{\ell}$ any $Q \times 1$ vector such that

$$\begin{pmatrix} \mathbf{k} \\ \boldsymbol{\ell} \end{pmatrix}' \begin{pmatrix} \mathbf{X}'\mathbf{X} & \mathbf{A}' \\ \mathbf{A} & \mathbf{0} \end{pmatrix} = \mathbf{0}.$$

Then,

$$\mathbf{k}'\mathbf{X}'\mathbf{X} = -\boldsymbol{\ell}'\mathbf{A} \quad \text{and} \quad \mathbf{Ak} = \mathbf{0}, \tag{1.17}$$

implying that

$$(\mathbf{Xk})'\mathbf{Xk} = \mathbf{k}'\mathbf{X}'\mathbf{Xk} = -\boldsymbol{\ell}'\mathbf{Ak} = 0$$

and hence (in light of Corollary 2.3.3) that

$$\mathbf{Xk} = \mathbf{0}. \tag{1.18}$$

Moreover, since (by assumption) $\mathbf{d} \in \mathcal{C}(\mathbf{A})$, there exists a vector $\mathbf{h}$ such that $\mathbf{d} = \mathbf{Ah}$, implying [in light of results (1.17) and (1.18)] that

$$\boldsymbol{\ell}'\mathbf{d} = \boldsymbol{\ell}'\mathbf{Ah} = -\mathbf{k}'\mathbf{X}'\mathbf{Xh} = -(\mathbf{Xk})'\mathbf{Xh} = 0. \tag{1.19}$$

Together, results (1.18) and (1.19) imply that

$$\begin{pmatrix} \mathbf{k} \\ \boldsymbol{\ell} \end{pmatrix}' \begin{pmatrix} \mathbf{X}'\mathbf{y} \\ \mathbf{d} \end{pmatrix} = (\mathbf{Xk})'\mathbf{y} + \boldsymbol{\ell}'\mathbf{d} = 0.$$

Thus, the constrained normal equations are compatible.

Minimizing values. For $q(\mathbf{b})$ to attain its minimum value at a point $\tilde{\mathbf{b}}$ (subject to the constraint $\mathbf{Ab} = \mathbf{d}$), it is necessary that $\tilde{\mathbf{b}}$ form the first (P-dimensional) part of some solution to the constrained normal equations. Is that condition sufficient as well as necessary? The answer is yes!

For purposes of confirming that the answer is yes, let $\tilde{\mathbf{b}}$ represent the first (P-dimensional) part and $\tilde{\mathbf{r}}$ the second (Q-dimensional) part of any solution to the constrained normal equations. Then, upon reexpressing $q(\mathbf{b})$ in the form

$$q(\mathbf{b}) = [\mathbf{y} - \mathbf{X}\tilde{\mathbf{b}} - \mathbf{X}(\mathbf{b} - \tilde{\mathbf{b}})]'[\mathbf{y} - \mathbf{X}\tilde{\mathbf{b}} - \mathbf{X}(\mathbf{b} - \tilde{\mathbf{b}})]$$

and upon observing that (for $\mathbf{b}$ such that $\mathbf{Ab} = \mathbf{d}$)

$$[(\mathbf{y} - \mathbf{X}\tilde{\mathbf{b}})'\mathbf{X}(\mathbf{b} - \tilde{\mathbf{b}})]' = (\mathbf{b} - \tilde{\mathbf{b}})'\mathbf{X}'(\mathbf{y} - \mathbf{X}\tilde{\mathbf{b}}) = (\mathbf{b} - \tilde{\mathbf{b}})'\mathbf{A}'\tilde{\mathbf{r}} = (\mathbf{Ab} - \mathbf{A}\tilde{\mathbf{b}})'\tilde{\mathbf{r}} = 0$$

and observing also that $[\mathbf{X}(\mathbf{b} - \tilde{\mathbf{b}})]'[\mathbf{X}(\mathbf{b} - \tilde{\mathbf{b}})]$ is the sum of squares of the elements of the vector $\mathbf{X}(\mathbf{b} - \tilde{\mathbf{b}})$, we find that (for $\mathbf{b}$ such that $\mathbf{Ab} = \mathbf{d}$)

$$q(\mathbf{b}) = q(\tilde{\mathbf{b}}) + [\mathbf{X}(\mathbf{b} - \tilde{\mathbf{b}})]'\mathbf{X}(\mathbf{b} - \tilde{\mathbf{b}}) \geq q(\tilde{\mathbf{b}}). \tag{1.20}$$

And it follows that $q(\mathbf{b})$ attains its minimum value (for $\mathbf{b}$ such that $\mathbf{Ab} = \mathbf{d}$) at $\tilde{\mathbf{b}}$. Moreover, for $\mathbf{b}$ such that $\mathbf{Ab} = \mathbf{d}$,

$$\begin{aligned} q(\mathbf{b}) = q(\tilde{\mathbf{b}}) \quad &\Leftrightarrow \quad [\mathbf{X}(\mathbf{b} - \tilde{\mathbf{b}})]'\mathbf{X}(\mathbf{b} - \tilde{\mathbf{b}}) = 0 \\ &\Leftrightarrow \quad \mathbf{X}(\mathbf{b} - \tilde{\mathbf{b}}) = \mathbf{0} \\ &\Leftrightarrow \quad \mathbf{Xb} = \mathbf{X}\tilde{\mathbf{b}} \\ &\Leftrightarrow \quad \mathbf{X}'\mathbf{Xb} = \mathbf{X}'\mathbf{X}\tilde{\mathbf{b}} \quad \text{(recalling Corollary 2.3.4)} \\ &\Leftrightarrow \quad \mathbf{X}'\mathbf{Xb} + \mathbf{A}'\tilde{\mathbf{r}} = \mathbf{X}'\mathbf{X}\tilde{\mathbf{b}} + \mathbf{A}'\tilde{\mathbf{r}} \\ &\Rightarrow \quad \mathbf{X}'\mathbf{Xb} + \mathbf{A}'\tilde{\mathbf{r}} = \mathbf{X}'\underline{\mathbf{y}}, \end{aligned}$$

In summary,

(1) $q(\mathbf{b})$ attains its minimum value (subject to the constraint $\mathbf{Ab} = \mathbf{d}$) at a point $\tilde{\mathbf{b}}$ if and only if $\tilde{\mathbf{b}}$ is the first (P-dimensional) part of some solution to the constrained normal equations; and

(2) $\mathbf{X}\tilde{\mathbf{b}} = \mathbf{X}\mathbf{b}^*$ for any two points $\tilde{\mathbf{b}}$ and $\mathbf{b}^*$ at which $q(\mathbf{b})$ attains its minimum value (subject to the constraint $\mathbf{Ab} = \mathbf{d}$) or, equivalently, that form the first (P-dimensional) parts of solutions to the constrained normal equations.

Constrained least squares estimators and the constrained conjugate normal equations. Let $\boldsymbol{\lambda}'\boldsymbol{\beta}$ represent any linear combination of the elements of $\boldsymbol{\beta}$ that is estimable (under the constrained G–M, Aitken, or general linear model). And let $q(\mathbf{b}) = (\mathbf{y}-\mathbf{Xb})'(\mathbf{y}-\mathbf{Xb})$. By definition, the constrained least squares estimator of $\boldsymbol{\lambda}'\boldsymbol{\beta}$ is $\boldsymbol{\lambda}'\tilde{\mathbf{b}}$, where (for each value of the observable random vector $\mathbf{y}$) $\tilde{\mathbf{b}}$ is any value of the vector $\mathbf{b}$ for which $q(\mathbf{b})$ attains its minimum value (subject to the constraint $\mathbf{Ab} = \mathbf{d}$). This estimator is well-defined in the sense that it does not vary with the choice of $\tilde{\mathbf{b}}$ [as is evident upon recalling from Subsection a that $\boldsymbol{\lambda}' = \mathbf{a}'\mathbf{X} + \mathbf{r}'\mathbf{A}$ for some vectors $\mathbf{a}$ and $\mathbf{r}$ and upon recalling from the preceding part of the present subsection that $\mathbf{X}\tilde{\mathbf{b}} = \mathbf{X}\mathbf{b}^*$ for any two points $\tilde{\mathbf{b}}$ and $\mathbf{b}^*$ at which $q(\mathbf{b})$ attains its minimum value (subject to the constraint $\mathbf{Ab} = \mathbf{d}$)].

The value of the constrained least squares estimator of $\boldsymbol{\lambda}'\boldsymbol{\beta}$ can be determined from a solution to the constrained normal equations; the constrained least squares estimate of $\boldsymbol{\lambda}'\boldsymbol{\beta}$ equals $\boldsymbol{\lambda}'\tilde{b}$, where $\tilde{\mathbf{b}}$ is the first (P-dimensional) part of any solution to those equations. Alternatively, it can be determined from a solution to the linear system

$$\begin{pmatrix} \mathbf{X}'\mathbf{X} & \mathbf{A}' \\ \mathbf{A} & \mathbf{0} \end{pmatrix} \begin{pmatrix} \mathbf{s} \\ \mathbf{t} \end{pmatrix} = \begin{pmatrix} \boldsymbol{\lambda} \\ \mathbf{0} \end{pmatrix}, \tag{1.21}$$

comprising $P+Q$ equations in a $(P+Q)$-dimensional vector of unknowns with first (P-dimensional) part $\mathbf{s}$ and second (Q-dimensional) part $\mathbf{t}$.

Let us refer to the equations forming linear system (1.21) (and to the linear system itself) as the *constrained conjugate normal equations*. Those equations are consistent. To see that, observe that there exist a P-dimensional vector $\mathbf{s}^*$ and a Q-dimensional vector $\mathbf{t}^*$ such that

$$\begin{pmatrix} \mathbf{X}'\mathbf{X} & \mathbf{A}' \\ \mathbf{A} & \mathbf{0} \end{pmatrix} \begin{pmatrix} \mathbf{s}^* \\ \mathbf{t}^* \end{pmatrix} = \begin{pmatrix} \mathbf{X}'\mathbf{a} \\ \mathbf{0} \end{pmatrix}$$

(as is evident from the consistency of the constrained normal equations), and observe further that a solution to linear system (1.21) can be obtained by taking $\mathbf{s} = \mathbf{s}^*$ and $\mathbf{t} = \mathbf{t}^* + \mathbf{r}$.

For any solution $\begin{pmatrix} \tilde{\mathbf{b}} \\ \tilde{\mathbf{r}} \end{pmatrix}$ to the constrained normal equations and any solution $\begin{pmatrix} \tilde{\mathbf{s}} \\ \tilde{\mathbf{t}} \end{pmatrix}$ to the constrained conjugate normal equations, we find that

$$\begin{aligned} \boldsymbol{\lambda}'\tilde{\mathbf{b}} &= (\mathbf{X}'\mathbf{X}\tilde{\mathbf{s}} + \mathbf{A}'\tilde{\mathbf{t}})'\tilde{\mathbf{b}} \\ &= \tilde{\mathbf{s}}'\mathbf{X}'\mathbf{X}\tilde{\mathbf{b}} + \tilde{\mathbf{t}}'\mathbf{A}\tilde{\mathbf{b}} \\ &= \tilde{\mathbf{s}}'(\mathbf{X}'\underline{\mathbf{y}} - \mathbf{A}'\tilde{\mathbf{r}}) + \tilde{\mathbf{t}}'\mathbf{d} \\ &= \tilde{\mathbf{s}}'\mathbf{X}'\underline{\mathbf{y}} - (\mathbf{A}\tilde{\mathbf{s}})'\tilde{\mathbf{r}} + \tilde{\mathbf{t}}'\mathbf{d} \\ &= \tilde{\mathbf{s}}'\mathbf{X}'\underline{\mathbf{y}} + \tilde{\mathbf{t}}'\mathbf{d}. \end{aligned} \tag{1.22}$$

It is clear from expression (1.22) that the constrained least squares estimator of $\boldsymbol{\lambda}'\boldsymbol{\beta}$ is a linear estimator. It is also clear that it satisfies conditions (1.3) and (1.12). Thus, the constrained least squares estimator of $\boldsymbol{\lambda}'\boldsymbol{\beta}$ is a linear, translation-equivariant, unbiased estimator.

More on estimability. As a generalization of the result that $\mathcal{R}(\mathbf{X}'\mathbf{X}) = \mathcal{R}(\mathbf{X})$, we have that

$$\mathcal{R}\begin{pmatrix} \mathbf{X}'\mathbf{X} \\ \mathbf{A} \end{pmatrix} = \mathcal{R}\begin{pmatrix} \mathbf{X} \\ \mathbf{A} \end{pmatrix}, \tag{1.23}$$

as can be readily verified by observing that

$$\begin{pmatrix} \mathbf{X}'\mathbf{X} \\ \mathbf{A} \end{pmatrix} = \begin{pmatrix} \mathbf{X}' & \mathbf{0} \\ \mathbf{0} & \mathbf{I} \end{pmatrix} \begin{pmatrix} \mathbf{X} \\ \mathbf{A} \end{pmatrix} \quad \text{and} \quad \begin{pmatrix} \mathbf{X} \\ \mathbf{A} \end{pmatrix} = \begin{pmatrix} \mathbf{X}(\mathbf{X}'\mathbf{X})^- & \mathbf{0} \\ \mathbf{0} & \mathbf{I} \end{pmatrix} \begin{pmatrix} \mathbf{X}'\mathbf{X} \\ \mathbf{A} \end{pmatrix}.$$

Thus, for $\boldsymbol{\lambda}'\boldsymbol{\beta}$ to be estimable under the constrained G–M, Aitken, or general linear model, it is necessary and sufficient that

$$\boldsymbol{\lambda}' \in \mathcal{R}\begin{pmatrix} \mathbf{X}'\mathbf{X} \\ \mathbf{A} \end{pmatrix} \tag{1.24}$$

or, equivalently, that

$$\boldsymbol{\lambda}' = \mathbf{s}'\mathbf{X}'\mathbf{X} + \mathbf{t}'\mathbf{A} \tag{1.25}$$

for some column vectors $\mathbf{s}$ and $\mathbf{t}$. Note that equality (1.25) can be reexpressed in the form

$$\boldsymbol{\lambda} = \mathbf{X}'\mathbf{X}\mathbf{s} + \mathbf{A}'\mathbf{t} \tag{1.26}$$

or in the form

$$(\mathbf{X}'\mathbf{X},\ \mathbf{A}')\begin{pmatrix} \mathbf{s} \\ \mathbf{t} \end{pmatrix} = \boldsymbol{\lambda}. \tag{1.27}$$

In the preceding part of the present subsection, it was shown that for $\boldsymbol{\lambda}'\boldsymbol{\beta}$ to be estimable under the constrained G–M, Aitken, or general linear model, it is necessary that the constrained conjugate normal equations be consistent. The consistency of the constrained conjugate normal equations is also a sufficient condition (for the estimability of $\boldsymbol{\lambda}'\boldsymbol{\beta}$), as is evident from the sufficiency of the existence of vectors $\mathbf{s}$ and $\mathbf{t}$ for which equality (1.27) is satisfied.

Note that the existence of vectors $\mathbf{s}$ and $\mathbf{t}$ for which equality (1.27) is satisfied can be recharacterized in terms of the consistency of a linear system. Summarizing, $\boldsymbol{\lambda}'\boldsymbol{\beta}$ is estimable under the constrained G–M, Aitken, or general linear model if and only if the linear system with coeficient matrix $(\mathbf{X}'\mathbf{X},\ \mathbf{A}')$ and right side $\boldsymbol{\lambda}$ is consistent and if and only if the constrained conjugate normal equations are consistent.

Other conditions that are necessary and sufficient for the estimability of $\boldsymbol{\lambda}'\boldsymbol{\beta}$ can be readily devised by making use of any of the various conditions that are necessary and sufficient for the consistency of a linear system. Previously (in Part 1 of Section 5.4c), such an approach was used in devising various conditions that are necessary and sufficient for the estimability of $\boldsymbol{\lambda}'\boldsymbol{\beta}$ under an unconstrained G–M, Aitken, or general linear model. It could be used in similar fashion to obtain generalizations of those conditions that are applicable when $\boldsymbol{\beta}$ is subject to the constraint $\mathbf{A}\boldsymbol{\beta} = \mathbf{d}$.

Some results on partitioned matrices. Before proceeding, it is convenient to establish some additional results on partitioned matrices, starting with the following result.

Lemma 8.1.1. For any $N \times P$ matrix $\mathbf{X}$ and $Q \times P$ matrix $\mathbf{A}$, $\mathcal{C}\begin{pmatrix}\mathbf{X}'\mathbf{X}\\ \mathbf{A}\end{pmatrix}$ and $\mathcal{C}\begin{pmatrix}\mathbf{A}'\\ \mathbf{0}\end{pmatrix}$ are essentially disjoint, that is, have only the null vector in common.

Proof. Let $\mathbf{u}$ represent a $(P+Q)$-dimensional column vector, and suppose that

$$\mathbf{u} \in \mathcal{C}\begin{pmatrix}\mathbf{X}'\mathbf{X}\\ \mathbf{A}\end{pmatrix} \cap \mathcal{C}\begin{pmatrix}\mathbf{A}'\\ \mathbf{0}\end{pmatrix}.$$

Then, there exist a P-dimensional column vector $\mathbf{s}$ and a Q-dimensional column vector $\mathbf{t}$ such that

$$\mathbf{u} = \begin{pmatrix}\mathbf{X}'\mathbf{X}\\ \mathbf{A}\end{pmatrix}\mathbf{s} \quad \text{and} \quad \mathbf{u} = \begin{pmatrix}\mathbf{A}'\\ \mathbf{0}\end{pmatrix}\mathbf{t},$$

in which case

$$\mathbf{X}'\mathbf{X}\mathbf{s} = \mathbf{A}'\mathbf{t} \quad \text{and} \quad \mathbf{A}\mathbf{s} = \mathbf{0}.$$

And it follows that

$$(\mathbf{X}\mathbf{s})'\mathbf{X}\mathbf{s} = \mathbf{s}'\mathbf{X}'\mathbf{X}\mathbf{s} = \mathbf{s}'\mathbf{A}'\mathbf{t} = (\mathbf{A}\mathbf{s})'\mathbf{t} = 0,$$

implying that $\mathbf{X}\mathbf{s} = \mathbf{0}$ and hence that

$$\mathbf{u} = \begin{pmatrix}\mathbf{X}'\mathbf{X}\mathbf{s}\\ \mathbf{A}\mathbf{s}\end{pmatrix} = \begin{pmatrix}\mathbf{X}'\mathbf{0}\\ \mathbf{0}\end{pmatrix} = \mathbf{0}.$$

Thus, $\mathcal{C}\begin{pmatrix}\mathbf{X}'\mathbf{X}\\ \mathbf{A}\end{pmatrix}$ and $\mathcal{C}\begin{pmatrix}\mathbf{A}'\\ \mathbf{0}\end{pmatrix}$ have only the null vector in common. Q.E.D.

As a corollary of Lemma 8.1.1, we have [upon recalling Theorem 2.4.25 and result (1.23)] the following result.

Corollary 8.1.2. For any $N \times P$ matrix $\mathbf{X}$ and $Q \times P$ matrix $\mathbf{A}$,

$$\operatorname{rank}\begin{pmatrix}\mathbf{X}'\mathbf{X} & \mathbf{A}'\\ \mathbf{A} & \mathbf{0}\end{pmatrix} = \operatorname{rank}\begin{pmatrix}\mathbf{X}'\mathbf{X}\\ \mathbf{A}\end{pmatrix} + \operatorname{rank}(\mathbf{A}) = \operatorname{rank}\begin{pmatrix}\mathbf{X}\\ \mathbf{A}\end{pmatrix} + \operatorname{rank}(\mathbf{A}).$$

The following result on the generalized inverses of partitioned matrices is applicable in particular to matrices of the form of the coefficient matrix of the constrained normal equations and the constrained conjugate normal equations.

Lemma 8.1.3. Let $\mathbf{A}$ represent a $P \times K$ matrix that has been partitioned as $\mathbf{A} = (\mathbf{A}_1,\ \mathbf{A}_2)$, and let $\mathbf{G}$ represent a generalized inverse of $\mathbf{A}$ that has been partitioned conformally as $\mathbf{G} = \begin{pmatrix}\mathbf{G}_1\\ \mathbf{G}_2\end{pmatrix}$ (so that $\mathbf{G}_1$ has the same number of rows as $\mathbf{A}_1$ has columns). Further, suppose that $\mathcal{C}(\mathbf{A}_1)$ and $\mathcal{C}(\mathbf{A}_2)$ are essentially disjoint (i.e., have only the null vector in common). Then, $\mathbf{A}_1\mathbf{G}_1\mathbf{A}_1 = \mathbf{A}_1$ (i.e., $\mathbf{G}_1$ is a generalized inverse of $\mathbf{A}_1$) and $\mathbf{A}_2\mathbf{G}_2\mathbf{A}_1 = \mathbf{0}$. And, similarly, $\mathbf{A}_2\mathbf{G}_2\mathbf{A}_2 = \mathbf{A}_2$ (i.e., $\mathbf{G}_2$ is a generalized inverse of $\mathbf{A}_2$) and $\mathbf{A}_1\mathbf{G}_1\mathbf{A}_2 = \mathbf{0}$.

Proof. Clearly,

$$\begin{aligned}(\mathbf{A}_1\mathbf{G}_1\mathbf{A}_1+\mathbf{A}_2\mathbf{G}_2\mathbf{A}_1,\ \mathbf{A}_1\mathbf{G}_1\mathbf{A}_2+\mathbf{A}_2\mathbf{G}_2\mathbf{A}_2)&\\ = (\mathbf{A}_1,\ \mathbf{A}_2)\begin{pmatrix}\mathbf{G}_1\\ \mathbf{G}_2\end{pmatrix}(\mathbf{A}_1,\ \mathbf{A}_2) = \mathbf{A}\mathbf{G}\mathbf{A} = \mathbf{A} = (\mathbf{A}_1,\ \mathbf{A}_2),&\end{aligned}$$

implying that

$$\mathbf{A}_1 - \mathbf{A}_1\mathbf{G}_1\mathbf{A}_1 = \mathbf{A}_2\mathbf{G}_2\mathbf{A}_1 \tag{1.28}$$

and that

$$\mathbf{A}_2 - \mathbf{A}_2\mathbf{G}_2\mathbf{A}_2 = \mathbf{A}_1\mathbf{G}_1\mathbf{A}_2. \tag{1.29}$$

And upon observing that each column of the left side of equality (1.28) is contained in $\mathcal{C}(\mathbf{A}_1)$ and each column of the right side is contained in $\mathcal{C}(\mathbf{A}_2)$, we conclude that the columns of the left side and the corresponding columns of the right side are null vectors and hence that $\mathbf{A}_1 - \mathbf{A}_1\mathbf{G}_1\mathbf{A}_1 = \mathbf{0}$, or equivalently $\mathbf{A}_1\mathbf{G}_1\mathbf{A}_1 = \mathbf{A}_1$, and that $\mathbf{A}_2\mathbf{G}_2\mathbf{A}_1 = \mathbf{0}$. Similarly, upon observing that each column of the left side of equality (1.29) is contained in $\mathcal{C}(\mathbf{A}_2)$ and each column of the right side is contained in $\mathcal{C}(\mathbf{A}_1)$, we conclude that the columns of the left side and the corresponding columns of the right side are null vectors and hence that $\mathbf{A}_2 - \mathbf{A}_2\mathbf{G}_2\mathbf{A}_2 = \mathbf{0}$, or equivalently $\mathbf{A}_2\mathbf{G}_2\mathbf{A}_2 = \mathbf{A}_2$, and that $\mathbf{A}_1\mathbf{G}_1\mathbf{A}_2 = \mathbf{0}$. Q.E.D.

As a corollary of Lemma 8.1.3, we have the following result.

Corollary 8.1.4. Let $\mathbf{B}$ represent a partitioned matrix of the form

$$\mathbf{B} = \begin{pmatrix} \mathbf{X}'\mathbf{X} & \mathbf{A}' \\ \mathbf{A} & \mathbf{0} \end{pmatrix},$$

and let $\mathbf{G}$ represent a generalized inverse (of $\mathbf{B}$) that has been partitioned as

$$\mathbf{G} = \begin{pmatrix} \mathbf{G}_{11} & \mathbf{G}_{12} \\ \mathbf{G}_{21} & \mathbf{G}_{22} \end{pmatrix},$$

conformally to the partitioning of $\mathbf{B}$ (so that the dimensions of $\mathbf{G}_{11}$ are the same as those of $\mathbf{X}'\mathbf{X}$). Then,

$$\mathbf{A}\mathbf{G}_{12}\mathbf{A} = \mathbf{A}\mathbf{G}_{21}'\mathbf{A} = \mathbf{A} \tag{1.30}$$

(i.e., $\mathbf{G}_{12}$ and $\mathbf{G}_{21}'$ are both generalized inverses of $\mathbf{A}$).

Proof. In light of Lemma 8.1.1, it follows from Lemma 8.1.3 that $(\mathbf{G}_{21}, \mathbf{G}_{22})$ is a generalized inverse of $\begin{pmatrix} \mathbf{A}' \\ \mathbf{0} \end{pmatrix}$ and hence that

$$\begin{pmatrix} \mathbf{A}' \\ \mathbf{0} \end{pmatrix} (\mathbf{G}_{21}, \mathbf{G}_{22}) \begin{pmatrix} \mathbf{A}' \\ \mathbf{0} \end{pmatrix} = \begin{pmatrix} \mathbf{A}' \\ \mathbf{0} \end{pmatrix}. \tag{1.31}$$

And equality (1.31) implies that

$$\mathbf{A}'\mathbf{G}_{21}\mathbf{A}' = \mathbf{A}'$$

and hence that

$$\mathbf{A}\mathbf{G}_{21}'\mathbf{A} = (\mathbf{A}'\mathbf{G}_{21}\mathbf{A}')' = (\mathbf{A}')' = \mathbf{A}.$$

Moreover, upon observing (in light of the symmetry of $\mathbf{B}$) that $\mathbf{G}' = \begin{pmatrix} \mathbf{G}_{11}' & \mathbf{G}_{21}' \\ \mathbf{G}_{12}' & \mathbf{G}_{22}' \end{pmatrix}$ (like $\mathbf{G}$ itself) is a generalized inverse of $\mathbf{B}$, we find [upon replacing $\mathbf{G}_{21}'$ in the equality $\mathbf{A}\mathbf{G}_{21}'\mathbf{A} = \mathbf{A}$ with $(\mathbf{G}_{12}')' = \mathbf{G}_{12}$] that $\mathbf{A}\mathbf{G}_{12}\mathbf{A} = \mathbf{A}$. Q.E.D.

Variances and covariances of constrained least squares estimators. Let us add to the results on constrained least squares estimation obtained earlier in the present subsection. Accordingly, suppose that $\mathbf{y}$ is an $N \times 1$ observable random vector that follows a constrained G–M, Aitken, or general linear model (with model equation $\mathbf{y} = \mathbf{X}\boldsymbol{\beta} + \mathbf{e}$ and constraint $\mathbf{A}\boldsymbol{\beta} = \mathbf{d}$). And recall that the constrained least squares estimator of an estimable linear combination $\boldsymbol{\lambda}'\boldsymbol{\beta}$ of the elements of $\boldsymbol{\beta}$ is expressible as

$$\tilde{\mathbf{s}}'\mathbf{X}'\mathbf{y} + \tilde{\mathbf{t}}'\mathbf{d} = (\mathbf{X}\tilde{\mathbf{s}})'\mathbf{y} + \tilde{\mathbf{t}}'\mathbf{d}, \tag{1.32}$$

where $\tilde{\mathbf{s}}$ and $\tilde{\mathbf{t}}$ are any values of $\mathbf{s}$ abd $\mathbf{t}$, respectively, that satisfy the onstrained conjugate normal equations $\begin{pmatrix} \mathbf{X}'\mathbf{X} & \mathbf{A}' \\ \mathbf{A} & \mathbf{0} \end{pmatrix} \begin{pmatrix} \mathbf{s} \\ \mathbf{t} \end{pmatrix} = \begin{pmatrix} \boldsymbol{\lambda} \\ \mathbf{0} \end{pmatrix}$—that the constrained least squares estimator is expresible in the form (1.32) is a consequence of result (1.22).

Let $\mathbf{G}$ represent any generalized inverse of $\begin{pmatrix} \mathbf{X}'\mathbf{X} & \mathbf{A}' \\ \mathbf{A} & \mathbf{0} \end{pmatrix}$, and partition $\mathbf{G}$ as $\mathbf{G} = \begin{pmatrix} \mathbf{G}_{11} & \mathbf{G}_{12} \\ \mathbf{G}_{21} & \mathbf{G}_{22} \end{pmatrix}$ (where $\mathbf{G}_{11}$ is of the same dimensions as $\mathbf{X}'\mathbf{X}$). Then, among the solutions to the constrained conjugate normal equations is the vector $\mathbf{G}\begin{pmatrix}\boldsymbol{\lambda}\\ \mathbf{0}\end{pmatrix}$, for which $\mathbf{s} = \mathbf{G}_{11}\boldsymbol{\lambda}$ and $\mathbf{t} = \mathbf{G}_{21}\boldsymbol{\lambda}$. Thus, the constrained least squares estimator (1.32) can be expressed in terms of $\mathbf{G}_{11}$ and $\mathbf{G}_{21}$ as follows:

$$(\mathbf{G}_{11}\boldsymbol{\lambda})'\mathbf{X}'\mathbf{y} + (\mathbf{G}_{21}\boldsymbol{\lambda})'\mathbf{d} = (\mathbf{X}\mathbf{G}_{11}\boldsymbol{\lambda})'\mathbf{y} + (\mathbf{G}_{21}\boldsymbol{\lambda})'\mathbf{d}. \tag{1.33}$$

Making use of expression (1.33), we find that under the constrained general linear model, the variance of the constrained least squares estimator of $\boldsymbol{\lambda}'\boldsymbol{\beta}$ is expressible as

$$\text{var}(\tilde{\mathbf{s}}'\mathbf{X}'\mathbf{y} + \tilde{\mathbf{t}}'\mathbf{d}) = \tilde{\mathbf{s}}'\mathbf{X}'\mathbf{V}(\boldsymbol{\theta})\mathbf{X}\tilde{\mathbf{s}} = (\mathbf{G}_{11}\boldsymbol{\lambda})'\mathbf{X}'\mathbf{V}(\boldsymbol{\theta})\mathbf{X}\mathbf{G}_{11}\boldsymbol{\lambda}. \tag{1.34}$$

More generally, suppose that $\boldsymbol{\lambda}_1'\boldsymbol{\beta}$ and $\boldsymbol{\lambda}_2'\boldsymbol{\beta}$ are any two estimable linear combinations of the elements of $\boldsymbol{\beta}$. Then, the constrained least squares estimator of $\boldsymbol{\lambda}_1'\boldsymbol{\beta}$ is $\tilde{\mathbf{s}}_1'\mathbf{X}'\mathbf{y} + \tilde{\mathbf{t}}_1'\mathbf{d}$, where $\tilde{\mathbf{s}}_1$ is any $P\times 1$ vector and $\tilde{\mathbf{t}}_1$ any $Q\times 1$ vector for which $\begin{pmatrix}\mathbf{X}'\mathbf{X} & \mathbf{A}'\\ \mathbf{A} & \mathbf{0}\end{pmatrix}\begin{pmatrix}\tilde{\mathbf{s}}_1\\ \tilde{\mathbf{t}}_1\end{pmatrix} = \begin{pmatrix}\boldsymbol{\lambda}_1\\ \mathbf{0}\end{pmatrix}$; and, similarly, the constrained least squares estimator of $\boldsymbol{\lambda}_2'\boldsymbol{\beta}$ is $\tilde{\mathbf{s}}_2'\mathbf{X}'\mathbf{y} + \tilde{\mathbf{t}}_2'\mathbf{d}$, where $\tilde{\mathbf{s}}_2$ is any $P\times 1$ vector and $\tilde{\mathbf{t}}_2$ any $Q\times 1$ vector for which $\begin{pmatrix}\mathbf{X}'\mathbf{X} & \mathbf{A}'\\ \mathbf{A} & \mathbf{0}\end{pmatrix}\begin{pmatrix}\tilde{\mathbf{s}}_2\\ \tilde{\mathbf{t}}_2\end{pmatrix} = \begin{pmatrix}\boldsymbol{\lambda}_2\\ \mathbf{0}\end{pmatrix}$. And under the constrained general linear model,

$$\text{cov}(\tilde{\mathbf{s}}_1'\mathbf{X}'\mathbf{y} + \tilde{\mathbf{t}}_1'\mathbf{d},\ \tilde{\mathbf{s}}_2'\mathbf{X}'\mathbf{y} + \tilde{\mathbf{t}}_2'\mathbf{d}) = \tilde{\mathbf{s}}_1'\mathbf{X}'\mathbf{V}(\boldsymbol{\theta})\mathbf{X}\tilde{\mathbf{s}}_2 = (\mathbf{G}_{11}\boldsymbol{\lambda}_1)'\mathbf{X}'\mathbf{V}(\boldsymbol{\theta})\mathbf{X}\mathbf{G}_{11}\boldsymbol{\lambda}_2. \tag{1.35}$$

In the special case of the constrained Aitken model, result (1.34) "simplifies" to

$$\text{var}(\tilde{\mathbf{s}}'\mathbf{X}'\mathbf{y} + \tilde{\mathbf{t}}'\mathbf{d}) = \sigma^2\tilde{\mathbf{s}}'\mathbf{X}'\mathbf{H}\mathbf{X}\tilde{\mathbf{s}} = \sigma^2(\mathbf{G}_{11}\boldsymbol{\lambda})'\mathbf{X}'\mathbf{H}\mathbf{X}\mathbf{G}_{11}\boldsymbol{\lambda} \tag{1.36}$$

and result (1.35) to

$$\text{cov}(\tilde{\mathbf{s}}_1'\mathbf{X}'\mathbf{y} + \tilde{\mathbf{t}}_1'\mathbf{d},\ \tilde{\mathbf{s}}_2'\mathbf{X}'\mathbf{y} + \tilde{\mathbf{t}}_2'\mathbf{d}) = \sigma^2\tilde{\mathbf{s}}_1'\mathbf{X}'\mathbf{H}\mathbf{X}\tilde{\mathbf{s}}_2 = \sigma^2(\mathbf{G}_{11}\boldsymbol{\lambda}_1)'\mathbf{X}'\mathbf{H}\mathbf{X}\mathbf{G}_{11}\boldsymbol{\lambda}_2. \tag{1.37}$$

And in the further special case of the constrained G–M model, we find that

$$\text{var}(\tilde{\mathbf{s}}'\mathbf{X}'\mathbf{y} + \tilde{\mathbf{t}}'\mathbf{d}) = \sigma^2\tilde{\mathbf{s}}'\mathbf{X}'\mathbf{X}\tilde{\mathbf{s}} = \sigma^2\tilde{\mathbf{s}}'(\boldsymbol{\lambda} - \mathbf{A}'\tilde{\mathbf{t}}) = \sigma^2\tilde{\mathbf{s}}'\boldsymbol{\lambda} = \sigma^2\boldsymbol{\lambda}'\tilde{\mathbf{s}} = \sigma^2\boldsymbol{\lambda}'\mathbf{G}_{11}\boldsymbol{\lambda} \tag{1.38}$$

and, similarly, that

$$\begin{aligned}&\text{cov}(\tilde{\mathbf{s}}_1'\mathbf{X}'\mathbf{y} + \tilde{\mathbf{t}}_1'\mathbf{d},\ \tilde{\mathbf{s}}_2'\mathbf{X}'\mathbf{y} + \tilde{\mathbf{t}}_2'\mathbf{d})\\ &\quad = \sigma^2\tilde{\mathbf{s}}_1'\mathbf{X}'\mathbf{X}\tilde{\mathbf{s}}_2 = \sigma^2\tilde{\mathbf{s}}_2'\mathbf{X}'\mathbf{X}\tilde{\mathbf{s}}_1 = \sigma^2\tilde{\mathbf{s}}_2'(\boldsymbol{\lambda}_1 - \mathbf{A}'\tilde{\mathbf{t}}_1) = \sigma^2\tilde{\mathbf{s}}_2'\boldsymbol{\lambda}_1 = \sigma^2\boldsymbol{\lambda}_1'\tilde{\mathbf{s}}_2 = \sigma^2\boldsymbol{\lambda}_1'\mathbf{G}_{11}\boldsymbol{\lambda}_2.\end{aligned} \tag{1.39}$$

The various formulas for the variances and covariances of the constrained least squares estimators can be regarded as generalizations of those presented (in the final part of Section 5.4c) for the variances and covariances of unconstrained least squares estimators. In these generalizations, the role of solutions to the (unconstrained) conjugate normal equations is assumed by the first (P-dimensional) parts of solutions to the constrained conjugate normal equations, and the role of $(\mathbf{X}'\mathbf{X})^-$ is assumed by $\mathbf{G}_{11}$.

"Vectorization". Suppose further that $\boldsymbol{\lambda}_1, \boldsymbol{\lambda}_2, \ldots, \boldsymbol{\lambda}_M$ are P-dimensional column vectors of constrants. And let $\boldsymbol{\Lambda} = (\boldsymbol{\lambda}_1, \boldsymbol{\lambda}_2, \ldots, \boldsymbol{\lambda}_M)$. Then, $\boldsymbol{\lambda}_1'\boldsymbol{\beta}, \boldsymbol{\lambda}_2'\boldsymbol{\beta}, \ldots, \boldsymbol{\lambda}_M'\boldsymbol{\beta}$ are linear combinations of the elements of $\boldsymbol{\beta}$, and $\boldsymbol{\Lambda}'\boldsymbol{\beta}$ is the M-dimensional column vector whose elements are $\boldsymbol{\lambda}_1'\boldsymbol{\beta}, \boldsymbol{\lambda}_2'\boldsymbol{\beta}, \ldots, \boldsymbol{\lambda}_M'\boldsymbol{\beta}$.

Now, suppose that all M of the linear combinations $\boldsymbol{\lambda}_1'\boldsymbol{\beta}, \boldsymbol{\lambda}_2'\boldsymbol{\beta}, \ldots, \boldsymbol{\lambda}_M'\boldsymbol{\beta}$ are estimable, in which case $\boldsymbol{\Lambda}'\boldsymbol{\beta}$ is said to be estimable—saying that $\boldsymbol{\Lambda}'\boldsymbol{\beta}$ is estimable when all of its elements are estimable is consistent with the convention adopted (in Section 5.6) for an unconstrained G–M, Aitken, or general linear model. As is evident from the results of Subsection a, $\boldsymbol{\Lambda}'\boldsymbol{\beta}$ is estimable if and only if $\boldsymbol{\Lambda}' = \mathbf{K}'\mathbf{X} + \mathbf{R}'\mathbf{A}$ for some matrices $\mathbf{K}$ and $\mathbf{R}$. And as is evident from the results of Parts 4 and 5 of the present subsection, $\boldsymbol{\Lambda}'\boldsymbol{\beta}$ is estimable if and only if $\boldsymbol{\Lambda}' = \mathbf{S}'\mathbf{X}'\mathbf{X} + \mathbf{T}'\mathbf{A}$ for some matrices $\mathbf{S}$ and $\mathbf{T}$ and is estimable if and only if the linear system

$$\begin{pmatrix}\mathbf{X}'\mathbf{X} & \mathbf{A}'\\ \mathbf{A} & \mathbf{0}\end{pmatrix}\begin{pmatrix}\mathbf{S}\\ \mathbf{T}\end{pmatrix} = \begin{pmatrix}\boldsymbol{\Lambda}\\ \mathbf{0}\end{pmatrix} \tag{1.40}$$

(where $\mathbf{S}$ is a $P \times M$ matrix of unknowns and $\mathbf{T}$ a $Q \times M$ matrix of unknowns) is consistent—linear system (1.40) is a "vectorized" version of the constrained conjugate normal equations (1.21).

The M-dimensional random column vector whose elements are the least squares estimators of the corresponding elements $\boldsymbol{\lambda}_1'\boldsymbol{\beta}, \boldsymbol{\lambda}_2'\boldsymbol{\beta}, \ldots, \boldsymbol{\lambda}_M'\boldsymbol{\beta}$ of the vector $\boldsymbol{\Lambda}'\boldsymbol{\beta}$ is referred to as the least squares estimator of $\boldsymbol{\Lambda}'\boldsymbol{\beta}$. And letting $\underline{\mathbf{y}}$ represent the observed value of $\mathbf{y}$, the constrained least squares estimate of $\boldsymbol{\Lambda}'\boldsymbol{\beta}$ (i.e., the value of the least squares estimator when $\mathbf{y} = \underline{\mathbf{y}}$) is expressible as $\boldsymbol{\Lambda}'\tilde{\mathbf{b}}$, where $\tilde{\mathbf{b}}$ is any solution to the constrained normal equation (1.16). Alternatively, the least squares estimate is expressible as $\tilde{\mathbf{S}}'\mathbf{X}'\underline{\mathbf{y}} + \tilde{\mathbf{T}}'\mathbf{d}$, where $\begin{pmatrix} \tilde{\mathbf{S}} \\ \tilde{\mathbf{T}} \end{pmatrix}$ is any solution to linear system (1.40), or as

$$(\mathbf{G}_{11}\boldsymbol{\Lambda})'\mathbf{X}'\underline{\mathbf{y}} + (\mathbf{G}_{21}\boldsymbol{\Lambda})'\mathbf{d},$$

where $\mathbf{G}_{11}$ and $\mathbf{G}_{21}$ are the matrices obtained upon partitioning any generalized inverse $\mathbf{G}$ of $\begin{pmatrix} \mathbf{X}'\mathbf{X} & \mathbf{A}' \\ \mathbf{A} & \mathbf{0} \end{pmatrix}$ as $\mathbf{G} = \begin{pmatrix} \mathbf{G}_{11} & \mathbf{G}_{12} \\ \mathbf{G}_{21} & \mathbf{G}_{22} \end{pmatrix}$.

Since the elements of $\tilde{\mathbf{S}}'\mathbf{X}'\mathbf{y} + \tilde{\mathbf{T}}'\mathbf{d}$ estimate the corresponding elements of $\boldsymbol{\Lambda}'\boldsymbol{\beta}$ unbiasedly,

$$\mathrm{E}(\tilde{\mathbf{S}}'\mathbf{X}'\mathbf{y} + \tilde{\mathbf{T}}'\mathbf{d}) = \boldsymbol{\Lambda}'\boldsymbol{\beta}. \tag{1.41}$$

And exxpressions for the variance-covariance matrix of the constrained least squares estimator of $\boldsymbol{\Lambda}'\boldsymbol{\beta}$ can be obtained from expressions (1.34), (1.35), (1.36), (1.37), (1.38), and (1.39). Under the constrained general linear model,

$$\mathrm{var}(\tilde{\mathbf{S}}'\mathbf{X}'\mathbf{y} + \tilde{\mathbf{T}}'\mathbf{d}) = \tilde{\mathbf{S}}'\mathbf{X}'\mathbf{V}(\boldsymbol{\theta})\mathbf{X}\tilde{\mathbf{S}} = (\mathbf{G}_{11}\boldsymbol{\Lambda})'\mathbf{X}'\mathbf{V}(\boldsymbol{\theta})\mathbf{X}\mathbf{G}_{11}\boldsymbol{\Lambda}; \tag{1.42}$$

under the constrained Aitken model,

$$\mathrm{var}(\tilde{\mathbf{S}}'\mathbf{X}'\mathbf{y} + \tilde{\mathbf{T}}'\mathbf{d}) = \sigma^2\tilde{\mathbf{S}}'\mathbf{X}'\mathbf{H}\mathbf{X}\tilde{\mathbf{S}} = \sigma^2(\mathbf{G}_{11}\boldsymbol{\Lambda})'\mathbf{X}'\mathbf{H}\mathbf{X}\mathbf{G}_{11}\boldsymbol{\Lambda}; \tag{1.43}$$

and under the constrained G–M model,

$$\mathrm{var}(\tilde{\mathbf{S}}'\mathbf{X}'\mathbf{y} + \tilde{\mathbf{T}}'\mathbf{d}) = \sigma^2\tilde{\mathbf{S}}'\mathbf{X}'\mathbf{X}\tilde{\mathbf{S}} = \sigma^2\tilde{\mathbf{S}}'(\boldsymbol{\Lambda} - \mathbf{A}'\tilde{\mathbf{T}}) = \sigma^2\tilde{\mathbf{S}}'\boldsymbol{\Lambda} = \sigma^2\boldsymbol{\Lambda}'\tilde{\mathbf{S}} = \sigma^2\boldsymbol{\Lambda}'\mathbf{G}_{11}\boldsymbol{\Lambda}. \tag{1.44}$$

The constrained least squares estimator of the mean vector and the constrained least squares residuals. Let us continue to suppose that $\mathbf{y}$ is an $N \times 1$ observable random vector that follows a constrained G–M, Aitken, or general linear model, with model equation $\mathbf{y} = \mathbf{X}\boldsymbol{\beta} + \mathbf{e}$ and constraint $\mathbf{A}\boldsymbol{\beta} = \mathbf{d}$. Further, let $\boldsymbol{\mu} = \mathrm{E}(\mathbf{y})$, and observe that $\boldsymbol{\mu} = \mathbf{X}\boldsymbol{\beta}$. Among the potential estimators of $\boldsymbol{\mu}$ is the constrained least squares estimator of $\mathbf{X}\boldsymbol{\beta}$. Let us denote this estimator by $\hat{\boldsymbol{\mu}}$.

Take $\tilde{\mathbf{b}}$ to be any value of the $P \times 1$ vector $\mathbf{b}$ and $\tilde{\mathbf{r}}$ any value of the $Q \times 1$ vector $\mathbf{r}$ that form a solution to the linear system

$$\begin{pmatrix} \mathbf{X}'\mathbf{X} & \mathbf{A}' \\ \mathbf{A} & \mathbf{0} \end{pmatrix} \begin{pmatrix} \mathbf{b} \\ \mathbf{r} \end{pmatrix} = \begin{pmatrix} \mathbf{X}'\mathbf{y} \\ \mathbf{d} \end{pmatrix}.$$

And take $\tilde{\mathbf{S}}$ to be any value of the $P \times N$ matrix $\mathbf{S}$ and $\tilde{\mathbf{T}}$ any value of the $Q \times N$ matrix $\mathbf{T}$ that form a solution to the linear system

$$\begin{pmatrix} \mathbf{X}'\mathbf{X} & \mathbf{A}' \\ \mathbf{A} & \mathbf{0} \end{pmatrix} \begin{pmatrix} \mathbf{S} \\ \mathbf{T} \end{pmatrix} = \begin{pmatrix} \mathbf{X}' \\ \mathbf{0} \end{pmatrix}$$

[which is linear system (1.40) in the special case where $\boldsymbol{\Lambda} = \mathbf{X}'$]. Then, $\hat{\boldsymbol{\mu}}$ can be expressed either in terms of $\tilde{\mathbf{b}}$ or in terms of $\tilde{\mathbf{S}}$ and $\tilde{\mathbf{T}}$. As is evident from the discussion in the preceding part of the present subsection,

$$\hat{\boldsymbol{\mu}} = \mathbf{X}\tilde{\mathbf{b}}, \tag{1.45}$$

and also

$$\hat{\boldsymbol{\mu}} = \tilde{\mathbf{S}}'\mathbf{X}'\mathbf{y} + \tilde{\mathbf{T}}'\mathbf{d}. \tag{1.46}$$

Let $\mathbf{W} = \mathbf{X}(\mathbf{I} - \mathbf{A}^-\mathbf{A})$. And note that $\mathbf{A}\tilde{\mathbf{S}} = \mathbf{0}$ and hence that

$$\tilde{\mathbf{S}} = (\mathbf{I} - \mathbf{A}^-\mathbf{A})\mathbf{U} \quad \text{for some matrix } \mathbf{U}.$$

Note also that

$$\mathbf{X}'\mathbf{X}(\mathbf{I}-\mathbf{A}^{-}\mathbf{A})\mathbf{U} + \mathbf{A}'\tilde{\mathbf{T}} = \mathbf{X}', \tag{1.47}$$

implying [upon premultiplying both sides of equality (1.47) by $(\mathbf{I}-\mathbf{A}^{-}\mathbf{A})'$] that

$$\mathbf{W}'\mathbf{W}\mathbf{U} = \mathbf{W}'.$$

Thus,

$$\mathbf{X}\tilde{\mathbf{S}} = \mathbf{X}(\mathbf{I}-\mathbf{A}^{-}\mathbf{A})\mathbf{U} = \mathbf{W}\mathbf{U} = \mathbf{P}_{\mathbf{W}}. \tag{1.48}$$

Moreover,

$$\mathbf{A}'\tilde{\mathbf{T}} = \mathbf{X}' - \mathbf{X}'\mathbf{X}\tilde{\mathbf{S}} = \mathbf{X}'(\mathbf{I}-\mathbf{P}_{\mathbf{W}}). \tag{1.49}$$

Now, let $\underline{\mathbf{b}}$ represent any $P\times 1$ vector such that $\mathbf{d}=\mathbf{A}\underline{\mathbf{b}}$—that $\mathbf{d}\in\mathbf{A}$ implies the existence of such a vector. Then, expression (1.46) is reexpressible as

$$\hat{\boldsymbol{\mu}} = (\mathbf{X}\tilde{\mathbf{S}})'\mathbf{y} + (\mathbf{A}'\tilde{\mathbf{T}})'\underline{\mathbf{b}},$$

and upon substituting expressions (1.48) and (1.49), we find that

$$\hat{\boldsymbol{\mu}} = \mathbf{P}_{\mathbf{W}}\mathbf{y} + (\mathbf{I}-\mathbf{P}_{\mathbf{W}})\mathbf{X}\underline{\mathbf{b}}. \tag{1.50}$$

Corresponding to the constrained least squares estimator $\hat{\boldsymbol{\mu}}$ of the mean vector $\boldsymbol{\mu}$ is the vector $\mathbf{y}-\hat{\boldsymbol{\mu}}$, whose elements consist of what might be referred to as the constrained least squares residuals. And corresponding to expressions (1.45), (1.46), and (1.50) for $\hat{\boldsymbol{\mu}}$ are the following expressions for $\mathbf{y}-\hat{\boldsymbol{\mu}}$:

$$\mathbf{y}-\hat{\boldsymbol{\mu}} = \mathbf{y} - \mathbf{X}\tilde{\mathbf{b}}, \tag{1.51}$$

$$\mathbf{y}-\hat{\boldsymbol{\mu}} = (\mathbf{I}-\mathbf{X}\tilde{\mathbf{S}})\mathbf{y} - \tilde{\mathbf{T}}'\mathbf{d}, \tag{1.52}$$

and

$$\mathbf{y}-\hat{\boldsymbol{\mu}} = (\mathbf{I}-\mathbf{P}_{\mathbf{W}})(\mathbf{y} - \mathbf{X}\underline{\mathbf{b}}). \tag{1.53}$$

Under the constrained model, $\hat{\boldsymbol{\mu}}$ is an unbiased estimator of $\boldsymbol{\mu}$, as is evident from the preceding parts of the present subsection and as can be verified directly by observing [in light of equality (1.46)] that (when $\mathbf{A}\boldsymbol{\beta}=\mathbf{d}$)

$$\mathrm{E}(\hat{\boldsymbol{\mu}}) = \tilde{\mathbf{S}}'\mathbf{X}'\mathbf{X}\boldsymbol{\beta} + \tilde{\mathbf{T}}'\mathbf{d} = (\mathbf{X}'\mathbf{X}\tilde{\mathbf{S}})'\boldsymbol{\beta} + \tilde{\mathbf{T}}'\mathbf{d} = (\mathbf{X}'-\mathbf{A}'\tilde{\mathbf{T}})'\boldsymbol{\beta} + \tilde{\mathbf{T}}'\mathbf{d} = \boldsymbol{\mu} - \tilde{\mathbf{T}}'\mathbf{d} + \tilde{\mathbf{T}}'\mathbf{d} = \boldsymbol{\mu}.$$

And (under the constrained model)

$$\mathrm{E}(\mathbf{y}-\hat{\boldsymbol{\mu}}) = \boldsymbol{\mu} - \boldsymbol{\mu} = \mathbf{0}.$$

Some results on matrices of the form $\mathbf{X}(\mathbf{I}-\mathbf{A}^{-}\mathbf{A})$. Before proceeding, it is convenient to establish some basic results pertaining to matrices of the form of the matrix $\mathbf{W}$ (from the preceding part of the present subsection). Accordingly, take $\mathbf{X}$ to be an $N\times P$ matrix and $\mathbf{A}$ a $Q\times P$ matrix, and define $\mathbf{W} = \mathbf{X}(\mathbf{I}-\mathbf{A}^{-}\mathbf{A})$. Then,

$$\mathcal{R}\begin{pmatrix}\mathbf{W}\\ \mathbf{A}\end{pmatrix} = \mathcal{R}\begin{pmatrix}\mathbf{X}\\ \mathbf{A}\end{pmatrix}, \tag{1.54}$$

as is evident upon observing that

$$\begin{pmatrix}\mathbf{W}\\ \mathbf{A}\end{pmatrix} = \begin{pmatrix}\mathbf{I} & -\mathbf{X}\mathbf{A}^{-}\\ \mathbf{0} & \mathbf{I}\end{pmatrix}\begin{pmatrix}\mathbf{X}\\ \mathbf{A}\end{pmatrix} \quad\text{and}\quad \begin{pmatrix}\mathbf{X}\\ \mathbf{A}\end{pmatrix} = \begin{pmatrix}\mathbf{I} & \mathbf{X}\mathbf{A}^{-}\\ \mathbf{0} & \mathbf{I}\end{pmatrix}\begin{pmatrix}\mathbf{W}\\ \mathbf{A}\end{pmatrix}.$$

And as a corollary of result (1.54), we have that

$$\operatorname{rank}\begin{pmatrix}\mathbf{W}\\ \mathbf{A}\end{pmatrix} = \operatorname{rank}\begin{pmatrix}\mathbf{X}\\ \mathbf{A}\end{pmatrix}. \tag{1.55}$$

Moreover,

$$\mathcal{R}(\mathbf{I}-\mathbf{A}^{-}\mathbf{A}) \cap \mathcal{R}(\mathbf{A}) = \{\mathbf{0}\}$$

—to see this, observe that the existence of vectors $\mathbf{r}$ and $\mathbf{s}$ for which $\mathbf{r}'(\mathbf{I}-\mathbf{A}^{-}\mathbf{A}) = \mathbf{s}'\mathbf{A}$ implies that $\mathbf{s}'\mathbf{A} = \mathbf{s}'\mathbf{A}\mathbf{A}^{-}\mathbf{A} = \mathbf{r}'(\mathbf{I}-\mathbf{A}^{-}\mathbf{A})\mathbf{A}^{-}\mathbf{A} = \mathbf{0}$—and [since $\mathcal{R}(\mathbf{W}) \subset \mathcal{R}(\mathbf{I}-\mathbf{A}^{-}\mathbf{A})$] it follows that

$$\mathcal{R}(\mathbf{W}) \cap \mathcal{R}(\mathbf{A}) = \{\mathbf{0}\}. \tag{1.56}$$

Together, results (1.55) and (1.56) imply (in light of Theorem 2.4.25) that

$$\text{rank}(\mathbf{W}) + \text{rank}(\mathbf{A}) = \text{rank}\begin{pmatrix}\mathbf{X}\\ \mathbf{A}\end{pmatrix}$$

or, equivalently, that

$$\text{rank}(\mathbf{W}) = \text{rank}\begin{pmatrix}\mathbf{X}\\ \mathbf{A}\end{pmatrix} - \text{rank}(\mathbf{A}). \tag{1.57}$$

Result (1.57) can be restated in terms that are more geometrically meaningful and that more directly relate the rank of $\mathbf{W}$ to the rank of $\mathbf{X}$. Such a restatement can be achieved by making use of a general result on the dimension of a sum of two subspaces—refer, e.g., to Harville (1997, theorem 17.4.1) for a statement of this result. When (in the present context) applied to the sum of $\mathcal{R}(\mathbf{X})$ and $\mathcal{R}(\mathbf{A})$, this result can be stated as follows:

$$\dim\left[\mathcal{R}\begin{pmatrix}\mathbf{X}\\ \mathbf{A}\end{pmatrix}\right] = \dim[\mathcal{R}(\mathbf{X})] + \dim[\mathcal{R}(\mathbf{A})] - \dim[\mathcal{R}(\mathbf{X}) \cap \mathcal{R}(\mathbf{A})]. \tag{1.58}$$

Or equivalently (since the rank of a matrix equals the dimension of its row space)

$$\text{rank}\begin{pmatrix}\mathbf{X}\\ \mathbf{A}\end{pmatrix} = \text{rank}(\mathbf{X}) + \text{rank}(\mathbf{A}) - \dim[\mathcal{R}(\mathbf{X}) \cap \mathcal{R}(\mathbf{A})]. \tag{1.59}$$

And upon substituting expression (1.59) in expression (1.57), we find that

$$\text{rank}(\mathbf{W}) = \text{rank}(\mathbf{X}) - \dim[\mathcal{R}(\mathbf{X}) \cap \mathcal{R}(\mathbf{A})]. \tag{1.60}$$

Vectors in the column space of the matrix $\mathbf{W}$ are related to those in the null space of the matrix $\mathbf{A}$ as follows:

$$\mathbf{u} \in \mathcal{C}(\mathbf{W}) \;\;\Leftrightarrow\;\; \mathbf{u} = \mathbf{X}\mathbf{v} \text{ for some } \mathbf{v} \in \mathcal{N}(\mathbf{A}), \tag{1.61}$$

as is evident upon observing that

$$\begin{aligned} \mathbf{u} \in \mathcal{C}(\mathbf{W}) \;\;&\Leftrightarrow\;\; \mathbf{u} = \mathbf{X}(\mathbf{I}-\mathbf{A}^{-}\mathbf{A})\mathbf{b} \text{ for some column vector } \mathbf{b} \\ &\Leftrightarrow\;\; \mathbf{u} = \mathbf{X}\mathbf{v} \text{ for some } \mathbf{v} \in \mathcal{C}(\mathbf{I}-\mathbf{A}^{-}\mathbf{A}) \end{aligned}$$

and recalling (in light of Corollary 2.11.4) that $\mathcal{N}(\mathbf{A}) = \mathcal{C}(\mathbf{I}-\mathbf{A}^{-}\mathbf{A})$.

Note (in light of Lemma 2.12.3) that the premultiplication of $\mathbf{W}$ by $\mathbf{X}'$ does not affect its row space or its rank. That is,

$$\mathcal{R}[\mathbf{X}'\mathbf{X}(\mathbf{I}-\mathbf{A}^{-}\mathbf{A})] = \mathcal{R}[\mathbf{X}(\mathbf{I}-\mathbf{A}^{-}\mathbf{A})] \;\text{ and }\; \text{rank}[\mathbf{X}'\mathbf{X}(\mathbf{I}-\mathbf{A}^{-}\mathbf{A})] = \text{rank}[\mathbf{X}(\mathbf{I}-\mathbf{A}^{-}\mathbf{A})]. \tag{1.62}$$

Residual sum of squares and the estimation of variances and covariances. Let us resume the "primary development." Accordingly, suppose that $\mathbf{y}$ is an $N \times 1$ observable random vector that follows a constrained G–M, Aitken, or general linear model with model equation $\mathbf{y} = \mathbf{X}\boldsymbol{\beta} + \mathbf{e}$, where $\boldsymbol{\beta}$ is a $P \times 1$ vector of unknown parameters that [for some $Q \times P$ matrix $\mathbf{A}$ and $Q \times 1$ vector $\mathbf{d} \in \mathcal{C}(\mathbf{A})$] is subject to the constraint $\mathbf{A}\boldsymbol{\beta} = \mathbf{d}$. Further, take $\underline{\mathbf{b}}$ to be any $P \times 1$ vector for which $\mathbf{d} = \mathbf{A}\underline{\mathbf{b}}$, and let $\boldsymbol{\mu} = \mathbf{X}\boldsymbol{\beta}$ [$= \text{E}(\mathbf{y})$]. And denote by $\hat{\boldsymbol{\mu}}$ the constrained least squares estimator of $\boldsymbol{\mu}$, and define $\mathbf{W} = \mathbf{X}(\mathbf{I}-\mathbf{A}^{-}\mathbf{A})$.

The quantity $(\mathbf{y}-\hat{\boldsymbol{\mu}})'(\mathbf{y}-\hat{\boldsymbol{\mu}})$ is referred to as the (constrained) residual sum of squares or sometimes as the (constrained) error sum of squares. Its value equals the sum of the squared values of the constrained least squares residuals. Thus, whether its value is small or large is indicative of how well or how poorly the model "fits" the data.

As previously indicated, $\text{E}(\mathbf{y}-\hat{\boldsymbol{\mu}}) = \mathbf{0}$. Moreover,

$$\text{var}(\mathbf{y}-\hat{\boldsymbol{\mu}}) = (\mathbf{I}-\mathbf{P}_{\mathbf{W}})\,\text{var}(\mathbf{y})\,(\mathbf{I}-\mathbf{P}_{\mathbf{W}})',$$

as is evident from result (1.53). And upon observing that $\mathbf{P}_{\mathbf{W}}$ is symmetric and idempotent, we find that

$$\text{E}[(\mathbf{y}-\hat{\boldsymbol{\mu}})'(\mathbf{y}-\hat{\boldsymbol{\mu}})] = \text{E}\{\text{tr}[(\mathbf{y}-\hat{\boldsymbol{\mu}})(\mathbf{y}-\hat{\boldsymbol{\mu}})']\} = \text{tr}[\text{var}(\mathbf{y}-\hat{\boldsymbol{\mu}})] = \text{tr}[(\mathbf{I}-\mathbf{P}_{\mathbf{W}})\,\text{var}(\mathbf{y})]. \tag{1.63}$$

In the special case of the constrained G–M model, the expression for $\mathrm{E}[(\mathbf{y}-\hat{\boldsymbol{\mu}})'(\mathbf{y}-\hat{\boldsymbol{\mu}})]$ simplifies [upon observing that $\mathrm{tr}(\mathbf{P_W}) = \mathrm{rank}(\mathbf{P_W}) = \mathrm{rank}(\mathbf{W})$] as follows:

$$\mathrm{E}[(\mathbf{y}-\hat{\boldsymbol{\mu}})'(\mathbf{y}-\hat{\boldsymbol{\mu}})] = \mathrm{tr}[(\mathbf{I}-\mathbf{P_W})(\sigma^2\mathbf{I})] = \sigma^2[N-\mathrm{rank}(\mathbf{W})]. \tag{1.64}$$

And upon dividing $(\mathbf{y}-\hat{\boldsymbol{\mu}})'(\mathbf{y}-\hat{\boldsymbol{\mu}})$ by $N-\mathrm{rank}(\mathbf{W})$, we obtain the quantity $\hat{\sigma}^2$ defined as follows:

$$\hat{\sigma}^2 = \frac{(\mathbf{y}-\hat{\boldsymbol{\mu}})'(\mathbf{y}-\hat{\boldsymbol{\mu}})}{N-\mathrm{rank}(\mathbf{W})}. \tag{1.65}$$

Under the constrained G–M model, $\hat{\sigma}^2$ is an unbiased estimator of σ^2. Moreover, by taking advantage of the results of the preceding part of the present subsection, $\hat{\sigma}^2$ can be reexpressed in either of the following two forms:

$$\hat{\sigma}^2 = \frac{(\mathbf{y}-\hat{\boldsymbol{\mu}})'(\mathbf{y}-\hat{\boldsymbol{\mu}})}{N-\left[\mathrm{rank}\begin{pmatrix}\mathbf{X}\\ \mathbf{A}\end{pmatrix}-\mathrm{rank}(\mathbf{A})\right]} \quad \text{or} \quad \hat{\sigma}^2 = \frac{(\mathbf{y}-\hat{\boldsymbol{\mu}})'(\mathbf{y}-\hat{\boldsymbol{\mu}})}{N-\mathrm{rank}(\mathbf{X})+\dim[\mathcal{R}(\mathbf{X})\cap\mathcal{R}(\mathbf{A})]}. \tag{1.66}$$

Now, let $\tilde{\mathbf{S}}$ represent the first [$(P\times N)$-dimensional] part of any solution to the linear system $\begin{pmatrix}\mathbf{X'X} & \mathbf{A'}\\ \mathbf{A} & \mathbf{0}\end{pmatrix}\begin{pmatrix}\mathbf{S}\\ \mathbf{T}\end{pmatrix} = \begin{pmatrix}\mathbf{X'}\\ \mathbf{0}\end{pmatrix}$, $\tilde{\mathbf{b}}$ represent the first (P-dimensional) part and $\tilde{\mathbf{r}}$ the second (Q-dimensional) part of any solution to the linear system $\begin{pmatrix}\mathbf{X'X} & \mathbf{A'}\\ \mathbf{A} & \mathbf{0}\end{pmatrix}\begin{pmatrix}\mathbf{b}\\ \mathbf{r}\end{pmatrix} = \begin{pmatrix}\mathbf{X'y}\\ \mathbf{d}\end{pmatrix}$, and $\mathbf{G}_{11}$ represent the submatrix of any generalized inverse $\mathbf{G}$ of $\begin{pmatrix}\mathbf{X'X} & \mathbf{A'}\\ \mathbf{A} & \mathbf{0}\end{pmatrix}$ obtained upon partitioning $\mathbf{G}$ conformally as $\mathbf{G} = \begin{pmatrix}\mathbf{G}_{11} & \mathbf{G}_{12}\\ \mathbf{G}_{21} & \mathbf{G}_{22}\end{pmatrix}$. Then, the constrained residual sum of squares is expressible in terms of the symmetric idempotent matrix $\mathbf{P_W}$ or alternatively in terms of $\tilde{\mathbf{S}}$, $\tilde{\mathbf{b}}$, or $\mathbf{G}_{11}$. As is evident from results (1.53) and (1.48) (and from observing that the choices for $\tilde{\mathbf{S}}$ include the matrix $\mathbf{G}_{11}\mathbf{X}'$),

$$\begin{aligned}(\mathbf{y}-\hat{\boldsymbol{\mu}})'(\mathbf{y}-\hat{\boldsymbol{\mu}}) &= (\mathbf{y}-\mathbf{X}\underline{\mathbf{b}})'(\mathbf{I}-\mathbf{P_W})(\mathbf{y}-\mathbf{X}\underline{\mathbf{b}}) &(1.67)\\ &= (\mathbf{y}-\mathbf{X}\underline{\mathbf{b}})'(\mathbf{y}-\mathbf{X}\underline{\mathbf{b}}) - (\mathbf{y}-\mathbf{X}\underline{\mathbf{b}})'\mathbf{P_W}(\mathbf{y}-\mathbf{X}\underline{\mathbf{b}}), &(1.68)\end{aligned}$$

$$\begin{aligned}(\mathbf{y}-\hat{\boldsymbol{\mu}})'(\mathbf{y}-\hat{\boldsymbol{\mu}}) &= (\mathbf{y}-\mathbf{X}\underline{\mathbf{b}})'(\mathbf{I}-\mathbf{X}\tilde{\mathbf{S}})(\mathbf{y}-\mathbf{X}\underline{\mathbf{b}}) &(1.69)\\ &= (\mathbf{y}-\mathbf{X}\underline{\mathbf{b}})'(\mathbf{y}-\mathbf{X}\underline{\mathbf{b}}) - (\mathbf{y}-\mathbf{X}\underline{\mathbf{b}})'\mathbf{X}\tilde{\mathbf{S}}(\mathbf{y}-\mathbf{X}\underline{\mathbf{b}}), &(1.70)\end{aligned}$$

and

$$\begin{aligned}(\mathbf{y}-\hat{\boldsymbol{\mu}})'(\mathbf{y}-\hat{\boldsymbol{\mu}}) &= (\mathbf{y}-\mathbf{X}\underline{\mathbf{b}})'(\mathbf{I}-\mathbf{X}\mathbf{G}_{11}\mathbf{X}')(\mathbf{y}-\mathbf{X}\underline{\mathbf{b}}) &(1.71)\\ &= (\mathbf{y}-\mathbf{X}\underline{\mathbf{b}})'(\mathbf{y}-\mathbf{X}\underline{\mathbf{b}}) - (\mathbf{y}-\mathbf{X}\underline{\mathbf{b}})'\mathbf{X}\mathbf{G}_{11}\mathbf{X}'(\mathbf{y}-\mathbf{X}\underline{\mathbf{b}}). &(1.72)\end{aligned}$$

And as is evident from result (1.51),

$$(\mathbf{y}-\hat{\boldsymbol{\mu}})'(\mathbf{y}-\hat{\boldsymbol{\mu}}) = (\mathbf{y}-\mathbf{X}\tilde{\mathbf{b}})'(\mathbf{y}-\mathbf{X}\tilde{\mathbf{b}}). \tag{1.73}$$

Moreover, upon reexpressing $\mathbf{y}-\mathbf{X}\tilde{\mathbf{b}}$ as $\mathbf{y}-\mathbf{X}\tilde{\mathbf{b}} = \mathbf{y}-\mathbf{X}\underline{\mathbf{b}} - \mathbf{X}(\tilde{\mathbf{b}}-\underline{\mathbf{b}})$, we find that

$$\begin{aligned}(\mathbf{y}-\mathbf{X}\tilde{\mathbf{b}})'(\mathbf{y}-\mathbf{X}\tilde{\mathbf{b}}) &= (\mathbf{y}-\mathbf{X}\tilde{\mathbf{b}})'(\mathbf{y}-\mathbf{X}\underline{\mathbf{b}}) - (\mathbf{y}-\mathbf{X}\tilde{\mathbf{b}})'\mathbf{X}(\tilde{\mathbf{b}}-\underline{\mathbf{b}})\\ &= (\mathbf{y}-\mathbf{X}\tilde{\mathbf{b}})'(\mathbf{y}-\mathbf{X}\underline{\mathbf{b}}) - (\mathbf{A}'\tilde{\mathbf{r}})'(\tilde{\mathbf{b}}-\underline{\mathbf{b}})\\ &= (\mathbf{y}-\mathbf{X}\tilde{\mathbf{b}})'(\mathbf{y}-\mathbf{X}\underline{\mathbf{b}}) - \tilde{\mathbf{r}}'(\mathbf{d}-\mathbf{d})\\ &= (\mathbf{y}-\mathbf{X}\underline{\mathbf{b}})'(\mathbf{y}-\mathbf{X}\underline{\mathbf{b}}) - (\tilde{\mathbf{b}}-\underline{\mathbf{b}})'\mathbf{X}'(\mathbf{y}-\mathbf{X}\underline{\mathbf{b}}),\end{aligned}$$

so that as an alternative version of expression (1.73) for $(\mathbf{y}-\hat{\boldsymbol{\mu}})'(\mathbf{y}-\hat{\boldsymbol{\mu}})$—a version in which [as in expressions (1.68), (1.70), and (1.72)] $(\mathbf{y}-\hat{\boldsymbol{\mu}})'(\mathbf{y}-\hat{\boldsymbol{\mu}})$ is expressed as a deviation from $(\mathbf{y}-\mathbf{X}\underline{\mathbf{b}})'(\mathbf{y}-\mathbf{X}\underline{\mathbf{b}})$—we have that

$$(\mathbf{y}-\hat{\boldsymbol{\mu}})'(\mathbf{y}-\hat{\boldsymbol{\mu}}) = (\mathbf{y}-\mathbf{X}\underline{\mathbf{b}})'(\mathbf{y}-\mathbf{X}\underline{\mathbf{b}}) - (\tilde{\mathbf{b}}-\underline{\mathbf{b}})'\mathbf{X}'(\mathbf{y}-\mathbf{X}\underline{\mathbf{b}}). \tag{1.74}$$

The constrained residual sum of squares can also be expressed in terms of $\tilde{\mathbf{b}}$ and $\tilde{\mathbf{r}}$ as follows:

$$\begin{aligned}(\mathbf{y}-\hat{\boldsymbol{\mu}})'(\mathbf{y}-\hat{\boldsymbol{\mu}}) &= (\mathbf{y}-\mathbf{X}\tilde{\mathbf{b}})'\mathbf{y} - (\mathbf{y}-\mathbf{X}\tilde{\mathbf{b}})'\mathbf{X}\tilde{\mathbf{b}} \\ &= (\mathbf{y}-\mathbf{X}\tilde{\mathbf{b}})'\mathbf{y} - (\mathbf{X}'\mathbf{y}-\mathbf{X}'\mathbf{X}\tilde{\mathbf{b}})'\tilde{\mathbf{b}} \\ &= (\mathbf{y}-\mathbf{X}\tilde{\mathbf{b}})'\mathbf{y} - (\mathbf{A}'\tilde{\mathbf{r}})'\tilde{\mathbf{b}} \\ &= (\mathbf{y}-\mathbf{X}\tilde{\mathbf{b}})'\mathbf{y} - \tilde{\mathbf{r}}'\mathbf{A}\tilde{\mathbf{b}} \\ &= \mathbf{y}'\mathbf{y} - \tilde{\mathbf{b}}'\mathbf{X}'\mathbf{y} - \tilde{\mathbf{r}}'\mathbf{d}. \end{aligned} \tag{1.75}$$

—either this expression or expression (1.74) could be regarded as the counterpart of expression (5.4.21).

Interpretation of the Lagrange multipliers. As on various previous occasions, let $q(\mathbf{b}) = (\underline{\mathbf{y}} - \mathbf{X}\mathbf{b})'(\underline{\mathbf{y}}-\mathbf{X}\mathbf{b})$ (where $\underline{\mathbf{y}} \in \mathbb{R}^N$). And let us consider further the problem of minimizing $q(\mathbf{b})$ subject to the constraint $\mathbf{A}\mathbf{b} = \mathbf{d}$ [where $\mathbf{d} \in \mathcal{C}(\mathbf{A})$]. Specifically, let us consider the dependence of the solution to this problem on the value of $\mathbf{d}$ as quantified by the function $h(\cdot)$ defined [for $\mathbf{d} \in \mathcal{C}(\mathbf{A})$] as follows:

$$h(\mathbf{d}) = \min_{\{\mathbf{b}\,:\,\mathbf{A}\mathbf{b}=\mathbf{d}\}} q(\mathbf{b}).$$

Take $\tilde{\mathbf{b}}$ to be any $P\times 1$ vector and $\tilde{\mathbf{r}}$ any $Q\times 1$ vector that simultaneously satisfy the equality

$$\begin{pmatrix} \mathbf{X}'\mathbf{X} & \mathbf{A}' \\ \mathbf{A} & \mathbf{0} \end{pmatrix} \begin{pmatrix} \tilde{\mathbf{b}} \\ \tilde{\mathbf{r}} \end{pmatrix} = \begin{pmatrix} \mathbf{X}'\underline{\mathbf{y}} \\ \mathbf{d} \end{pmatrix}. \tag{1.76}$$

Further, let $\mathbf{G}$ represent any generalized inverse of $\begin{pmatrix} \mathbf{X}'\mathbf{X} & \mathbf{A}' \\ \mathbf{A} & \mathbf{0} \end{pmatrix}$, partition $\mathbf{G}$ as $\mathbf{G} = \begin{pmatrix} \mathbf{G}_{11} & \mathbf{G}_{12} \\ \mathbf{G}_{21} & \mathbf{G}_{22} \end{pmatrix}$ (where $\mathbf{G}_{11}$ is of dimensions $P\times P$), and recall that $\mathbf{G}' = \begin{pmatrix} \mathbf{G}_{11}' & \mathbf{G}_{21}' \\ \mathbf{G}_{12}' & \mathbf{G}_{22}' \end{pmatrix}$, like $\mathbf{G}$ itself, is a generalized inverse of $\begin{pmatrix} \mathbf{X}'\mathbf{X} & \mathbf{A}' \\ \mathbf{A} & \mathbf{0} \end{pmatrix}$. And observe (in light of the results of Part 3 of the present subsection) that the value of $\mathbf{X}\tilde{\mathbf{b}}$ is uniquely determined by equality (1.76) and that as a consequence the value of $\mathbf{A}'\tilde{\mathbf{r}}$ is also uniquely determined (i.e., the values of $\mathbf{X}\tilde{\mathbf{b}}$ and $\mathbf{A}'\tilde{\mathbf{r}}$ do not vary with the choice of $\tilde{\mathbf{b}}$ and $\tilde{\mathbf{r}}$).

Among the choices for $\tilde{\mathbf{b}}$ and $\tilde{\mathbf{r}}$ [that satisfy equality (1.76)] are those obtained by taking

$$\begin{pmatrix} \tilde{\mathbf{b}} \\ \tilde{\mathbf{r}} \end{pmatrix} = \begin{pmatrix} \mathbf{G}_{11} & \mathbf{G}_{12} \\ \mathbf{G}_{21} & \mathbf{G}_{22} \end{pmatrix} \begin{pmatrix} \mathbf{X}'\underline{\mathbf{y}} \\ \mathbf{d} \end{pmatrix} \tag{1.77}$$

and those obtained by taking

$$\begin{pmatrix} \tilde{\mathbf{b}} \\ \tilde{\mathbf{r}} \end{pmatrix} = \begin{pmatrix} \mathbf{G}_{11}' & \mathbf{G}_{21}' \\ \mathbf{G}_{12}' & \mathbf{G}_{22}' \end{pmatrix} \begin{pmatrix} \mathbf{X}'\underline{\mathbf{y}} \\ \mathbf{d} \end{pmatrix}. \tag{1.78}$$

And in light of result (1.75),

$$h(\mathbf{d}) = \underline{\mathbf{y}}'\underline{\mathbf{y}} - \tilde{\mathbf{b}}'\mathbf{X}'\underline{\mathbf{y}} - \tilde{\mathbf{r}}'\mathbf{d}. \tag{1.79}$$

Since $\tilde{\mathbf{b}}$ and $\tilde{\mathbf{r}}$ depend on $\mathbf{d}$, expression (1.79) for $h(\mathbf{d})$ depends on $\mathbf{d}$ implicitly as well as explicitly. To obtain an expression for $h(\mathbf{d})$ in which the dependence on $\mathbf{d}$ is altogether explicit, it suffices to start with expression (1.79) and to substitute for $\tilde{\mathbf{b}}$ and $\tilde{\mathbf{r}}$ from expression (1.77) or (1.78). When the substitution is from expression (1.78), we obtain the expression

$$h(\mathbf{d}) = \underline{\mathbf{y}}'\underline{\mathbf{y}} - (\mathbf{X}'\underline{\mathbf{y}})'\mathbf{G}_{11}\mathbf{X}'\underline{\mathbf{y}} - \mathbf{d}'\mathbf{G}_{21}\mathbf{X}'\underline{\mathbf{y}} - (\mathbf{G}_{12}'\mathbf{X}'\underline{\mathbf{y}})'\mathbf{d} - \mathbf{d}'\mathbf{G}_{22}\mathbf{d}. \tag{1.80}$$

Now, suppose that $\operatorname{rank}(\mathbf{A}) = Q$. Then, the rows of $\mathbf{A}$ are linearly independent and hence the columns of $\mathbf{A}'$ are linearly independent. And since (as noted earlier) the value of $\mathbf{A}'\tilde{\mathbf{r}}$ is uniquely determined by equality (1.76), it follows from the linear independence of the columns of $\mathbf{A}'$ that the value of $\tilde{\mathbf{r}}$ is also uniquely determined. Moreover, this value is such that

$$\frac{\partial h}{\partial \mathbf{d}} = -2\tilde{\mathbf{r}}. \tag{1.81}$$

Result (1.81) can be established by applying well-known results on Lagrange multipliers [described, e.g., by Magnus and Neudecker (1988, sec. 7.16)]. Alternatively, it can be established via the direct differentiation of expression (1.80). Making use of formulas (5.4.7) and (5.4.8) [and observing that $\mathbf{d}'\mathbf{G}_{21}\mathbf{X}'\underline{\mathbf{y}} = (\mathbf{G}_{21}\mathbf{X}'\underline{\mathbf{y}})'\mathbf{d}$], we find that

$$\begin{aligned}\frac{\partial h}{\partial \mathbf{d}} &= -\mathbf{G}_{21}\mathbf{X}'\underline{\mathbf{y}} - \mathbf{G}_{12}'\mathbf{X}'\underline{\mathbf{y}} - (\mathbf{G}_{22}+\mathbf{G}_{22}')\mathbf{d} \\ &= -(\mathbf{G}_{21}\mathbf{X}'\underline{\mathbf{y}} + \mathbf{G}_{22}\mathbf{d}) - (\mathbf{G}_{12}'\mathbf{X}'\underline{\mathbf{y}} + \mathbf{G}_{22}'\mathbf{d}).\end{aligned}$$

And upon observing that $\tilde{\mathbf{r}}$ is expressible as $\tilde{\mathbf{r}} = \mathbf{G}_{21}\mathbf{X}'\underline{\mathbf{y}} + \mathbf{G}_{22}\mathbf{d}$ and also as $\tilde{\mathbf{r}} = \mathbf{G}_{12}'\mathbf{X}'\underline{\mathbf{y}} + \mathbf{G}_{22}'\mathbf{d}$, we conclude that

$$\frac{\partial h}{\partial \mathbf{d}} = -\tilde{\mathbf{r}} - \tilde{\mathbf{r}} = -2\tilde{\mathbf{r}}.$$

The significance of result (1.81) is that it characterizes the vector $\tilde{\mathbf{r}}$ in a meaningful way; the elements of the vector $-2\tilde{\mathbf{r}}$ can be used to gauge the sensitivity of the residual sum of squares $h(\mathbf{d})$ to small changes in the elements of $\mathbf{d}$.

The result (1.81) can be extended to the more general case where rank$(\mathbf{A})$ equals some number, say Q_*, that may be less than Q. Such an extension can be achieved by identifying Q_* rows of the $Q\times P$ matrix $\mathbf{A}$ that form a linearly independent set, say its $i_1, i_2, \ldots, i_{Q_*}$th rows, and by restricting the choice of $\tilde{\mathbf{b}}$ and $\tilde{\mathbf{r}}$ so that all of the elements of $\tilde{\mathbf{r}}$ save its $i_1, i_2, \ldots, i_{Q_*}$th elements equal 0. Then, in lieu of result (1.81), we have that

$$\frac{\partial h}{\partial \mathbf{d}_*} = -2\tilde{\mathbf{r}}_*, \tag{1.82}$$

where $\mathbf{d}_*$ is the subvector of $\mathbf{d}$ and $\tilde{\mathbf{r}}_*$ the subvector of $\tilde{\mathbf{r}}$ formed by their $i_1, i_2, \ldots, i_{Q_*}$th elements—in regard to the interpretation of result (1.82), it is assumed that any changes in the $i_1, i_2, \ldots, i_{Q_*}$th elements of $\mathbf{d}$ are accompanied by "corresponding changes" in the other elements of $\mathbf{d}$.

e. The F test

Procedures for constructing confidence intervals (or sets) and for testing hypotheses are described and discussed in Chapter 7. The coverage therein is for the case where an unobservable random vector $\mathbf{y}$ follows an unconstrained G–M model. By making use of various of the results in the preceding subsection of the present section, that coverage can (at least in principle) be extended to the case where $\mathbf{y}$ follows a constrained G–M model. Let us consider the use of such an approach in extending the coverage of the F test.

Accordingly, suppose that $\mathbf{y}$ is an $N\times 1$ observable random vector that follows a constrained G–M model with model equation $\mathbf{y} = \mathbf{X}\boldsymbol{\beta} + \mathbf{e}$, where $\boldsymbol{\beta}$ is a $P\times 1$ vector of unknown parameters that [for a $Q\times P$ matrix $\mathbf{A}$ and a $Q\times 1$ vector $\mathbf{d}\in\mathcal{C}(\mathbf{A})$] satisfies the constraint $\mathbf{A}\boldsymbol{\beta} = \mathbf{d}$. Further, let $\boldsymbol{\tau} = \boldsymbol{\Lambda}'\boldsymbol{\beta}$, where $\boldsymbol{\Lambda}$ is a $P\times M$ matrix of constants. And suppose that τ is estimable, and consider the problem of testing the null hypothesis $H_0: \tau = \tau^{(0)}$ versus the alternative hypothesis $H_1: \tau \neq \tau^{(0)}$.

Let us assume that the hypothesized value $\tau^{(0)}$ is such that $\begin{pmatrix}\tau^{(0)}\\ \mathbf{d}\end{pmatrix} \in \mathcal{C}\begin{pmatrix}\boldsymbol{\Lambda}'\\ \mathbf{A}\end{pmatrix}$ and hence that $\begin{pmatrix}\tau^{(0)}\\ \mathbf{d}\end{pmatrix} = \begin{pmatrix}\boldsymbol{\Lambda}'\\ \mathbf{A}\end{pmatrix}\underline{\mathbf{b}}$ (or, equivalently, $\tau^{(0)} = \boldsymbol{\Lambda}'\underline{\mathbf{b}}$ and $\mathbf{d} = \mathbf{A}\underline{\mathbf{b}}$) for some vector $\underline{\mathbf{b}}$—if $\begin{pmatrix}\tau^{(0)}\\ \mathbf{d}\end{pmatrix} \notin \mathcal{C}\begin{pmatrix}\boldsymbol{\Lambda}'\\ \mathbf{A}\end{pmatrix}$, there would be no values of $\boldsymbol{\beta}$ that satisfy both the constraint and the null hypothesis. Assume also that $\operatorname{rank}\begin{pmatrix}\boldsymbol{\Lambda}'\\ \mathbf{A}\end{pmatrix} > \operatorname{rank}(\mathbf{A})$—if $\operatorname{rank}\begin{pmatrix}\boldsymbol{\Lambda}'\\ \mathbf{A}\end{pmatrix} = \operatorname{rank}(\mathbf{A})$ [and $\begin{pmatrix}\tau^{(0)}\\ \mathbf{d}\end{pmatrix} \in \mathcal{C}\begin{pmatrix}\boldsymbol{\Lambda}'\\ \mathbf{A}\end{pmatrix}$], then $\mathbf{A}\boldsymbol{\beta} = \mathbf{d} \Rightarrow \boldsymbol{\Lambda}'\boldsymbol{\beta} = \tau^{(0)}$. And observe that (since $\boldsymbol{\tau}$ is estimable) there exists an $(N+Q)\times M$ matrix $\mathbf{L}$ such that $\boldsymbol{\Lambda}' = \mathbf{L}'\begin{pmatrix}\mathbf{X}\\ \mathbf{A}\end{pmatrix}$ or equivalently, upon partitioning $\mathbf{L}$ as $\mathbf{L} = \begin{pmatrix}\mathbf{L}_1\\ \mathbf{L}_2\end{pmatrix}$ (where $\mathbf{L}_1$ has N rows), such that $\boldsymbol{\Lambda}' = \mathbf{L}_1'\mathbf{X} + \mathbf{L}_2'\mathbf{A}$. Further, let $\mathbf{W} = \mathbf{X}(\mathbf{I}-\mathbf{A}^-\mathbf{A})$, take $\mathbf{G}_{11}$ to be the $P\times P$ matrix obtained upon partitioning any generalized inverse $\mathbf{G}$ of $\begin{pmatrix}\mathbf{X}'\mathbf{X} & \boldsymbol{\Lambda}'\\ \mathbf{A} & \mathbf{0}\end{pmatrix}$ conformally as $\mathbf{G} = \begin{pmatrix}\mathbf{G}_{11} & \mathbf{G}_{12}\\ \mathbf{G}_{21} & \mathbf{G}_{22}\end{pmatrix}$, and denote by $\hat{\boldsymbol{\tau}}$ the constrained least squares estimator of $\boldsymbol{\tau}$ and by $\hat{\boldsymbol{\mu}}$ the constrained least squares estimator of $\boldsymbol{\mu}$ $(= \mathbf{X}\boldsymbol{\beta})$.

***F* statistic and test (in the case of the constrained G–M model).** Let

$$F = \frac{(\hat{\boldsymbol{\tau}} - \boldsymbol{\tau}^{(0)})'\mathbf{B}^{-}(\hat{\boldsymbol{\tau}} - \boldsymbol{\tau}^{(0)})/M_*}{(\mathbf{y} - \hat{\boldsymbol{\mu}})'(\mathbf{y} - \hat{\boldsymbol{\mu}})/(N - P_*)},$$

where $\mathbf{B} = \boldsymbol{\Lambda}'\mathbf{G}_{11}\boldsymbol{\Lambda}$, $M_* = \text{rank}\begin{pmatrix}\boldsymbol{\Lambda}'\\ \mathbf{A}\end{pmatrix} - \text{rank}(\mathbf{A})$, and $P_* = \text{rank}\begin{pmatrix}\mathbf{X}\\ \mathbf{A}\end{pmatrix} - \text{rank}(\mathbf{A})$. And consider the test of the null hypothesis H_0 (versus the alternative hypothesis H_1) with the following critical region:

$$\{\mathbf{y} \,:\, F > \bar{F}_{\dot{\gamma}}(M_*,\, N - P_*)\}. \tag{1.83}$$

This test can be regarded as an extension of what in the special case of the unconstrained G–M model is referred to as the ($\dot{\gamma}$-level) F test. And the statistic F on which it is based is sometimes (at least in that special case) referred to as the F statistic.

As is evident from result (1.44), the matrix $\mathbf{B}$ has various alternative representations and is such that

$$\text{var}(\hat{\boldsymbol{\tau}}) = \sigma^2\mathbf{B}.$$

And as is evident from result (1.59),

$$P_* = \text{rank}(\mathbf{X}) - \dim[\mathcal{R}(\mathbf{X}) \cap \mathcal{R}(\mathbf{A})]$$

and [upon applying result (1.59) with $\boldsymbol{\Lambda}'$ in place of $\mathbf{X}$]

$$M_* = \text{rank}(\boldsymbol{\Lambda}) - \dim[\mathcal{R}(\boldsymbol{\Lambda}') \cap \mathcal{R}(\mathbf{A})].$$

Distribution of the* F *statistic [when* $\mathbf{e} \sim N(\mathbf{0}, \sigma^2\mathbf{I})$*]. If the distribution of the vector $\mathbf{e}$ of residual effects in the G–M model is MVN (with mean $\mathbf{0}$ and variance-covariance matrix $\sigma^2\mathbf{I}$), then

$$F \sim SF(M_*,\, N - P_*,\, \lambda), \tag{1.84}$$

where $\lambda = (1/\sigma^2)(\boldsymbol{\tau} - \boldsymbol{\tau}^{(0)})'\mathbf{B}^{-}(\boldsymbol{\tau} - \boldsymbol{\tau}^{(0)})$.

Let us verify result (1.84). For that purpose, it is convenient to reexpress F in the form

$$F = \frac{(1/\sigma^2)(\hat{\boldsymbol{\tau}} - \boldsymbol{\tau}^{(0)})'\mathbf{B}^{-}(\hat{\boldsymbol{\tau}} - \boldsymbol{\tau}^{(0)})/M_*}{(1/\sigma^2)(\mathbf{y} - \hat{\boldsymbol{\mu}})'(\mathbf{y} - \hat{\boldsymbol{\mu}})/(N - P_*)}. \tag{1.85}$$

Note that $\mathbf{A}(\boldsymbol{\beta} - \underline{\mathbf{b}}) = \mathbf{d} - \mathbf{d} = \mathbf{0}$, so that $\boldsymbol{\beta} - \underline{\mathbf{b}} = (\mathbf{I} - \mathbf{A}^{-}\mathbf{A})\mathbf{u}$ for some column vector $\mathbf{u}$, in which case

$$(\mathbf{I} - \mathbf{P}_{\mathbf{W}})\mathbf{X}(\boldsymbol{\beta} - \underline{\mathbf{b}}) = (\mathbf{I} - \mathbf{P}_{\mathbf{W}})\mathbf{W}\mathbf{u} = \mathbf{0}.$$

Thus, making use of expression (1.53), we find that

$$\mathbf{y} - \hat{\boldsymbol{\mu}} = (\mathbf{I} - \mathbf{P}_{\mathbf{W}})\mathbf{X}(\boldsymbol{\beta} - \underline{\mathbf{b}}) + (\mathbf{I} - \mathbf{P}_{\mathbf{W}})\mathbf{e} = (\mathbf{I} - \mathbf{P}_{\mathbf{W}})\mathbf{e}. \tag{1.86}$$

And letting $\tilde{\mathbf{b}}$ represent the first (P-dimensional) part of any solution to the constrained normal equations $\begin{pmatrix}\mathbf{X}'\mathbf{X} & \mathbf{A}'\\ \mathbf{A} & \mathbf{0}\end{pmatrix}\begin{pmatrix}\mathbf{b}\\ \mathbf{r}\end{pmatrix} = \begin{pmatrix}\mathbf{X}'\mathbf{y}\\ \mathbf{d}\end{pmatrix}$ and making use of expression (1.50), we find also that

$$\begin{aligned}\hat{\boldsymbol{\tau}} - \boldsymbol{\tau}^{(0)} = \boldsymbol{\Lambda}'\tilde{\mathbf{b}} - \boldsymbol{\Lambda}'\underline{\mathbf{b}} &= \mathbf{L}_1'\mathbf{X}\tilde{\mathbf{b}} + \mathbf{L}_2'\mathbf{A}\tilde{\mathbf{b}} - \mathbf{L}_1'\mathbf{X}\underline{\mathbf{b}} - \mathbf{L}_2'\mathbf{A}\underline{\mathbf{b}}\\ &= \mathbf{L}_1'\hat{\boldsymbol{\mu}} - \mathbf{L}_1'\mathbf{X}\underline{\mathbf{b}} + \mathbf{L}_2'(\mathbf{d} - \mathbf{d})\\ &= \mathbf{L}_1'\mathbf{P}_{\mathbf{W}}(\mathbf{X}\boldsymbol{\beta} + \mathbf{e}) + \mathbf{L}_1'(\mathbf{I} - \mathbf{P}_{\mathbf{W}})\mathbf{X}\underline{\mathbf{b}} - \mathbf{L}_1'\mathbf{X}\underline{\mathbf{b}}\\ &= \mathbf{L}_1'\mathbf{P}_{\mathbf{W}}\mathbf{X}(\boldsymbol{\beta} - \underline{\mathbf{b}}) + \mathbf{L}_1'\mathbf{P}_{\mathbf{W}}\mathbf{e}.\end{aligned} \tag{1.87}$$

Based on results (1.87) and (1.86), we find that

$$\text{E}(\hat{\boldsymbol{\tau}} - \boldsymbol{\tau}^{(0)}) = \mathbf{L}_1'\mathbf{P}_{\mathbf{W}}\mathbf{X}(\boldsymbol{\beta} - \underline{\mathbf{b}}) \tag{1.88}$$

and (since $\mathbf{P_W}$ is symmetric and idempotent) that

$$\operatorname{cov}(\hat{\boldsymbol{\tau}}-\boldsymbol{\tau}^{(0)},\ \mathbf{y}-\hat{\boldsymbol{\mu}}) = \operatorname{cov}[\mathbf{L}_1'\mathbf{P_W}\mathbf{e},\ (\mathbf{I}-\mathbf{P_W})\mathbf{e}] = \mathbf{L}_1'\mathbf{P_W}(\sigma^2\mathbf{I})(\mathbf{I}-\mathbf{P_W}) = \mathbf{0} \tag{1.89}$$

and

$$\operatorname{var}(\hat{\boldsymbol{\tau}}-\boldsymbol{\tau}^{(0)}) = \operatorname{var}(\mathbf{L}_1'\mathbf{P_W}\mathbf{e}) = \sigma^2\mathbf{L}_1'\mathbf{P_W}\mathbf{L}_1. \tag{1.90}$$

Note that [since $\mathrm{E}(\hat{\boldsymbol{\tau}})=\boldsymbol{\tau}$] result (1.88) implies that $\boldsymbol{\tau}-\boldsymbol{\tau}^{(0)}$ is expressible as

$$\boldsymbol{\tau}-\boldsymbol{\tau}^{(0)} = \mathbf{L}_1'\mathbf{P_W}\mathbf{X}(\boldsymbol{\beta}-\underline{\mathbf{b}}), \tag{1.91}$$

and [since $\operatorname{var}(\hat{\boldsymbol{\tau}}-\boldsymbol{\tau}^{(0)}) = \sigma^2\mathbf{B}$] result (1.90) implies that the matrix $\mathbf{B}$ is expressible as

$$\mathbf{B} = \mathbf{L}_1'\mathbf{P_W}\mathbf{L}_1 = (\mathbf{P_W}\mathbf{L}_1)'\mathbf{P_W}\mathbf{L}_1. \tag{1.92}$$

Now, suppose that $\mathbf{e} \sim N(\mathbf{0}, \sigma^2\mathbf{I})$. Then, it follows from result (1.89) that $\hat{\boldsymbol{\tau}}-\boldsymbol{\tau}^{(0)}$ and $\mathbf{y}-\hat{\boldsymbol{\mu}}$ are statistically independent and hence that $(1/\sigma^2)(\hat{\boldsymbol{\tau}}-\tau^{(0)})'\mathbf{B}^-(\hat{\boldsymbol{\tau}}-\tau^{(0)})$ and $(1/\sigma^2)(\mathbf{y}-\hat{\boldsymbol{\mu}})'(\mathbf{y}-\hat{\boldsymbol{\mu}})$ are statistically independent. Moreover, upon observing that $(1/\sigma^2)(\mathbf{y}-\hat{\boldsymbol{\mu}})'(\mathbf{y}-\hat{\boldsymbol{\mu}}) = \mathbf{e}'[(1/\sigma^2)(\mathbf{I}-\mathbf{P_W})]\mathbf{e}$ and observing that $(1/\sigma^2)(\mathbf{I}-\mathbf{P_W})\operatorname{var}(\mathbf{e}) = \mathbf{I}-\mathbf{P_W}$, upon observing [in light of Lemma 2.8.4 and result (1.57)] that

$$\operatorname{rank}[(1/\sigma^2)(\mathbf{I}-\mathbf{P_W})] = \operatorname{rank}(\mathbf{I}-\mathbf{P_W}) = N-\operatorname{rank}(\mathbf{P_W}) = N-\operatorname{rank}(\mathbf{W}) = N-P_*,$$

and upon applying Corollary 6.6.4, we find that

$$(1/\sigma^2)(\mathbf{y}-\hat{\boldsymbol{\mu}})'(\mathbf{y}-\hat{\boldsymbol{\mu}}) \sim \chi^2(N-P_*). \tag{1.93}$$

Turning now to the distribution of $(1/\sigma^2)(\hat{\boldsymbol{\tau}}-\tau^{(0)})'\mathbf{B}^-(\hat{\boldsymbol{\tau}}-\tau^{(0)})$, let $\mathbf{K} = \mathbf{P_W}\mathbf{L}_1$, and observe [in light of results (1.87) and (1.92)] that

$$\begin{aligned}(1/\sigma^2)&(\hat{\boldsymbol{\tau}}-\tau^{(0)})'\mathbf{B}^-(\hat{\boldsymbol{\tau}}-\tau^{(0)}) \\ &= \sigma^{-2}(\boldsymbol{\beta}-\underline{\mathbf{b}})'\mathbf{X}'\mathbf{P_K}\mathbf{X}(\boldsymbol{\beta}-\underline{\mathbf{b}}) + 2\sigma^{-1}[\mathbf{P_K}\mathbf{X}(\boldsymbol{\beta}-\underline{\mathbf{b}})]'(\sigma^{-1}\mathbf{e}) + (\sigma^{-1}\mathbf{e})'\mathbf{P_K}(\sigma^{-1}\mathbf{e}).\end{aligned} \tag{1.94}$$

Then, upon applying Theorem 6.6.1 to expression (1.94) (which is a second-degree polynomial in $\sigma^{-1}\mathbf{e}$), we find that

$$(1/\sigma^2)(\hat{\boldsymbol{\tau}}-\tau^{(0)})'\mathbf{B}^-(\hat{\boldsymbol{\tau}}-\tau^{(0)}) \sim \chi^2[\operatorname{rank}(\mathbf{P_K}),\ \sigma^{-2}(\boldsymbol{\beta}-\underline{\mathbf{b}})'\mathbf{X}'\mathbf{P_K}\mathbf{X}(\boldsymbol{\beta}-\underline{\mathbf{b}})]. \tag{1.95}$$

Moreover, in light of result (1.91),

$$(\boldsymbol{\beta}-\underline{\mathbf{b}})'\mathbf{X}'\mathbf{P_K}\mathbf{X}(\boldsymbol{\beta}-\underline{\mathbf{b}}) = (\boldsymbol{\beta}-\underline{\mathbf{b}})'\mathbf{X}'\mathbf{K}\mathbf{B}^-\mathbf{K}'\mathbf{X}(\boldsymbol{\beta}-\underline{\mathbf{b}}) = (\boldsymbol{\tau}-\boldsymbol{\tau}^{(0)})'\mathbf{B}^-(\boldsymbol{\tau}-\boldsymbol{\tau}^{(0)}). \tag{1.96}$$

And upon observing that $\operatorname{rank}(\mathbf{L}_1'\mathbf{P_W}) = (\operatorname{rank}(\mathbf{L}_1'\mathbf{W})$ (as can be readily verified) and upon observing that

$$\boldsymbol{\Lambda}'(\mathbf{I}-\boldsymbol{\Lambda}^-\boldsymbol{\Lambda}) = (\mathbf{L}_1'\mathbf{X}+\mathbf{L}_2'\boldsymbol{\Lambda})(\mathbf{I}-\boldsymbol{\Lambda}^-\boldsymbol{\Lambda}) = \mathbf{L}_1'\mathbf{W},$$

we find that

$$\operatorname{rank}(\mathbf{P_K}) = \operatorname{rank}(\mathbf{K}) = \operatorname{rank}(\mathbf{K}') = \operatorname{rank}(\mathbf{L}_1'\mathbf{P_W}) = \operatorname{rank}(\mathbf{L}_1'\mathbf{W}) = \operatorname{rank}[\boldsymbol{\Lambda}'(\mathbf{I}-\boldsymbol{\Lambda}^-\boldsymbol{\Lambda})]$$

and hence [upon applying result (1.57) with $\boldsymbol{\Lambda}'$ in place of $\mathbf{X}$] that

$$\operatorname{rank}(\mathbf{P_K}) = \operatorname{rank}\begin{pmatrix}\boldsymbol{\Lambda}' \\ \boldsymbol{\Lambda}\end{pmatrix} - \operatorname{rank}(\boldsymbol{\Lambda}). \tag{1.97}$$

In summary, $(1/\sigma^2)(\hat{\boldsymbol{\tau}}-\tau^{(0)})'\mathbf{B}^-(\hat{\boldsymbol{\tau}}-\tau^{(0)}) \sim \chi^2(M_*, \lambda)$, $(1/\sigma^2)(\mathbf{y}-\hat{\boldsymbol{\mu}})'(\mathbf{y}-\hat{\boldsymbol{\mu}}) \sim \chi^2(N-P_*)$, and $(1/\sigma^2)(\hat{\boldsymbol{\tau}}-\tau^{(0)})'\mathbf{B}^-(\hat{\boldsymbol{\tau}}-\tau^{(0)})$ and $(1/\sigma^2)(\mathbf{y}-\hat{\boldsymbol{\mu}})'(\mathbf{y}-\hat{\boldsymbol{\mu}})$ are statistically independent, leading [in light of expression (1.85)] to the conclusion that $F \sim SF(M_*, N-P_*, \lambda)$.

Assumptions about the form of the distribution of the vector* e *of residual effects needed to insure the "validity" of the F test. Whether or not the test with critical region (1.83) is of level $\dot{\gamma}$ depends on the form of the distribution of the vector $\mathbf{e}$. If this distribution is such that under the null hypothesis H_0 (i.e., when $\boldsymbol{\tau} = \boldsymbol{\tau}^{(0)}$) $F \sim SF(M_*, N-P_*)$, then clearly the test with critical region (1.83) is of level $\dot{\gamma}$. As is evident from result (1.84), $F \sim SF(M_*, N-P_*)$ under H_0 when the distribution of $\mathbf{e}$ is MVN. More generally, $F \sim SF(M_*, N-P_*)$ under H_0 when $\mathbf{e}$ has an absolutely continuous spherical distribution, as is demonstrated in what follows [thereby extending the results obtained (in Section 7.3b) for the unconstrained G–M model].

Let us make use of the results established in the preceding part of the present subsection and adopt the notation introduced therein. And let us take $\mathbf{O}_1$ and $\mathbf{O}_2$ to be $N\times N$ orthogonal matrices such that

$$\mathbf{O}_1'\mathbf{P}_{\mathbf{K}}\mathbf{O}_1 = \operatorname{diag}(\mathbf{I}_{M_*},\, \mathbf{0}) \quad \text{and} \quad \mathbf{O}_2'(\mathbf{I}-\mathbf{P}_{\mathbf{W}})\mathbf{O}_2 = \operatorname{diag}(\mathbf{I}_{N-P_*},\, \mathbf{0})$$

—the existence of such matrices is evident from the results of Section 6.7a upon observing that $\mathbf{P}_{\mathbf{K}}$ is symmetric and idempotent of rank M_* and that $\mathbf{I}-\mathbf{P}_{\mathbf{W}}$ is symmetric and idempotent of rank $N-P_*$. Further, let us partition $\mathbf{O}_1$ as $\mathbf{O}_1 = (\mathbf{O}_{11},\, \mathbf{O}_{12})$ and $\mathbf{O}_2$ as $\mathbf{O}_2 = (\mathbf{O}_{21},\, \mathbf{O}_{22})$ [where $\mathbf{O}_{11}$ is of dimensions $N\times M_*$ and $\mathbf{O}_{21}$ of dimensions $N\times(N-P_*)$], so that

$$\mathbf{P}_{\mathbf{K}} = \mathbf{O}_1 \operatorname{diag}(\mathbf{I}_{M_*},\, \mathbf{0})\mathbf{O}_1' = \mathbf{O}_{11}\mathbf{O}_{11}' \quad \text{and} \quad \mathbf{I}-\mathbf{P}_{\mathbf{W}} = \mathbf{O}_2 \operatorname{diag}(\mathbf{I}_{N-P_*},\, \mathbf{0})\mathbf{O}_2' = \mathbf{O}_{21}\mathbf{O}_{21}'$$

and [since $\mathbf{O}_{21}\mathbf{O}_{21}'\mathbf{O}_{11}\mathbf{O}_{11}' = (\mathbf{I}-\mathbf{P}_{\mathbf{W}})\mathbf{P}_{\mathbf{K}} = (\mathbf{I}-\mathbf{P}_{\mathbf{W}})\mathbf{P}_{\mathbf{W}}\mathbf{L}_1(\mathbf{K}'\mathbf{K})^-\mathbf{K}' = \mathbf{0}$]

$$\mathbf{O}_{21}'\mathbf{O}_{11} = \mathbf{0};$$

and let us take $\mathbf{O} = (\mathbf{O}_{11},\, \mathbf{O}_{21},\, \mathbf{O}_3)$, where $\mathbf{O}_3$ is an $N\times(P_*-M_*)$ matrix whose columns form an orthonormal basis for the null space of the matrix $(\mathbf{O}_{11},\, \mathbf{O}_{21})'$—the existence of such an orthonormal basis follows from Theorem 2.4.23. Clearly, $\mathbf{O}$ is an orthogonal matrix.

Now, let $\mathbf{z} = \mathbf{O}'\mathbf{e}$, and suppose that $\mathbf{e}$ has an absolutely continuous spherical distribution. Then, $\mathbf{z}$ has an absolutely continuous spherical distribution, and the marginal distribution of the (M_*+N-P_*)-dimensional vector consisting of the first through M_*+N-P_* elements of $\mathbf{z}$, say $z_1, z_2, \ldots, z_{M_*+N-P_*}$, is an absolutely continuous spherical distribution.

The constrained residual sum of squares $(\mathbf{y}-\hat{\boldsymbol{\mu}})'(\mathbf{y}-\hat{\boldsymbol{\mu}})$ is reexpressible as

$$(\mathbf{y}-\hat{\boldsymbol{\mu}})'(\mathbf{y}-\hat{\boldsymbol{\mu}}) = \mathbf{e}'(\mathbf{I}-\mathbf{P}_{\mathbf{W}})\mathbf{e} = \mathbf{e}'\mathbf{O}_{21}\mathbf{O}_{21}'\mathbf{e} = \textstyle\sum_{i=M_*+1}^{M_*+N-P_*} z_i^2,$$

and, under H_0, $(\hat{\boldsymbol{\tau}}-\boldsymbol{\tau}^{(0)})'\mathbf{B}^-(\hat{\boldsymbol{\tau}}-\boldsymbol{\tau}^{(0)})$ is reexpressible as

$$(\hat{\boldsymbol{\tau}}-\boldsymbol{\tau}^{(0)})'\mathbf{B}^-(\hat{\boldsymbol{\tau}}-\boldsymbol{\tau}^{(0)}) = \mathbf{e}'\mathbf{P}_{\mathbf{K}}\mathbf{e} = \mathbf{e}'\mathbf{O}_{11}\mathbf{O}_{11}'\mathbf{e} = \textstyle\sum_{i=1}^{M_*} z_i^2.$$

Thus, under H_0, the F statistic is reexpressible as

$$F = \frac{\sum_{i=1}^{M_*} z_i^2/M_*}{\sum_{i=M_*+1}^{M_*+N-P_*} z_i^2/(N-P_*)}, \tag{1.98}$$

leading (on the basis of the discussion in Part 1 of Section 6.3b) to the conclusion that, under H_0,

$$F \sim SF(M_*,\, N-P_*). \tag{1.99}$$

f. An alternative approach: transformation to an "equivalent" unconstrained model

Suppose (as in Subsections d and e) that $\mathbf{y}$ is an $N\times 1$ observable random vector that follows a constrained G–M, Aitken, or general linear model (with model equation $\mathbf{y} = \mathbf{X}\boldsymbol{\beta}+\mathbf{e}$ and constraint $\mathbf{A}\boldsymbol{\beta} = \mathbf{d}$). And observe (in light of Theorem 2.11.7) that a $P\times 1$ vector $\underline{\mathbf{b}}$ is a solution to the (consistent) linear system $\mathbf{A}\mathbf{b} = \mathbf{d}$ (in the $P\times 1$ vector $\mathbf{b}$) if and only if

$$\underline{\mathbf{b}} = \mathbf{A}^-\mathbf{d} + (\mathbf{I}-\mathbf{A}^-\mathbf{A})\mathbf{t} \tag{1.100}$$

for some $P\times 1$ vector $\mathbf{t}$. Accordingly, an $N\times 1$ vector, say $\mathbf{m}$, is expressible in the form $\mathbf{m} = \mathbf{X}\underline{\mathbf{b}}$ for some solution $\underline{\mathbf{b}}$ to $\mathbf{A}\mathbf{b} = \mathbf{d}$ if and only if

$$\mathbf{m} = \mathbf{X}\mathbf{A}^-\mathbf{d} + \mathbf{X}(\mathbf{I}-\mathbf{A}^-\mathbf{A})\mathbf{t} \tag{1.101}$$

for some $P\times 1$ vector $\mathbf{t}$. This result suggests that we replace $\mathbf{X}\boldsymbol{\beta}$ in the model equation $\mathbf{y} = \mathbf{X}\boldsymbol{\beta}+\mathbf{e}$ with the expression

$$\mathbf{X}\mathbf{A}^-\mathbf{d} + \mathbf{W}\boldsymbol{\tau},$$

where $\mathbf{W} = \mathbf{X}(\mathbf{I}-\mathbf{A}^-\mathbf{A})$ and where $\boldsymbol{\tau}$ is a $P\times 1$ vector. Then,

$$\mathbf{y} = \mathbf{X}\mathbf{A}^-\mathbf{d} + \mathbf{W}\boldsymbol{\tau} + \mathbf{e} \tag{1.102}$$

or, equivalently,

$$\mathbf{z} = \mathbf{W}\boldsymbol{\tau} + \mathbf{e}, \tag{1.103}$$

where $\mathbf{z} = \mathbf{y} - \mathbf{X}\mathbf{A}^-\mathbf{d}$.

Let us regard equality (1.103) as the model equation for an (unconstrained) G–M, Aitken, or general linear model. This model is one in which it is the transformed vector $\mathbf{z}$ (rather than the vector $\mathbf{y}$) that plays the role of the observable random vector (the realization of which is the data vector), in which $\mathbf{W}$ is the model matrix, and in which $\boldsymbol{\tau}$ represents an (unconstrained) vector of parameters.

The distribution of $\mathbf{y}$ (or $\mathbf{z}$) under the constrained G–M, Aitken, or general linear model (with model equation $\mathbf{y} = \mathbf{X}\boldsymbol{\beta} + \mathbf{e}$ and constraint $\mathbf{A}\boldsymbol{\beta} = \mathbf{d}$) can be readily determined from its distribution under the corresponding unconstrained G–M, Aitken, or general linear model (with model equation $\mathbf{z} = \mathbf{W}\boldsymbol{\tau} + \mathbf{e}$). It is simply a matter of observing that the latter distribution depends on the parameter vector $\boldsymbol{\tau}$ only through the value of $\mathbf{W}\boldsymbol{\tau}$ and hence only through the value of $(\mathbf{I}-\mathbf{A}^-\mathbf{A})\boldsymbol{\tau}$ and of reexpressing that distribution in terms of the parameter vector $\boldsymbol{\beta}$ by replacing $(\mathbf{I}-\mathbf{A}^-\mathbf{A})\boldsymbol{\tau}$ with $\boldsymbol{\beta}-\mathbf{A}^-\mathbf{d}$—when $(\mathbf{I}-\mathbf{A}^-\mathbf{A})\boldsymbol{\tau} = \boldsymbol{\beta}-\mathbf{A}^-\mathbf{d}$,

$$\mathbf{y} = \mathbf{X}\mathbf{A}^-\mathbf{d} + \mathbf{z} = \mathbf{X}\mathbf{A}^-\mathbf{d} + \mathbf{X}(\boldsymbol{\beta}-\mathbf{A}^-\mathbf{d}) + \mathbf{e} = \mathbf{X}\boldsymbol{\beta} + \mathbf{e}$$

and

$$\mathbf{A}\boldsymbol{\beta} = \mathbf{A}[\mathbf{A}^-\mathbf{d} + (\mathbf{I}-\mathbf{A}^-\mathbf{A})\boldsymbol{\tau}] = \mathbf{d}.$$

Moreover, results obtained under the unconstrained G–M, Aitken, or general linear model for the transformed observable random vector $\mathbf{z}$ can be readily translated into results obtainable under the constrained G–M, Aitken, or general linear model for the observable random vector $\mathbf{y}$, as is demonstrated in what follows.

Estimability. A linear combination $\boldsymbol{\lambda}'\boldsymbol{\beta}$ of the elements of $\boldsymbol{\beta}$ is estimable under the constrained G–M, Aitken, or general linear model for $\mathbf{y}$ if and only if the linear combination $\boldsymbol{\lambda}'(\mathbf{I}-\mathbf{A}^-\mathbf{A})\boldsymbol{\tau}$ (of the elements of $\boldsymbol{\tau}$) is estimable under the corresponding unconstrained G–M, Aitken, or general linear model for the transformed vector $\mathbf{z}$.

For purposes of verification, recall (from Subsection a) that $\boldsymbol{\lambda}'\boldsymbol{\beta}$ is estimable under the constrained G–M, Aitken, or general linear model for $\mathbf{y}$ if and only if $\boldsymbol{\lambda}' = \mathbf{a}'\mathbf{X} + \mathbf{r}'\mathbf{A}$ for some column vectors $\mathbf{a}$ and $\mathbf{r}$. Moreover, $\boldsymbol{\lambda}' = \mathbf{a}'\mathbf{X} + \mathbf{r}'\mathbf{A}$ for some column vectors $\mathbf{a}$ and $\mathbf{r}$ if and only if $\boldsymbol{\lambda}'(\mathbf{I}-\mathbf{A}^-\mathbf{A}) = \boldsymbol{\ell}'\mathbf{W}$ for some column vector $\boldsymbol{\ell}$ (as can be readily verified) and hence if and only if $\boldsymbol{\lambda}'(\mathbf{I}-\mathbf{A}^-\mathbf{A})\boldsymbol{\tau}$ is estimable under the unconstrained G–M, Aitken, or general linear model for $\mathbf{z}$ (with model equation $\mathbf{z} = \mathbf{W}\boldsymbol{\tau} + \mathbf{e}$).

Constrained and unconstrained normal equations and least squares estimators. Let $\underline{\mathbf{b}}$ represent any value of $\mathbf{b}$ and $\underline{\mathbf{r}}$ any value of $\mathbf{r}$ for which $\begin{pmatrix}\underline{\mathbf{b}}\\ \underline{\mathbf{r}}\end{pmatrix}$ constitutes a solution to the constrained normal equations $\begin{pmatrix}\mathbf{X}'\mathbf{X} & \mathbf{A}'\\ \mathbf{A} & \mathbf{0}\end{pmatrix}\begin{pmatrix}\mathbf{b}\\ \mathbf{r}\end{pmatrix} = \begin{pmatrix}\mathbf{X}'\mathbf{y}\\ \mathbf{d}\end{pmatrix}$ (for the constrained G–M, Aitken, or general linear model for $\mathbf{y}$), and let $\underline{\mathbf{t}}$ represent any solution to the normal equations $\mathbf{W}'\mathbf{W}\mathbf{t} = \mathbf{W}'\mathbf{z}$ (for the corresponding unconstrained G–M, Aitken, or general linear model for the transformed vector $\mathbf{z}$). Then, for any linear combination $\boldsymbol{\lambda}'\boldsymbol{\beta}$ of the elements of $\boldsymbol{\beta}$ that is estimable (under the constrained G–M, Aitken, or general linear model), $\boldsymbol{\lambda}'\underline{\mathbf{b}}$ (which is the constrained least squares estimator of $\boldsymbol{\lambda}'\boldsymbol{\beta}$) is reexpressible in terms of $\underline{\mathbf{t}}$ (and in terms of the least squares estimator $\boldsymbol{\lambda}'(\mathbf{I}-\mathbf{A}^-\mathbf{A})\underline{\mathbf{t}}$ of $\boldsymbol{\lambda}'(\mathbf{I}-\mathbf{A}^-\mathbf{A})\boldsymbol{\tau}$) as

$$\boldsymbol{\lambda}'\underline{\mathbf{b}} = \boldsymbol{\lambda}'\mathbf{A}^-\mathbf{d} + \boldsymbol{\lambda}'(\mathbf{I}-\mathbf{A}^-\mathbf{A})\underline{\mathbf{t}}. \tag{1.104}$$

Let us verify equality (1.104). Since $\mathbf{A}\underline{\mathbf{b}} = \mathbf{d}$,

$$\underline{\mathbf{b}} = \mathbf{A}^-\mathbf{d} + (\mathbf{I}-\mathbf{A}^-\mathbf{A})\mathbf{t}^* \tag{1.105}$$

for some vector $\mathbf{t}^*$. And since $\mathbf{X}'\mathbf{X}\underline{\mathbf{b}} + \mathbf{A}'\underline{\mathbf{r}} = \mathbf{X}'\mathbf{y}$,

$$\mathbf{X}'\mathbf{W}\mathbf{t}^* + \mathbf{A}'\underline{\mathbf{r}} = \mathbf{X}'\mathbf{z}. \tag{1.106}$$

Moreover, equality (1.106) implies that

$$\mathbf{W}'\mathbf{W}\mathbf{t}^* = \mathbf{W}'\mathbf{z},$$

as becomes evident upon premultiplying both sides of equality (1.106) by $(\mathbf{I}-\mathbf{A}^-\mathbf{A})'$.

Now, upon premultiplying expression (1.105) by $\boldsymbol{\lambda}'$, we obtain the expression

$$\boldsymbol{\lambda}'\underline{\mathbf{b}} = \boldsymbol{\lambda}'\mathbf{A}^-\mathbf{d} + \boldsymbol{\lambda}'(\mathbf{I}-\mathbf{A}^-\mathbf{A})\mathbf{t}^*,$$

which indicates that equality (1.104) holds for some solution to $\mathbf{W}'\mathbf{W}\mathbf{t} = \mathbf{W}'\mathbf{z}$. To confirm that it holds for every solution, observe (in light of the estimability of $\boldsymbol{\lambda}'\boldsymbol{\beta}$) that $\boldsymbol{\lambda}' = \mathbf{a}'\mathbf{X} + \mathbf{r}'\mathbf{A}$ for some column vectors $\mathbf{a}$ and $\mathbf{r}$ and hence that $\boldsymbol{\lambda}'(\mathbf{I}-\mathbf{A}^-\mathbf{A}) = \mathbf{a}'\mathbf{W}$ for some column vector $\mathbf{a}$, and recall that $\mathbf{W}\mathbf{t}$ has the same value for every solution to $\mathbf{W}'\mathbf{W}\mathbf{t} = \mathbf{W}'\mathbf{z}$.

Residual vector and the residual sum of squares. In the case of the (unconstrained) G–M, Aitken, or general linear model for the transformed observable random vector $\mathbf{z}$ (with model equation $\mathbf{z} = \mathbf{W}\boldsymbol{\tau} + \mathbf{e}$),

$$\mathrm{E}(\mathbf{z}) = \mathbf{W}\boldsymbol{\tau};$$

the residual vector (i.e., the vector whose elements are the least squares residuals) is

$$\mathbf{z} - \mathbf{W}\underline{\mathbf{t}},$$

where $\underline{\mathbf{t}}$ is any solution to $\mathbf{W}'\mathbf{W}\mathbf{t} = \mathbf{W}'\mathbf{z}$; the residual sum of squares is

$$(\mathbf{z}-\mathbf{W}\underline{\mathbf{t}})'(\mathbf{z}-\mathbf{W}\underline{\mathbf{t}});$$

and (in the special case of the G–M model)

$$(\mathbf{z}-\mathbf{W}\underline{\mathbf{t}})'(\mathbf{z}-\mathbf{W}\underline{\mathbf{t}})/[N-\mathrm{rank}(\mathbf{W})]$$

is an unbiased estimator of σ^2.

In the case of the constrained G–M, Aitken, or general linear model for the observable random vector $\mathbf{y}$ (with model equation $\mathbf{y} = \mathbf{X}\boldsymbol{\beta} + \mathbf{e}$ and constraint $\mathbf{A}\boldsymbol{\beta} = \mathbf{d}$),

$$\mathrm{E}(\mathbf{y}) = \mathbf{X}\boldsymbol{\beta};$$

the residual vector is

$$\mathbf{y}-\mathbf{X}\underline{\mathbf{b}},$$

where $\underline{\mathbf{b}}$ is the $P\times 1$ vector whose elements are the first P elements of any solution to the constrained normal equations $\begin{pmatrix}\mathbf{X}'\mathbf{X} & \mathbf{A}' \\ \mathbf{A} & \mathbf{0}\end{pmatrix}\begin{pmatrix}\mathbf{b} \\ \mathbf{r}\end{pmatrix} = \begin{pmatrix}\mathbf{X}'\mathbf{y} \\ \mathbf{d}\end{pmatrix}$; the residual sum of squares is

$$(\mathbf{y}-\mathbf{X}\underline{\mathbf{b}})'(\mathbf{y}-\mathbf{X}\underline{\mathbf{b}});$$

and (in the special case of the constrained G–M model)

$$\frac{(\mathbf{y}-\mathbf{X}\underline{\mathbf{b}})'(\mathbf{y}-\mathbf{X}\underline{\mathbf{b}})}{N - [\mathrm{rank}\begin{pmatrix}\mathbf{X} \\ \mathbf{A}\end{pmatrix} - \mathrm{rank}(\mathbf{A})]} \quad \left(= \frac{(\mathbf{y}-\mathbf{X}\underline{\mathbf{b}})'(\mathbf{y}-\mathbf{X}\underline{\mathbf{b}})}{N - \mathrm{rank}(\mathbf{W})}\right) \tag{1.107}$$

is an unbiased estimator of σ^2.

It follows from result (1.104) that

$$\mathbf{X}\underline{\mathbf{b}} = \mathbf{X}\mathbf{A}^-\mathbf{d} + \mathbf{W}\underline{\mathbf{t}}. \tag{1.108}$$

Thus,

$$\mathbf{z} - \mathbf{W}\underline{\mathbf{t}} = \mathbf{y} - \mathbf{X}\mathbf{A}^-\mathbf{d} - \mathbf{W}\underline{\mathbf{t}} = \mathbf{y}-\mathbf{X}\underline{\mathbf{b}} \tag{1.109}$$

and

$$(\mathbf{z}-\mathbf{W}\underline{\mathbf{t}})'(\mathbf{z}-\mathbf{W}\underline{\mathbf{t}}) = (\mathbf{y}-\mathbf{X}\underline{\mathbf{b}})'(\mathbf{y}-\mathbf{X}\underline{\mathbf{b}}). \tag{1.110}$$

That is, the residual vector and the residual sum of squares obtained on the basis of the unconstrained G–M, Aitken, or general linear model (with model equation $\mathbf{z} = \mathbf{W}\boldsymbol{\tau} + \mathbf{e}$) are the same as those obtained on the basis of the constrained G–M, Aitken, or general linear model (with model equation $\mathbf{y} = \mathbf{X}\boldsymbol{\beta} + \mathbf{e}$ and constraint $\mathbf{A}\boldsymbol{\beta} = \mathbf{d}$). Moreover, the two estimators of σ^2 are also the same.

A generalization. The results and the discussion of the preceding parts of the present subsection (Subsection f) can be generalized. Let $\tilde{\mathbf{b}}$ represent any value of the $P \times 1$ vector $\mathbf{b}$ that satisfies the equality $\mathbf{Ab} = \mathbf{d}$. Further, let $\mathbf{L}$ represent any matrix (with P rows) whose columns span $\mathcal{N}(\mathbf{A})$, and denote by K the number of columns in $\mathbf{L}$—necessarily, $K \geq P - \text{rank}(\mathbf{A})$. Then, as generalizations of results (1.100) and (1.101), we have that

$$\underline{\mathbf{b}} = \tilde{\mathbf{b}} + \mathbf{L}\underline{\mathbf{t}} \tag{1.111}$$

and

$$\mathbf{m} = \mathbf{X}\tilde{\mathbf{b}} + \mathbf{XLt} \tag{1.112}$$

for some $K \times 1$ vector $\mathbf{t}$—results (1.100) and (1.101) are the special cases of results (1.111) and (1.112) where $\tilde{\mathbf{b}} = \mathbf{A}^{-}\mathbf{d}$ and $\mathbf{L} = \mathbf{I} - \mathbf{A}^{-}\mathbf{A}$ (and where $K = P$). And upon letting $\boldsymbol{\tau}$ represent a $K \times 1$ vector and upon replacing $\mathbf{X}\boldsymbol{\beta}$ in the model equation $\mathbf{y} = \mathbf{X}\boldsymbol{\beta} + \mathbf{e}$ with the expression

$$\mathbf{X}\tilde{\mathbf{b}} + \mathbf{W}\boldsymbol{\tau},$$

where $\mathbf{W} = \mathbf{XL}$, we obtain [as a generalization of expression (1.102)]

$$\mathbf{y} = \mathbf{X}\tilde{\mathbf{b}} + \mathbf{W}\boldsymbol{\tau} + \mathbf{e} \tag{1.113}$$

or equivalently [as a generalization of equality (1.103)]

$$\mathbf{z} = \mathbf{W}\boldsymbol{\tau} + \mathbf{e}, \tag{1.114}$$

where $\mathbf{z} = \mathbf{y} - \mathbf{X}\tilde{\mathbf{b}}$.

Now, regard $\mathbf{z} = \mathbf{y} - \mathbf{X}\tilde{\mathbf{b}}$ as an observable random vector that follows an (unconstrained) G–M, Aitken, or general linear model with model equation (1.114). Then, as in the special case where $\tilde{\mathbf{b}} = \mathbf{A}^{-}\mathbf{d}$ and $\mathbf{L} = \mathbf{I} - \mathbf{A}^{-}\mathbf{A}$, the distribution of $\mathbf{z}$ depends on the parameter vector $\boldsymbol{\tau}$ only through the value of $\mathbf{L}\boldsymbol{\tau}$, and the distribution of $\mathbf{y}$ under the constrained G–M, Aitken, or general linear model for $\mathbf{y}$ (with model equation $\mathbf{y} = \mathbf{X}\boldsymbol{\beta} + \mathbf{e}$ and constraint $\mathbf{A}\boldsymbol{\beta} = \mathbf{d}$) can be determined from the distribution of $\mathbf{z}$ simply by replacing $\mathbf{L}\boldsymbol{\tau}$ with $\boldsymbol{\beta} - \tilde{\mathbf{b}}$. And (as in the special case where $\tilde{\mathbf{b}} = \mathbf{A}^{-}\mathbf{d}$ and $\mathbf{L} = \mathbf{I} - \mathbf{A}^{-}\mathbf{A}$) inferences from $\mathbf{y}$ about $\boldsymbol{\lambda}'\boldsymbol{\beta}$ under the constrained G–M, Aitken, or general linear model are related to inferences from $\mathbf{z}$ about $\boldsymbol{\lambda}'\mathbf{L}\boldsymbol{\tau}$ as follows:

(1) $\boldsymbol{\lambda}'\boldsymbol{\beta}$ is estimable if and only if $\boldsymbol{\lambda}'\mathbf{L}\boldsymbol{\tau}$ is estimable;

(2) if $\boldsymbol{\lambda}'\boldsymbol{\beta}$ is estimable, then for any solution $\begin{pmatrix} \underline{\mathbf{b}} \\ \underline{\mathbf{r}} \end{pmatrix}$ to the constrained normal equations $\begin{pmatrix} \mathbf{X}'\mathbf{X} & \mathbf{A}' \\ \mathbf{A} & \mathbf{0} \end{pmatrix} \begin{pmatrix} \mathbf{b} \\ \mathbf{r} \end{pmatrix} = \begin{pmatrix} \mathbf{X}'\mathbf{y} \\ \mathbf{d} \end{pmatrix}$ and any solution $\underline{\mathbf{t}}$ to the unconstrained normal equations $\mathbf{W}'\mathbf{Wt} = \mathbf{W}'\mathbf{z}$,

$$\boldsymbol{\lambda}'\underline{\mathbf{b}} = \boldsymbol{\lambda}'\tilde{\mathbf{b}} + \boldsymbol{\lambda}'\mathbf{L}\underline{\mathbf{t}}, \tag{1.115}$$

so that the (constrained) least squares estimator of an estimable linear combination $\boldsymbol{\lambda}'\boldsymbol{\beta}$ equals a constant $\boldsymbol{\lambda}'\tilde{\mathbf{b}}$ plus the (unconstrained) least squares estimator of $\boldsymbol{\lambda}'\mathbf{L}\boldsymbol{\tau}$.

The validity of results (1) and (2) in the special case where $\tilde{\mathbf{b}} = \mathbf{A}^{-}\mathbf{d}$ and $\mathbf{L} = \mathbf{I} - \mathbf{A}^{-}\mathbf{A}$ is evident from what was established earlier (in preceding parts of the present subsection). And the validity of result (1) in that special case can be used to establish its validity in the general case. Specifically, upon observing that

$$\mathbf{L} = (\mathbf{I} - \mathbf{A}^{-}\mathbf{A})\mathbf{S} \quad \text{and} \quad \mathbf{I} - \mathbf{A}^{-}\mathbf{A} = \mathbf{LT} \quad \text{for some matrices } \mathbf{S} \text{ and } \mathbf{T}, \tag{1.116}$$

the validity of result (1) can be extended to the general case by observing that

$$\begin{aligned} \boldsymbol{\lambda}'\mathbf{L}\boldsymbol{\tau} \text{ is estimable in the general case} &\Leftrightarrow \boldsymbol{\lambda}'\mathbf{L} = \mathbf{a}'\mathbf{XL} \text{ for some vector } \mathbf{a} \\ &\Leftrightarrow (\boldsymbol{\lambda}' - \mathbf{a}'\mathbf{X})\mathbf{L} = \mathbf{0} \text{ for some vector } \mathbf{a} \\ &\Leftrightarrow (\boldsymbol{\lambda}' - \mathbf{a}'\mathbf{X})(\mathbf{I} - \mathbf{A}^{-}\mathbf{A}) = \mathbf{0} \text{ for some vector } \mathbf{a} \\ &\Leftrightarrow \boldsymbol{\lambda}'(\mathbf{I} - \mathbf{A}^{-}\mathbf{A}) = \mathbf{a}'\mathbf{X}(\mathbf{I} - \mathbf{A}^{-}\mathbf{A}) \text{ for some vector } \mathbf{a} \\ &\Leftrightarrow \boldsymbol{\lambda}'(\mathbf{I} - \mathbf{A}^{-}\mathbf{A})\boldsymbol{\tau} \text{ is estimable in the special case.} \end{aligned}$$

Further, the validity of result (2) can be extended to the general case by observing that (as in the special case where $\tilde{\mathbf{b}} = \mathbf{A}^{-}\mathbf{d}$ and $\mathbf{L} = \mathbf{I}-\mathbf{A}^{-}\mathbf{A}$) $\underline{\mathbf{b}} = \tilde{\mathbf{b}} + \mathbf{L}\mathbf{t}^{*}$ for some vector $\mathbf{t}^{*}$ and that $\mathbf{AL} = \mathbf{0}$ and by proceeding in essentially the same way as in establishing the special case of result (2).

As in the special case where $\tilde{\mathbf{b}} = \mathbf{A}^{-}\mathbf{d}$ and $\mathbf{L} = \mathbf{I}-\mathbf{A}^{-}\mathbf{A}$, the transformed observable random vector $\mathbf{z} = \mathbf{y} - \mathbf{X}\tilde{\mathbf{b}}$ (which follows a G–M, Aitken, or general linear model with model equation $\mathbf{z} = \mathbf{W}\boldsymbol{\tau} + \mathbf{e}$, where $\mathbf{W} = \mathbf{XL}$) is such that

$$\mathrm{E}(\mathbf{z}) = \mathbf{W}\boldsymbol{\tau};$$

the vector of least squares residuals is

$$\mathbf{z} - \mathbf{W}\underline{\mathbf{t}},$$

where $\underline{\mathbf{t}}$ is any solution to $\mathbf{W}'\mathbf{W}\mathbf{t} = \mathbf{W}'\mathbf{z}$; the residual sum of squares is

$$(\mathbf{z}-\mathbf{W}\underline{\mathbf{t}})'(\mathbf{z}-\mathbf{W}\underline{\mathbf{t}});$$

and (in the special case of the G–M model)

$$(\mathbf{z}-\mathbf{W}\underline{\mathbf{t}})'(\mathbf{z}-\mathbf{W}\underline{\mathbf{t}})/[N-\mathrm{rank}(\mathbf{W})] \tag{1.117}$$

is an unbiased estimator of σ^2. And upon letting $\underline{\mathbf{b}}$ represent the $P\times 1$ vector whose elements are the first P elements of any solution to the constrained normal equations $\begin{pmatrix} \mathbf{X}'\mathbf{X} & \mathbf{A}' \\ \mathbf{A} & \mathbf{0} \end{pmatrix} \begin{pmatrix} \mathbf{b} \\ \mathbf{r} \end{pmatrix} = \begin{pmatrix} \mathbf{X}'\mathbf{y} \\ \mathbf{d} \end{pmatrix}$ and upon making use of result (1.115), we find that (as in the special case where $\tilde{\mathbf{b}} = \mathbf{A}^{-}\mathbf{d}$ and $\mathbf{L} = \mathbf{I}-\mathbf{A}^{-}\mathbf{A}$)

$$\mathbf{X}\underline{\mathbf{b}} = \mathbf{X}\tilde{\mathbf{b}} + \mathbf{W}\underline{\mathbf{t}} \tag{1.118}$$

and hence that

$$\mathbf{z} - \mathbf{W}\underline{\mathbf{t}} = \mathbf{y} - \mathbf{X}\tilde{\mathbf{b}} - \mathbf{W}\underline{\mathbf{t}} = \mathbf{y}-\mathbf{X}\underline{\mathbf{b}} \tag{1.119}$$

and

$$(\mathbf{z}-\mathbf{W}\underline{\mathbf{t}})'(\mathbf{z}-\mathbf{W}\underline{\mathbf{t}}) = (\mathbf{y}-\mathbf{X}\underline{\mathbf{b}})'(\mathbf{y}-\mathbf{X}\underline{\mathbf{b}}). \tag{1.120}$$

Moreover, $\mathrm{rank}(\mathbf{W})$ is the same as in the special case where $\mathbf{L} = \mathbf{I}-\mathbf{A}^{-}\mathbf{A}$ [as is evident from result(1.116)], so that (as in that special case)

$$\mathrm{rank}(\mathbf{W}) = \mathrm{rank}\begin{pmatrix} \mathbf{X} \\ \mathbf{A} \end{pmatrix} - \mathrm{rank}(\mathbf{A}) \tag{1.121}$$

and (as in the special case where $\tilde{\mathbf{b}} = \mathbf{A}^{-}\mathbf{d}$ and $\mathbf{L} = \mathbf{I}-\mathbf{A}^{-}\mathbf{A}$)

$$\frac{(\mathbf{z}-\mathbf{W}\underline{\mathbf{t}})'(\mathbf{z}-\mathbf{W}\underline{\mathbf{t}})}{N-\mathrm{rank}(\mathbf{W})} = \frac{(\mathbf{y}-\mathbf{X}\underline{\mathbf{b}})'(\mathbf{y}-\mathbf{X}\underline{\mathbf{b}})}{N - [\mathrm{rank}\begin{pmatrix} \mathbf{X} \\ \mathbf{A} \end{pmatrix} - \mathrm{rank}(\mathbf{A})]}.$$

The QR decomposition of $\mathbf{A}'$ ***as a basis for choosing*** $\tilde{\mathbf{b}}$ ***and*** $\mathbf{L}$. Assume that the constraint $\mathbf{A}\boldsymbol{\beta} = \mathbf{d}$ is such that $\mathbf{A}$ is of full row rank—this assumption can be made without loss of generality in the sense that if $\mathrm{rank}(\mathbf{A}) = Q_{*} < Q$, then $\mathbf{A}$ contains Q_{*} linearly independent rows, and the constraint $\mathbf{A}\boldsymbol{\beta} = \mathbf{d}$ can be replaced by the equivalent constraint $\mathbf{A}_{*}\boldsymbol{\beta} = \mathbf{d}_{*}$, where $\mathbf{A}_{*}$ is the $Q_{*}\times P$ matrix (of full row rank) whose rows are Q_{*} linearly independent rows of $\mathbf{A}$ and where $\mathbf{d}_{*}$ is the $Q_{*}\times 1$ vector whose elements are the Q_{*} elements of $\mathbf{d}$ corresponding to those Q_{*} linearly independent rows of $\mathbf{A}$. And observe that there exist a $P\times P$ orthogonal matrix $\mathbf{O}$ and a $Q\times Q$ upper triangular matrix $\mathbf{U}_1$ (with strictly positive diagonal elements) such that

$$\mathbf{A}' = \mathbf{O}\mathbf{U} = \mathbf{O}_1\mathbf{U}_1, \tag{1.122}$$

where $\mathbf{U} = \begin{pmatrix} \mathbf{U}_1 \\ \mathbf{0} \end{pmatrix}$ and $\mathbf{O} = (\mathbf{O}_1, \mathbf{O}_2)$ [with $\mathbf{O}_1$ being of dimensions $P\times Q$ and $\mathbf{O}_2$ of dimensions $P\times(P-Q)$]. The decomposition (1.122) is the QR decomposition of the matrix $\mathbf{A}'$—refer to Section 5.4e for a discussion of QR decompositions.

The equalities (1.122) are reexpressible in the form

$$\mathbf{A} = \mathbf{U}'\mathbf{O}' = \mathbf{U}_1'\mathbf{O}_1'.$$

Thus,

$$\mathbf{A}\mathbf{O}_2 = \mathbf{U}_1'\mathbf{O}_1'\mathbf{O}_2 = \mathbf{0},$$

so that one choice for the matrix $\mathbf{L}$ [with $\mathcal{C}(\mathbf{L}) = \mathcal{N}(\mathbf{A})$] is

$$\mathbf{L} = \mathbf{O}_2. \tag{1.123}$$

Corresponding to the constrained G–M, Aitken, or general linear model for $\mathbf{y}$ (with model equation $\mathbf{y} = \mathbf{X}\boldsymbol{\beta} + \mathbf{e}$ and constraint $\mathbf{A}\boldsymbol{\beta} = \mathbf{d}$) is the unconstrained G–M, Aitken, or general linear model for the transformed observable random vector $\mathbf{z} = \mathbf{y} - \mathbf{X}\tilde{\mathbf{b}}$ with model equation $\mathbf{z} = \mathbf{W}\boldsymbol{\tau} + \mathbf{e}$. When $\mathbf{L} = \mathbf{O}_2$,

$$\mathbf{W} = \mathbf{X}\mathbf{L} = \mathbf{X}\mathbf{O}_2. \tag{1.124}$$

Moreover,

$$\mathbf{A}\mathbf{b} = \mathbf{d} \quad \Leftrightarrow \quad \mathbf{U}_1'\mathbf{O}_1'\mathbf{b} = \mathbf{d},$$

so that $\tilde{\mathbf{b}}$ can be recharacterized as an arbitrary solution to the linear system

$$\mathbf{U}_1'\mathbf{O}_1'\mathbf{b} = \mathbf{d}. \tag{1.125}$$

g. Optimality properties

Many of the statistical procedures that might be applied to an observable random vector $\mathbf{y}$ that follows an (unconstrained) G–M model (with model equation $\mathbf{y} = \mathbf{X}\boldsymbol{\beta} + \mathbf{e}$) depend on $\mathbf{y}$ only through the least squares estimates of various estimable linear combinations of the elements of $\boldsymbol{\beta}$ and through the residual sum of squares. Those procedures can be extended for use when $\mathbf{y}$ follows a constrained G–M model (in which $\mathbf{A}\boldsymbol{\beta} = \mathbf{d}$). It is simply a matter of replacing the least squares estimates and the residual sum of squares devised on the basis of the unconstrained model with those devised on the basis of the constrained model.

How might the extension of the various procedures be "legitimized" and to what extent do any optimality properties carry over to the extended procedures? One way to address those issues would be to undertake a more general version of the development undertaken earlier in establishing the legitimacy and optimality properties of the procedures under an unconstrained G–M model. Alternatively, we can attempt to build on what has already been established by exploiting the various relationships between constrained least squares estimation from $\mathbf{y}$ under a constrained G–M model (with model equation $\mathbf{y} = \mathbf{X}\boldsymbol{\beta} + \mathbf{e}$ and constraint $\mathbf{A}\boldsymbol{\beta} = \mathbf{d}$) and unconstrained least squares estimation from the transformed vector $\mathbf{z} = \mathbf{y} - \mathbf{X}\mathbf{A}^-\mathbf{d}$ under the corresponding unconstrained G–M model [i.e., the unconstrained G–M model with model equation $\mathbf{z} = \mathbf{W}\boldsymbol{\tau} + \mathbf{e}$, where $\mathbf{W} = \mathbf{X}(\mathbf{I} - \mathbf{A}^-\mathbf{A})$]. More specifically, we can reexpress the extended procedures in terms of $\mathbf{z}$ and in terms related to the unconstrained G–M model with model equation $\mathbf{z} = \mathbf{W}\boldsymbol{\tau} + \mathbf{e}$, apply any of the results obtained earlier under an unconstrained G–M model that are applicable to the reexpressed procedures, and then translate those results into terms of $\mathbf{y}$ and in terms related to the constrained G–M model. In what follows, that approach is illustrated by using it to extend (to constrained least squares estimators) the results (covered by the Gauss-Markov theorem) on the optimality of least squares estimators in the absence of any constraints.

Best linear unbiased estimation. Let $\mathbf{y}$ represent an $N \times 1$ observable random vector that follows a constrained G–M, Aitken, or general linear model with model equation $\mathbf{y} = \mathbf{X}\boldsymbol{\beta} + \mathbf{e}$ and constraint $\mathbf{A}\boldsymbol{\beta} = \mathbf{d}$. Further, let $\boldsymbol{\lambda}'\boldsymbol{\beta}$ represent any estimable linear combination of the elements of $\boldsymbol{\beta}$ (so that $\boldsymbol{\lambda}' = \mathbf{a}'\mathbf{X} + \mathbf{r}'\mathbf{A}$ for some $P \times 1$ vector $\mathbf{a}$ and some $Q \times 1$ vector $\mathbf{r}$). And let $\mathbf{z} = \mathbf{y} - \mathbf{X}\mathbf{A}^-\mathbf{d}$ and $\mathbf{W} = \mathbf{X}(\mathbf{I} - \mathbf{A}^-\mathbf{A})$, and take $\mathbf{t}$ to be any $P \times 1$ vector for which $\mathbf{W}'\mathbf{W}\mathbf{t} = \mathbf{W}'\mathbf{z}$.

Now, observe (in light of the results of Subsection f) that the constrained least squares estimator of $\boldsymbol{\lambda}'\boldsymbol{\beta}$ is expressible in terms of the transformed vector $\mathbf{z}$ as

$$\boldsymbol{\lambda}'\mathbf{A}^-\mathbf{d} + \boldsymbol{\lambda}'(\mathbf{I} - \mathbf{A}^-\mathbf{A})\mathbf{t}. \tag{1.126}$$

And suppose that $\mathbf{z}$ were assumed to follow an unconstrained G–M, Aitken, or general linear model with model equation $\mathbf{z} = \mathbf{W}\boldsymbol{\tau} + \mathbf{e}$. Then, $\boldsymbol{\lambda}'(\mathbf{I}-\mathbf{A}^{-}\mathbf{A})\boldsymbol{\tau}$ would be an estimable linear combination of the elements of $\boldsymbol{\tau}$, $\boldsymbol{\lambda}'(\mathbf{I}-\mathbf{A}^{-}\mathbf{A})\mathbf{t}$ would be the least squares estimator of $\boldsymbol{\lambda}'(\mathbf{I}-\mathbf{A}^{-}\mathbf{A})\boldsymbol{\tau}$, and the quantity (1.126) would be an unbiased estimator of

$$\boldsymbol{\lambda}'\mathbf{A}^{-}\mathbf{d} + \boldsymbol{\lambda}'(\mathbf{I}-\mathbf{A}^{-}\mathbf{A})\boldsymbol{\tau}. \tag{1.127}$$

Moreover, the quantity (1.126) is expressible in the form $k + \boldsymbol{\ell}'\mathbf{z}$ (where k is a constant and $\boldsymbol{\ell}$ a vector of constants) and hence would be a linear estimator of the parametric function (1.127). In fact, among all linear unbiased estimators of the parametric function (1.127), it would (in the special case of the G–M model) be optimal in the sense that its variance would be smaller than that of any other linear unbiased estimator (as is evident from Theorem 5.5.1, i.e., the Gauss-Markov theorem).

Let us translate this result (which pertains to $\mathbf{z}$ when $\mathbf{z}$ is assumed to follow an unconstrained G–M model with model equation $\mathbf{z} = \mathbf{W}\boldsymbol{\tau} + \mathbf{e}$) into terms that apply to $\mathbf{y}$ and the constrained G–M model (in which $\mathbf{y} = \mathbf{X}\boldsymbol{\beta} + \mathbf{e}$ and $\mathbf{A}\boldsymbol{\beta} = \mathbf{d}$). Corresponding to each scalar c and each $N\times 1$ vector $\mathbf{a}$ is an expression of the form

$$c + \mathbf{a}'\mathbf{y}. \tag{1.128}$$

And corresponding to each expression of the form (1.128) is an expression of the form $k + \boldsymbol{\ell}'\mathbf{z}$ defined by the one-to-one mapping $k = c + \mathbf{a}'\mathbf{X}\mathbf{A}^{-}\mathbf{d}$ and $\boldsymbol{\ell} = \mathbf{a}$ from the collection of all values of c and $\mathbf{a}$ onto the collection of all values of k and $\boldsymbol{\ell}$. When $k = c + \mathbf{a}'\mathbf{X}\mathbf{A}^{-}\mathbf{d}$ and $\boldsymbol{\ell} = \mathbf{a}$,

$$k + \boldsymbol{\ell}'\mathbf{z} \equiv c + \mathbf{a}'\mathbf{y}.$$

The quantity $c + \mathbf{a}'\mathbf{y}$ is an unbiased estimator of $\boldsymbol{\lambda}'\boldsymbol{\beta}$ (under a constrained G–M, Aitken, or general linear model with $\mathbf{y} = \mathbf{X}\boldsymbol{\beta} + \mathbf{e}$ and $\mathbf{A}\boldsymbol{\beta} = \mathbf{d}$) if and only if

$$c = \mathbf{r}'\mathbf{d} \quad \text{and} \quad \mathbf{a}'\mathbf{X} + \mathbf{r}'\mathbf{A} = \boldsymbol{\lambda}' \quad \text{for some } Q\times 1 \text{ vector } \mathbf{r} \tag{1.129}$$

(as is evident upon recalling the results of Subsection a). And (when $\mathbf{z}$ is assumed to follow an unconstrained G–M, Aitken, or general linear model with model equation $\mathbf{z} = \mathbf{W}\boldsymbol{\tau} + \mathbf{e}$) $k + \boldsymbol{\ell}'\mathbf{z}$ is an unbiased estimator of $\boldsymbol{\lambda}'\mathbf{A}^{-}\mathbf{d} + \boldsymbol{\lambda}'(\mathbf{I}-\mathbf{A}^{-}\mathbf{A})\boldsymbol{\tau}$ if and only if

$$k = \boldsymbol{\lambda}'\mathbf{A}^{-}\mathbf{d} \qquad \text{and} \qquad \boldsymbol{\ell}'\mathbf{W} = \boldsymbol{\lambda}'(\mathbf{I}-\mathbf{A}^{-}\mathbf{A}). \tag{1.130}$$

Moreover, when $k = c + \mathbf{a}'\mathbf{X}\mathbf{A}^{-}\mathbf{d}$ and $\boldsymbol{\ell} = \mathbf{a}$, condition (1.130) is equivalent to condition (1.129), as can be readily verified. Thus, the quantity $c + \mathbf{a}'\mathbf{y}$ is an unbiased estimator of $\boldsymbol{\lambda}'\boldsymbol{\beta}$ (when $\mathbf{y}$ follows a G–M, Aitken, or general linear model with $\mathbf{y} = \mathbf{X}\boldsymbol{\beta} + \mathbf{e}$ and $\mathbf{A}\boldsymbol{\beta} = \mathbf{d}$) if and only if the corresponding quantity $k + \boldsymbol{\ell}'\mathbf{z}$ is an unbiased estimator of $\boldsymbol{\lambda}'\mathbf{A}^{-}\mathbf{d} + \boldsymbol{\lambda}'(\mathbf{I}-\mathbf{A}^{-}\mathbf{A})\boldsymbol{\tau}$ (when $\mathbf{z}$ is assumed to follow an unconstrained G–M, Aitken, or general linear model with model equation $\mathbf{z} = \mathbf{W}\boldsymbol{\tau} + \mathbf{e}$).

When $\mathbf{y}$ follows a constrained G–M, Aitken, or general linear model (with $\mathbf{y} = \mathbf{X}\boldsymbol{\beta} + \mathbf{e}$ and $\mathbf{A}\boldsymbol{\beta} = \mathbf{d}$) and $\mathbf{z}$ is assumed to follow an unconstrained G–M, Aitken, or general linear model with model equation $\mathbf{z} = \mathbf{W}\boldsymbol{\tau} + \mathbf{e}$, the quantity $c + \mathbf{a}'\mathbf{y}$ has the same variance as the corresponding quantity $k + \boldsymbol{\ell}'\mathbf{z}$ $(= c + \mathbf{a}'\mathbf{X}\mathbf{A}^{-}\mathbf{d} + \mathbf{a}'\mathbf{z})$. And the constrained least squares estimator of $\boldsymbol{\lambda}'\boldsymbol{\beta}$ (under the constrained G–M, Aitken, or general linear model for $\mathbf{y}$) is of the form $c + \mathbf{a}'\mathbf{y}$ and is equal to $\boldsymbol{\lambda}'\mathbf{A}^{-}\mathbf{d} + \boldsymbol{\lambda}'(\mathbf{I}-\mathbf{A}^{-}\mathbf{A})\mathbf{t}$ (where $\mathbf{W}'\mathbf{W}\mathbf{t} = \mathbf{W}'\mathbf{z}$), which is the corresponding quantity of the form $k + \boldsymbol{\ell}'\mathbf{z}$ and which (under the unconstrained G–M, Aitken, or general linear model for $\mathbf{z}$ with model equation $\mathbf{z} = \mathbf{W}\boldsymbol{\tau} + \mathbf{e}$) is the least squares estimator of $\boldsymbol{\lambda}'\mathbf{A}^{-}\mathbf{d} + \boldsymbol{\lambda}'(\mathbf{I}-\mathbf{A}^{-}\mathbf{A})\boldsymbol{\tau}$. These observations lead to the conclusion that when $\mathbf{y}$ follows a constrained G–M model (with $\mathbf{y} = \mathbf{X}\boldsymbol{\beta} + \mathbf{e}$ and $\mathbf{A}\boldsymbol{\beta} = \mathbf{d}$), the constrained least squares estimator of $\boldsymbol{\lambda}'\boldsymbol{\beta}$ is the best linear unbiased estimator (in the sense that it has minimum variance among all unbiased estimators of the form $c + \mathbf{a}'\mathbf{y}$).

Best translation-equivariant (linear) estimation. Let us consider how the results of the preceding part of the present subsection (on the estimation of $\boldsymbol{\lambda}'\boldsymbol{\beta}$ from an observable random vector $\mathbf{y}$ that follows a constrained G–M, Aitken, or general linear model with $\mathbf{y} = \mathbf{X}\boldsymbol{\beta} + \mathbf{e}$ and $\mathbf{A}\boldsymbol{\beta} = \mathbf{d}$) are affected when the criterion of unbiasedness is replaced by one of translation equivariance.

Suppose that $\mathbf{z}$ $(= \mathbf{y} - \mathbf{X}\mathbf{A}^{-}\mathbf{d})$ were assumed to follow an unconstrained G–M, Aitken, or general linear model with model equation $\mathbf{z} = \mathbf{W}\boldsymbol{\tau} + \mathbf{e}$ [where $\mathbf{W} = \mathbf{X}(\mathbf{I} - \mathbf{A}^{-}\mathbf{A})$]. Then, for $k + \boldsymbol{\ell}'\mathbf{z}$ to be a translation-equivariant estimator of $\boldsymbol{\lambda}'(\mathbf{I} - \mathbf{A}^{-}\mathbf{A})\boldsymbol{\tau}$, it would be necessary and sufficient that

$$\boldsymbol{\ell}'\mathbf{W} = \boldsymbol{\lambda}'(\mathbf{I} - \mathbf{A}^{-}\mathbf{A}). \tag{1.131}$$

Condition (1.131) would likewise be necessary and sufficient for $k + \boldsymbol{\ell}'\mathbf{z}$ to be a translation-equivariant estimator of $\boldsymbol{\lambda}'\mathbf{A}^{-}\mathbf{d} + \boldsymbol{\lambda}'(\mathbf{I} - \mathbf{A}^{-}\mathbf{A})\boldsymbol{\tau}$.

Among the estimators of $\boldsymbol{\lambda}'\mathbf{A}^{-}\mathbf{d} + \boldsymbol{\lambda}'(\mathbf{I} - \mathbf{A}^{-}\mathbf{A})\boldsymbol{\tau}$ that would be expressible in the form $k + \boldsymbol{\ell}'\mathbf{z}$ and that would satisfy condition (1.131) would be the estimator (1.126). In fact, in the case of an unconstrained G–M model with model equation $\mathbf{z} = \mathbf{W}\boldsymbol{\tau} + \mathbf{e}$, the estimator (1.126) would be the best linear translation-equivariant estimator of $\boldsymbol{\lambda}'\mathbf{A}^{-}\mathbf{d} + \boldsymbol{\lambda}'(\mathbf{I} - \mathbf{A}^{-}\mathbf{A})\boldsymbol{\tau}$ [in the sense that among all estimators of $\boldsymbol{\lambda}'\mathbf{A}^{-}\mathbf{d} + \boldsymbol{\lambda}'(\mathbf{I} - \mathbf{A}^{-}\mathbf{A})\boldsymbol{\tau}$ that are expressible in the form $k + \boldsymbol{\ell}'\mathbf{z}$ and that are translation equivariant, it would have the smallest mean squared error]—this result can be readily verified by making use of Corollary 5.5.2.

Let us translate these results into terms applicable to the estimation of $\boldsymbol{\lambda}'\boldsymbol{\beta}$ from the observable random vector $\mathbf{y}$ (when $\mathbf{y}$ follows a constrained G–M, Aitken, or general linear model with $\mathbf{y} = \mathbf{X}\boldsymbol{\beta} + \mathbf{e}$ and $\mathbf{A}\boldsymbol{\beta} = \mathbf{d}$). For the quantity $c + \mathbf{a}'\mathbf{y}$ to be a translation-equivariant estimator of $\boldsymbol{\lambda}'\boldsymbol{\beta}$ (under the constrained G–M, Aitken, or general linear model), it is necessary and sufficient that

$$(\boldsymbol{\lambda}' - \mathbf{a}'\mathbf{X})(\mathbf{I} - \mathbf{A}^{-}\mathbf{A}) = \mathbf{0}. \tag{1.132}$$

And condition (1.132) is reexpressible in the form

$$\mathbf{a}'\mathbf{W} = \boldsymbol{\lambda}'(\mathbf{I} - \mathbf{A}^{-}\mathbf{A}). \tag{1.133}$$

Thus, the quantity $c + \mathbf{a}'\mathbf{y}$ is a translation-equivariant estimator of $\boldsymbol{\lambda}'\boldsymbol{\beta}$ (when $\mathbf{y}$ follows a constrained G–M, Aitken, or general linear model with $\mathbf{y} = \mathbf{X}\boldsymbol{\beta} + \mathbf{e}$ and $\mathbf{A}\boldsymbol{\beta} = \mathbf{d}$) if and only if the corresponding quantity $k + \boldsymbol{\ell}'\mathbf{z}$ $(= c + \mathbf{a}'\mathbf{X}\mathbf{A}^{-}\mathbf{d} + \mathbf{a}'\mathbf{z})$ is a translation-equivariant estimator of $\boldsymbol{\lambda}'\mathbf{A}^{-}\mathbf{d} + \boldsymbol{\lambda}'(\mathbf{I} - \mathbf{A}^{-}\mathbf{A})\boldsymbol{\tau}$ (when $\mathbf{z}$ is assumed to follow an unconstrained G–M, Aitken, or general linear model with model equation $\mathbf{z} = \mathbf{W}\boldsymbol{\tau} + \mathbf{e}$), in which case the two quantities have the same bias (as can be verified via a relatively straightforward exercise) as well as the same variance and hence have the same mean squared error.

Upon recalling that the constrained least squares estimator of $\boldsymbol{\lambda}'\boldsymbol{\beta}$ (under the constrained G–M, Aitken, or general linear model for $\mathbf{y}$) is of the form $c + \mathbf{a}'\mathbf{y}$ and is translation equivariant and that it equals $\boldsymbol{\lambda}'\mathbf{A}^{-}\mathbf{d} + \boldsymbol{\lambda}'(\mathbf{I} - \mathbf{A}^{-}\mathbf{A})\mathbf{t}$ (where $\mathbf{W}'\mathbf{W}\mathbf{t} = \mathbf{W}'\mathbf{z}$), we conclude that when $\mathbf{y}$ follows a constrained G–M model (with $\mathbf{y} = \mathbf{X}\boldsymbol{\beta} + \mathbf{e}$ and $\mathbf{A}\boldsymbol{\beta} = \mathbf{d}$), the constrained least squares estimator of $\boldsymbol{\lambda}'\boldsymbol{\beta}$ is the best linear translation-equivariant estimator (in the sense that it has minimum mean squared error among all translation-equivariant estimators of the form $c + \mathbf{a}'\mathbf{y}$).

h. Inequality constraints

The constraint $\mathbf{A}\boldsymbol{\beta} = \mathbf{d}$ in the constrained G–M, Aitken, or general linear model consists of Q individual equality constraints

$$\mathbf{a}_i'\boldsymbol{\beta} = d_i \quad (i = 1, 2, \ldots, Q) \tag{1.134}$$

(where $\mathbf{a}_i'$ is the ith row of $\mathbf{A}$ and d_i the ith element of $\mathbf{d}$). A broader class of constrained linear models can be obtained by allowing for the possibility that some or all of the individual constraints are inequality constraints rather than equality constraints. Accordingly, suppose that the Q individual constraints in the constrained G–M, Aitken, or general linear model are

$$\mathbf{a}_i'\boldsymbol{\beta} = d_i \quad (i \in I) \tag{1.135}$$

and

$$\mathbf{a}_i'\boldsymbol{\beta} \geq d_i \quad (i \in \bar{I}), \tag{1.136}$$

where I and $\bar{I}$ are mutually exclusive and exhaustive subsets of $\{1, 2, \ldots, Q\}$. Further, let

$$B = \{\mathbf{b} \in \mathbb{R}^P : \ \mathbf{a}_i'\mathbf{b} = d_i \ (i \in I), \ \ \mathbf{a}_i'\mathbf{b} \geq d_i \ (i \in \bar{I})\}.$$

And drop the assumption that $\mathbf{d} \in \mathcal{C}(\mathbf{A})$ in favor of the less restrictive assumption that the set B is nonempty.

In connection with the inequalities (1.136), note that for any $Q \times 1$ vector $\mathbf{a}$ and any scalar d,

$$\mathbf{a}'\boldsymbol{\beta} \le d \quad \Leftrightarrow \quad (-\mathbf{a})'\boldsymbol{\beta} \ge -d.$$

Consequently, there is no essential loss of generality in not including among the inequalities (1.136) any inequalities of the form $\mathbf{a}'\boldsymbol{\beta} \le d$.

When the constraints on $\boldsymbol{\beta}$ consist of the equality constraints (1.135) and the inequality constraints (1.136), the parameter space for the constrained G–M or Aitken model is

$$\{\boldsymbol{\beta}, \sigma \;:\; \boldsymbol{\beta} \in B,\; \sigma > 0\}.$$

And the parameter space for the constrained general linear model is

$$\{\boldsymbol{\beta}, \boldsymbol{\theta} \;:\; \boldsymbol{\beta} \in B,\; \boldsymbol{\theta} \in \boldsymbol{\Omega}\}.$$

Least squares estimation. Let $\mathbf{A}_*$ represent the matrix whose rows are $\mathbf{a}_i'$ $(i \in I)$. And let $\boldsymbol{\lambda}$ represent any $P \times 1$ vector such that $\boldsymbol{\lambda}' \in \mathcal{R}\begin{pmatrix} \mathbf{X} \\ \mathbf{A}_* \end{pmatrix}$. Then, conceptually, the constrained least squares estimation of $\boldsymbol{\lambda}'\boldsymbol{\beta}$ in the presence of the inequality constraints (1.136) [as well as the equality constraints (1.135)] is relatively simple. By definition, the constrained least squares estimator of $\boldsymbol{\lambda}'\boldsymbol{\beta}$ is $\boldsymbol{\lambda}'\tilde{\mathbf{b}}$, where $\tilde{\mathbf{b}}$ is any value of the $P \times 1$ vector $\mathbf{b}$ at which $(\mathbf{y}-\mathbf{X}\mathbf{b})'(\mathbf{y}-\mathbf{X}\mathbf{b})$ attains its minimum value subject to the constraints

$$\mathbf{a}_i'\mathbf{b} = d_i \;\; (i \in I) \quad \text{and} \quad \mathbf{a}_i'\mathbf{b} \ge d_i \;\; (i \in \bar{I}) \tag{1.137}$$

—it is worth noting that the value of $\boldsymbol{\lambda}'\tilde{\mathbf{b}}$ does not vary with the choice of $\tilde{\mathbf{b}}$.

For some discussion of the constrained least squares estimator $\boldsymbol{\lambda}'\tilde{\mathbf{b}}$ in a relatively simple special case [that where the minimization of $(\mathbf{y}-\mathbf{X}\mathbf{b})'(\mathbf{y}-\mathbf{X}\mathbf{b})$ is subject to the constraints $d_0 \ge \mathbf{a}'\mathbf{b} \ge d_1$, or equivalently $\mathbf{a}'\mathbf{b} \ge d_1$ and $(-\mathbf{a})'\mathbf{b} \ge -d_0$, and where $\mathbf{a} = \boldsymbol{\lambda}$], refer to Escobar and Skarpness (1986) and to Ohtani (1987).

Confidence intervals or sets. How might one account for the inequality constraints in constructing confidence intervals (or 1-dimensional confidence sets) for one or more linear combinations of the elements of $\boldsymbol{\beta}$ or in constructing multidimensional confidence sets for one or more vectors of such linear combinations? Perhaps the simplest way of doing so would be to first construct confidence intervals or sets that would be suitable in the absence of the inequality constraints and to subsequently retain only those members of each interval or set that are "compatible" with the inequality constraints. Thus, in the case of a confidence set for a linear combination or vector of linear combinations $\boldsymbol{\alpha} = \boldsymbol{\Lambda}'\boldsymbol{\beta}$, a value $\tilde{\boldsymbol{\alpha}}$ is retained only if $\tilde{\boldsymbol{\alpha}} = \boldsymbol{\Lambda}'\mathbf{b}$ for some vector $\mathbf{b}$ such that $\mathbf{a}_i'\mathbf{b} \ge d_i$ $(i \in \bar{I})$. Clearly, the intervals or sets formed by those members of the original intervals or sets that are retained have the same probability of coverage (both one-at-a-time and simultaneous) as the original intervals or sets, and they may be smaller. Note that in "extreme" cases, none of the values of a linear combination or vector of linear combinations may be retained, in which case the resultant interval or set is the empty set.

Hypothesis tests and multiple comparisons. How might one account for the inequality constraints in testing the null hypothesis that a linear combination of the elements of $\boldsymbol{\beta}$ equals a hypothesized value, in testing multiple null hypotheses of this kind, or in testing the null hypothesis that a vector of linear combinations equals a hypothesized value? As in the absence of the inequality constraints, confidence intervals or sets can be used as a basis for devising such tests: the null hypothesis is rejected when the observed value of $\mathbf{y}$ is such that the confidence interval or set does not include the hypothesized value. When the confidence intervals or sets that would have been used for this purpose in the absence of the inequality constraints are modified to account for those constraints, the associated tests of hypotheses may produce additional rejections, none of which are false rejections.

Slack variables. Each of the inequality constraints $\mathbf{a}_i'\mathbf{b} \geq d_i$ $(i \in \bar{I})$ can be converted into an equality constraint by introducing an additional variable called a *slack variable* or a surplus variable. For instance, consider (for any i in $\bar{I}$) the inequality constraint

$$\mathbf{a}_i'\mathbf{b} \geq d_i. \tag{1.138}$$

Upon introducing a slack variable $s_i \geq 0$, constraint (1.138) can be replaced by the equality constraint

$$\mathbf{a}_i'\mathbf{b} = d_i + s_i. \tag{1.139}$$

Clearly, a value of $\mathbf{b}$ satisfies the inequality constraint (1.138) if and only if it satisfies the equality constraint (1.139) for some nonnegative value of s_i. Moreover, if (for some j in $\bar{I}$) $\mathbf{a}_j = -\mathbf{a}_i$ (with $-d_j \geq d_i$) and $s_j \geq 0$ is the slack variable corresponding to the inequality

$$\mathbf{a}_j'\mathbf{b} \geq d_j, \tag{1.140}$$

then $d_j + s_j = -(d_i + s_i)$, in which case

$$s_j \geq 0 \quad \Leftrightarrow \quad s_i \leq -d_j - d_i,$$

so that a value of $\mathbf{b}$ satisfies both of inequalities (1.138) and (1.140) if and only if it satisfies equality (1.139) for some value of s_i in the interval

$$0 \leq s_i \leq -d_j - d_i.$$

Now, as is evident from the preceding discussion, there exists a subset $\tilde{I}$ of $\bar{I}$ for which none of the vectors $\mathbf{a}_i$ $(i \in \tilde{I})$ is a scalar multiple of any of the others and that has the following property: a value of $\mathbf{b}$ satisfies the inequality constraints $\mathbf{a}_i'\mathbf{b} \geq d_i$ $(i \in \bar{I})$ if and only if

$$\mathbf{a}_i'\mathbf{b} = d_i + s_i \qquad (i \in \tilde{I})$$

for some scalars s_i $(i \in \tilde{I})$ with

$$0 \leq s_i \leq u_i \qquad (i \in \tilde{I}),$$

where (for $i \in \tilde{I}$) u_i is either a finite upper bound that is determinable from d_i and another of the d_j's or is $+\infty$. Thus, the problem of minimizing $(\mathbf{y}-\mathbf{X}\mathbf{b})'(\mathbf{y}-\mathbf{X}\mathbf{b})$ with respect to $\mathbf{b}$ subject to the constraints (1.137) is equivalent to the problem of minimizing $(\mathbf{y}-\mathbf{X}\mathbf{b})'(\mathbf{y}-\mathbf{X}\mathbf{b})$ with respect to $\mathbf{b}$ and s_i $(i \in \tilde{I})$ subject to the constraints

$$\mathbf{a}_i'\mathbf{b} = d_i \ (i \in I) \quad \text{and} \quad \mathbf{a}_i'\mathbf{b} = d_i + s_i \ (i \in \tilde{I}) \tag{1.141}$$

and the constraints

$$0 \leq s_i \leq u_i \qquad (i \in \tilde{I}). \tag{1.142}$$

An equivalent minimization problem. Denote by $\mathbf{s}$ the vector whose elements are the slack variables s_i $(i \in \tilde{I})$, and observe (in light of Theorem 2.11.1) that there exists a value of the vector $\mathbf{b}$ for which all of the equalities (1.141) are simultaneously satisfied if and only if

$$\left[\begin{pmatrix} \mathbf{I} & \mathbf{0} \\ \mathbf{0} & \mathbf{I} \end{pmatrix} - \begin{pmatrix} \mathbf{A}_1 \\ \mathbf{A}_2 \end{pmatrix}\begin{pmatrix} \mathbf{A}_1 \\ \mathbf{A}_2 \end{pmatrix}^-\right]\begin{pmatrix} \mathbf{d}_1 \\ \mathbf{d}_2 + \mathbf{s} \end{pmatrix} = \begin{pmatrix} \mathbf{0} \\ \mathbf{0} \end{pmatrix}, \tag{1.143}$$

where $\mathbf{A}_1$ is the matrix with rows $\mathbf{a}_i'$ $(i \in I)$ and $\mathbf{A}_2$ the matrix with rows $\mathbf{a}_i'$ $(i \in \tilde{I})$ and where $\mathbf{d}_1$ is the column vector with elements d_i $(i \in I)$ and $\mathbf{d}_2$ the column vector with elements d_i $(i \in \tilde{I})$. Further, let $\mathbf{G}$ represent any generalized inverse of the partitioned matrix $\begin{pmatrix} \mathbf{X}'\mathbf{X} & \mathbf{A}_1' & \mathbf{A}_2' \\ \mathbf{A}_1 & \mathbf{0} & \mathbf{0} \\ \mathbf{A}_2 & \mathbf{0} & \mathbf{0} \end{pmatrix}$, and partition $\mathbf{G}$ conformally with $\begin{pmatrix} \mathbf{X}'\mathbf{X} & \mathbf{A}_1' & \mathbf{A}_2' \\ \mathbf{A}_1 & \mathbf{0} & \mathbf{0} \\ \mathbf{A}_2 & \mathbf{0} & \mathbf{0} \end{pmatrix}$ as $\begin{pmatrix} \mathbf{G}_{11} & \mathbf{G}_{12} & \mathbf{G}_{13} \\ \mathbf{G}_{21} & \mathbf{G}_{22} & \mathbf{G}_{23} \\ \mathbf{G}_{31} & \mathbf{G}_{32} & \mathbf{G}_{33} \end{pmatrix}$. And observe that for any value of $\mathbf{s}$ that satisfies condition (1.143), $(\mathbf{y}-\mathbf{X}\mathbf{b})'(\mathbf{y}-\mathbf{X}\mathbf{b})$ attains a minimum value [with respect to $\mathbf{b}$ subject to condition (1.141)] of

$$[\mathbf{y}-\mathbf{X}\mathbf{G}_{11}\mathbf{X}'\mathbf{y}-\mathbf{X}\mathbf{G}_{12}\mathbf{d}_1-\mathbf{X}\mathbf{G}_{13}(\mathbf{d}_2+\mathbf{s})]'[\mathbf{y}-\mathbf{X}\mathbf{G}_{11}\mathbf{X}'\mathbf{y}-\mathbf{X}\mathbf{G}_{12}\mathbf{d}_1-\mathbf{X}\mathbf{G}_{13}(\mathbf{d}_2+\mathbf{s})] \tag{1.144}$$

and does so at

$$\mathbf{b} = \mathbf{G}_{11}\mathbf{X}'\mathbf{y} + \mathbf{G}_{12}\mathbf{d}_1 + \mathbf{G}_{13}(\mathbf{d}_2+\mathbf{s}). \tag{1.145}$$

It is now clear that the problem of finding a value of $\mathbf{b}$ that minimizes $(\mathbf{y}-\mathbf{X}\mathbf{b})'(\mathbf{y}-\mathbf{X}\mathbf{b})$ with respect to $\mathbf{b}$ subject to the constraints (1.137) can be solved indirectly by minimizing the quantity (1.144) with respect to $\mathbf{s}$ subject to the constraints (1.142) and (1.143) and by obtaining the value of $\mathbf{b}$ from expression (1.145). This approach can be advantageous in applications where the number of inequality constraints is small relative to the number P of elements in $\boldsymbol{\beta}$.

Computational considerations. Both the problem of minimizing $(\mathbf{y}-\mathbf{X}\mathbf{b})'(\mathbf{y}-\mathbf{X}\mathbf{b})$ with respect to $\mathbf{b}$ subject to the constraints (1.137) and that of minimizing the quantity (1.144) with respect to $\mathbf{s}$ subject to the constraints (1.142) and (1.143) are in essence quadratic programming problems. For a thorough discussion of these kinds of problems and of various algorithms for solving them, refer to Nocedal and Wright (2006).

8.2 Constrained Models: Jointly Nonestimable Constraints

Suppose that $\mathbf{y}$ is an $N \times 1$ observable random vector that follows a G–M, Aitken, or general linear model with model equation $\mathbf{y} = \mathbf{X}\boldsymbol{\beta} + \mathbf{e}$, where $\boldsymbol{\beta}$ is a $P \times 1$ vector of unknown parameters. Let us consider further the situation where [for some $Q \times P$ matrix $\mathbf{A}$ and some vector $\mathbf{d} \in \mathcal{C}(\mathbf{A})$] the parameter vector $\boldsymbol{\beta}$ is subject to the equality constraint $\mathbf{A}\boldsymbol{\beta} = \mathbf{d}$. A number of results for that situation were presented previously (in Section 8.1); those results were general in nature, that is, applicable to any such situation.

In the present section, the focus is on the special case where some or all of the elements of the vector $\mathbf{A}\boldsymbol{\beta}$ have a property (defined in the context of the unconstrained model) that is referred to as joint nonestimability. This property is described and discussed in Subsection a, thereby setting the stage for the main development (which is contained in Subsection b). As discussed in Subsection c, results obtained for the constrained linear model in the special case where the elements of $\mathbf{A}\boldsymbol{\beta}$ are jointly nonestimable may be of intereest even in the absence of any real constraints.

a. Joint nonestimability

Definition. For purposes of defining joint nonestimability, suppose that $\mathbf{y}$ is an $N \times 1$ observable random vector that follows an (unconstrained) G–M, Aitken, or general linear model with model equation $\mathbf{y} = \mathbf{X}\boldsymbol{\beta}+\mathbf{e}$ and $N\times P$ model matrix $\mathbf{X}$ [where $\boldsymbol{\beta} = (\beta_1, \beta_2, \ldots, \beta_P)'$ is a vector of unknown and unconstrained parameters]. Further, let $\boldsymbol{\lambda}_1, \boldsymbol{\lambda}_2, \ldots, \boldsymbol{\lambda}_M$ represent $P\times 1$ vectors of constants and define $\boldsymbol{\Lambda} = (\boldsymbol{\lambda}_1, \boldsymbol{\lambda}_2, \ldots, \boldsymbol{\lambda}_M)$, so that $\boldsymbol{\lambda}_1'\boldsymbol{\beta}, \boldsymbol{\lambda}_2'\boldsymbol{\beta}, \ldots, \boldsymbol{\lambda}_M'\boldsymbol{\beta}$ represent linear combinations of the elements of $\boldsymbol{\beta}$ and $\boldsymbol{\Lambda}'\boldsymbol{\beta} = (\boldsymbol{\lambda}_1'\boldsymbol{\beta}, \boldsymbol{\lambda}_2'\boldsymbol{\beta}, \ldots, \boldsymbol{\lambda}_M'\boldsymbol{\beta})'$ represents an M-dimensional column vector of such linear combinations. And take $\boldsymbol{\lambda} = \sum_{i=1}^M \ell_i\boldsymbol{\lambda}_i$, where $\ell_1, \ell_2, \ldots, \ell_M$ represent arbitrary constants, or equivalently take $\boldsymbol{\lambda} = \boldsymbol{\Lambda}\boldsymbol{\ell}$, where $\boldsymbol{\ell} = (\ell_1, \ell_2, \ldots, \ell_M)'$ is an M-dimensional column vector of constants, so that $\boldsymbol{\lambda}'\boldsymbol{\beta}$ represents an arbitrary linear combination of $\boldsymbol{\lambda}_1'\boldsymbol{\beta}, \boldsymbol{\lambda}_2'\boldsymbol{\beta}, \ldots, \boldsymbol{\lambda}_M'\boldsymbol{\beta}$.

If $\ell_1, \ell_2, \ldots, \ell_M$ are such that $\boldsymbol{\lambda} = \mathbf{0}$ (or, equivalently, such that $\sum_{i=1}^M \ell_i\boldsymbol{\lambda}_i = \mathbf{0}$ or, also equivalently, such that $\boldsymbol{\Lambda}\boldsymbol{\ell} = \mathbf{0}$), then the linear combination of $\boldsymbol{\lambda}_1'\boldsymbol{\beta}, \boldsymbol{\lambda}_2'\boldsymbol{\beta}, \ldots, \boldsymbol{\lambda}_M'\boldsymbol{\beta}$ is said to be trivial—the linear combination is trivial if the vector $\boldsymbol{\ell} = (\ell_1, \ell_2, \ldots, \ell_M)'$ is an M-dimensional null vector and is trivial for some nonnull values of $\boldsymbol{\ell}$ if and only if $\boldsymbol{\lambda}_1, \boldsymbol{\lambda}_2, \ldots, \boldsymbol{\lambda}_M$ are linearly dependent or, equivalently, if and only if $\operatorname{rank}(\boldsymbol{\Lambda}) < M$. Otherwise (if $\ell_1, \ell_2, \ldots, \ell_M$ are such that $\boldsymbol{\lambda}$ is nonnull), the linear combination of $\boldsymbol{\lambda}_1'\boldsymbol{\beta}, \boldsymbol{\lambda}_2'\boldsymbol{\beta}, \ldots, \boldsymbol{\lambda}_M'\boldsymbol{\beta}$ is said to be nontrivial.

If $\boldsymbol{\lambda}_1'\boldsymbol{\beta}, \boldsymbol{\lambda}_2'\boldsymbol{\beta}, \dots, \boldsymbol{\lambda}_M'\boldsymbol{\beta}$ are all estimable (in which case the vector $\boldsymbol{\Lambda}'\boldsymbol{\beta}$ is said to be estimable), then $\boldsymbol{\lambda}'\boldsymbol{\beta}$ is also estimable—refer to Sections 5.3 and 5.6. It may be tempting to conclude that a similar result holds for nonestimable linear combinations of the elements of $\boldsymbol{\beta}$; that is, to conclude that if $\boldsymbol{\lambda}_1'\boldsymbol{\beta}, \boldsymbol{\lambda}_2'\boldsymbol{\beta}, \dots, \boldsymbol{\lambda}_M'\boldsymbol{\beta}$ are all nonestimable, then $\boldsymbol{\lambda}'\boldsymbol{\beta}$ is also nonestimable. Such a conclusion would be erroneous and would not necessarily be correct even if the case or cases where $\ell_1, \ell_2, \dots, \ell_M$ are such that $\boldsymbol{\lambda} = \mathbf{0}$ were excluded. In general, linear combinations of nonestimable linear combinations (of the elements of $\boldsymbol{\beta}$) are not necessarily nonestimable. As an extremely simple example, consider the case where the $N \times P$ model matrix $\mathbf{X}$ is the 1×2 matrix $\mathbf{X} = (1, 1)$; in this case, β_1 and β_2 are both nonestimable, but $\beta_1 + \beta_2$ is estimable. Nor (aside from the special case where $M = 1$) is the "converse" necessarily true; that is, unless $M = 1$, $\boldsymbol{\lambda}'\boldsymbol{\beta}$ being nonestimable does not necessarily imply that $\boldsymbol{\lambda}_i'\boldsymbol{\beta}$ is nonestimable, not even for a value of i for which $\ell_i \neq 0$.

The cases in which one or more of the linear combinations $\boldsymbol{\lambda}_1'\boldsymbol{\beta}, \boldsymbol{\lambda}_2'\boldsymbol{\beta}, \dots, \boldsymbol{\lambda}_M'\boldsymbol{\beta}$ are nonestimable can be divided into three categories that can be characterized as follows:

(1) at least one of the M linear combinations $\boldsymbol{\lambda}_1'\boldsymbol{\beta}, \boldsymbol{\lambda}_2'\boldsymbol{\beta}, \dots, \boldsymbol{\lambda}_M'\boldsymbol{\beta}$ is nonestimable and at least one is estimable;

(2) each of the M linear combinations $\boldsymbol{\lambda}_1'\boldsymbol{\beta}, \boldsymbol{\lambda}_2'\boldsymbol{\beta}, \dots, \boldsymbol{\lambda}_M'\boldsymbol{\beta}$ is nonestimable, but some nontrivial linear combination of $\boldsymbol{\lambda}_1'\boldsymbol{\beta}, \boldsymbol{\lambda}_2'\boldsymbol{\beta}, \dots, \boldsymbol{\lambda}_M'\boldsymbol{\beta}$ is estimable;

(3) each of the M linear combinations $\boldsymbol{\lambda}_1'\boldsymbol{\beta}, \boldsymbol{\lambda}_2'\boldsymbol{\beta}, \dots, \boldsymbol{\lambda}_M'\boldsymbol{\beta}$ is nonestimable, and every nontrivial linear combination of $\boldsymbol{\lambda}_1'\boldsymbol{\beta}, \boldsymbol{\lambda}_2'\boldsymbol{\beta}, \dots, \boldsymbol{\lambda}_M'\boldsymbol{\beta}$ is nonestimable.

When the linear combinations $\boldsymbol{\lambda}_1'\boldsymbol{\beta}, \boldsymbol{\lambda}_2'\boldsymbol{\beta}, \dots, \boldsymbol{\lambda}_M'\boldsymbol{\beta}$ have the characteristics identified with the third of these three categories, they are said to be jointly nonestimable. And when $\boldsymbol{\lambda}_1'\boldsymbol{\beta}, \boldsymbol{\lambda}_2'\boldsymbol{\beta}, \dots, \boldsymbol{\lambda}_M'\boldsymbol{\beta}$ have those characteristics, the vector $\boldsymbol{\Lambda}'\boldsymbol{\beta}$ may likewise be referred to as jointly nonestimable (or at the risk of some possible confusion, may be referred to simply as nonestimable). Thus, the linear combinations $\boldsymbol{\lambda}_1'\boldsymbol{\beta}, \boldsymbol{\lambda}_2'\boldsymbol{\beta}, \dots, \boldsymbol{\lambda}_M'\boldsymbol{\beta}$ are jointly nonestimable (and the vector $\boldsymbol{\Lambda}'\boldsymbol{\beta}$ of linear combinations is jointly nonestimable) if each of the linear combinations $\boldsymbol{\lambda}_1'\boldsymbol{\beta}, \boldsymbol{\lambda}_2'\boldsymbol{\beta}, \dots, \boldsymbol{\lambda}_M'\boldsymbol{\beta}$ is nonestimable and if in addition every nontrivial linear combination of $\boldsymbol{\lambda}_1'\boldsymbol{\beta}, \boldsymbol{\lambda}_2'\boldsymbol{\beta}, \dots, \boldsymbol{\lambda}_M'\boldsymbol{\beta}$ is nonestimable. Note that (since $0 = \mathbf{0}'\boldsymbol{\beta}$ is estimable) the limitation to nontrivial linear combinations (in the requirement that linear combinations of $\boldsymbol{\lambda}_1'\boldsymbol{\beta}, \boldsymbol{\lambda}_2'\boldsymbol{\beta}, \dots, \boldsymbol{\lambda}_M'\boldsymbol{\beta}$ be nonestimable) is a critical part of the definition of joint nonestimability.

For purposes of illustrating the concept of joint nonestimability, consider a very simple example in which $N = 1$, $P = 3$, and $\mathbf{X} = (1, 1, 1)$. If $\boldsymbol{\lambda}_1 = (1, 0, 0)'$ and $\boldsymbol{\lambda}_2 = (0, 1, 0)'$, then the two linear combinations $\boldsymbol{\lambda}_1'\boldsymbol{\beta}$ and $\boldsymbol{\lambda}_2'\boldsymbol{\beta}$ are jointly nonestimable—$\boldsymbol{\lambda}_1'\boldsymbol{\beta}$ and $\boldsymbol{\lambda}_2'\boldsymbol{\beta}$ are both nonestimable and every nontrivial linear combination of $\boldsymbol{\lambda}_1'\boldsymbol{\beta}$ and $\boldsymbol{\lambda}_2'\boldsymbol{\beta}$ is also nonestimable, as can be readily verified. However, if $\boldsymbol{\lambda}_1 = (1, 0, 0)'$ and $\boldsymbol{\lambda}_2 = (0, 1, 1)'$, then $\boldsymbol{\lambda}_1'\boldsymbol{\beta}$ and $\boldsymbol{\lambda}_2'\boldsymbol{\beta}$ are not jointly nonestimable—they are both nonestimable, but $\boldsymbol{\lambda}_1'\boldsymbol{\beta} + \boldsymbol{\lambda}_2'\boldsymbol{\beta}$ is clearly estimable.

Necessary and sufficient conditions (for joint nonestimability). Let us develop further the concepts introduced in Part 1 of the present subsection. And in doing so, let us adopt the same framework and employ the same notation as in Part 1.

Clearly, any linear combination of $\boldsymbol{\lambda}_1'\boldsymbol{\beta}, \boldsymbol{\lambda}_2'\boldsymbol{\beta}, \dots, \boldsymbol{\lambda}_M'\boldsymbol{\beta}$ is expressible in the form $\boldsymbol{\ell}'\boldsymbol{\Lambda}'\boldsymbol{\beta}$ (for some $M \times 1$ vector $\boldsymbol{\ell}$), and $\boldsymbol{\ell}'\boldsymbol{\Lambda}'\boldsymbol{\beta}$ is nonestimable if and only if $\boldsymbol{\ell}'\boldsymbol{\Lambda}' \notin \mathcal{R}(\mathbf{X})$. Thus, $\boldsymbol{\lambda}_1'\boldsymbol{\beta}, \boldsymbol{\lambda}_2'\boldsymbol{\beta}, \dots, \boldsymbol{\lambda}_M'\boldsymbol{\beta}$ are jointly nonestimable if and only if the matrix $\boldsymbol{\Lambda}' = (\boldsymbol{\lambda}_1, \boldsymbol{\lambda}_2, \dots, \boldsymbol{\lambda}_M)'$ is such that

$$\mathcal{R}(\boldsymbol{\Lambda}') \cap \mathcal{R}(\mathbf{X}) = \{\mathbf{0}\}, \tag{2.1}$$

that is, if and only if $\boldsymbol{\Lambda}$ is such that $\mathcal{R}(\boldsymbol{\Lambda}')$ and $\mathcal{R}(\mathbf{X})$ are essentially disjoint—refer to Section 2.4 for the definition of essential disjointness. It is also the case that $\boldsymbol{\lambda}_1'\boldsymbol{\beta}, \boldsymbol{\lambda}_2'\boldsymbol{\beta}, \dots, \boldsymbol{\lambda}_M'\boldsymbol{\beta}$ are jointly nonestimable if and only if $\boldsymbol{\Lambda}$ is such that

$$\operatorname{rank}\begin{pmatrix} \mathbf{X} \\ \boldsymbol{\Lambda}' \end{pmatrix} = \operatorname{rank}(\mathbf{X}) + \operatorname{rank}(\boldsymbol{\Lambda}), \tag{2.2}$$

as is evident from Theorem 2.4.25.

Number of linearly independent jointly nonestimable linear combinations. How many essentially different linear combinations can be included in an $M \times 1$ vector $\boldsymbol{\Lambda}'\boldsymbol{\beta} = (\boldsymbol{\lambda}_1'\boldsymbol{\beta}, \boldsymbol{\lambda}_2'\boldsymbol{\beta}, \ldots, \boldsymbol{\lambda}_M'\boldsymbol{\beta})'$ of jointly nonestimable linear combinations $\boldsymbol{\lambda}_1'\boldsymbol{\beta}, \boldsymbol{\lambda}_2'\boldsymbol{\beta}, \ldots, \boldsymbol{\lambda}_M'\boldsymbol{\beta}$? Let us take essentially different to mean linearly independent. Making use of equality (2.2), we find that

$$\text{rank}(\boldsymbol{\Lambda}) = \text{rank}\begin{pmatrix} \mathbf{X} \\ \boldsymbol{\Lambda}' \end{pmatrix} - \text{rank}(\mathbf{X}) \leq P - \text{rank}(\mathbf{X}).$$

Thus, at most, the vector $\boldsymbol{\Lambda}'\boldsymbol{\beta}$ of jointly nonestimable linear combinations includes $P - \text{rank}(\mathbf{X})$ linearly independent linear combinations—note that any nonempty subset of jointly nonestimable linear combinations is itself jointly nonestimable.

Can $P - \text{rank}(\mathbf{X})$ linearly independent jointly nonestimable linear combinations always be found? The answer is yes. For purposes of verification, let $P_* = \text{rank}(\mathbf{X})$, and take $\mathbf{X}_*$ to be a $P_* \times P$ submatrix of $\mathbf{X}$, the rows of which form a basis for $\mathcal{R}(\mathbf{X})$. And (supposing that $P_* < P$), take $\boldsymbol{\lambda}_1, \boldsymbol{\lambda}_2, \ldots, \boldsymbol{\lambda}_{P-P_*}$ to be a sequence of $P \times 1$ vectors defined recursively by taking (for $i = 1, 2, \ldots, P - P_*$) $\boldsymbol{\lambda}_i$ to be any $P \times 1$ vector that satisfies the condition

$$\text{rank}(\mathbf{X}_*', \boldsymbol{\lambda}_1, \boldsymbol{\lambda}_2, \ldots, \boldsymbol{\lambda}_i) = P_* + i$$

—that such a vector exists is evident upon observing that otherwise the $P_* + i - 1$ columns of the matrix $(\mathbf{X}_*', \boldsymbol{\lambda}_1, \boldsymbol{\lambda}_2, \ldots, \boldsymbol{\lambda}_{i-1})$ would form a basis for $\mathcal{R}^P$ consisting of fewer than P vectors. For example, if the $N \times P$ model matrix $\mathbf{X}$ were the 2×3 matrix $\mathbf{X} = \begin{pmatrix} 1 & 1 & 1 \\ 2 & 2 & 2 \end{pmatrix}$, then $P_* = 1$ and we could take $\mathbf{X}_* = (1, 1, 1)$, $\boldsymbol{\lambda}_1 = (1, 0, 0)'$, and $\boldsymbol{\lambda}_2 = (0, 1, 0)'$.

b. Constrained normal equations in the special case of jointly nonestimable constraints

Suppose that $\mathbf{y}$ is an $N \times 1$ observable random vector that follows a constrained G–M, Aitken, or general linear model with model equation $\mathbf{y} = \mathbf{X}\boldsymbol{\beta} + \mathbf{e}$, where $\boldsymbol{\beta}$ is a $P \times 1$ vector of unknown parameters that [for some $Q \times P$ matrix $\mathbf{A}$ and some $Q \times 1$ vector $\mathbf{d} \in \mathcal{C}(\mathbf{A})$] is subject to the equality constraint $\mathbf{A}\boldsymbol{\beta} = \mathbf{d}$. Suppose further that $\mathbf{A}$ is such that (in the absence of the constraint) $\mathbf{A}\boldsymbol{\beta}$ would be jointly nonestimable or, equivalently, that $\mathbf{A}$ is such that $\mathcal{R}(\mathbf{X}) \cap \mathcal{R}(\mathbf{A}) = \{\mathbf{0}\}$ or, also equivalently, that $\mathbf{A}$ is such that $\text{rank}\begin{pmatrix} \mathbf{X} \\ \mathbf{A} \end{pmatrix} = \text{rank}(\mathbf{X}) + \text{rank}(\mathbf{A})$. And letting $\underline{\mathbf{y}}$ represent the observed value of $\mathbf{y}$, consider the solution of the constrained normal equations

$$\begin{pmatrix} \mathbf{X}'\mathbf{X} & \mathbf{A}' \\ \mathbf{A} & \mathbf{0} \end{pmatrix} \begin{pmatrix} \mathbf{b} \\ \mathbf{r} \end{pmatrix} = \begin{pmatrix} \mathbf{X}'\underline{\mathbf{y}} \\ \mathbf{d} \end{pmatrix} \tag{2.3}$$

in the special case where $\mathbf{A}$ satisfies any of those three equivalent conditions.

Augmented normal equations. Let $\tilde{\mathbf{b}}$ represent the "$\mathbf{b}$-component" and $\tilde{\mathbf{r}}$ the "$\mathbf{r}$-component" of an arbitrary solution to the constrained normal equations (2.3). Then, in the special case under consideration,

$$\tilde{\mathbf{r}}'\mathbf{A} = \mathbf{0} \quad \text{and} \quad (\underline{\mathbf{y}} - \mathbf{X}\tilde{\mathbf{b}})'\mathbf{X} = \mathbf{0},$$

as is evident upon observing that

$$\tilde{\mathbf{r}}'\mathbf{A} = (\mathbf{A}'\tilde{\mathbf{r}})' = (\mathbf{X}'\underline{\mathbf{y}} - \mathbf{X}'\mathbf{X}\tilde{\mathbf{b}})' = (\underline{\mathbf{y}} - \mathbf{X}\tilde{\mathbf{b}})'\mathbf{X}$$

and that $\mathcal{R}(\mathbf{A})$ and $\mathcal{R}(\mathbf{X})$ are essentially disjoint. And it follows that $\mathbf{A}'\tilde{\mathbf{r}} = (\tilde{\mathbf{r}}'\mathbf{A})' = \mathbf{0}$ and that $\mathbf{X}'\mathbf{X}\tilde{\mathbf{b}} = (\tilde{\mathbf{b}}'\mathbf{X}'\mathbf{X})' = (\underline{\mathbf{y}}'\mathbf{X})' = \mathbf{X}'\underline{\mathbf{y}}$ and hence that

$$\begin{pmatrix} \mathbf{X}'\mathbf{X} \\ \mathbf{A} \end{pmatrix} \tilde{\mathbf{b}} = \begin{pmatrix} \mathbf{X}'\underline{\mathbf{y}} \\ \mathbf{d} \end{pmatrix}. \tag{2.4}$$

Thus, in the special case under consideration, the first (P-dimensional) part of any solution to the constrained normal equations (2.3) is a solution to the "augmented normal equations"

$$\begin{pmatrix} \mathbf{X}'\mathbf{X} \\ \mathbf{A} \end{pmatrix} \mathbf{b} = \begin{pmatrix} \mathbf{X}'\underset{\sim}{\mathbf{y}} \\ \mathbf{d} \end{pmatrix} \tag{2.5}$$

(in the $P \times 1$ vector $\mathbf{b}$)—these equations are those obtained by augmenting the normal equations $\mathbf{X}'\mathbf{X}\mathbf{b} = \mathbf{X}'\underset{\sim}{\mathbf{y}}$ for the unconstrained model with the constraint $\mathbf{A}\mathbf{b} = \mathbf{d}$.

Since the constrained normal equations are consistent, the augmented normal equations are [in the special case under consideration and as is evident from equality (2.4)] consistent. Moreover, if $\tilde{\mathbf{b}}$ is any solution to the augmented normal equations and if $\tilde{\mathbf{r}}$ is the $Q \times 1$ null vector (or any other $Q \times 1$ vector for which $\mathbf{A}'\tilde{\mathbf{r}} = \mathbf{0}$), then the vector $\begin{pmatrix} \tilde{\mathbf{b}} \\ \tilde{\mathbf{r}} \end{pmatrix}$ is a solution to the constrained normal equations, so that (in the special case under consideration) the constrained normal equations and the augmented normal equations are "interchangeable."

Ranks of the coefficient matrices of the conjugate normal equations and the augmented normal equations. Note that $\mathcal{R}(\mathbf{X}'\mathbf{X}) \subset \mathcal{R}(\mathbf{X})$—in fact, $\mathcal{R}(\mathbf{X}'\mathbf{X}) = \mathcal{R}(\mathbf{X})$. Thus, in the special case under consideration, $\mathcal{R}(\mathbf{X}'\mathbf{X})$ and $\mathcal{R}(\mathbf{A})$ are essentially disjoint and hence (in that special case) the rank of $\begin{pmatrix} \mathbf{X}'\mathbf{X} \\ \mathbf{A} \end{pmatrix}$ (which is the coefficient matrix of the augmented normal equations) is expressible as

$$\operatorname{rank}\begin{pmatrix} \mathbf{X}'\mathbf{X} \\ \mathbf{A} \end{pmatrix} = \operatorname{rank}(\mathbf{X}'\mathbf{X}) + \operatorname{rank}(\mathbf{A}) = \operatorname{rank}(\mathbf{X}) + \operatorname{rank}(\mathbf{A}) \tag{2.6}$$

and the rank of the matrix $\begin{pmatrix} \mathbf{X}'\mathbf{X} & \mathbf{A}' \\ \mathbf{A} & \mathbf{0} \end{pmatrix}$ (which is the coefficient matrix of the constrained normal equations) is (upon recalling Corollary 8.1.2) expressible as

$$\operatorname{rank}\begin{pmatrix} \mathbf{X}'\mathbf{X} & \mathbf{A}' \\ \mathbf{A} & \mathbf{0} \end{pmatrix} = \operatorname{rank}\begin{pmatrix} \mathbf{X}'\mathbf{X} \\ \mathbf{A} \end{pmatrix} + \operatorname{rank}(\mathbf{A}) = \operatorname{rank}(\mathbf{X}) + 2\operatorname{rank}(\mathbf{A}). \tag{2.7}$$

c. Addition of constraints to an unconstrained model (as a matter of "convenience")

Suppose that $\mathbf{y}$ is an $N \times 1$ observable random vector that follows an (unconstrained) G–M, Aitken, or general linear model with model equation $\mathbf{y} = \mathbf{X}\boldsymbol{\beta} + \mathbf{e}$, where $\boldsymbol{\beta}$ is a $P \times 1$ vector of unknown (and unconstrained) parameters. Suppose further that for some $P \times 1$ vector $\boldsymbol{\lambda}$ for which $\boldsymbol{\lambda}' \in \mathcal{R}(\mathbf{X})$, we wish to make inferences about the (estimable) linear combination $\boldsymbol{\lambda}'\boldsymbol{\beta}$. And observe that the least squares estimate of $\boldsymbol{\lambda}'\boldsymbol{\beta}$ along with various related quantities for making inferences about $\boldsymbol{\lambda}'\boldsymbol{\beta}$ are determinable from a solution to the (unconstrained) normal equations $\mathbf{X}'\mathbf{X}\mathbf{b} = \mathbf{X}'\underset{\sim}{\mathbf{y}}$ (where $\underset{\sim}{\mathbf{y}}$ is the observed value of $\mathbf{y}$), from a solution to the conjugate normal equations $\mathbf{X}'\mathbf{X}\mathbf{s} = \boldsymbol{\lambda}$, and/or from a generalized inverse of $\mathbf{X}'\mathbf{X}$.

Now, suppose that $\operatorname{rank}(\mathbf{X}) < P$ and hence that $\operatorname{rank}(\mathbf{X}'\mathbf{X}) < P$, in which case the solution to the normal equations or conjugate normal equations is nonunique and $\mathbf{X}'\mathbf{X}$ has multiple generalized inverses. And take $\mathbf{A}$ to be a $Q \times P$ matrix of full row rank, where $Q = P - \operatorname{rank}(\mathbf{X})$ and $\mathbf{A}$ is such that $\mathbf{A}\boldsymbol{\beta}$ is jointly nonestimable—the exixtence of such a matrix was established (and a procedure for constructing such a matrix outlined) in the final part of Subsection a. Then, the nonuniqueness of the solution to the normal equations could (if desired) be eliminated by subjecting the solution to the restriction $\mathbf{A}\mathbf{b} = \mathbf{d}$ (where $\mathbf{d}$ can be any $Q \times 1$ vector) and hence by taking $\mathbf{b}$ to be the unique solution to the augmented normal equations

$$\begin{pmatrix} \mathbf{X}'\mathbf{X} \\ \mathbf{A} \end{pmatrix} \mathbf{b} = \begin{pmatrix} \mathbf{X}'\underset{\sim}{\mathbf{y}} \\ \mathbf{d} \end{pmatrix} \tag{2.8}$$

—the existence and uniqueness of the solution to these equations follows from the results of Subsection b. Alternatively, the same solution to the normal equations is obtainable as the first (P-dimensional) part of the unique solution to the constrained normal equations

$$\begin{pmatrix} \mathbf{X}'\mathbf{X} & \mathbf{A}' \\ \mathbf{A} & \mathbf{0} \end{pmatrix} \begin{pmatrix} \mathbf{b} \\ \mathbf{r} \end{pmatrix} = \begin{pmatrix} \mathbf{X}'\underset{\sim}{\mathbf{y}} \\ \mathbf{d} \end{pmatrix} \tag{2.9}$$

—the solution to these equations is unique and its first part is identical to the solution to the augmented normal equations (2.8), as is evident upon making further use of the results of Subsection b.

Since $\boldsymbol{\lambda}' \in \mathcal{R}(\mathbf{X})$, $\boldsymbol{\lambda}' = \mathbf{a}'\mathbf{X}$ (for some vector $\mathbf{a}$) or equivalently $\boldsymbol{\lambda} = \mathbf{X}'\mathbf{a}$, so that the conjugate normal equations are of the same general form as the normal equations and hence the results relating the solution of the normal equations to the solution of linear system (2.8) and the solution of linear system (2.9) can be readily extended to the solution of the conjugate normal equations—it's a simple matter of replacing $\underset{\sim}{\mathbf{y}}$ with $\mathbf{a}$. Accordingly, the "augmented conjugate normal equations"

$$\begin{pmatrix} \mathbf{X}'\mathbf{X} \\ \mathbf{A} \end{pmatrix} \mathbf{s} = \begin{pmatrix} \boldsymbol{\lambda} \\ \mathbf{0} \end{pmatrix}, \tag{2.10}$$

are consistent with a unique solution (that is a particular one of the solutions to the unconstrained conjugate normal equations $\mathbf{X}'\mathbf{X}\mathbf{s} = \boldsymbol{\lambda}$). And analogous to linear system (2.9), we have the linear system

$$\begin{pmatrix} \mathbf{X}'\mathbf{X} & \mathbf{A}' \\ \mathbf{A} & \mathbf{0} \end{pmatrix} \begin{pmatrix} \mathbf{s} \\ \mathbf{t} \end{pmatrix} = \begin{pmatrix} \boldsymbol{\lambda} \\ \mathbf{0} \end{pmatrix}, \tag{2.11}$$

which is consistent and has a unique solution, the first (P-dimensional) part of which is a particular one of the solutions to $\mathbf{X}'\mathbf{X}\mathbf{s} = \boldsymbol{\lambda}$ [the same one that is the unique solution to linear system (2.10)].

Turning our attention to generalized inverses of the $P \times P$ matrix $\mathbf{X}'\mathbf{X}$ (which is the coefficient matrix of the normal equations $\mathbf{X}'\mathbf{X}\mathbf{b} = \mathbf{X}'\underset{\sim}{\mathbf{y}}$ and of the conjugate normal equations $\mathbf{X}'\mathbf{X}\mathbf{s} = \boldsymbol{\lambda}$), observe that the coefficient matrix $\begin{pmatrix} \mathbf{X}'\mathbf{X} \\ \mathbf{A} \end{pmatrix}$ of linear systems (2.8) and (2.10) is of dimensions $(P+Q) \times P$ and of rank P, in which case it has a left inverse, say $\mathbf{G}$. Observe also that the coefficient matrix $\begin{pmatrix} \mathbf{X}'\mathbf{X} & \mathbf{A}' \\ \mathbf{A} & \mathbf{0} \end{pmatrix}$ of linear systems (2.9) and (2.11) is a nonsingular matrix (of order $P+Q$) and hence has an ordinary inverse, say $\mathbf{B}$. Further, take $\mathbf{G}_1$ to be the $P \times P$ submatrix of $\mathbf{G}$ and $\mathbf{B}_{11}$ the $P \times P$ submatrix of $\mathbf{B}$ defined by the partitionings $\mathbf{G} = (\mathbf{G}_1, \mathbf{G}_2)$ and $\mathbf{B} = \begin{pmatrix} \mathbf{B}_{11} & \mathbf{B}_{12} \\ \mathbf{B}_{21} & \mathbf{B}_{22} \end{pmatrix}$.

The submatrix $\mathbf{G}_1$ of the left inverse $\mathbf{G}$ of the matrix $\begin{pmatrix} \mathbf{X}'\mathbf{X} \\ \mathbf{A} \end{pmatrix}$ is a generalized inverse of $\mathbf{X}'\mathbf{X}$. To see this, observe that $\mathbf{G}$ is a left inverse of $\begin{pmatrix} \mathbf{X}'\mathbf{X} \\ \mathbf{A} \end{pmatrix}$ if and only if $\mathbf{G}'$ [$= \begin{pmatrix} \mathbf{G}_1' \\ \mathbf{G}_2' \end{pmatrix}$] is a right inverse of $(\mathbf{X}'\mathbf{X}, \mathbf{A}')$, and observe also that $\mathcal{C}(\mathbf{X}'\mathbf{X})$ and $\mathcal{C}(\mathbf{A}')$ are essentially disjoint. Then, as is evident upon applying Lemma 8.1.3, $\mathbf{X}'\mathbf{X}\mathbf{G}_1'\mathbf{X}'\mathbf{X} = \mathbf{X}'\mathbf{X}$, thereby implying that $\mathbf{G}_1'$ is a generalized inverse of $\mathbf{X}'\mathbf{X}$ and hence (since $\mathbf{X}'\mathbf{X}$ is symmetric) that G_1 [$= (\mathbf{G}_1')'$] is a generalized inverse of $\mathbf{X}'\mathbf{X}$.

Like the submatrix $\mathbf{G}_1$ of the left inverse $\mathbf{G}$ of the matrix $\begin{pmatrix} \mathbf{X}'\mathbf{X} \\ \mathbf{A} \end{pmatrix}$, the submatrix $\mathbf{B}_{11}$ of the ordinary inverse $\mathbf{B}$ of the matrix $\begin{pmatrix} \mathbf{X}'\mathbf{X} & \mathbf{A}' \\ \mathbf{A} & \mathbf{0} \end{pmatrix}$ is a generalized inverse of the matrix $\mathbf{X}'\mathbf{X}$. That $\mathbf{B}_{11}$ is a generalized inverse of $\mathbf{X}'\mathbf{X}$ can be verified by making further use of Lemma 8.1.3. Upon recalling (from Lemma 8.1.1) that $\mathcal{C}\begin{pmatrix} \mathbf{X}'\mathbf{X} \\ \mathbf{A} \end{pmatrix}$ and $\mathcal{C}\begin{pmatrix} \mathbf{A}' \\ \mathbf{0} \end{pmatrix}$ are essentially disjoint, it follows from Lemma 8.1.3 that $(\mathbf{B}_{11}, \mathbf{B}_{12})$ is a generalized inverse of $\begin{pmatrix} \mathbf{X}'\mathbf{X} \\ \mathbf{A} \end{pmatrix}$ and hence [since $\begin{pmatrix} \mathbf{X}'\mathbf{X} \\ \mathbf{A} \end{pmatrix}$ is of full column rank] that $(\mathbf{B}_{11}, \mathbf{B}_{12})$ is a left inverse of $\begin{pmatrix} \mathbf{X}'\mathbf{X} \\ \mathbf{A} \end{pmatrix}$. And upon applying [with $(\mathbf{B}_{11}, \mathbf{B}_{12})$ in place of $\mathbf{G} = (\mathbf{G}_1, \mathbf{G}_2)$] the (already verified) result that the submatrix $\mathbf{G}_1$ of the matrix $\mathbf{G}$ is a generalized inverse of $\mathbf{X}'\mathbf{X}$, we conclude that $\mathbf{B}_{11}$ is a generalized inverse of $\mathbf{X}'\mathbf{X}$.

d. Extensions

Subsections a, b, and c introduce the notion of jointly nonestimable constraints and present various results related to invoking such constraints as a matter of convenience. The coverage in those subsections pertains to situations in which an $N \times 1$ observable random vector $\mathbf{y}$ is assumed to follow an unconstrained G–M, Aitken, or general linear model with model equation $\mathbf{y} = \mathbf{X}\boldsymbol{\beta} + \mathbf{e}$, where $\boldsymbol{\beta}$ is a $P \times 1$ vector of unknown (and unconstrained) parameters. As demonstrated in what follows, that coverage can be readily extended to situations in which $\mathbf{y}$ is assumed to follow a constrained G–M, Aitken, or general linear model, where $\boldsymbol{\beta}$ is subject to linear equality constraints.

Joint nonestimability. Suppose that $\mathbf{y}$ is an $N \times 1$ observable random vector that follows a constrained G–M, Aitken, or general linear model with model equation $\mathbf{y} = \mathbf{X}\boldsymbol{\beta} + \mathbf{e}$, where $\boldsymbol{\beta}$ is a $P \times 1$ vector of unknown parameters that [for some $Q \times P$ matrix $\mathbf{A}$ and $Q \times 1$ vector $\mathbf{d} \in \mathcal{C}(\mathbf{A})$] is subject to the constraint $\mathbf{A}\boldsymbol{\beta} = \mathbf{d}$. And let $\boldsymbol{\lambda}_1, \boldsymbol{\lambda}_2, \ldots, \boldsymbol{\lambda}_M$ represent $P \times 1$ vectors of constants and let $\boldsymbol{\Lambda} = (\boldsymbol{\lambda}_1, \boldsymbol{\lambda}_2, \ldots, \boldsymbol{\lambda}_M)$, in which case $\boldsymbol{\lambda}_1'\boldsymbol{\beta}, \boldsymbol{\lambda}_2'\boldsymbol{\beta}, \ldots, \boldsymbol{\lambda}_M'\boldsymbol{\beta}$ are linear combinations of the elements of $\boldsymbol{\beta}$ and $\boldsymbol{\Lambda}'\boldsymbol{\beta} = (\boldsymbol{\lambda}_1'\boldsymbol{\beta}, \boldsymbol{\lambda}_2'\boldsymbol{\beta}, \ldots, \boldsymbol{\lambda}_M'\boldsymbol{\beta})'$ is an $M \times 1$ vector of such linear combinations.

As in the case of an unconstrained model, the linear combinations $\boldsymbol{\lambda}_1'\boldsymbol{\beta}, \boldsymbol{\lambda}_2'\boldsymbol{\beta}, \ldots, \boldsymbol{\lambda}_M'\boldsymbol{\beta}$ are said to be jointly nonestimable [and the vector $\boldsymbol{\Lambda}'\boldsymbol{\beta} = (\boldsymbol{\lambda}_1'\boldsymbol{\beta}, \boldsymbol{\lambda}_2'\boldsymbol{\beta}, \ldots, \boldsymbol{\lambda}_M'\boldsymbol{\beta})'$ is said to be jointly nonestimable] if each of the linear combinations $\boldsymbol{\lambda}_1'\boldsymbol{\beta}, \boldsymbol{\lambda}_2'\boldsymbol{\beta}, \ldots, \boldsymbol{\lambda}_M'\boldsymbol{\beta}$ is nonestimable and if in addition every nontrivial linear combination of $\boldsymbol{\lambda}_1'\boldsymbol{\beta}, \boldsymbol{\lambda}_2'\boldsymbol{\beta}, \ldots, \boldsymbol{\lambda}_M'\boldsymbol{\beta}$ is nonestimable. However, the constrained case differs from the unconstrained case in that a linear combination, say $\boldsymbol{\lambda}'\boldsymbol{\beta}$, is nonestimable if and only if $\boldsymbol{\lambda}' \notin \mathcal{R}\begin{pmatrix}\mathbf{X}\\ \mathbf{A}\end{pmatrix}$ [whereas in the unconstrained case, $\boldsymbol{\lambda}'\boldsymbol{\beta}$ is nonestimable if and only if $\boldsymbol{\lambda}' \notin \mathcal{R}(\mathbf{X})$], so that some linear combinations that are nonestimable under the unconstrained model may be estimable under the constrained model. Moreover, in the case of the constrained model, $\boldsymbol{\Lambda}'\boldsymbol{\beta}$ is jointly nonestimable if and only if

$$\mathcal{R}\begin{pmatrix}\mathbf{X}\\ \mathbf{A}\end{pmatrix} \cap \mathcal{R}(\boldsymbol{\Lambda}') = \{\mathbf{0}\}$$

and is jointly nonestimable if and only if

$$\operatorname{rank}\begin{pmatrix}\mathbf{X}\\ \mathbf{A}\\ \boldsymbol{\Lambda}'\end{pmatrix} = \operatorname{rank}\begin{pmatrix}\mathbf{X}\\ \mathbf{A}\end{pmatrix} + \operatorname{rank}(\boldsymbol{\Lambda}).$$

Jointly nonestimable constraints. Suppose that $\mathbf{y}$ is an $N \times 1$ observable random vector that follows a constrained G–M, Aitken, or general linear model with model equation $\mathbf{y} = \mathbf{X}\boldsymbol{\beta} + \mathbf{e}$, where $\boldsymbol{\beta}$ is a $P \times 1$ vector of unknown parameters that [for some $Q \times P$ matrix $\mathbf{A}$ and $Q \times 1$ vector $\mathbf{d} \in \mathcal{C}(\mathbf{A})$] is subject to the constraint $\mathbf{A}\boldsymbol{\beta} = \mathbf{d}$. Suppose further that $\mathbf{A} = \begin{pmatrix}\mathbf{A}_1\\ \mathbf{A}_2\end{pmatrix}$ for some $Q_1 \times P$ matrix $\mathbf{A}_1$ and some $Q_2 \times P$ matrix $\mathbf{A}_2$ (with $Q_2 = Q - Q_1$) for which

$$\mathcal{R}\begin{pmatrix}\mathbf{X}\\ \mathbf{A}_1\end{pmatrix} \cap \mathcal{R}(\mathbf{A}_2) = \{\mathbf{0}\}. \tag{2.12}$$

Then, upon partitioning $\mathbf{d}$ conformally as $\mathbf{d} = \begin{pmatrix}\mathbf{d}_1\\ \mathbf{d}_2\end{pmatrix}$ and denoting by $\underline{\mathbf{y}}$ the observed value of $\mathbf{y}$, the constrained normal equations are expressible as

$$\begin{pmatrix}\mathbf{X}'\mathbf{X} & \mathbf{A}_1' & \mathbf{A}_2'\\ \mathbf{A}_1 & \mathbf{0} & \mathbf{0}\\ \mathbf{A}_2 & \mathbf{0} & \mathbf{0}\end{pmatrix}\begin{pmatrix}\mathbf{b}\\ \mathbf{r}_1\\ \mathbf{r}_2\end{pmatrix} = \begin{pmatrix}\mathbf{X}'\underline{\mathbf{y}}\\ \mathbf{d}_1\\ \mathbf{d}_2\end{pmatrix}. \tag{2.13}$$

This setting can be regarded as an extension of that considered in Subsection b. In this extension, the role of the augmented normal equations is played by the equations

$$\begin{pmatrix}\mathbf{X}'\mathbf{X} & \mathbf{A}_1'\\ \mathbf{A} & \mathbf{0}\end{pmatrix}\begin{pmatrix}\mathbf{b}\\ \mathbf{r}_1\end{pmatrix} = \begin{pmatrix}\mathbf{X}'\underline{\mathbf{y}}\\ \mathbf{d}\end{pmatrix}. \tag{2.14}$$

These equations are related to the equations (2.13) in a way that can be discerned by employing reasoning similar to that employed in Subsection b in relating the equations (2.5) to the equations

(2.3). The first [$(P+Q_1)$-dimensional] part of any solution to the equations (2.13) is a solution to the equations (2.14) [implying in particular that the equations (2.14) are consistent]. And, conversely, if $\begin{pmatrix} \tilde{\mathbf{b}} \\ \tilde{\mathbf{r}}_1 \end{pmatrix}$ is any solution to the equations (2.14) and if $\tilde{\mathbf{r}}_2$ is the $Q_2 \times 1$ null vector or any other $Q_2 \times 1$ vector for which $\mathbf{A}_2'\tilde{\mathbf{r}}_2 = \mathbf{0}$, then $\begin{pmatrix} \tilde{\mathbf{b}} \\ \tilde{\mathbf{r}}_1 \\ \tilde{\mathbf{r}}_2 \end{pmatrix}$ is a solution to the equations (2.13).

Note that $\mathcal{R}\begin{pmatrix} \mathbf{X}'\mathbf{X} \\ \mathbf{A}_1 \end{pmatrix} = \mathcal{R}\begin{pmatrix} \mathbf{X} \\ \mathbf{A}_1 \end{pmatrix}$ [as is evident from result (1.23)], so that condition (2.12) is equivalent to the condition

$$\mathcal{R}\begin{pmatrix} \mathbf{X}'\mathbf{X} \\ \mathbf{A}_1 \end{pmatrix} \cap \mathcal{R}(\mathbf{A}_2) = \{\mathbf{0}\}. \tag{2.15}$$

Note also that

$$\mathcal{R}\begin{pmatrix} \mathbf{X}'\mathbf{X} & \mathbf{A}_1' \\ \mathbf{A}_1 & \mathbf{0} \end{pmatrix} \cap \mathcal{R}(\mathbf{A}_2,\ \mathbf{0}) = \{\mathbf{0}\} \tag{2.16}$$

and that

$$\mathcal{R}\begin{pmatrix} \mathbf{X}'\mathbf{X} & \mathbf{A}' \\ \mathbf{A}_1 & \mathbf{0} \end{pmatrix} \cap \mathcal{R}(\mathbf{A}_2,\ \mathbf{0}) = \{\mathbf{0}\}. \tag{2.17}$$

Thus, making use of Theorem 2.4.25 and Corollary 8.1.2, we find that the rank of the coefficient matrix of linear system (2.14) is expressible as

$$\begin{aligned} \operatorname{rank}\begin{pmatrix} \mathbf{X}'\mathbf{X} & \mathbf{A}_1' \\ \mathbf{A} & \mathbf{0} \end{pmatrix} &= \operatorname{rank}\begin{pmatrix} \mathbf{X}'\mathbf{X} & \mathbf{A}_1' \\ \mathbf{A}_1 & \mathbf{0} \end{pmatrix} + \operatorname{rank}(\mathbf{A}_2,\ \mathbf{0}) \\ &= \operatorname{rank}\begin{pmatrix} \mathbf{X} \\ \mathbf{A}_1 \end{pmatrix} + \operatorname{rank}(\mathbf{A}_1) + \operatorname{rank}(\mathbf{A}_2) \end{aligned} \tag{2.18}$$

and that the rank of the coefficient matrix of linear system (2.13) is expressible as

$$\begin{aligned} \operatorname{rank}\begin{pmatrix} \mathbf{X}'\mathbf{X} & \mathbf{A}_1' & \mathbf{A}_2' \\ \mathbf{A}_1 & \mathbf{0} & \mathbf{0} \\ \mathbf{A}_2 & \mathbf{0} & \mathbf{0} \end{pmatrix} &= \operatorname{rank}\begin{pmatrix} \mathbf{X}'\mathbf{X} & \mathbf{A}' \\ \mathbf{A}_1 & \mathbf{0} \end{pmatrix} + \operatorname{rank}\,(\mathbf{A}_2,\ \mathbf{0}) \\ &= \operatorname{rank}\left[\begin{pmatrix} \mathbf{X}'\mathbf{X} & \mathbf{A}' \\ \mathbf{A}_1 & \mathbf{0} \end{pmatrix}'\right] + \operatorname{rank}(\mathbf{A}_2) \\ &= \operatorname{rank}\begin{pmatrix} \mathbf{X}'\mathbf{X} & \mathbf{A}_1' \\ \mathbf{A} & \mathbf{0} \end{pmatrix} + \operatorname{rank}(\mathbf{A}_2) \qquad (2.19) \\ &= \operatorname{rank}\begin{pmatrix} \mathbf{X} \\ \mathbf{A}_1 \end{pmatrix} + \operatorname{rank}(\mathbf{A}_1) + 2\operatorname{rank}(\mathbf{A}_2). \qquad (2.20) \end{aligned}$$

Additional constraints. Suppose that $\mathbf{y}$ is an $N{\times}1$ observable random vector that follows a constrained G–M, Aitken, or general linear model with model equation $\mathbf{y} = \mathbf{X}\boldsymbol{\beta} + \mathbf{e}$, where $\boldsymbol{\beta}$ is a $P \times 1$ vector of unknown parameters that [for some $Q_1 \times P$ matrix $\mathbf{A}_1$ and $Q_1 \times 1$ vector $\mathbf{d}_1 \in \mathcal{C}(\mathbf{A}_1)$] is subject to the constraint $\mathbf{A}_1\boldsymbol{\beta} = \mathbf{d}_1$. Then, for purposes of making inferences about an estimable linear combination (of the elements of $\boldsymbol{\beta}$), say $\boldsymbol{\lambda}'\boldsymbol{\beta}$, we may wish to solve the constrained normal equations $\begin{pmatrix} \mathbf{X}'\mathbf{X} & \mathbf{A}_1' \\ \mathbf{A}_1 & \mathbf{0} \end{pmatrix}\begin{pmatrix} \mathbf{b} \\ \mathbf{r}_1 \end{pmatrix} = \begin{pmatrix} \mathbf{X}'\underline{\mathbf{y}} \\ \mathbf{d}_1 \end{pmatrix}$ (where $\underline{\mathbf{y}}$ is the observed value of $\mathbf{y}$), to solve the constrained conjugate normal equations $\begin{pmatrix} \mathbf{X}'\mathbf{X} & \mathbf{A}_1' \\ \mathbf{A}_1 & \mathbf{0} \end{pmatrix}\begin{pmatrix} \mathbf{s} \\ \mathbf{t}_1 \end{pmatrix} = \begin{pmatrix} \boldsymbol{\lambda} \\ \mathbf{0} \end{pmatrix}$, and/or to obtain a generalized inverse of the matrix $\begin{pmatrix} \mathbf{X}'\mathbf{X} & \mathbf{A}_1' \\ \mathbf{A}_1 & \mathbf{0} \end{pmatrix}$.

The matrix $\begin{pmatrix} \mathbf{X}'\mathbf{X} & \mathbf{A}_1' \\ \mathbf{A}_1 & \mathbf{0} \end{pmatrix}$ is a symmetric matrix of order $P+Q_1$. And (as is evident from Corollary 8.1.2),

$$\operatorname{rank}\begin{pmatrix} \mathbf{X}'\mathbf{X} & \mathbf{A}_1' \\ \mathbf{A}_1 & \mathbf{0} \end{pmatrix} = \operatorname{rank}\begin{pmatrix} \mathbf{X} \\ \mathbf{A}_1 \end{pmatrix} + \operatorname{rank}(\mathbf{A}_1) \le P + Q_1.$$

Let us assume that $\mathbf{A}_1$ is of full row rank (i.e., of rank Q_1)—this assumption can be made without loss of generality in the sense that if $\mathbf{A}_*$ is a matrix whose rows are rank($\mathbf{A}_1$) linearly independent rows of $\mathbf{A}_1$ and $\mathbf{d}_*$ is a vector whose elements are the corresponding elements of $\mathbf{d}_1$, then the constraint $\mathbf{A}_*\boldsymbol{\beta} = \mathbf{d}_*$ is equivalent to the constraint $\mathbf{A}_1\boldsymbol{\beta} = \mathbf{d}_1$. Then, the matrix $\begin{pmatrix} \mathbf{X}'\mathbf{X} & \mathbf{A}_1' \\ \mathbf{A}_1 & \mathbf{0} \end{pmatrix}$ is nonsingular (i.e., of rank $P+Q_1$) if the matrix $\begin{pmatrix} \mathbf{X} \\ \mathbf{A}_1 \end{pmatrix}$ is of full column rank P, but not otherwise.

By introducing additional constraints of a certain kind, one can obtain constrained normal equations and constrained conjugate normal equations for which the coefficient matrix is nonsingular and can do so in such a way that the inferences are unaffected. More specifically, take $\mathbf{A}_2$ to be a $Q_2 \times P$ matrix of full row rank, where $Q_2 = P - \text{rank}\begin{pmatrix} \mathbf{X} \\ \mathbf{A}_1 \end{pmatrix}$—the existence of such a matrix can be established via a procedure outlined in the final part of Subsection a. And let $\mathbf{A} = \begin{pmatrix} \mathbf{A}_1 \\ \mathbf{A}_2 \end{pmatrix}$ and $\mathbf{d} = \begin{pmatrix} \mathbf{d}_1 \\ \mathbf{d}_2 \end{pmatrix}$, where $\mathbf{d}_2$ is any $Q_2 \times 1$ vector of constants—the assumption that $\mathbf{A}_1$ is of full row rank together with the assumptions about $\mathbf{A}_2$ imply that $\mathbf{A}$ is of full row rank and insure that $\mathbf{d} \in \mathbf{A}$.

When $\boldsymbol{\beta}$ is subject to the constraint $\mathbf{A}_2\boldsymbol{\beta} = \mathbf{d}_2$ as well as the constraint $\mathbf{A}_1\boldsymbol{\beta} = \mathbf{d}_1$, the constrained normal equations are

$$\begin{pmatrix} \mathbf{X}'\mathbf{X} & \mathbf{A}_1' & \mathbf{A}_2' \\ \mathbf{A}_1 & \mathbf{0} & \mathbf{0} \\ \mathbf{A}_2 & \mathbf{0} & \mathbf{0} \end{pmatrix} \begin{pmatrix} \mathbf{b} \\ \mathbf{r}_1 \\ \mathbf{r}_2 \end{pmatrix} = \begin{pmatrix} \mathbf{X}'\underline{\mathbf{y}} \\ \mathbf{d}_1 \\ \mathbf{d}_2 \end{pmatrix}, \tag{2.21}$$

and the role played by the augmented normal equations "when $Q_1 = 0$" is assumed [for purposes of making inferences about $\boldsymbol{\lambda}'\boldsymbol{\beta}$ when $\boldsymbol{\lambda}' \in \mathcal{R}\begin{pmatrix} \mathbf{X} \\ \mathbf{A}_1 \end{pmatrix}$] by the equations

$$\begin{pmatrix} \mathbf{X}'\mathbf{X} & \mathbf{A}_1' \\ \mathbf{A} & \mathbf{0} \end{pmatrix} \begin{pmatrix} \mathbf{b} \\ \mathbf{r}_1 \end{pmatrix} = \begin{pmatrix} \mathbf{X}'\underline{\mathbf{y}} \\ \mathbf{d} \end{pmatrix}. \tag{2.22}$$

And upon applying results (2.18) and (2.19), we find that

$$\text{rank}\begin{pmatrix} \mathbf{X}'\mathbf{X} & \mathbf{A}_1' \\ \mathbf{A} & \mathbf{0} \end{pmatrix} = \text{rank}\begin{pmatrix} \mathbf{X} \\ \mathbf{A}_1 \end{pmatrix} + Q_1 + P - \text{rank}\begin{pmatrix} \mathbf{X} \\ \mathbf{A}_1 \end{pmatrix} = P + Q_1$$

and that

$$\text{rank}\begin{pmatrix} \mathbf{X}'\mathbf{X} & \mathbf{A}_1' & \mathbf{A}_2' \\ \mathbf{A}_1 & \mathbf{0} & \mathbf{0} \\ \mathbf{A}_2 & \mathbf{0} & \mathbf{0} \end{pmatrix} = P + Q_1 + Q_2$$

[so that the coefficient matrix of linear system (2.22) is of full column rank and the coefficient matrix of linear system (2.21) is nonsingular]. Thus, each of the linear systems (2.21) and (2.22) has a unique solution and (as is evident from the preceding part of the present subsection) the first [$(P+Q_1)$-dimensional] part of the solution to linear system (2.21) equals the solution to linear system (2.22). Moreover, the solution to linear system (2.22) is among the solutions to the linear system $\begin{pmatrix} \mathbf{X}'\mathbf{X} & \mathbf{A}_1' \\ \mathbf{A}_1 & \mathbf{0} \end{pmatrix} \begin{pmatrix} \mathbf{b} \\ \mathbf{r}_1 \end{pmatrix} = \begin{pmatrix} \mathbf{X}'\underline{\mathbf{y}} \\ \mathbf{d}_1 \end{pmatrix}$.

In the case of the constrained conjugate normal equations for making inferences about $\boldsymbol{\lambda}'\boldsymbol{\beta}$ [when $\boldsymbol{\lambda}' \in \mathcal{R}\begin{pmatrix} \mathbf{X} \\ \mathbf{A}_1 \end{pmatrix}$], the counterpart of the linear system (2.21) is the linear system

$$\begin{pmatrix} \mathbf{X}'\mathbf{X} & \mathbf{A}_1' & \mathbf{A}_2' \\ \mathbf{A}_1 & \mathbf{0} & \mathbf{0} \\ \mathbf{A}_2 & \mathbf{0} & \mathbf{0} \end{pmatrix} \begin{pmatrix} \mathbf{s} \\ \mathbf{t}_1 \\ \mathbf{t}_2 \end{pmatrix} = \begin{pmatrix} \boldsymbol{\lambda} \\ \mathbf{0} \\ \mathbf{0} \end{pmatrix} \tag{2.23}$$

and the counterpart of linear system (2.22) is the linear system

$$\begin{pmatrix} \mathbf{X}'\mathbf{X} & \mathbf{A}_1' \\ \mathbf{A} & \mathbf{0} \end{pmatrix} \begin{pmatrix} \mathbf{s} \\ \mathbf{t}_1 \end{pmatrix} = \begin{pmatrix} \boldsymbol{\lambda} \\ \mathbf{0} \end{pmatrix}. \tag{2.24}$$

Analogous to the results for linear systems (2.21) and (2.22), each of linear systems (2.23) and (2.24) has a unique solution, the first $[(P+Q_1)$-dimensional] part of the solution to linear system (2.23) equals the solution to linear system (2.24), and the solution to linear system (2.24) is among the solutions to the linear system $\begin{pmatrix} \mathbf{X}'\mathbf{X} & \mathbf{A}_1' \\ \mathbf{A}_1 & \mathbf{0} \end{pmatrix} \begin{pmatrix} \mathbf{s} \\ \mathbf{t}_1 \end{pmatrix} = \begin{pmatrix} \boldsymbol{\lambda} \\ \mathbf{0} \end{pmatrix}$.

Now, let $\mathbf{G}$ represent the inverse of the matrix $\begin{pmatrix} \mathbf{X}'\mathbf{X} & \mathbf{A}_1' & \mathbf{A}_2' \\ \mathbf{A}_1 & \mathbf{0} & \mathbf{0} \\ \mathbf{A}_2 & \mathbf{0} & \mathbf{0} \end{pmatrix}$ and partition $\mathbf{G}$ as $\mathbf{G} = \begin{pmatrix} \mathbf{G}_1 \\ \mathbf{G}_2 \end{pmatrix}$, where $\mathbf{G}_1$ is of dimensions $(P+Q_1) \times (P+Q_1+Q_2)$. Then, upon observing [in light of result (2.17)] that $\mathcal{C}\begin{pmatrix} \mathbf{X}'\mathbf{X} & \mathbf{A}_1' \\ \mathbf{A} & \mathbf{0} \end{pmatrix}$ and $\mathcal{C}\begin{pmatrix} \mathbf{A}_2' \\ \mathbf{0} \end{pmatrix}$ are essentially disjoint and upon applying Lemma 8.1.3, we find that $\mathbf{G}_1$ is a generalized inverse of $\begin{pmatrix} \mathbf{X}'\mathbf{X} & \mathbf{A}_1' \\ \mathbf{A} & \mathbf{0} \end{pmatrix}$ [and hence is a left inverse of $\begin{pmatrix} \mathbf{X}'\mathbf{X} & \mathbf{A}_1' \\ \mathbf{A} & \mathbf{0} \end{pmatrix}$].

Further, let $\mathbf{G}_1$ represent any left inverse of $\begin{pmatrix} \mathbf{X}'\mathbf{X} & \mathbf{A}_1' \\ \mathbf{A} & \mathbf{0} \end{pmatrix}$, and partition $\mathbf{G}_1$ as $\mathbf{G}_1 = (\mathbf{G}_{11}, \mathbf{G}_{12})$, where $\mathbf{G}_{11}$ is of dimensions $(P+Q_1) \times (P+Q_1)$. Then, upon observing [in light of result (2.16)] that $\mathcal{C}\begin{pmatrix} \mathbf{X}'\mathbf{X} & \mathbf{A}_1' \\ \mathbf{A}_1 & \mathbf{0} \end{pmatrix}$ and $\mathcal{C}\begin{pmatrix} \mathbf{A}_2' \\ \mathbf{0} \end{pmatrix}$ are essentially disjoint and observing that $\mathbf{G}_1' = \begin{pmatrix} \mathbf{G}_{11}' \\ \mathbf{G}_{12}' \end{pmatrix}$ is a generalized inverse of $\begin{pmatrix} \mathbf{X}'\mathbf{X} & \mathbf{A}' \\ \mathbf{A}_1 & \mathbf{0} \end{pmatrix}$ and upon applying Lemma 8.1.3,we find that $\mathbf{G}_{11}'$ is a generalized inverse of $\begin{pmatrix} \mathbf{X}'\mathbf{X} & \mathbf{A}_1' \\ \mathbf{A}_1 & \mathbf{0} \end{pmatrix}$ and hence that $\mathbf{G}_{11} = (\mathbf{G}_{11}')'$ is a generalized inverse of $\begin{pmatrix} \mathbf{X}'\mathbf{X} & \mathbf{A}_1' \\ \mathbf{A}_1 & \mathbf{0} \end{pmatrix}$.

8.3 Some Results on the Testing of Hypotheses

a. Unconstrained models

Suppose that $\mathbf{y}$ is an $N \times 1$ observable random vector that follows an (unconstrained) G–M model with model equation $\mathbf{y} = \mathbf{X}\boldsymbol{\beta} + \mathbf{e}$, where $\boldsymbol{\beta}$ is a $P \times 1$ vector of unknown (and unconstrained) parameters and where $\mathrm{E}(\mathbf{e}) = \mathbf{0}$ and $\mathrm{var}(\mathbf{e}) = \sigma^2\mathbf{I}$. And let $\boldsymbol{\tau} = \boldsymbol{\Lambda}'\boldsymbol{\beta}$ represent an M-dimensional column vector of estimable linear combinations of the elements of $\boldsymbol{\beta}$, take $\tilde{\mathbf{b}}$ to be any solution to the normal equations $\mathbf{X}'\mathbf{X}\mathbf{b} = \mathbf{X}'\mathbf{y}$, and define $\hat{\boldsymbol{\tau}} = \boldsymbol{\Lambda}'\tilde{\mathbf{b}}$ and $\mathbf{C} = \boldsymbol{\Lambda}'(\mathbf{X}'\mathbf{X})^-\boldsymbol{\Lambda}$ [so that $\hat{\boldsymbol{\tau}}$ is the least squares estimator of $\boldsymbol{\tau}$ and $\mathrm{var}(\hat{\boldsymbol{\tau}}) = \sigma^2\mathbf{C}$]. Further, let $P_* = \mathrm{rank}(\mathbf{X})$ and $M_* = \mathrm{rank}(\boldsymbol{\Lambda})$, and denote by $\boldsymbol{\tau}^{(0)}$ $[\in \mathcal{C}(\boldsymbol{\Lambda}')]$ a hypothesized value of $\boldsymbol{\tau}$. Finally, recall (from Chapter 7) that (under suitable assumptions about the distribution of $\mathbf{e}$) a size-$\dot{\gamma}$ test of the testable null hypothesis $H_0 : \boldsymbol{\tau} = \boldsymbol{\tau}^{(0)}$ (versus the alternative hypothesis $H_1 : \boldsymbol{\tau} \neq \boldsymbol{\tau}^{(0)}$) is provided by the test (known as the F test) that rejects H_0 whenever the statistic

$$F = \frac{(\hat{\boldsymbol{\tau}} - \boldsymbol{\tau}^{(0)})'\mathbf{C}^-(\hat{\boldsymbol{\tau}} - \boldsymbol{\tau}^{(0)})/M_*}{(\mathbf{y} - \mathbf{X}\tilde{\mathbf{b}})'(\mathbf{y} - \mathbf{X}\tilde{\mathbf{b}})/(N - P_*)} \tag{3.1}$$

(known as the F statistic) exceeds $\bar{F}_{\dot{\gamma}}(M_*,\, N - P_*)$.

As discussed in some detail in what follows, various of the results on the constrained G–M model are of relevance in the testing of the null hypothesis H_0 (even though the actual model is such that $\boldsymbol{\beta}$ is unconstrained). These results are ones for a constrained G–M model in which the constraint is taken to be

$$\boldsymbol{\Lambda}'\boldsymbol{\beta} = \boldsymbol{\tau}^{(0)}$$

(corresponding to an assumption that the null hypothesis is satisfied).

An alternative expression for the F ***statistic.*** For $\mathbf{b} \in \mathbb{R}^P$, let $q(\mathbf{b}) = (\mathbf{y} - \mathbf{X}\mathbf{b})'(\mathbf{y} - \mathbf{X}\mathbf{b})$. Then, as will subsequently be shown, the quantity $(\hat{\boldsymbol{\tau}} - \boldsymbol{\tau}^{(0)})'\mathbf{C}^-(\hat{\boldsymbol{\tau}} - \boldsymbol{\tau}^{(0)})$, which appears in the numerator of the F statistic (3.1), is reexpressible as follows:

$$(\hat{\boldsymbol{\tau}}-\boldsymbol{\tau}^{(0)})'\mathbf{C}^{-}(\hat{\boldsymbol{\tau}}-\boldsymbol{\tau}^{(0)}) = \min_{\mathbf{b}\,:\,\boldsymbol{\Lambda}'\mathbf{b}=\boldsymbol{\tau}^{(0)}} q(\mathbf{b}) - \min_{\mathbf{b}\in\mathcal{R}^P} q(\mathbf{b}). \tag{3.2}$$

Moreover, $q(\mathbf{b})$ attains its minimum value for $\mathbf{b} \in \mathcal{R}^P$ at $\mathbf{b} = \bar{\mathbf{b}}$ (where $\bar{\mathbf{b}}$ is any solution to the unconstrained normal equations), and (in light of the results of Section 8.1d) $q(\mathbf{b})$ attains its minimum value for $\mathbf{b}$ such that $\boldsymbol{\Lambda}'\mathbf{b} = \boldsymbol{\tau}^{(0)}$ at $\tilde{\mathbf{b}}$, where $\tilde{\mathbf{b}}$ is any value of the $P\times 1$ vector $\mathbf{b}$ and $\tilde{\mathbf{r}}$ any value of the $M\times 1$ vector $\mathbf{r}$ that together form a solution to the constrained normal equations

$$\begin{pmatrix} \mathbf{X}'\mathbf{X} & \boldsymbol{\Lambda} \\ \boldsymbol{\Lambda}' & \mathbf{0} \end{pmatrix} \begin{pmatrix} \mathbf{b} \\ \mathbf{r} \end{pmatrix} = \begin{pmatrix} \mathbf{X}'\mathbf{y} \\ \boldsymbol{\tau}^{(0)} \end{pmatrix}. \tag{3.3}$$

Thus, as an alternative version of expression (3.2), we have that

$$(\hat{\boldsymbol{\tau}}-\boldsymbol{\tau}^{(0)})'\mathbf{C}^{-}(\hat{\boldsymbol{\tau}}-\boldsymbol{\tau}^{(0)}) = q(\tilde{\mathbf{b}}) - q(\bar{\mathbf{b}}). \tag{3.4}$$

Clearly, by making use of result (3.4), the F statistic (3.1) is reexpressible in the following form:

$$F = \frac{[q(\tilde{\mathbf{b}}) - q(\bar{\mathbf{b}})]/M_*}{q(\bar{\mathbf{b}})/(N-P_*)}. \tag{3.5}$$

Let us verify result (3.2) or, equivalently, result (3.4). As a first step in the verification, observe that among the choices for the vectors $\tilde{\mathbf{b}}$ and $\tilde{\mathbf{r}}$ (that together form a solution to the constrained normal equations) are those obtained by taking $\tilde{\mathbf{r}}$ to be any solution to the linear system

$$\mathbf{C}\mathbf{r} = \hat{\boldsymbol{\tau}}-\boldsymbol{\tau}^{(0)} \tag{3.6}$$

(in $\mathbf{r}$) and taking $\tilde{\mathbf{b}} = \bar{\mathbf{b}} - (\mathbf{X}'\mathbf{X})^{-}\boldsymbol{\Lambda}\tilde{\mathbf{r}}$—that linear system (3.6) has a solution is evident upon observing that $\hat{\boldsymbol{\tau}}-\boldsymbol{\tau}^{(0)} \in \mathcal{C}(\boldsymbol{\Lambda}')$ and upon recalling [from result (7.3.1)] that $\text{rank}(\mathbf{C}) = \text{rank}(\boldsymbol{\Lambda}')$ and hence [since $\mathcal{C}(\mathbf{C}) \subset \mathcal{C}(\boldsymbol{\Lambda}')$] that $\mathcal{C}(\mathbf{C}) = \mathcal{C}(\boldsymbol{\Lambda}')$. That $\tilde{\mathbf{b}}$ and $\tilde{\mathbf{r}}$ can be chosen thusly can be readily confirmed.

Accordingly, suppose that $\tilde{\mathbf{r}}$ is a solution to linear system (3.6) and that $\tilde{\mathbf{b}} = \bar{\mathbf{b}} - (\mathbf{X}'\mathbf{X})^{-}\boldsymbol{\Lambda}\tilde{\mathbf{r}}$. Then, upon making use of result (5.4.16) (and upon observing that $\boldsymbol{\Lambda}'[(\mathbf{X}'\mathbf{X})^{-}]'\mathbf{X}'\mathbf{X} = \boldsymbol{\Lambda}'$), we find that

$$\begin{aligned} q(\tilde{\mathbf{b}}) - q(\bar{\mathbf{b}}) &= [\mathbf{X}(\tilde{\mathbf{b}}-\bar{\mathbf{b}})]'\mathbf{X}(\tilde{\mathbf{b}}-\bar{\mathbf{b}}) \\ &= \tilde{\mathbf{r}}'\boldsymbol{\Lambda}'[(\mathbf{X}'\mathbf{X})^{-}]'\mathbf{X}'\mathbf{X}(\mathbf{X}'\mathbf{X})^{-}\boldsymbol{\Lambda}\tilde{\mathbf{r}} \\ &= \tilde{\mathbf{r}}'\mathbf{C}\tilde{\mathbf{r}} = (\mathbf{C}\tilde{\mathbf{r}})'\mathbf{C}^{-}\mathbf{C}\tilde{\mathbf{r}} \\ &= (\hat{\boldsymbol{\tau}}-\boldsymbol{\tau}^{(0)})'\mathbf{C}^{-}(\hat{\boldsymbol{\tau}}-\boldsymbol{\tau}^{(0)}), \end{aligned}$$

thereby completing the verification of result (3.4) [and in doing so serving also to verify results (3.2) and (3.5)].

Likelihood ratio test. Let us consider further the problem of testing the testable null hypothesis $H_0: \boldsymbol{\tau} = \boldsymbol{\tau}^{(0)}$ versus the alternative hypothesis $H_1: \boldsymbol{\tau} \neq \boldsymbol{\tau}^{(0)}$ [where $\boldsymbol{\tau} = \boldsymbol{\Lambda}'\boldsymbol{\beta}$ and $\boldsymbol{\tau}^{(0)} \in \mathcal{C}(\boldsymbol{\Lambda}')$]. Specifically, let us consider a method for devising tests of hypotheses known as the likelihood ratio method as applied to the problem of testing H_0—refer, e.g., to Casella and Berger (2002, sec. 8.2) for a general description and a discussion and illustration of the likelihood ratio method. For that purpose, suppose that the distribution of the vector $\mathbf{e}$ of residual effects in the G–M model is MVN, in which case $\mathbf{y} \sim N(\mathbf{X}\boldsymbol{\beta}, \sigma^2\mathbf{I})$.

As in Part 1 of the present subsection, take $q(\mathbf{b})$ to be the function defined (for $\mathbf{b} \in \mathcal{R}^P$) as follows: $q(\mathbf{b}) = (\mathbf{y}-\mathbf{X}\mathbf{b})'(\mathbf{y}-\mathbf{X}\mathbf{b})$. Then, the likelihood function is the function, say $L(\boldsymbol{\beta}, \sigma)$, of $\boldsymbol{\beta}$ and σ that is expressible as

$$L(\boldsymbol{\beta}, \sigma) = (2\pi\sigma^2)^{-N/2}\exp[-q(\boldsymbol{\beta})/(2\sigma^2)]. \tag{3.7}$$

And the likelihood ratio test of H_0 is the test that rejects H_0 whenever the ratio

$$\frac{\max_{\boldsymbol{\beta},\sigma\,:\,\boldsymbol{\Lambda}'\boldsymbol{\beta}=\boldsymbol{\tau}^{(0)},\ \sigma>0} L(\boldsymbol{\beta}, \sigma)}{\max_{\boldsymbol{\beta}\in\mathcal{R}^P,\ \sigma>0} L(\boldsymbol{\beta}, \sigma)} \tag{3.8}$$

(which is referred to as the likelihood ratio test statistic) is less than a constant c $(0<c<1)$, where c is chosen so as to determine the size of the test.

Now, consider the maximization of $L(\boldsymbol{\beta}, \sigma)$ with respect to σ (for any particular value of $\boldsymbol{\beta}$). Unless $\mathbf{y} \in \mathcal{C}(\mathbf{X})$ (which is an event of probability 0), $\log L(\boldsymbol{\beta}, \sigma)$ is (for "fixed" $\boldsymbol{\beta}$) expressible as a function of σ of the form (5.9.4). Thus, it follows from the results of Section 5.9a that [unless $\mathbf{y} \in \mathcal{C}(\mathbf{X})$] $L(\boldsymbol{\beta}, \sigma)$ attains its maximum value with respect to σ (for fixed β) when $\sigma^2 = q(\boldsymbol{\beta})/N$.

Continuing, take (as in Part 1 of the present subsection) $\bar{\mathbf{b}}$ to be any solution to the (unconstrained) normal equations $\mathbf{X}'\mathbf{X}\mathbf{b} = \mathbf{X}'\mathbf{y}$ and take $\tilde{\mathbf{b}}$ to be the first (P-dimensional) part of any solution to the constrained normal equations $\begin{pmatrix} \mathbf{X}'\mathbf{X} & \boldsymbol{\Lambda} \\ \boldsymbol{\Lambda}' & \mathbf{0} \end{pmatrix} \begin{pmatrix} \mathbf{b} \\ \mathbf{r} \end{pmatrix} = \begin{pmatrix} \mathbf{X}'\mathbf{y} \\ \boldsymbol{\tau}^{(0)} \end{pmatrix}$, in which case $q(\mathbf{b})$ attains its minimum value (in the absence of any constraints) at the point $\bar{\mathbf{b}}$ and attains its minimum value when $\mathbf{b}$ is subject to the constraint $\boldsymbol{\Lambda}'\mathbf{b} = \boldsymbol{\tau}^{(0)}$ at the point $\tilde{\mathbf{b}}$. And let $\bar{\sigma} = [q(\bar{\mathbf{b}})/N]^{1/2}$ and $\tilde{\sigma} = [q(\tilde{\mathbf{b}})/N]^{1/2}$. Then, clearly,

$$L(\bar{\mathbf{b}}, \bar{\sigma}) = \max_{\boldsymbol{\beta} \in \mathbb{R}^P,\ \sigma>0} L(\boldsymbol{\beta}, \sigma) \quad \text{and} \quad L(\tilde{\mathbf{b}}, \tilde{\sigma}) = \max_{\boldsymbol{\beta}, \sigma:\, \boldsymbol{\Lambda}'\boldsymbol{\beta} = \boldsymbol{\tau}^{(0)},\ \sigma>0} L(\boldsymbol{\beta}, \sigma),$$

so that the likelihood ratio test statistic is reexpressible as

$$\frac{L(\tilde{\mathbf{b}}, \tilde{\sigma})}{L(\bar{\mathbf{b}}, \bar{\sigma})} = \left(\frac{\bar{\sigma}^2}{\tilde{\sigma}^2}\right)^{N/2} = \left[\frac{q(\bar{\mathbf{b}})}{q(\tilde{\mathbf{b}})}\right]^{N/2}. \tag{3.9}$$

Moreover,

$$\left[\frac{q(\bar{\mathbf{b}})}{q(\tilde{\mathbf{b}})}\right]^{N/2} < c \;\Leftrightarrow\; \frac{q(\bar{\mathbf{b}})}{q(\tilde{\mathbf{b}})} < c^{2/N} \;\Leftrightarrow\; \frac{q(\tilde{\mathbf{b}})}{q(\bar{\mathbf{b}})} > c^{-2/N} \;\Leftrightarrow\; \frac{[q(\tilde{\mathbf{b}}) - q(\bar{\mathbf{b}})]/M_*}{q(\bar{\mathbf{b}})/(N-P_*)} > k, \tag{3.10}$$

where $k = [(N-P_*)/M_*](c^{-2/N} - 1)$.

Upon recalling the results of Part 1 (of the present subsection), it becomes clear that when $k = \bar{F}_{\dot{\gamma}}(M_*,\, N-P_*)$, the critical region defined by the last of the four equivalent inequalities encompassed in result (3.10) is that of the size-$\dot{\gamma}$ F test, leading to the conclusion that the size-$\dot{\gamma}$ F test is the equivalent of a size-$\dot{\gamma}$ likelihood ratio test. Alternatively, the equivalence of the two tests could be established by making use of the canonical form of the G–M model. The latter approach was considered in Chapter 7 (in the form of Exercise 7.8).

b. Constrained models

The results of Subsection a are applicable to the testing of hypotheses under an unconstrained G–M model. Let us extend those results to the testing of hypotheses under a constrained G–M model.

Accordingly, suppose that $\mathbf{y}$ is an $N \times 1$ observable random vector that follows a constrained G–M model with model equation $\mathbf{y} = \mathbf{X}\boldsymbol{\beta} + \mathbf{e}$, where $\boldsymbol{\beta}$ is a $P \times 1$ vector of unknown parameters that [for a $Q \times P$ matrix $\mathbf{A}$ and a $Q \times 1$ vector $\mathbf{d} \in \mathcal{C}(\mathbf{A})$] is subject to the constraint $\mathbf{A}\boldsymbol{\beta} = \mathbf{d}$ and where $\mathrm{E}(\mathbf{e}) = \mathbf{0}$ and $\mathrm{var}(\mathbf{e}) = \sigma^2\mathbf{I}$. And let $\boldsymbol{\tau} = \boldsymbol{\Lambda}'\boldsymbol{\beta}$ represent an M-dimensional column vector of estimable linear combinations of the elements of $\boldsymbol{\beta}$ [so that $\boldsymbol{\Lambda}' \in \mathcal{R}\begin{pmatrix} \mathbf{X} \\ \mathbf{A} \end{pmatrix}$], take $\bar{\mathbf{b}}$ to be any value of $\mathbf{b}$ and $\bar{\mathbf{r}}$ any value of $\mathbf{r}$ that satisfy the constrained normal equations

$$\begin{pmatrix} \mathbf{X}'\mathbf{X} & \mathbf{A}' \\ \mathbf{A} & \mathbf{0} \end{pmatrix} \begin{pmatrix} \mathbf{b} \\ \mathbf{r} \end{pmatrix} = \begin{pmatrix} \mathbf{X}'\mathbf{y} \\ \mathbf{d} \end{pmatrix} \tag{3.11}$$

(in which $\mathbf{b}$ is a $P \times 1$ vector and $\mathbf{r}$ a $Q \times 1$ vector of unknowns), take $\mathbf{G}$ to be any generalized inverse of the matrix $\begin{pmatrix} \mathbf{X}'\mathbf{X} & \mathbf{A}' \\ \mathbf{A} & \mathbf{0} \end{pmatrix}$, partition $\mathbf{G}$ conformally as $\begin{pmatrix} \mathbf{G}_{11} & \mathbf{G}_{12} \\ \mathbf{G}_{21} & \mathbf{G}_{22} \end{pmatrix}$, and define $\hat{\boldsymbol{\tau}} = \boldsymbol{\Lambda}'\bar{\mathbf{b}}$ and $\mathbf{B} = \boldsymbol{\Lambda}'\mathbf{G}_{11}\boldsymbol{\Lambda}$ [so that $\hat{\boldsymbol{\tau}}$ is the constrained least squares estimator of $\boldsymbol{\tau}$ and $\mathrm{var}(\hat{\boldsymbol{\tau}}) = \sigma^2\mathbf{B}$]. Further, let $P_* = \mathrm{rank}\begin{pmatrix} \mathbf{X} \\ \mathbf{A} \end{pmatrix} - \mathrm{rank}(\mathbf{A})$ and $M_* = \mathrm{rank}\begin{pmatrix} \boldsymbol{\Lambda}' \\ \mathbf{A} \end{pmatrix} - \mathrm{rank}(\mathbf{A})$, and denote by $\boldsymbol{\tau}^{(0)}$ an

$M \times 1$ vector [for which $\begin{pmatrix} \mathbf{d} \\ \boldsymbol{\tau}^{(0)} \end{pmatrix} \in \mathcal{C}\begin{pmatrix} \mathbf{A} \\ \boldsymbol{\Lambda}' \end{pmatrix}$] that is the hypothesized value of $\boldsymbol{\tau}$. Finally, recall (from Section 8.1e) that (under suitable assumptions about the distribution of $\mathbf{e}$) a size-$\dot{\gamma}$ test of the testable null hypothesis $H_0 : \boldsymbol{\tau} = \boldsymbol{\tau}^{(0)}$ (versus the alternative hypothesis $H_1 : \boldsymbol{\tau} \neq \boldsymbol{\tau}^{(0)}$) is provided by the test (known as the F test) that rejects H_0 whenever the statistic

$$F = \frac{(\hat{\boldsymbol{\tau}} - \boldsymbol{\tau}^{(0)})'\mathbf{B}^-(\hat{\boldsymbol{\tau}} - \boldsymbol{\tau}^{(0)})/M_*}{(\mathbf{y} - \mathbf{X}\bar{\mathbf{b}})'(\mathbf{y} - \mathbf{X}\bar{\mathbf{b}})/(N - P_*)} \tag{3.12}$$

(known as the F statistic) exceeds $\bar{F}_{\dot{\gamma}}(M_*,\, N - P_*)$.

An alternative expression for the F statistic. For those $\mathbf{b} \in \mathcal{R}^P$ for which $\mathbf{Ab} = \mathbf{d}$, let $q(\mathbf{b}) = (\mathbf{y} - \mathbf{Xb})'(\mathbf{y} - \mathbf{Xb})$. Then, as will subsequently be shown, the quantity $(\hat{\boldsymbol{\tau}} - \boldsymbol{\tau}^{(0)})'\mathbf{B}^-(\hat{\boldsymbol{\tau}} - \boldsymbol{\tau}^{(0)})$, which appears in the numerator of the F statistic (3.12), is reexpressible as follows:

$$(\hat{\boldsymbol{\tau}} - \boldsymbol{\tau}^{(0)})'\mathbf{B}^-(\hat{\boldsymbol{\tau}} - \boldsymbol{\tau}^{(0)}) = \min_{\mathbf{b}\,:\,\mathbf{Ab}=\mathbf{d},\ \boldsymbol{\Lambda}'\mathbf{b}=\boldsymbol{\tau}^{(0)}} q(\mathbf{b}) - \min_{\mathbf{b}\,:\,\mathbf{Ab}=\mathbf{d}} q(\mathbf{b}). \tag{3.13}$$

This result provides an extension (to constrained models) of result (3.2). Further, letting $\tilde{\mathbf{b}}$, $\tilde{\mathbf{r}}$, and $\tilde{\mathbf{u}}$ represent any values of $\mathbf{b}$, $\mathbf{r}$, and $\mathbf{u}$, respectively, that together form a solution to the "further constrained" normal equations

$$\begin{pmatrix} \mathbf{X}'\mathbf{X} & \mathbf{A}' & \boldsymbol{\Lambda} \\ \mathbf{A} & \mathbf{0} & \mathbf{0} \\ \boldsymbol{\Lambda}' & \mathbf{0} & \mathbf{0} \end{pmatrix} \begin{pmatrix} \mathbf{b} \\ \mathbf{r} \\ \mathbf{u} \end{pmatrix} = \begin{pmatrix} \mathbf{X}'\mathbf{y} \\ \mathbf{d} \\ \boldsymbol{\tau}^{(0)} \end{pmatrix} \tag{3.14}$$

(where $\mathbf{b}$ is of dimension P, $\mathbf{r}$ of dimension Q, and $\mathbf{u}$ of dimension M), we have [as an alternative version of expression (3.13)] the following extension of result (3.4):

$$(\hat{\boldsymbol{\tau}} - \boldsymbol{\tau}^{(0)})'\mathbf{B}^-(\hat{\boldsymbol{\tau}} - \boldsymbol{\tau}^{(0)}) = q(\tilde{\mathbf{b}}) - q(\bar{\mathbf{b}}). \tag{3.15}$$

And as is evident from result (3.15), the F statistic (3.12) is reexpressible as

$$F = \frac{[q(\tilde{\mathbf{b}}) - q(\bar{\mathbf{b}})]/M_*}{q(\bar{\mathbf{b}})/(N - P_*)}; \tag{3.16}$$

result (3.16) is the counterpart of result (3.5).

Let us verify result (3.13) or, equivalently, result (3.15). For that purpose, let $\bar{\mathbf{u}}$ represent any solution to the linear system

$$\mathbf{Bu} = \hat{\boldsymbol{\tau}} - \boldsymbol{\tau}^{(0)} \tag{3.17}$$

(in $\mathbf{u}$)—that linear system (3.17) is consistent is evident upon observing that together results (1.92) and (1.87) imply that $\hat{\boldsymbol{\tau}} - \boldsymbol{\tau}^{(0)} \in \mathcal{C}(\mathbf{B})$. And observe that among the choices for $\tilde{\mathbf{b}}$, $\tilde{\mathbf{r}}$, and $\tilde{\mathbf{u}}$ are those obtained by taking

$$\begin{pmatrix} \tilde{\mathbf{b}} \\ \tilde{\mathbf{r}} \end{pmatrix} = \begin{pmatrix} \bar{\mathbf{b}} \\ \bar{\mathbf{r}} \end{pmatrix} - \begin{pmatrix} \bar{\mathbf{S}} \\ \bar{\mathbf{T}} \end{pmatrix}\bar{\mathbf{u}} \quad \text{and} \quad \tilde{\mathbf{u}} = \bar{\mathbf{u}}, \tag{3.18}$$

where $\bar{\mathbf{S}} = \mathbf{G}_{11}\boldsymbol{\Lambda}$ and $\bar{\mathbf{T}} = \mathbf{G}_{21}\boldsymbol{\Lambda}$ [in which case $\begin{pmatrix} \bar{\mathbf{S}} \\ \bar{\mathbf{T}} \end{pmatrix}$ is a solution to the linear system

$$\begin{pmatrix} \mathbf{X}'\mathbf{X} & \mathbf{A}' \\ \mathbf{A} & \mathbf{0} \end{pmatrix} \begin{pmatrix} \mathbf{S} \\ \mathbf{T} \end{pmatrix} = \begin{pmatrix} \boldsymbol{\Lambda} \\ \mathbf{0} \end{pmatrix}$$

(in the $P \times M$ matrix $\mathbf{S}$ and $Q \times M$ matrix $\mathbf{T}$)]—that a solution to linear system (3.14) can be obtained from the solution to linear system (3.11) and the solution to linear system (3.17) in accordance with formula (3.18) can be readily confirmed.

Now, upon making use of result (1.20), we find that

$$\begin{aligned}
q(\tilde{\mathbf{b}}) - q(\bar{\mathbf{b}}) &= [\mathbf{X}(\tilde{\mathbf{b}}-\bar{\mathbf{b}})]'\mathbf{X}(\tilde{\mathbf{b}}-\bar{\mathbf{b}}) \\
&= (\mathbf{X}\bar{\mathbf{S}}\bar{\mathbf{u}})'\mathbf{X}\bar{\mathbf{S}}\bar{\mathbf{u}} \\
&= \bar{\mathbf{u}}'(\mathbf{X}'\mathbf{X}\bar{\mathbf{S}})'\bar{\mathbf{S}}\bar{\mathbf{u}} \\
&= \bar{\mathbf{u}}'(\boldsymbol{\Lambda} - \mathbf{A}'\bar{\mathbf{T}})'\bar{\mathbf{S}}\bar{\mathbf{u}} \\
&= \bar{\mathbf{u}}'(\boldsymbol{\Lambda}'\bar{\mathbf{S}} - \bar{\mathbf{T}}'\mathbf{A}\bar{\mathbf{S}})\bar{\mathbf{u}} \\
&= \bar{\mathbf{u}}'(\mathbf{B} - \bar{\mathbf{T}}'\mathbf{0})\bar{\mathbf{u}} \\
&= \bar{\mathbf{u}}'\mathbf{B}\bar{\mathbf{u}} = (\mathbf{B}\bar{\mathbf{u}})'\mathbf{B}^{-}\mathbf{B}\bar{\mathbf{u}} = (\hat{\boldsymbol{\tau}}-\boldsymbol{\tau}^{(0)})'\mathbf{B}^{-}(\hat{\boldsymbol{\tau}}-\boldsymbol{\tau}^{(0)}),
\end{aligned}$$

thereby completing the verification of result (3.15) [and in doing so, serving also to verify results (3.16) and (3.13)].

Likelihood ratio test. Suppose that the distribution of the vector $\mathbf{e}$ of residual effects in the constrained G–M model is MVN, in which case $\mathbf{y} \sim N(\mathbf{X}\boldsymbol{\beta}, \sigma^2\mathbf{I})$ (where $\mathbf{A}\boldsymbol{\beta} = \mathbf{d}$). And as in Part 1 (of the present subsection), let (for those $\mathbf{b} \in \mathbb{R}^P$ for which $\mathbf{A}\mathbf{b} = \mathbf{d}$) $q(\mathbf{b}) = (\mathbf{y}-\mathbf{X}\mathbf{b})'(\mathbf{y}-\mathbf{X}\mathbf{b})$. Further, for those values of $\boldsymbol{\beta}$ for which $\mathbf{A}\boldsymbol{\beta} = \mathbf{d}$ and for $\sigma > 0$, let $L(\boldsymbol{\beta}, \sigma)$ represent the likelihood function [which (as in the case of the unconstrained G–M model) is expressible in the form (3.7)]. Then, under the constrained G–M model, the likelihood ratio test statistic (for testing the null hypothesis H_0 versus the alternative hypothesis H_1) is

$$\frac{\max_{\boldsymbol{\beta},\sigma\,:\,\mathbf{A}\boldsymbol{\beta}=\mathbf{d},\ \boldsymbol{\Lambda}'\boldsymbol{\beta}=\boldsymbol{\tau}^{(0)},\ \sigma>0} L(\boldsymbol{\beta}, \sigma)}{\max_{\boldsymbol{\beta},\sigma\,:\,\mathbf{A}\boldsymbol{\beta}=\mathbf{d},\ \sigma>0} L(\boldsymbol{\beta}, \sigma)}. \tag{3.19}$$

Now, take $\bar{\mathbf{b}}$ and $\tilde{\mathbf{b}}$ to be as defined in Part 1 (of the present subsection). Then, by proceeding in the same fashion as in the derivation of expression (3.9) in the case of the unconstrained G–M model, it can be shown that the likelihood ratio test statistic is reexpressible as

$$\left[\frac{q(\bar{\mathbf{b}})}{q(\tilde{\mathbf{b}})}\right]^{N/2}. \tag{3.20}$$

And as is evident upon extending result (3.10) to the case where the likelihood ratio test statistic is that for the constrained G–M model and upon recalling the results of Part 1 of the present subsection, the size-$\dot{\gamma}$ F test of the null hypothesis H_0 (versus the alternative hypothesis H_1) is (in the case of the constrained G–M model as in the case of the unconstrained G–M model) the equivalent of a size-$\dot{\gamma}$ likelihood ratio test.

8.4 Model "Sum of Squares"

Suppose that $\mathbf{y}$ is an $N\times 1$ observable random vector that follows a constrained or unconstrained G–M model with model equation $\mathbf{y}=\mathbf{X}\boldsymbol{\beta}+\mathbf{e}$, where $\boldsymbol{\beta}$ is a $P\times 1$ vector of unknown parameters [that for some $Q\times P$ matrix $\mathbf{A}$ and $Q\times 1$ vector $\mathbf{d}\in\mathcal{C}(\mathbf{A})$ may be subject to the constraint $\mathbf{A}\boldsymbol{\beta}=\mathbf{d}$]. Let us adopt the custom of referring to the quantity $\mathbf{y}'\mathbf{y}$ as the total sum of squares or simply as the total SS, and let us denote that quantity by the symbol $TotSS$. The magnitude of the difference between $\mathbf{y}$ and the constrained or unconstrained least squares estimator $\hat{\boldsymbol{\mu}}$ of its expected value $\boldsymbol{\mu}=\mathbf{X}\boldsymbol{\beta}$ is reflected by the constrained or unconstrained residual sum of squares $(\mathbf{y}-\hat{\boldsymbol{\mu}})'(\mathbf{y}-\hat{\boldsymbol{\mu}})$; let us optionally refer to that quantity simply as the residual SS, and let us denote it by the symbol $ResSS$.

Let us write $ModSS$ for the difference $TotSS - ResSS$ between the total sum of squares and the residual sum of squares, and let us refer to that difference as the model SS. The model SS represents the part of the total SS that is "accounted for" or "explained by" the model.

It is worth noting that there are special cases of the constrained G–M model where for some values of $\mathbf{y}$ the residual SS may actually be larger than the total SS and hence where the model SS may be negative (so that the use of the term model SS comes with the caveat that in general it does not stand for model sum of squares). In fact, some such values of $\mathbf{y}$ necessarily exist when the constraint $\mathbf{A}\boldsymbol{\beta}=\mathbf{d}$ is one where $\mathbf{k}'\mathbf{d}\neq 0$ for some $Q\times 1$ vector $\mathbf{k}$ for which $\mathbf{k}'\mathbf{A}\in\mathcal{R}(\mathbf{X})$. As an extremely simple example, suppose that $\mathbf{A}$ is the $1\times P$ matrix whose only row equals a nonnull row of $\mathbf{X}$, that $\mathbf{d}=(1)$, and that $\mathbf{y}=\mathbf{0}$ (in which case $TotSS=0$ and $ResSS\geq 1$).

a. Alternative expressions for the F statistic

Continue to suppose that $\mathbf{y}$ is an $N\times 1$ observable random vector that follows a constrained or unconstrained G–M model with model equation $\mathbf{y}=\mathbf{X}\boldsymbol{\beta}+\mathbf{e}$, where $\boldsymbol{\beta}$ is a $P\times 1$ vector of unknown parameters [that for some $Q\times P$ matrix $\mathbf{A}$ and $Q\times 1$ vector $\mathbf{d}\in\mathcal{C}(\mathbf{A})$ may be subject to the constraint $\mathbf{A}\boldsymbol{\beta}=\mathbf{d}$]. Let us revisit some of the results presented in Section 8.3. Accordingly, take F to be the F statistic for testing the testable null hypothesis $H_0: \boldsymbol{\tau}=\boldsymbol{\tau}^{(0)}$ (where $\boldsymbol{\tau}=\boldsymbol{\Lambda}'\boldsymbol{\beta}$) versus the alternative hypothesis $H_1: \boldsymbol{\tau}\neq\boldsymbol{\tau}^{(0)}$.

Recall (from Section 8.3) that F is expressible as in result (3.5) or as in result (3.16), depending on whether the underlying model is the unconstrained model or the constrained model; in both cases, the numerator of the F statistic is expressed in terms of the difference between two residual sums of squares. These expressions can be presented in a common framework and can be recharacterized by adopting the notation and the nomenclature introduced in the present section.

Depending on which of the two models the test is based, let $ResSS_1$ represent the residual sum of squares for the unconstrained G–M model or that for the G–M model with constraint $\mathbf{A}\boldsymbol{\beta}=\mathbf{d}$. And let $ResSS_0$ represent the residual sum of squares for the model obtained by augmenting the base model with the constraint (or additional constraint) $\boldsymbol{\Lambda}'\boldsymbol{\beta}=\boldsymbol{\tau}^{(0)}$. Further, denote by $ModSS_1 = TotSS - ResSS_1$ and by $ModSS_0 = TotSS - ResSS_0$ the corresponding model SS's. Then, the expressions for the F statistic given by result (3.5) or that given by result (3.16) can be restated as

$$F=\frac{(ResSS_0 - ResSS_1)/M_*}{ResSS_1/(N-P_*)}, \tag{4.1}$$

where either $M_*=\operatorname{rank}(\boldsymbol{\Lambda})$ and $P_*=\operatorname{rank}(\mathbf{X})$ or $M_*=\operatorname{rank}\begin{pmatrix}\boldsymbol{\Lambda}'\\ \mathbf{A}\end{pmatrix}-\operatorname{rank}(\mathbf{A})$ and $P_*=\operatorname{rank}\begin{pmatrix}\mathbf{X}\\ \mathbf{A}\end{pmatrix}-\operatorname{rank}(\mathbf{A})$ (depending on whether it is the unconstrained model or the constrained model that is the base model). Moreover, the numerator of the F statistic can be reexpressed in terms of the model SS's, in which case

$$F=\frac{(ModSS_1 - ModSS_0)/M_*}{ResSS_1/(N-P_*)}. \tag{4.2}$$

b. Model SS (and residual SS) for an unconstrained G–M model: some specifics

Suppose that $\mathbf{y}$ is an $N\times 1$ observable random vector that follows an unconstrained G–M model with model equation $\mathbf{y}=\mathbf{X}\boldsymbol{\beta}+\mathbf{e}$, where $\boldsymbol{\beta}$ is a $P\times 1$ vector of unknown (and unconstrained) parameters. Further, let $\boldsymbol{\mu} = E(\mathbf{y}) = \mathbf{X}\boldsymbol{\beta}$, and let $\hat{\boldsymbol{\mu}}$ represent the least squares estimator of $\boldsymbol{\mu}$. And observe that $\hat{\boldsymbol{\mu}}$ can be expressed in terms of a solution, say $\mathbf{b}=\tilde{\mathbf{b}}$, to the normal equations $\mathbf{X}'\mathbf{X}\mathbf{b} = \mathbf{X}'\mathbf{y}$, a solution, say $\mathbf{S} = \tilde{\mathbf{S}}$, to the conjugage normal equations $\mathbf{X}'\mathbf{X}\mathbf{S} = \mathbf{X}'$, or a generalized inverse $(\mathbf{X}'\mathbf{X})^-$ of the $P\times P$ matrix $\mathbf{X}'\mathbf{X}$; or it can be expressed in terms of the projection matrix $\mathbf{P}_{\mathbf{X}}$. Specifically,

$$\hat{\boldsymbol{\mu}} = \mathbf{X}\tilde{\mathbf{b}} = \mathbf{X}\tilde{\mathbf{S}}\mathbf{y} = \mathbf{X}(\mathbf{X}'\mathbf{X})^-\mathbf{X}'\mathbf{y} = \mathbf{P}_{\mathbf{X}}\mathbf{y}. \tag{4.3}$$

Corresponding to $\hat{\boldsymbol{\mu}}$ and to the various expressions for $\hat{\boldsymbol{\mu}}$ are the following expressions for the vector of least squares residuals:

$$\mathbf{y}-\hat{\boldsymbol{\mu}} = \mathbf{y}-\mathbf{X}\tilde{\mathbf{b}} = (\mathbf{I}-\mathbf{X}\tilde{\mathbf{S}})\mathbf{y} = [\mathbf{I}-\mathbf{X}(\mathbf{X}'\mathbf{X})^-\mathbf{X}']\mathbf{y} = (\mathbf{I}-\mathbf{P}_{\mathbf{X}})\mathbf{y}. \tag{4.4}$$

Thus, the residual sum of squares can be expressed as

$$(\mathbf{y}-\hat{\boldsymbol{\mu}})'(\mathbf{y}-\hat{\boldsymbol{\mu}}) = \mathbf{y}'(\mathbf{y}-\hat{\boldsymbol{\mu}}) = \mathbf{y}'(\mathbf{y}-\mathbf{X}\tilde{\mathbf{b}}) = \mathbf{y}'(\mathbf{I}-\mathbf{X}\tilde{\mathbf{S}})\mathbf{y} = \mathbf{y}'[\mathbf{I}-\mathbf{X}(\mathbf{X}'\mathbf{X})^-\mathbf{X}']\mathbf{y} = \mathbf{y}'(\mathbf{I}-\mathbf{P}_{\mathbf{X}})\mathbf{y}, \tag{4.5}$$

implying in particular that

$$ResSS = TotSS - \mathbf{y}'\mathbf{P}_{\mathbf{X}}\mathbf{y}, \tag{4.6}$$

or equivalently that

$$TotSS = ResSS + \mathbf{y}'\mathbf{P}_{\mathbf{X}}\mathbf{y}$$

and hence that the model SS is expressible as

$$ModSS = \mathbf{y}'\mathbf{P}_{\mathbf{X}}\mathbf{y}. \tag{4.7}$$

Moreover, since $\mathbf{P}_{\mathbf{X}}$ is symmetric and idempotent, the model SS is also expressible as follows:

$$ModSS = \mathbf{y}'\hat{\boldsymbol{\mu}} = \hat{\boldsymbol{\mu}}'\mathbf{y} = \hat{\boldsymbol{\mu}}'\hat{\boldsymbol{\mu}}. \tag{4.8}$$

Note that in the special case of the unconstrained G–M model, the model SS is in fact a sum of squares. As is evident from result (4.8), it is expressible in particular as the sum of squares of the least squares estimators of the elements of $\boldsymbol{\mu}$.

c. Model SS (and residual SS) for a constrained G–M model: some specifics

Suppose that $\mathbf{y}$ is an $N\times 1$ observable random vector that follows a constrained G–M model with model equation $\mathbf{y}=\mathbf{X}\boldsymbol{\beta}+\mathbf{e}$, where $\boldsymbol{\beta}$ is a $P\times 1$ vector of unknown parameters that [for some $Q\times P$ matrix $\mathbf{A}$ and $Q\times 1$ vector $\mathbf{d}\in\mathcal{C}(\mathbf{A})$] is confined to the set $\{\boldsymbol{\beta} : \mathbf{A}\boldsymbol{\beta}=\mathbf{d}\}$. Further, let $\hat{\boldsymbol{\mu}}$ represent the constrained least squares estimator of the vector $\boldsymbol{\mu} = E(\mathbf{y}) = \mathbf{X}\boldsymbol{\beta}$. And defining $\mathbf{W} = \mathbf{X}(\mathbf{I}-\mathbf{A}^-\mathbf{A})$ and taking $\underline{\mathbf{b}} = \mathbf{A}^-\mathbf{d}$ (or taking $\underline{\mathbf{b}}$ to be any other $P\times 1$ vector for which $\mathbf{A}\mathbf{b}=\mathbf{d}$), recall (from Section 8.1d) that $\hat{\boldsymbol{\mu}}$ is expressible in terms of a solution, say that with $\mathbf{b}=\tilde{\mathbf{b}}$ and $\mathbf{r}=\tilde{\mathbf{r}}$, to the constrained normal equations $\begin{pmatrix}\mathbf{X}'\mathbf{X} & \mathbf{A}'\\ \mathbf{A} & \mathbf{0}\end{pmatrix}\begin{pmatrix}\mathbf{b}\\ \mathbf{r}\end{pmatrix} = \begin{pmatrix}\mathbf{X}'\mathbf{y}\\ \mathbf{d}\end{pmatrix}$, in terms of a solution, say that with $\mathbf{S} = \tilde{\mathbf{S}}$ and $\mathbf{T} = \tilde{\mathbf{T}}$, to the constrained conjugate normal equations $\begin{pmatrix}\mathbf{X}'\mathbf{X} & \mathbf{A}'\\ \mathbf{A} & \mathbf{0}\end{pmatrix}\begin{pmatrix}\mathbf{S}\\ \mathbf{T}\end{pmatrix} = \begin{pmatrix}\mathbf{X}'\\ \mathbf{0}\end{pmatrix}$, or in terms of the projection matrix $\mathbf{P}_{\mathbf{W}}$ as follows:

$$\hat{\boldsymbol{\mu}} = \mathbf{X}\tilde{\mathbf{b}} = \mathbf{X}\tilde{\mathbf{S}}\mathbf{y} + \tilde{\mathbf{T}}'\mathbf{d} = \mathbf{P}_{\mathbf{W}}\mathbf{y} + (\mathbf{I}-\mathbf{P}_{\mathbf{W}})\mathbf{X}\underline{\mathbf{b}}. \tag{4.9}$$

Corresponding to $\hat{\boldsymbol{\mu}}$ and to the various expressions for $\hat{\boldsymbol{\mu}}$ are the following expressions for the vector of constrained least squares residuals:

$$\mathbf{y}-\hat{\boldsymbol{\mu}} = \mathbf{y}-\mathbf{X}\tilde{\mathbf{b}} = (\mathbf{I}-\mathbf{X}\tilde{\mathbf{S}})\mathbf{y} - \tilde{\mathbf{T}}'\mathbf{d} = (\mathbf{I}-\mathbf{P}_{\mathbf{W}})(\mathbf{y}-\mathbf{X}\underline{\mathbf{b}}). \tag{4.10}$$

And by making use of various of these expressions, the residual sum of squares is expressible as

$$(\mathbf{y}-\hat{\boldsymbol{\mu}})'(\mathbf{y}-\hat{\boldsymbol{\mu}}) = (\mathbf{y}-\mathbf{X}\tilde{\mathbf{b}})'(\mathbf{y}-\mathbf{X}\tilde{\mathbf{b}}) = (\mathbf{y}-\mathbf{X}\underline{\mathbf{b}})'(\mathbf{I}-\mathbf{P_W})(\mathbf{y}-\mathbf{X}\underline{\mathbf{b}}). \tag{4.11}$$

Moreover, as is evident from result (1.75), it is also expressible in the form

$$ResSS = TotSS - \tilde{\mathbf{b}}'\mathbf{X}'\mathbf{y} - \tilde{\mathbf{r}}'\mathbf{d}. \tag{4.12}$$

Thus,

$$TotSS = ResSS + \tilde{\mathbf{b}}'\mathbf{X}'\mathbf{y} + \tilde{\mathbf{r}}'\mathbf{d},$$

and hence the model SS is expressible as

$$ModSS = \tilde{\mathbf{b}}'\mathbf{X}'\mathbf{y} + \tilde{\mathbf{r}}'\mathbf{d}. \tag{4.13}$$

Special case: $\mathbf{d} = \mathbf{0}$. Suppose that $\mathbf{d} = \mathbf{0}$, so that $\boldsymbol{\beta}$ is confined to the set $\{\boldsymbol{\beta} : \mathbf{A}\boldsymbol{\beta} = \mathbf{0}\}$. In that special case, the vector $\underline{\mathbf{b}}$ (for which $\mathbf{A}\underline{\mathbf{b}} = \mathbf{d}$) can be set to $\mathbf{0}$. Accordingly, when $\mathbf{d} = \mathbf{0}$, the model SS is expressible as

$$ModSS = \tilde{\mathbf{b}}'\mathbf{X}'\mathbf{y} = \hat{\boldsymbol{\mu}}'\mathbf{y} = \mathbf{y}'\hat{\boldsymbol{\mu}} = \mathbf{y}'\mathbf{P_W}\mathbf{y} = \hat{\boldsymbol{\mu}}'\hat{\boldsymbol{\mu}}. \tag{4.14}$$

Note that in the special case where $\mathbf{d} = \mathbf{0}$, the model SS is truly a sum of squares. It is expressible in particular as the sum of the squared elements of the vector $\hat{\boldsymbol{\mu}}$. Also, note the resemblance between the various formulas for $\hat{\boldsymbol{\mu}}$, $\mathbf{y}-\hat{\boldsymbol{\mu}}$, the residual SS, and the model SS in the special case (of the constrained G–M model) where $\mathbf{d} = \mathbf{0}$ and those presented in Subsection b for the unconstrained G–M model. When modified by taking $\tilde{\mathbf{b}}$ to be the "**b**-part" of the solution to the linear system $\begin{pmatrix}\mathbf{X}'\mathbf{X} & \mathbf{A}' \\ \mathbf{A} & \mathbf{0}\end{pmatrix}\begin{pmatrix}\mathbf{b} \\ \mathbf{r}\end{pmatrix} = \begin{pmatrix}\mathbf{X}'\mathbf{y} \\ \mathbf{0}\end{pmatrix}$ rather than the solution to the linear system $\mathbf{X}'\mathbf{X}\mathbf{b} = \mathbf{X}'\mathbf{y}$, by taking $\tilde{\mathbf{S}}$ to be the "**S**-part" of the solution to $\begin{pmatrix}\mathbf{X}'\mathbf{X} & \mathbf{A}' \\ \mathbf{A} & \mathbf{0}\end{pmatrix}\begin{pmatrix}\mathbf{S} \\ \mathbf{T}\end{pmatrix} = \begin{pmatrix}\mathbf{X}' \\ \mathbf{0}\end{pmatrix}$ rather than the solution to $\mathbf{X}'\mathbf{X}\mathbf{S} = \mathbf{X}'$, by replacing $(\mathbf{X}'\mathbf{X})^-$ with the $P \times P$ matrix $\mathbf{G}_{11}$ obtained upon partitioning any generalized inverse $\mathbf{G}$ of $\begin{pmatrix}\mathbf{X}'\mathbf{X} & \mathbf{A}' \\ \mathbf{A} & \mathbf{0}\end{pmatrix}$ conformally as $\mathbf{G} = \begin{pmatrix}\mathbf{G}_{11} & \mathbf{G}_{12} \\ \mathbf{G}_{21} & \mathbf{G}_{22}\end{pmatrix}$, and by replacing $\mathbf{P_X}$ with $\mathbf{P_W}$, the formulas for the unconstrained G–M model are applicable to the special case (where $\mathbf{d} = \mathbf{0}$) of the constrained G–M model.

Special case: jointly nonestimable constraints. Suppose that the matrix $\mathbf{A}$ is such that in the absence of the constraint $\mathbf{A}\boldsymbol{\beta} = \mathbf{d}$ (i.e., under the unconstrained G–M model), $\mathbf{A}\boldsymbol{\beta}$ would be jointly nonestimable. Or, equivalently, suppose that $\mathbf{A}$ is such that $\mathcal{R}(\mathbf{X}) \cap \mathcal{R}(\mathbf{A}) = \{\mathbf{0}\}$, or, also equivalently, suppose that $\mathbf{A}$ is such that $\operatorname{rank}\begin{pmatrix}\mathbf{X} \\ \mathbf{A}\end{pmatrix} = \operatorname{rank}(\mathbf{X}) + \operatorname{rank}(\mathbf{A})$. Then, as discussed in Section 8.2b, it would be the case that the "**b**-part" of any solution to the constrained normal equations $\begin{pmatrix}\mathbf{X}'\mathbf{X} & \mathbf{A}' \\ \mathbf{A} & \mathbf{0}\end{pmatrix}\begin{pmatrix}\mathbf{b} \\ \mathbf{r}\end{pmatrix} = \begin{pmatrix}\mathbf{X}'\mathbf{y} \\ \mathbf{d}\end{pmatrix}$ would be a solution to the augmented normal equations $\begin{pmatrix}\mathbf{X}'\mathbf{X} \\ \mathbf{A}\end{pmatrix}\mathbf{b} = \begin{pmatrix}\mathbf{X}'\mathbf{y} \\ \mathbf{d}\end{pmatrix}$ and hence a solution to the unconstrained normal equations $\mathbf{X}'\mathbf{X}\mathbf{b} = \mathbf{X}'\mathbf{y}$. Thus, the residual SS for the constrained model would be the same as that for the unconstrained model. And we conclude that (in this special case) the model SS for the constrained model would be the same as that for the unconstrained model.

8.5 Analysis of Variance (ANOVA)

Closely related to topics such as constrained (or augmented) linear models and the testing of hypotheses is something known as the analysis of variance and that is widely identified by the acronym ANOVA.

a. Analysis of variance in general

Suppose that $\mathbf{y}$ is an $N \times 1$ observable random vector that follows a G–M, Aitken, or general linear model with model equation $\mathbf{y} = \mathbf{X}\boldsymbol{\beta} + \mathbf{e}$, where $\boldsymbol{\beta}$ is a $P \times 1$ vector of unknown (and unconstrained) parameters. Then, the term analysis of variance (or acronym ANOVA) is used in referring to a decomposition $\mathbf{y}'\mathbf{y} = \mathbf{y}'\mathbf{A}_1\mathbf{y} + \mathbf{y}'\mathbf{A}_2\mathbf{y} + \cdots + \mathbf{y}'\mathbf{A}_K\mathbf{y}$ of the total SS [into some number K (≥ 2) of quadratic forms], where $\mathbf{A}_1, \mathbf{A}_2, \ldots, \mathbf{A}_K$ are $N \times N$ nonnull symmetric matrices of constants (that sum to $\mathbf{I}_N$). Or, more generally, letting $\mathbf{A}$ represent an $N \times N$ nonnull symmetric idempotent matrix of constants, this term (or acronym) is used in referring to a decomposition $\mathbf{y}'\mathbf{A}\mathbf{y} = \mathbf{y}'\mathbf{A}_1\mathbf{y} + \mathbf{y}'\mathbf{A}_2\mathbf{y} + \cdots + \mathbf{y}'\mathbf{A}_K\mathbf{y}$ of the quadratic form $\mathbf{y}'\mathbf{A}\mathbf{y}$ [into some number K (≥ 2) of quadratic forms], where $\mathbf{A}_1, \mathbf{A}_2, \ldots, \mathbf{A}_K$ are $N \times N$ nonnull symmetric matrices of constants (that sum to $\mathbf{A}$). For the decomposition $\mathbf{y}'\mathbf{y} = \sum_{j=1}^K \mathbf{y}'\mathbf{A}_j\mathbf{y}$ of the total SS to qualify as an analysis of variance or ANOVA, the matrices $\mathbf{A}_1, \mathbf{A}_2, \ldots, \mathbf{A}_K$ must satisfy the condition

$$N = \text{rank}(\mathbf{A}_1) + \text{rank}(\mathbf{A}_2) + \cdots + \text{rank}(\mathbf{A}_K).$$

And, more generally, for the decomposition $\mathbf{y}'\mathbf{A}\mathbf{y} = \sum_{j=1}^K \mathbf{y}'\mathbf{A}_j\mathbf{y}$ of $\mathbf{y}'\mathbf{A}\mathbf{y}$ (where $\mathbf{A}$ is a nonnull symmetric idempotent matrix) to qualify as an analysis of variance or ANOVA, the matrices $\mathbf{A}_1, \mathbf{A}_2, \ldots, \mathbf{A}_K$ must satisfy the condition

$$\text{rank}(\mathbf{A}) = \text{rank}(\mathbf{A}_1) + \text{rank}(\mathbf{A}_2) + \cdots + \text{rank}(\mathbf{A}_K).$$

Now, suppose that the quadratic form $\mathbf{y}'\mathbf{A}\mathbf{y}$ (where $\mathbf{A}$ is an $N \times N$ nonnull symmetric idempotent matrix) is decomposable as $\mathbf{y}'\mathbf{A}\mathbf{y} = \sum_{j=1}^K \mathbf{y}'\mathbf{A}_j\mathbf{y}$ for some number K (≥ 2) of $N \times N$ nonnull symmetric matrices $\mathbf{A}_1, \mathbf{A}_2, \ldots, \mathbf{A}_K$ (that sum to $\mathbf{A}$). Suppose also that $\mathbf{A}_1, \mathbf{A}_2, \ldots, \mathbf{A}_K$ are such that $\text{rank}(\mathbf{A}) = \sum_{j=1}^K \text{rank}(\mathbf{A}_j)$, so that the decomposition qualifies as an analysis of variance. And observe (in light of Theorem 6.8.7) that $\mathbf{A}_1, \mathbf{A}_2, \ldots, \mathbf{A}_K$ are idempotent, implying that (for $j = 1, 2, \ldots, K$) $\mathbf{y}'\mathbf{A}_j\mathbf{y} = (\mathbf{A}_j\mathbf{y})'\mathbf{A}_j\mathbf{y}$ and hence that $\mathbf{y}'\mathbf{A}_j\mathbf{y}$ is expressible as a sum of squares. Similarly, $\mathbf{y}'\mathbf{A}\mathbf{y}$ is expressible as a sum of squares. Accordingly, let us write SS_j for $\mathbf{y}'\mathbf{A}_j\mathbf{y}$ $(j = 1, 2, \ldots, K)$ and SS_+ for $\mathbf{y}'\mathbf{A}\mathbf{y}$, thereby allowing the decomposition $\mathbf{y}'\mathbf{A}\mathbf{y} = \sum_{j=1}^K \mathbf{y}'\mathbf{A}_j\mathbf{y}$ to be rewritten as $SS_+ = \sum_{j=1}^K SS_j$. Further, let us write r_j for $\text{rank}(\mathbf{A}_j)$ $(j = 1, 2, \ldots, K)$ and r_+ for $\text{rank}(\mathbf{A})$, in which case the condition $\text{rank}(\mathbf{A}) = \sum_{j=1}^K \text{rank}(\mathbf{A}_j)$ may be rewritten as $r_+ = \sum_{j=1}^K r_j$.

In the special case where $\mathbf{y}$ follows a G–M model and where the distribution of the vector $\mathbf{e}$ of the model's residual effects is MVN, $\sigma^{-1}\mathbf{y} \sim N(\sigma^{-1}\boldsymbol{\mu}, \mathbf{I}_N)$, where $\boldsymbol{\mu} = \text{E}(\mathbf{y}) = \mathbf{X}\boldsymbol{\beta}$, and the joint distribution of the sums of squares $SS_1, SS_2, \ldots, SS_K$ in the analysis of variance can be deduced from Corollary 6.8.11 (a form of Cochran's theorem). In that special case, $SS_1, SS_2, \ldots, SS_K$ are statistically independent, and (for $j = 1, 2, \ldots, K$) $SS_j/\sigma^2 \sim \chi^2(r_j, \lambda_j)$, where $\lambda_j = \boldsymbol{\mu}'\mathbf{A}_j\boldsymbol{\mu}/\sigma^2$. (And in that special case, $SS_+/\sigma^2 \sim \chi^2(r_+, \lambda_+)$, where $\lambda_+ = \boldsymbol{\mu}'\mathbf{A}\boldsymbol{\mu}/\sigma^2$.)

An analysis of variance is typically presented as a table with columns that may be labeled source of variation (or source), degrees of freedom (or df), sum of squares (or SS), and mean square (or MS) and possibly with additional columns including one labeled expected mean square (or EMS). This table is referred to as an analysis of variance (or ANOVA) table and generally takes the following form (or something similar):

Source	df	SS	MS	EMS
Source 1	r_1	SS_1	SS_1/r_1	$\sigma^2 + r_1^{-1}\boldsymbol{\mu}'\mathbf{A}_1\boldsymbol{\mu}$
Source 2	r_2	SS_2	SS_2/r_2	$\sigma^2 + r_2^{-1}\boldsymbol{\mu}'\mathbf{A}_2\boldsymbol{\mu}$
$\vdots$	$\vdots$	$\vdots$	$\vdots$	$\vdots$
Source K	r_K	SS_K	SS_K/r_K	$\sigma^2 + r_K^{-1}\boldsymbol{\mu}'\mathbf{A}_K\boldsymbol{\mu}$
Total	r_+	SS_+		

Clearly, the jth line in the table corresponds to the jth term in the decomposition $SS_+ = \sum_{j=1}^{K} SS_j$. And as is apparent from the entries in the fourth column of the table (the entries under the heading MS), the ratios SS_j/r_j $(j = 1,2,\ldots,K)$ are referred to as mean squares. Further, the expected values of the mean squares are referred to simply as expected mean squares. The expected mean squares depend on $\text{var}(\mathbf{y})$. The entries for the expected mean squares in the table are for the case where $\text{var}(\mathbf{y}) = \sigma^2\mathbf{I}$ (as under the G–M model); they are obtainable by observing that the trace of an idempotent matrix equals the rank and by applying formula (5.7.11), and they are related to $\lambda_1, \lambda_2, \ldots, \lambda_K$ in that (for $j = 1,2,\ldots,K$) $r_j^{-1}\boldsymbol{\mu}'\mathbf{A}_j\boldsymbol{\mu} = \sigma^2\lambda_j/r_j$ or, equivalently, $\lambda_j = r_j(r_j^{-1}\boldsymbol{\mu}'\mathbf{A}_j\boldsymbol{\mu})/\sigma^2$. In an application, the entries in the table for the degrees of freedom, the sums of squares, and the mean squares are replaced by their numerical values, the entries for the expected mean squares are replaced by algebraic expressions that are specific to the application, and the entries for the sources are replaced by meaningful words or phrases.

In many applications, the last (Kth) line in the ANOVA table would be that for the residual sum of squares obtained upon fitting the (unconstrained) model by least squares. In such an application, it would be the case that $r_K = N - \text{rank}(\mathbf{X})$, $\mathbf{A}_K = \mathbf{I} - \mathbf{P}_\mathbf{X}$, $\boldsymbol{\mu}'\mathbf{A}_K\boldsymbol{\mu} = 0$, and $\sigma^2 + r_K^{-1}\boldsymbol{\mu}'\mathbf{A}_K\boldsymbol{\mu} = \sigma^2$, that the entry on the Kth line under Source might be the word Residual or perhaps the word Error, and that the last (Kth) mean square would be what for the G–M model would be the usual unbiased estimator of σ^2.

Subjecting an ANOVA to certain basic kinds of modification. Suppose that the $N \times N$ nonnull symmetric idempotent matrix $\mathbf{A}$ differs from $\mathbf{I}_N$. And consider the decomposition

$$\mathbf{y}'\mathbf{y} = \mathbf{y}'(\mathbf{I}-\mathbf{A})\mathbf{y} + \textstyle\sum_{j=1}^{K}\mathbf{y}'\mathbf{A}_j\mathbf{y} \tag{5.1}$$

(of the total sum of squares) obtained upon adding the term $\mathbf{y}'(\mathbf{I}-\mathbf{A})\mathbf{y}$ to the decomposition $\mathbf{y}'\mathbf{A}\mathbf{y} = \sum_{j=1}^{K}\mathbf{y}'\mathbf{A}_j\mathbf{y}$ of $\mathbf{y}'\mathbf{A}\mathbf{y}$. Then, if the decomposition of $\mathbf{y}'\mathbf{A}\mathbf{y}$ qualifies as an ANOVA, decomposition (5.1) (of the total sum of squares) also qualifies as an ANOVA, as is evident upon observing (in light of Lemma 2.8.4) that $\text{rank}(\mathbf{I}_N - \mathbf{A}) = N - \text{rank}(\mathbf{A})$.

Now, consider the effect of deleting one or more terms from a decomposition $\mathbf{y}'\mathbf{A}\mathbf{y} = \sum_{j=1}^{K}\mathbf{y}'\mathbf{A}_j\mathbf{y}$ (of $\mathbf{y}'\mathbf{A}\mathbf{y}$) that qualifies as an ANOVA. Accordingly, suppose that all but K^* of the terms are deleted and that following the deletion it is the $j_1, j_2, \ldots, j_{K^*}$th terms that remain. Further, let $\mathbf{A}_* = \sum_{k=1}^{K^*}\mathbf{A}_{j_k}$. And observe (in light of Theorem 6.8.7) that $\mathbf{A}_i\mathbf{A}_j = \mathbf{0}$ for $j \neq i$ and that (as noted earlier) the $\mathbf{A}_i$'s are idempotent. Thus, $\mathbf{A}_*$ is idempotent (and is symmetric), and (as is evident upon making additional use of Theorem 6.8.7) $\text{rank}(\mathbf{A}_*) = \sum_{k=1}^{K^*}\text{rank}(\mathbf{A}_{j_k})$, leading to the conclusion that the decomposition $\mathbf{y}'\mathbf{A}_*\mathbf{y} = \sum_{k=1}^{K^*}\mathbf{y}'\mathbf{A}_{j_k}\mathbf{y}$ formed by deleting terms from the original decomposition $\mathbf{y}'\mathbf{A}\mathbf{y} = \sum_{j=1}^{K}\mathbf{y}'\mathbf{A}_j\mathbf{y}$ (like the original decomposition itself) qualifies as an ANOVA.

Alternatively, suppose that the $K - K^*$ terms of the decomposition $\mathbf{y}'\mathbf{A}\mathbf{y} = \sum_{j=1}^{K}\mathbf{y}'\mathbf{A}_j\mathbf{y}$ other than the $j_1, j_2, \ldots, j_{K^*}$th terms, say the $j_{K^*+1}, j_{K^*+2}, \ldots, j_K$th terms, were not deleted but rather were combined with the j_{K^*}th term. And consider the resultant decomposition

$$\mathbf{y}'\mathbf{A}\mathbf{y} = \textstyle\sum_{k=1}^{K^*-1}\mathbf{y}'\mathbf{A}_{j_k}\mathbf{y} + \mathbf{y}'\big(\sum_{k=K^*}^{K}\mathbf{A}_{j_k}\big)\mathbf{y}. \tag{5.2}$$

Since (as is being assumed) the original decomposition $\mathbf{y}'\mathbf{A}\mathbf{y} = \sum_{j=1}^{K}\mathbf{y}'\mathbf{A}_j\mathbf{y}$ qualifies as an ANOVA, $\text{rank}(\mathbf{A}) = \sum_{j=1}^{K}\text{rank}(\mathbf{A}_j)$, and (as noted earlier) $\mathbf{A}_i\mathbf{A}_j = \mathbf{0}$ for $j \neq i$ and the $\mathbf{A}_i$'s are idempotent. Then, $\sum_{k=K^*}^{K}\mathbf{A}_{j_k}$ is idempotent (and is symmetric), in which case $\text{rank}\big(\sum_{k=K^*}^{K}\mathbf{A}_{j_k}\big) = \sum_{k=K^*}^{K}\text{rank}(\mathbf{A}_{j_k})$ (as is evident from Theorem 6.8.7), implying that

$$\text{rank}(\mathbf{A}) = \textstyle\sum_{k=1}^{K}\text{rank}(\mathbf{A}_{j_k}) = \sum_{k=1}^{K^*-1}\text{rank}(\mathbf{A}_{j_k}) + \text{rank}\big(\sum_{k=K^*}^{K}\mathbf{A}_{j_k}\big)$$

and leading to the conclusion that decomposition (5.2) (like the original decomposition) qualifies as an ANOVA.

Note that the deletion or combination of terms in a decomposition that qualifies as an ANOVA or the expansion of the scope of such a decomposition (so as to allow for the inclusion of additional terms) corresponds to a deletion or combination of lines in an ANOVA table or an expansion of the scope of an ANOVA table (so as to allow for the inclusion of additional lines).

A connection to the Dirichlet distribution. Suppose that $\mathbf{A} \neq \mathbf{I}$ and that the decomposition $\mathbf{y}'\mathbf{A}\mathbf{y} = \sum_{j=1}^{K} \mathbf{y}'\mathbf{A}_j\mathbf{y}$ qualifies as an ANOVA. Further, suppose that $\mathbf{y}$ follows a G–M model and that the distribution of the vector $\mathbf{e}$ of residual effects is MVN. And let $\mathbf{z} = \sigma^{-1}\mathbf{e} = \sigma^{-1}(\mathbf{y} - \boldsymbol{\mu})$ [where $\boldsymbol{\mu} = \mathrm{E}(\mathbf{y}) = \mathbf{X}\boldsymbol{\beta}$], and observe that $\mathbf{z} \sim N(\mathbf{0}, \mathbf{I}_N)$. Then, the distribution of $\mathbf{z}'\mathbf{A}_1\mathbf{z}, \mathbf{z}'\mathbf{A}_2\mathbf{z}, \ldots, \mathbf{z}'\mathbf{A}_K\mathbf{z}$, and $\mathbf{z}'\mathbf{z}-\mathbf{z}'\mathbf{A}\mathbf{z}$ [$= \mathbf{z}'(\mathbf{I}-\mathbf{A})\mathbf{z}$] can be deduced from Cochran's theorem; $\mathbf{z}'\mathbf{A}_1\mathbf{z}, \mathbf{z}'\mathbf{A}_2\mathbf{z}, \ldots, \mathbf{z}'\mathbf{A}_K\mathbf{z}$, and $\mathbf{z}'\mathbf{z}-\mathbf{z}'\mathbf{A}\mathbf{z}$ are statistically independent, and $\mathbf{z}'\mathbf{A}_j\mathbf{z} \sim \chi^2(r_j)$ $(j = 1, 2, \ldots, K)$ and $\mathbf{z}'\mathbf{z}-\mathbf{z}'\mathbf{A}\mathbf{z} \sim \chi^2(N-r_+)$.

Now, consider the (joint) distribution of the ratios $\mathbf{z}'\mathbf{A}_1\mathbf{z}/\mathbf{z}'\mathbf{z}$, $\mathbf{z}'\mathbf{A}_2\mathbf{z}/\mathbf{z}'\mathbf{z}, \ldots,$ $\mathbf{z}'\mathbf{A}_K\mathbf{z}/\mathbf{z}'\mathbf{z}$. According to Theorem 6.8.15, $\mathbf{z}'\mathbf{A}_1\mathbf{z}/\mathbf{z}'\mathbf{z}$, $\mathbf{z}'\mathbf{A}_2\mathbf{z}/\mathbf{z}'\mathbf{z}, \ldots,$ $\mathbf{z}'\mathbf{A}_K\mathbf{z}/\mathbf{z}'\mathbf{z}$ have a $Di[r_1/2,\ r_2/2,\ \ldots,\ r_K/2,\ (N-r_+)/2;\ K]$ distribution. Moreover, the supposition that $\mathbf{e}$ has an MVN distribution is stronger than necessary to insure that $\mathbf{z}'\mathbf{A}_1\mathbf{z}/\mathbf{z}'\mathbf{z}$, $\mathbf{z}'\mathbf{A}_2\mathbf{z}/\mathbf{z}'\mathbf{z}, \ldots,$ $\mathbf{z}'\mathbf{A}_K\mathbf{z}/\mathbf{z}'\mathbf{z}$ have a $Di[r_1/2,\ r_2/2,\ \ldots,\ r_K/2,\ (N-r_+)/2;\ K]$ distribution. In light of the discussion at the end of Section 6.8c, it is sufficient that $\mathbf{e}$ (and hence $\mathbf{z}$) have an absolutely continuous spherical distribution.

Nonconforming use of terminology. The terms analysis of variance and analysis of variance table (or in abbreviated form, ANOVA and ANOVA table) are sometimes used elsewhere in ways that do not conform fully to the ways in which they are used here. One relatively common kind of nonconformance consists of selecting lines from what could be regarded as two different ANOVA tables and of forming a single table from the selected lines and then referring to the resultant table as an ANOVA table.

b. Two- or three-line ANOVA table for testing a single vector-valued homogeneous null hypothesis

Suppose that $\mathbf{y}$ is an $N \times 1$ observable random vector that follows a G–M model with model equation $\mathbf{y} = \mathbf{X}\boldsymbol{\beta} + \mathbf{e}$, where $\boldsymbol{\beta}$ is a $P \times 1$ vector of unknown (and unconstrained) parameters. And take $\mathbf{W}$ to be any matrix (with N rows) such that $\mathcal{C}(\mathbf{W}) \subset \mathcal{C}(\mathbf{X})$ and hence such that $\mathbf{W} = \mathbf{X}\mathbf{L}$ for some matrix $\mathbf{L}$. Further, let $TotSS = \mathbf{y}'\mathbf{y}$, and consider the decomposition

$$TotSS = SS_1 + SS_2 + SS_3 \tag{5.3}$$

of the total sum of squares into three components SS_1, SS_2, and SS_3 defined as follows:

$$SS_1 = \mathbf{y}'\mathbf{P_W}\mathbf{y}, \quad SS_2 = \mathbf{y}'(\mathbf{P_X}-\mathbf{P_W})\mathbf{y}, \quad \text{and} \quad SS_3 = \mathbf{y}'(\mathbf{I}-\mathbf{P_X})\mathbf{y}.$$

Let $r_1 = \operatorname{rank}(\mathbf{P_W})$, $r_2 = \operatorname{rank}(\mathbf{P_X}-\mathbf{P_W})$, and $r_3 = \operatorname{rank}(\mathbf{I}-\mathbf{P_X})$. And observe that $\mathbf{P_X}$ and $\mathbf{P_W}$ are symmetric and idempotent and that $\mathbf{P_X}\mathbf{W} = \mathbf{P_X}\mathbf{X}\mathbf{L} = \mathbf{X}\mathbf{L} = \mathbf{X}$, so that all three of the matrices $\mathbf{P_W}$, $\mathbf{P_X}-\mathbf{P_W}$, and $\mathbf{I}-\mathbf{P_X}$ are symmetric and idempotent. Clearly,

$$r_1 = \operatorname{rank}(\mathbf{W}) \qquad \text{and} \qquad r_3 = N - \operatorname{rank}(\mathbf{X}). \tag{5.4}$$

Moreover,

$$r_2 = \operatorname{rank}(\mathbf{X}) - \operatorname{rank}(\mathbf{W}), \tag{5.5}$$

as is evident from Theorem 6.8.7 (upon observing that $\mathbf{P_W} + (\mathbf{P_X}-\mathbf{P_W}) + (\mathbf{I}-\mathbf{P_X}) = \mathbf{I}$) or, more directly, upon recalling that the rank of an idempotent matrix equals its trace and observing that

$$\operatorname{rank}(\mathbf{P_X}-\mathbf{P_W}) = \operatorname{tr}(\mathbf{P_X}-\mathbf{P_W}) = \operatorname{tr}(\mathbf{P_X}) - \operatorname{tr}(\mathbf{P_W}) = \operatorname{rank}(\mathbf{P_X}) - \operatorname{rank}(\mathbf{P_W}) = \operatorname{rank}(\mathbf{X}) - \operatorname{rank}(\mathbf{W}).$$

Together, results (5.4) and (5.5) imply that

$$r_1 + r_2 + r_3 = N$$

and hence that decomposition (5.3) is an ANOVA.

Letting $\boldsymbol{\mu} = \mathrm{E}(\mathbf{y}) = \mathbf{X}\boldsymbol{\beta}$, the ANOVA (5.3) is expressible in tabular form as follows:

Source	df	SS	MS	EMS
Source 1	$r_1=\text{rank}(\mathbf{W})$	$SS_1=\mathbf{y}'\mathbf{P_W}\mathbf{y}$	SS_1/r_1	$\sigma^2+r_1^{-1}\boldsymbol{\mu}'\mathbf{P_W}\boldsymbol{\mu}$
Source 2	$r_2=\text{rank}(\mathbf{X})-\text{rank}(\mathbf{W})$	$SS_2=\mathbf{y}'(\mathbf{P_X}-\mathbf{P_W})\mathbf{y}$	SS_2/r_2	$\sigma^2+r_2^{-1}\boldsymbol{\mu}'(\mathbf{I}-\mathbf{P_W})\boldsymbol{\mu}$
Residual	$r_3=N-\text{rank}(\mathbf{X})$	$SS_3=\mathbf{y}'(\mathbf{I}-\mathbf{P_X})\mathbf{y}$	SS_3/r_3	σ^2
Total	N	$TotSS$		

Now, let $\boldsymbol{\tau} = \boldsymbol{\Lambda}'\boldsymbol{\beta}$ represent a vector of estimable linear combinations of the elements of $\boldsymbol{\beta}$ (where $\boldsymbol{\Lambda}$ is a nonnull matrix). Further, let $\hat{\boldsymbol{\tau}}$ represent the least squares estimator of $\boldsymbol{\tau}$. And recall that $\text{E}(\hat{\boldsymbol{\tau}}) = \boldsymbol{\tau}$ and that $\text{var}(\hat{\boldsymbol{\tau}}) = \sigma^2\mathbf{C}$, where $\mathbf{C} = \boldsymbol{\Lambda}'(\mathbf{X}'\mathbf{X})^-\boldsymbol{\Lambda}$. As is to be shown in what follows, there is an ANOVA of the form (5.3) that is intimately related to the formation of the F statistic for testing the homogeneous null hypothesis $H_0 : \boldsymbol{\tau} = \mathbf{0}$ versus the alternative hypothesis $H_1 : \boldsymbol{\tau} \neq \mathbf{0}$. Specifically, it is the special case of the ANOVA where $\mathbf{W} = \mathbf{X}[\mathbf{I} - (\boldsymbol{\Lambda}')^-\boldsymbol{\Lambda}']$ that is intimately related.

Accordingly, suppose that $\mathbf{W} = \mathbf{X}[\mathbf{I} - (\boldsymbol{\Lambda}')^-\boldsymbol{\Lambda}']$. Then, $\mathbf{y}'\mathbf{P_W}\mathbf{y}$ is the model SS for a constrained G–M model in which the constraint is that associated with the null hypothesis (i.e., in which the constraint is $\boldsymbol{\Lambda}'\boldsymbol{\beta} = \mathbf{0}$)—refer to result (4.14). And upon recalling that $\mathbf{y}'\mathbf{P_X}\mathbf{y}$ is the model SS for the unconstrained G–M model and upon reexpressing result (3.4) in terms of the constrained and unconstrained model SS's, we find that

$$\hat{\boldsymbol{\tau}}'\mathbf{C}^-\hat{\boldsymbol{\tau}} = \mathbf{y}'(\mathbf{P_X}-\mathbf{P_W})\mathbf{y}. \tag{5.6}$$

Moreover,

$$\text{rank}(\boldsymbol{\Lambda}) = \text{rank}(\mathbf{X}) - \text{rank}(\mathbf{W}), \tag{5.7}$$

as is verifiable by, for example, observing [in light of result (1.57) and the estimability of $\boldsymbol{\Lambda}'\boldsymbol{\beta}$] that

$$\text{rank}(\boldsymbol{\Lambda}) = \text{rank}(\boldsymbol{\Lambda}') = \text{rank}\begin{pmatrix}\mathbf{X}\\ \boldsymbol{\Lambda}'\end{pmatrix} - \text{rank}(\mathbf{W}) = \text{rank}(\mathbf{X}) - \text{rank}(\mathbf{W}).$$

Finally, note [in light of result (5.7.11)] that

$$\text{E}(\hat{\boldsymbol{\tau}}'\mathbf{C}^-\hat{\boldsymbol{\tau}}) = \text{tr}(\sigma^2\mathbf{C}^-\mathbf{C}) + \boldsymbol{\tau}'\mathbf{C}^-\boldsymbol{\tau}$$

and [in light of Corollary 2.8.3, Lemma 2.10.13, and result (7.3.1)] that

$$\text{tr}(\sigma^2\mathbf{C}^-\mathbf{C}) = \sigma^2\,\text{tr}(\mathbf{C}^-\mathbf{C}) = \sigma^2\,\text{rank}(\mathbf{C}^-\mathbf{C}) = \sigma^2\,\text{rank}(\mathbf{C}) = \sigma^2\,\text{rank}(\boldsymbol{\Lambda}),$$

so that

$$\text{E}(\hat{\boldsymbol{\tau}}'\mathbf{C}^-\hat{\boldsymbol{\tau}}) = \sigma^2\,\text{rank}(\boldsymbol{\Lambda}) + \boldsymbol{\tau}'\mathbf{C}^-\boldsymbol{\tau}. \tag{5.8}$$

Results (5.6), (5.7), and (5.8) serve to relate the special case of the ANOVA (5.3) where $\mathbf{W} = \mathbf{X}[\mathbf{I} - (\boldsymbol{\Lambda}')^-\boldsymbol{\Lambda}']$ to the formation of the F statistic for testing the null hypothesis $H_0 : \boldsymbol{\tau} = \mathbf{0}$ (versus the alternative hypothesis $H_1 : \boldsymbol{\tau} \neq \mathbf{0}$). When expressed in the form of a table and in terms related to the testing of H_0, that version of the ANOVA is [in light of results (5.6), (5.7), and (5.8)] as follows:

Source	df	SS	MS	EMS
Restricted model	$r_1=\text{rank}(\mathbf{W})$	$SS_1=\mathbf{y}'\mathbf{P_W}\mathbf{y}$		
Hypothesis	$r_2=\text{rank}(\boldsymbol{\Lambda})$	$SS_2=\hat{\boldsymbol{\tau}}'\mathbf{C}^-\hat{\boldsymbol{\tau}}$	SS_2/r_2	$\sigma^2+r_2^{-1}\boldsymbol{\tau}'\mathbf{C}^-\boldsymbol{\tau}$
Residual	$r_3=N-\text{rank}(\mathbf{X})$	$SS_3=\mathbf{y}'(\mathbf{I}-\mathbf{P_X})\mathbf{y}$	SS_3/r_3	σ^2
Total	N	$TotSS$		

The F statistic, say F, is expressible in terms of the mean squares SS_2/r_2 and SS_3/r_3 from the last two lines of this table; specifically,

$$F = \frac{SS_2/r_2}{SS_3/r_3}. \tag{5.9}$$

Note that the first two entries in the ANOVA table under the heading Source are now ones that are at least somewhat meaningful in the context of hypothesis testing. Note also that the entries in the first line of the ANOVA table under the headings MS and EMS have been omitted; these entries have little bearing on the testing of H_0. In fact, we might now choose to omit the entire first line of the table; upon doing so, we obtain the following ANOVA table:

Source	df	SS	MS	EMS
Hypothesis	$r_2 = \text{rank}(\mathbf{\Lambda})$	$SS_2 = \hat{\boldsymbol{\tau}}'\mathbf{C}^{-}\hat{\boldsymbol{\tau}}$	SS_2/r_2	$\sigma^2 + r_2^{-1}\boldsymbol{\tau}'\mathbf{C}^{-}\boldsymbol{\tau}$
Residual	$r_3 = N - \text{rank}(\mathbf{X})$	$SS_3 = \mathbf{y}'(\mathbf{I} - \mathbf{P_X})\mathbf{y}$	SS_3/r_3	σ^2
Total	$N - \text{rank}(\mathbf{W})$	$\mathbf{y}'(\mathbf{I} - \mathbf{P_W})\mathbf{y}$		

c. Regression sum of squares

A (linear) regression model can be regarded as a special case of a G–M model. A regression model may or may not have an "intercept term." In the "no-intercept" case, the entries in each column of the model matrix $\mathbf{X}$ are "interpretable" as the values of an "explanatory variable"—in general, some of the explanatory variables may be functionally related to others. In the more commonly encountered "with-intercept" case, the model matrix $\mathbf{X}$ is of the form $\mathbf{X} = (\mathbf{1},\ \mathbf{X}_*)$, so that $\mathbf{X}\boldsymbol{\beta} = \beta_1\mathbf{1} + \mathbf{X}_*\boldsymbol{\beta}_*$, where $\boldsymbol{\beta} = \begin{pmatrix}\beta_1\\ \boldsymbol{\beta}_*\end{pmatrix}$; and it is the entries in each column of $\mathbf{X}_*$ that are "interpretable" as "explanatory variables." In both cases, the values of the entries in the observable random vector $\mathbf{y}$ are "interpretable" as the values of a "response variable," and interest may center on how and to what extent (if any) the variability among the values of the response variable is "attributable" to the differences among the values of each of the explanatory variables. When in the no-intercept case all of the explanatory variables equal 0, the response variable is implicity assumed to center around 0, whereas, under the same circumstance in the with-intercept case, it is assumed to center around the unknown value of the parameter β_1.

Among the settings for which a (linear) regression model might be suitable are those where the N vectors formed by the values of a response variable and the corresponding values of a vector of explanatory variables can reasonably be regarded as a random sample (of size N) from the joint distribution of a random variable, say y, and a random vector, say $\mathbf{u}$. The settings of this kind for which a (linear) regression model would be suitable are those for which the conditional distribution of y given $\mathbf{u}$ qualifies as a G–M model (as it does when the joint distribution of y and $\mathbf{u}$ is MVN)—refer to Section 4.3 for some relevant discussion.

In what follows, the regression sum of squares is taken (by definition) to be the hypothesis sum of squares for testing the null hypothesis $H_0 : \mathbf{X}\boldsymbol{\beta} = \mathbf{0}$ versus the alternative hypothesis $H_1 : \mathbf{X}\boldsymbol{\beta} \neq \mathbf{0}$ (in the case of the no-intercept model) and that for testing the null hypothesis $H_0 : \mathbf{X}_*\boldsymbol{\beta}_* = \mathbf{0}$ versus the alternative hypothesis $H_1 : \mathbf{X}_*\boldsymbol{\beta}_* \neq \mathbf{0}$ (in the case of the with-intercept model). When $\text{rank}(\mathbf{X}) = P$ or $\text{rank}(\mathbf{X}_*) = P-1$, the null hypothesis is equivalent to $H_0 : \boldsymbol{\beta} = \mathbf{0}$ or $H_0 : \boldsymbol{\beta}_* = \mathbf{0}$ and the alternative hypothesis is equivalent to $H_1 : \boldsymbol{\beta} \neq \mathbf{0}$ or $H_1 : \boldsymbol{\beta}_* \neq \mathbf{0}$. Subsequently, the regression sum of squares is sometimes denoted by the symbol *RegSS*. And, on occasion, the term regression sum of squares is abbreviated to regression SS.

Regression SS: no-intercept case. Suppose that $\mathbf{y}$ is an $N \times 1$ observable random vector that follows a G–M model with model equation $\mathbf{y} = \mathbf{X}\boldsymbol{\beta} + \mathbf{e}$, where $\boldsymbol{\beta}$ is a $P \times 1$ vector of unknown (and unconstrained) parameters and where $\mathbf{e}$ is an unobservable random vector with $\text{E}(\mathbf{e}) = \mathbf{0}$ and $\text{var}(\mathbf{e}) = \sigma^2\mathbf{I}$. Further, let $\boldsymbol{\mu} = \text{E}(\mathbf{y}) = \mathbf{X}\boldsymbol{\beta}$. And consider the problem of testing the null hypothesis $H_0 : \boldsymbol{\mu} = \mathbf{0}$ (versus the alternative hypothesis $H_1 : \boldsymbol{\mu} \neq \mathbf{0}$) or "more generally" that of testing the null hypothesis $H_0 : \boldsymbol{\tau} = \mathbf{0}$ (versus the alternative hypothesis $H_1 : \boldsymbol{\tau} \neq \mathbf{0}$), where $\boldsymbol{\tau} = \mathbf{\Lambda}'\boldsymbol{\beta}$ for some matrix $\mathbf{\Lambda}$ (with P rows) for which $\mathcal{R}(\mathbf{\Lambda}') = \mathcal{R}(\mathbf{X})$ [so that $\boldsymbol{\tau}$ is estimable and $\text{rank}(\mathbf{\Lambda}) = \text{rank}(\mathbf{X})$]. (Actually, the two null hypotheses are equivalent in the sense that $\boldsymbol{\tau} = \mathbf{0} \Leftrightarrow \boldsymbol{\mu} = \mathbf{0}$.)

As is evident for example from results (5.6) and (5.7),

$$\mathbf{y}'\mathbf{P_X}\mathbf{y} = \hat{\boldsymbol{\tau}}'\mathbf{C}^{-}\hat{\boldsymbol{\tau}}$$

[where $\hat{\boldsymbol{\tau}}$ is the least squares estimator of $\boldsymbol{\tau}$ and where $\mathbf{C} = \boldsymbol{\Lambda}'(\mathbf{X}'\mathbf{X})^{-}\boldsymbol{\Lambda}$]. Thus, the F statistic for testing H_0 versus H_1 (under suitable assumptions about the distribution of $\mathbf{e}$) is expressible in terms of the mean squares from the following ANOVA table:

Source	df	SS	MS	EMS
Regression	$r_1 = \text{rank}(\mathbf{X})$	$SS_1 = \mathbf{y}'\mathbf{P_X}\mathbf{y}$	SS_1/r_1	$\sigma^2 + r_1^{-1}\boldsymbol{\mu}'\boldsymbol{\mu}$
Residual	$r_2 = N - \text{rank}(\mathbf{X})$	$SS_2 = \mathbf{y}'(\mathbf{I} - \mathbf{P_X})\mathbf{y}$	SS_2/r_2	σ^2
Total	N	$TotSS$		

Specifically, the F statistic, say F, for testing H_0 versus H_1 is expressible as

$$F = \frac{SS_1/r_1}{SS_2/r_2}.$$

And as indicated by the entry on line 1 of the ANOVA table under the heading Source, the regression SS is (in the no-intercept case) expressible as

$$RegSS = \mathbf{y}'\mathbf{P_X}\mathbf{y}.$$

Regression SS: with-intercept case. As in the preceding part of the present subsection, take $\mathbf{y}$ to be an $N\times 1$ observable random vector that follows a G–M model with model equation $\mathbf{y} = \mathbf{X}\boldsymbol{\beta} + \mathbf{e}$, where $\boldsymbol{\beta}$ is a $P\times 1$ vector of unknown (and unconstrained) parameters and where $\mathbf{e}$ is an unobservable random vector with $\text{E}(\mathbf{e}) = \mathbf{0}$ and $\text{var}(\mathbf{e}) = \sigma^2\mathbf{I}$. And continue to take $\boldsymbol{\mu} = \text{E}(\mathbf{y})$ $(= \mathbf{X}\boldsymbol{\beta})$.

In the with-intercept case, $\mathbf{X}\boldsymbol{\beta} = \beta_1\mathbf{1} + \mathbf{X}_*\boldsymbol{\beta}_*$. Assume that $\mathbf{1} \notin \mathcal{C}(\mathbf{X}_*)$ (as would ordinarily be the case) or, equivalently, that $\text{rank}(\mathbf{X}) = \text{rank}(\mathbf{X}_*) + 1$. Then, a linear combination $\lambda_1\beta_1 + \boldsymbol{\lambda}_*'\boldsymbol{\beta}_*$ of the elements of $\boldsymbol{\beta}$ is estimable provided that $\boldsymbol{\lambda}_*' \in \mathcal{R}(\mathbf{X}_*)$, as is evident upon observing (in light of the results of Section 5.3b) that (for a constant k_1 and vector of constants k_*)

$$\mathbf{1}k_1 + \mathbf{X}_*\mathbf{k}_* = \mathbf{0} \;\Rightarrow\; k_1 = 0 \text{ and } \mathbf{X}_*\mathbf{k}_* = \mathbf{0} \;\Rightarrow\; \lambda_1 k_1 + \boldsymbol{\lambda}_*'k_* = 0 \text{ [provided that } \boldsymbol{\lambda}_*' \in \mathcal{R}(\mathbf{X}_*)].$$

Thus, $\mathbf{X}_*\boldsymbol{\beta}_*$ is estimable (and β_1 is estimable).

Now, consider the problem of testing the null hypothesis $H_0 : \mathbf{X}_*\boldsymbol{\beta}_* = \mathbf{0}$ versus the alternative hypothesis $H_1 : \mathbf{X}_*\boldsymbol{\beta}_* \neq \mathbf{0}$. In addressing this problem, it is instructive to regard it as a special case of the problem of testing the null hypothesis $H_0 : \boldsymbol{\tau} = \mathbf{0}$ versus the alternative hypothesis $H_1 : \boldsymbol{\tau} \neq \mathbf{0}$, where $\boldsymbol{\tau} = \boldsymbol{\Lambda}'\boldsymbol{\beta}$ is a vector of estimable linear combinations of the elements of $\boldsymbol{\beta}$ and $\text{rank}(\boldsymbol{\Lambda}) = \text{rank}(\mathbf{X}) - 1$ [and where the model matrix $\mathbf{X}$ may or may not be of the form $\mathbf{X} = (\mathbf{1}, \mathbf{X}_*)$]. The latter problem is a special case of the problem considered in Subsection b and can be addressed accordingly.

Let $\mathbf{W} = \mathbf{X}[\mathbf{I} - (\boldsymbol{\Lambda}')^{-}\boldsymbol{\Lambda}']$, and observe [in light of result (5.7)] that [when $\text{rank}(\boldsymbol{\Lambda}) = \text{rank}(\mathbf{X}) - 1$]

$$\text{rank}(\mathbf{W}) = \text{rank}(\mathbf{X}) - \text{rank}(\boldsymbol{\Lambda}) = 1. \tag{5.10}$$

Result (5.10) implies that

$$\mathbf{W} = \mathbf{w}\boldsymbol{\ell}'$$

for some (nonnull) $N\times 1$ vector $\mathbf{w}$ and some (nonnull) $P\times 1$ vector $\boldsymbol{\ell}$. And it follows that

$$\mathbf{P_W} = \mathbf{W}(\mathbf{W}'\mathbf{W})^{-}\mathbf{W}' = \mathbf{w}\boldsymbol{\ell}'[(\mathbf{w}'\mathbf{w})^{-1}(\boldsymbol{\ell}'\boldsymbol{\ell})^{-1}\mathbf{I}_P]\boldsymbol{\ell}\mathbf{w}' = (\mathbf{w}'\mathbf{w})^{-1}\mathbf{w}\mathbf{w}' \tag{5.11}$$

and hence that

$$\mathbf{y}'\mathbf{P_W}\mathbf{y} = (\mathbf{w}'\mathbf{w})^{-1}(\mathbf{w}'\mathbf{y})^2.$$

Letting $\tilde{\mu} = (\mathbf{w}'\mathbf{w})^{-1}\mathbf{w}'\boldsymbol{\mu}$, the F statistic for testing $H_0 : \boldsymbol{\tau} = \mathbf{0}$ versus $H_1 : \boldsymbol{\tau} \neq \mathbf{0}$ [in the special case where $\text{rank}(\boldsymbol{\Lambda}) = \text{rank}(\mathbf{X}) - 1$] is expressible in terms of the mean squares from the following ANOVA table:

Source	df	SS	MS	EMS
CF	1	$CF = (\mathbf{w}'\mathbf{w})^{-1}(\mathbf{w}'\mathbf{y})^2$		
Hypothesis	$r_2 = \text{rank}(\mathbf{X}) - 1$	$SS_2 = \mathbf{y}'\mathbf{P_X}\mathbf{y} - CF$	SS_2/r_2	$\sigma^2 + r_2^{-1}(\boldsymbol{\mu} - \tilde{\mu}\mathbf{w})'(\boldsymbol{\mu} - \tilde{\mu}\mathbf{w})$
Residual	$r_3 = N - \text{rank}(\mathbf{X})$	$SS_3 = \mathbf{y}'(\mathbf{I} - \mathbf{P_X})\mathbf{y}$	SS_3/r_3	σ^2
Total	N	$TotSS$		

Specifically, the F statistic, say F, is expressible as

$$F = \frac{SS_2/r_2}{SS_3/r_3}.$$

These results on the testing of $H_0 : \boldsymbol{\tau} = \mathbf{0}$ versus $H_1 : \boldsymbol{\tau} \neq \mathbf{0}$ are obtainable from the results of Subsection b by making use of results (5.10) and (5.11). The sum of squares from the first line of the ANOVA table is referred to as the correction factor and has been identified by the acronym CF and denoted by the symbol CF —such usage of the term correction factor has become common practice.

Finally, suppose that $\mathbf{X} = (\mathbf{1}, \mathbf{X}_*)$ and that $\boldsymbol{\Lambda}' = (\mathbf{0}, \mathbf{X}_*)$, in which case $\boldsymbol{\tau} = (\mathbf{0}, \mathbf{X}_*)\boldsymbol{\beta} = \mathbf{X}_*\boldsymbol{\beta}_*$ [where $\boldsymbol{\beta} = \begin{pmatrix} \beta_1 \\ \boldsymbol{\beta}_* \end{pmatrix}$]. Then, $(\mathbf{0}, \mathbf{X}_*)'(\mathbf{X}_*\mathbf{X}_*')^-$ is a generalized inverse of $\boldsymbol{\Lambda}'$ and upon setting $(\boldsymbol{\Lambda}')^- = (\mathbf{0}, \mathbf{X}_*)'(\mathbf{X}_*\mathbf{X}_*')^-$, we find that

$$\mathbf{W} = \mathbf{X}[\mathbf{I} - (\mathbf{0}, \mathbf{X}_*)'(\mathbf{X}_*\mathbf{X}_*')^-(\mathbf{0}, \mathbf{X}_*)] = (\mathbf{1}, \mathbf{X}_*)\begin{pmatrix} 1 & \mathbf{0}' \\ \mathbf{0} & \mathbf{I} - \mathbf{X}_*'(\mathbf{X}_*\mathbf{X}_*')^-\mathbf{X}_* \end{pmatrix} = (\mathbf{1}, \mathbf{0}).$$

Thus, $\mathbf{W} = \mathbf{w}\boldsymbol{\ell}'$ when $\mathbf{w} = \mathbf{1}$ and $\boldsymbol{\ell}' = (1, \mathbf{0}')$. And when $\mathbf{w} = \mathbf{1}$, the ANOVA table is reexpressible (upon letting $\bar{y} = N^{-1}\mathbf{1}'\mathbf{y}$ and $\bar{\mu} = N^{-1}\mathbf{1}'\boldsymbol{\mu}$) as follows:

Source	df	SS	MS	EMS
CF	1	$CF = N\bar{y}^2$		
Regression	$r_2 = \text{rank}(\mathbf{X}) - 1$	$SS_2 = \mathbf{y}'\mathbf{P_X}\mathbf{y} - CF$	SS_2/r_2	$\sigma^2 + r_2^{-1}(\boldsymbol{\mu} - \bar{\mu}\mathbf{1})'(\boldsymbol{\mu} - \bar{\mu}\mathbf{1})$
Residual	$r_3 = N - \text{rank}(\mathbf{X})$	$SS_3 = \mathbf{y}'(\mathbf{I} - \mathbf{P_X})\mathbf{y}$	SS_3/r_3	σ^2
Total	N	$TotSS$		

The regression SS (in the with-intercept case) is the sum of squares from the second line of this ANOVA table (as indicated by the entry on the second line of the table under the heading Source). Thus, in the with-intercept case,

$$RegSS = \mathbf{y}'\mathbf{P_X}\mathbf{y} - N\bar{y}^2.$$

The first line of the ANOVA table (i.e., the line for the correction factor) is sometimes omitted. That results in an ANOVA table with two lines as follows:

Source	df	SS	MS	EMS
Regression	$r_1 = \text{rank}(\mathbf{X}) - 1$	$SS_1 = \mathbf{y}'\mathbf{P_X}\mathbf{y} - CF$	SS_1/r_1	$\sigma^2 + r_1^{-1}(\boldsymbol{\mu} - \bar{\mu}\mathbf{1})'(\boldsymbol{\mu} - \bar{\mu}\mathbf{1})$
Residual	$r_2 = N - \text{rank}(\mathbf{X})$	$SS_2 = \mathbf{y}'(\mathbf{I} - \mathbf{P_X})\mathbf{y}$	SS_2/r_2	σ^2
Total (corr.)	$N - 1$	$TotSS - CF$		

d. Coefficient of determination

Let us consider further the setting discussed in Subsection c. That setting is one in which inferences are to be made about how a so-called response variable is related to one or more explanatory variables and in which the inferences are to be based on a so-called regression model.

Let *RegSS* represent the regression SS (either that for the with-intercept version of the regression model or that for the no-intercept version), let *TotSS* represent the "uncorrected" total SS (i.e., $\mathbf{y}'\mathbf{y}$), and let CF represent the correction factor (i.e., $N\bar{y}^2$, where $\bar{y} = N^{-1}\mathbf{1}'\mathbf{y}$). And recall that $RegSS = \mathbf{y}'\mathbf{P_X}\mathbf{y} - CF$ in the case of the with-intercept version of the regression model, and $RegSS = \mathbf{y}'\mathbf{P_X}\mathbf{y}$ in the case of the no-intercept version. In the case of the with-intercept version, the ratio $\dfrac{RegSS}{TotSS - CF} = \dfrac{\mathbf{y}'\mathbf{P_X}\mathbf{y} - CF}{TotSS - CF}$ has come to be known as the coefficient of determination. In the case of the no-intercept version, it is the ratio $\dfrac{RegSS}{TotSS} = \dfrac{\mathbf{y}'\mathbf{P_X}\mathbf{y}}{TotSS}$ that is referred to as the coefficient of determination.

In the case of either the with-intercept version of the regression model or the no-intercept version, interest in the coefficient of determination centers on its interpretation as a measure of the extent to which the observed variability in the response variable is related to the observed variability in the explanatory variables. The coefficient of determination has a lower bound of 0 and an upper bound of 1. The extent of the relationship between the observed variability in the response variable and that in the explanatory variables increases from no relationship to complete relationship as the coefficient of determination increases from 0 to 1—in fact, a coefficient of determination "close to 0" is indicative of a relationship much less extensive than that expected simply by chance.

The definition of the coefficient of determination can be restated in a form that covers a broader range of settings. Suppose that $\mathbf{y}$ is an $N\times 1$ observable random vector that follows a G–M model with model equation $\mathbf{y} = \mathbf{X}\boldsymbol{\beta} + \mathbf{e}$, where $\boldsymbol{\beta}$ is a $P\times 1$ vector of unknown (and unconstrained) parameters. And as is customary (for reasons that will subsequently become apparent), denote the coefficient of determination by the symbol R^2, and (as a general definition) take

$$R^2 = \begin{cases} \dfrac{\mathbf{y}'\mathbf{P_X}\mathbf{y} - N\bar{y}^2}{\mathbf{y}'\mathbf{y} - N\bar{y}^2}, & \text{if } \mathbf{1} \in \mathcal{C}(\mathbf{X}), \\[2ex] \dfrac{\mathbf{y}'\mathbf{P_X}\mathbf{y}}{\mathbf{y}'\mathbf{y}}, & \text{if } \mathbf{1} \notin \mathcal{C}(\mathbf{X}) \end{cases} \tag{5.12}$$

(where, as before, $\bar{y} = N^{-1}\mathbf{1}'\mathbf{y}$).

Reexpression of the coefficient of determination as a product-moment correlation. Let us consider further the quantity R^2 defined by expression (5.12). Accordingly, continue to suppose that $\mathbf{y}$ is an $N\times 1$ observable random vector that follows a G–M model with model equation $\mathbf{y} = \mathbf{X}\boldsymbol{\beta} + \mathbf{e}$ and to take $\bar{y} = N^{-1}\mathbf{1}'\mathbf{y}$. Further, let $\boldsymbol{\mu} = \mathrm{E}(\mathbf{y}) = \mathbf{X}\boldsymbol{\beta}$. And take $\hat{\boldsymbol{\mu}} = \mathbf{P_X}\mathbf{y}$ (so that $\hat{\boldsymbol{\mu}}$ is the least squares estimator of $\boldsymbol{\mu}$), and define $\bar{\mu} = N^{-1}\mathbf{1}'\boldsymbol{\mu}$.

Consider first the special case where $\mathbf{1} \in \mathcal{C}(\mathbf{X})$. In that special case, $\mathbf{P_X}\mathbf{1} = \mathbf{1}$ and hence $\mathbf{1}'\mathbf{P_X} = \mathbf{1}'$. Thus,

$$N^{-1}\mathbf{1}'\hat{\boldsymbol{\mu}} = N^{-1}\mathbf{1}'\mathbf{P_X}\mathbf{y} = N^{-1}\mathbf{1}'\mathbf{y} = \bar{y}.$$

And it follows that $\hat{\boldsymbol{\mu}} - \bar{y}\mathbf{1}$ is the least squares estimator of the vector $\boldsymbol{\mu} - \bar{\mu}\mathbf{1}$.

Now, regard the elements of the vector $\mathbf{y} - \bar{y}\mathbf{1}$ as the N values of a variable and the elements of the vector $\hat{\boldsymbol{\mu}} - \bar{y}\mathbf{1}$ as the N values of a second variable, and denote by R the product-moment correlation between these two variables. Then, by definition,

$$R = \frac{(\mathbf{y} - \bar{y}\mathbf{1})'(\hat{\boldsymbol{\mu}} - \bar{y}\mathbf{1})}{[(\mathbf{y} - \bar{y}\mathbf{1})'(\mathbf{y} - \bar{y}\mathbf{1})]^{1/2}\,[(\hat{\boldsymbol{\mu}} - \bar{y}\mathbf{1})'(\hat{\boldsymbol{\mu}} - \bar{y}\mathbf{1})]^{1/2}}. \tag{5.13}$$

And as suggested by the notation, the coefficient of determination equals the square of the product-moment correlation (5.13)

The validity of that relationship can be established by observing that both of the quantities $(\mathbf{y} - \bar{y}\mathbf{1})'(\hat{\boldsymbol{\mu}} - \bar{y}\mathbf{1})$ and $(\hat{\boldsymbol{\mu}} - \bar{y}\mathbf{1})'(\hat{\boldsymbol{\mu}} - \bar{y}\mathbf{1})$ are equal to $\mathbf{y}'\mathbf{P_X}\mathbf{y} - N\bar{y}^2$ [as can be readily verified by making use of the basic properties (symmetry and idempotency) of the matrix $\mathbf{P_X}$ and by making use of the two equalities $\mathbf{P_X}\mathbf{1} = \mathbf{1}$ and $\mathbf{1}'\mathbf{P_X} = \mathbf{1}'$] and by observing that $(\mathbf{y} - \bar{y}\mathbf{1})'(\mathbf{y} - \bar{y}\mathbf{1}) = \mathbf{y}'\mathbf{y} - N\bar{y}^2$. From those two observations, it follows that

$$R = \frac{(\mathbf{y}'\mathbf{P_X}\mathbf{y} - N\bar{y}^2)^{1/2}}{(\mathbf{y}'\mathbf{y} - N\bar{y}^2)^{1/2}}, \tag{5.14}$$

the square of which is [in the special case where $\mathbf{1} \in \mathcal{C}(\mathbf{X})$] identical to expression (5.12) for the coefficient of determination. Note [in light of expression (5.14)] that the product-moment correlation R is necessarily nonnegative, so that R^2 conveys the same amount of information as R.

Turning to the other special case, that is, the special case where $\mathbf{1} \notin \mathcal{C}(\mathbf{X})$, regard the elements of the vector $\mathbf{y}$ as the N values of a variable and the elements of the vector $\hat{\boldsymbol{\mu}}$ as the N values of a second variable; and take R to be the product-moment correlation between these two variables, so that

$$R = \frac{\mathbf{y}'\hat{\boldsymbol{\mu}}}{(\mathbf{y}'\mathbf{y})^{1/2}(\hat{\boldsymbol{\mu}}'\hat{\boldsymbol{\mu}})^{1/2}}. \tag{5.15}$$

Then, clearly, $\mathbf{y}'\hat{\boldsymbol{\mu}} = \mathbf{y}'\mathbf{P_X}\mathbf{y}$ and $\hat{\boldsymbol{\mu}}'\hat{\boldsymbol{\mu}} = (\mathbf{P_X}\mathbf{y})'\mathbf{P_X}\mathbf{y} = \mathbf{y}'\mathbf{P_X}\mathbf{y}$. Thus, R is reexpressible as

$$R = \frac{(\mathbf{y}'\mathbf{P_X}\mathbf{y})^{1/2}}{(\mathbf{y}'\mathbf{y})^{1/2}}, \tag{5.16}$$

leading to the conclusion that in the special case where $\mathbf{1} \notin \mathcal{C}(\mathbf{X})$ it is the product-moment correlation (5.15) whose square equals the coefficient of determination. Note [in light of expression (5.16)] that the product-moment correlation (5.15), like the product-moment correlation (5.13), is inherently nonnegative.

Adjusted coefficient of determination. As defined by expression (5.12), the coefficient of determination could conceivably be employed as a criterion for comparing the "performance" of various models. However, its suitability for use in such a capacity would be limited to models for which the rank of the model matrix $\mathbf{X}$ is the same. That it would be unsuitable for use in comparing the performance of models for which $\text{rank}(\mathbf{X})$ is not the same is suggested by the following observation: if R_1^2 is the coefficient of determination for a model with model matrix $\mathbf{X}_1$ and R_2^2 that for any model with a model matrix $\mathbf{X}_2$ for which $\mathcal{C}(\mathbf{X}_1) \subset \mathcal{C}(\mathbf{X}_2)$, then necessarily $R_1^2 \le R_2^2$. Thus, in the case of a regression model, one can achieve a larger coefficient of determination simply by incorporating additional explanatory variables.

There is a variation on R^2 (the coefficient of determination) known as adjusted R^2 that (unlike the original version) takes into account the value of $\text{rank}(\mathbf{X})$. Let us first consider this variation in the special case where $\mathbf{1} \in \mathcal{C}(\mathbf{X})$.

Accordingly, suppose that $\mathbf{1} \in \mathcal{C}(\mathbf{X})$. Further, let $SS_1 = \mathbf{y}'\mathbf{P_X}\mathbf{y} - N\bar{y}^2$, $SS_2 = \mathbf{y}'(\mathbf{I}-\mathbf{P_X})\mathbf{y}$, and $SS_+ = SS_1 + SS_2 = \mathbf{y}'\mathbf{y} - N\bar{y}^2$; and let $r_1 = \text{rank}(\mathbf{X})-1$, $r_2 = N - \text{rank}(\mathbf{X})$, and $r_+ = r_1 + r_2 = N-1$. And observe that R^2 is reexpressible in the following form:

$$R^2 = \frac{SS_1}{SS_+} = \frac{SS_+ - SS_2}{SS_+} = 1 - \frac{SS_2}{SS_+}. \tag{5.17}$$

The adjustment that leads to adjusted R^2 consists of "adjusting" expression (5.17) by dividing the two sums of squares SS_2 and SS_+ by their respective degrees of freedom r_2 and r_+. Thus, in the special case where $\mathbf{1} \in \mathcal{C}(\mathbf{X})$, adjusted R^2 is the quantity $AdjR^2$ defined as follows:

$$AdjR^2 = 1 - \frac{SS_2/r_2}{SS_+/r_+} = 1 - \frac{r_+}{r_2}\frac{SS_2}{SS_+} = 1 - \frac{N-1}{N-\text{rank}(\mathbf{X})}\frac{\mathbf{y}'(\mathbf{I}-\mathbf{P_X})\mathbf{y}}{\mathbf{y}'\mathbf{y} - N\bar{y}^2}. \tag{5.18}$$

The quantity $AdjR^2$ is reexpressible in the form

$$AdjR^2 = R^2 - \frac{r_1}{r_2}(1-R^2) = R^2 - \frac{\text{rank}(\mathbf{X})-1}{N-\text{rank}(\mathbf{X})}(1-R^2), \tag{5.19}$$

as can be readily verified. Clearly, that quantity is smaller than R^2 by an amount proportional to $1-R^2$, with a constant of proportionality equal to $\dfrac{\text{rank}(\mathbf{X})-1}{N-\text{rank}(\mathbf{X})}$.

Turning to the other special case, that is, the special case where $\mathbf{1} \notin \mathcal{C}(\mathbf{X})$, take $SS_2 = \mathbf{y}'(\mathbf{I}-\mathbf{P_X})\mathbf{y}$

and $r_2 = N - \text{rank}(\mathbf{X})$ [the same as in the special case where $\mathbf{1} \in \mathcal{C}(\mathbf{X})$], and let $SS_1 = \mathbf{y}'\mathbf{P_X}\mathbf{y}$, $SS_+ = \mathbf{y}'\mathbf{y}$, $r_1 = \text{rank}(\mathbf{X})$, and $r_+ = N$. In this special case, the adjusted R^2 [that obtained by proceeding down the same path as in the special case where $\mathbf{1} \in \mathcal{C}(\mathbf{X})$] is the quantity $AdjR^2$ defined as follows:

$$AdjR^2 = 1 - \frac{SS_2/r_2}{SS_+/r_+} = 1 - \frac{r_+}{r_2}\frac{SS_2}{SS_+} = 1 - \frac{N}{N-\text{rank}(\mathbf{X})}\frac{\mathbf{y}'(\mathbf{I}-\mathbf{P_X})\mathbf{y}}{\mathbf{y}'\mathbf{y}}. \tag{5.20}$$

And analogous to result (5.19),

$$AdjR^2 = R^2 - \frac{r_1}{r_2}(1-R^2) = R^2 - \frac{\text{rank}(\mathbf{X})}{N-\text{rank}(\mathbf{X})}(1-R^2), \tag{5.21}$$

so that [as in the special case where $\mathbf{1} \in \mathcal{C}(\mathbf{X})$] adjusted R^2 is smaller than R^2 by an amount proportional to $1-R^2$—in the special case where $\mathbf{1} \notin \mathcal{C}(\mathbf{X})$, the constant of proportionality equals $\frac{\text{rank}(\mathbf{X})}{N-\text{rank}(\mathbf{X})}$.

e. Some extensions: K- or $(K–1)$-line ANOVA tables and their use in testing multiple vector-valued homogeneous null hypotheses

As in Subsection b, suppose that $\mathbf{y}$ is an $N \times 1$ observable random vector that follows a G–M model with model equation $\mathbf{y} = \mathbf{X}\boldsymbol{\beta} + \mathbf{e}$, where $\boldsymbol{\beta}$ is a $P \times 1$ vector of unknown (and unconstrained) parameters. And for purposes of extending the development in Subsection b, let $\mathbf{W}_1, \mathbf{W}_2, \ldots, \mathbf{W}_{K-1}$ represent a collection of $N \times N$ matrices for which $\mathbf{W}_{K-1} = \mathbf{X}$ and

$$\mathcal{C}(\mathbf{W}_1) \subset \mathcal{C}(\mathbf{W}_2) \subset \cdots \subset \mathcal{C}(\mathbf{W}_{K-2}) \subset \mathcal{C}(\mathbf{W}_{K-1}) \; [= \mathcal{C}(\mathbf{X})]. \tag{5.22}$$

Further, let $TotSS = \mathbf{y}'\mathbf{y}$, and consider the decomposition

$$TotSS = SS_1 + SS_2 + \cdots + SS_{K-2} + SS_{K-1} + SS_K \tag{5.23}$$

of the total sum of squares into K components $SS_1, SS_2, \ldots, SS_{K-2}, SS_{K-1}$, and SS_K defined as follows:

$$SS_1 = \mathbf{y}'\mathbf{P}_{\mathbf{W}_1}\mathbf{y}, \quad SS_j = \mathbf{y}'(\mathbf{P}_{\mathbf{W}_j} - \mathbf{P}_{\mathbf{W}_{j-1}})\mathbf{y} \;\; (j = 2, 3, \ldots, K–2, K–1), \quad \text{and} \quad SS_K = \mathbf{y}'(\mathbf{I}-\mathbf{P_X})\mathbf{y}.$$

Let $r_1 = \text{rank}(\mathbf{P}_{\mathbf{W}_1})$, $r_j = \text{rank}(\mathbf{P}_{\mathbf{W}_j} - \mathbf{P}_{\mathbf{W}_{j-1}})$ $(j = 2, 3, \ldots, K-2, K-1)$, and $r_K = \text{rank}(\mathbf{I}-\mathbf{P_X})$; and observe that $r_1 = \text{rank}(\mathbf{W}_1)$, that (for $j = 2, 3, \ldots, K-2, K-1$) $\mathbf{P}_{\mathbf{W}_j} - \mathbf{P}_{\mathbf{W}_{j-1}}$ (like $\mathbf{P}_{\mathbf{W}_j}$ and $\mathbf{P}_{\mathbf{W}_{j-1}}$) is idempotent and hence

$$r_j = \text{tr}(\mathbf{P}_{\mathbf{W}_j} - \mathbf{P}_{\mathbf{W}_{j-1}}) = \text{tr}(\mathbf{P}_{\mathbf{W}_j}) - \text{tr}(\mathbf{P}_{\mathbf{W}_{j-1}}) = \text{rank}(\mathbf{P}_{\mathbf{W}_j}) - \text{rank}(\mathbf{P}_{\mathbf{W}_{j-1}}) = \text{rank}(\mathbf{W}_j) - \text{rank}(\mathbf{W}_{j-1}),$$

and that $r_K = N - \text{rank}(\mathbf{X})$ $[= N - \text{rank}(\mathbf{W}_{K-1})]$. Then, clearly, $\sum_{j=1}^{K} r_j = N$, so that the decomposition (5.23) is an ANOVA. And letting $\boldsymbol{\mu} = \text{E}(\mathbf{y}) = \mathbf{X}\boldsymbol{\beta}$, this ANOVA is expressible in tabular form as follows:

Source	df	SS	MS	EMS
Source 1	r_1	$SS_1 = \mathbf{y}'\mathbf{P}_{\mathbf{W}_1}\mathbf{y}$	SS_1/r_1	$\sigma^2 + r_1^{-1}\boldsymbol{\mu}'\mathbf{P}_{\mathbf{W}_1}\boldsymbol{\mu}$
Source 2	r_2	$SS_2 = \mathbf{y}'(\mathbf{P}_{\mathbf{W}_2} - \mathbf{P}_{\mathbf{W}_1})\mathbf{y}$	SS_2/r_2	$\sigma^2 + r_2^{-1}\boldsymbol{\mu}'(\mathbf{P}_{\mathbf{W}_2} - \mathbf{P}_{\mathbf{W}_1})\boldsymbol{\mu}$
$\vdots$	$\vdots$	$\vdots$	$\vdots$	$\vdots$
Source K-2	r_{K-2}	$SS_{K-2} = \mathbf{y}'(\mathbf{P}_{\mathbf{W}_{K-2}} - \mathbf{P}_{\mathbf{W}_{K-3}})\mathbf{y}$	SS_{K-2}/r_{K-2}	$\sigma^2 + r_{K-2}^{-1}\boldsymbol{\mu}'(\mathbf{P}_{\mathbf{W}_{K-2}} - \mathbf{P}_{\mathbf{W}_{K-3}})\boldsymbol{\mu}$
Source K-1	r_{K-1}	$SS_{K-1} = \mathbf{y}'(\mathbf{P_X} - \mathbf{P}_{\mathbf{W}_{K-2}})\mathbf{y}$	SS_{K-1}/r_{K-1}	$\sigma^2 + r_{K-1}^{-1}\boldsymbol{\mu}'(\mathbf{I} - \mathbf{P}_{\mathbf{W}_{K-2}})\boldsymbol{\mu}$
Residual	r_K	$SS_K = \mathbf{y}'(\mathbf{I} - \mathbf{P_X})\mathbf{y}$	SS_K/r_K	σ^2
Total	N	$TotSS$		

Multi-part G-M model. Suppose now that the columns of the model matrix $\mathbf{X}$ of the G-M model (or, equivalently, the "independent variables" giving rise to those columns) are partitioned into some number of groups, say L, of sizes $P_1, P_2, \ldots, P_L$, respectively, so that

$$\mathbf{X} = (\mathbf{X}_1, \mathbf{X}_2, \ldots, \mathbf{X}_L) \tag{5.24}$$

where (for $j = 1, 2, \ldots, L$) $\mathbf{X}_j$ is the $N \times P_j$ matrix formed by the jth group of columns. And suppose that the vector $\boldsymbol{\beta}$ is partitioned conformally into subvectors $\boldsymbol{\beta}_1, \boldsymbol{\beta}_2, \ldots, \boldsymbol{\beta}_L$ of dimensions $P_1, P_2, \ldots, P_L$, respectively, so that $\boldsymbol{\beta}' = (\boldsymbol{\beta}_1', \boldsymbol{\beta}_2', \ldots, \boldsymbol{\beta}_L')$ and the model equation $\mathbf{y} = \mathbf{X}\boldsymbol{\beta} + \mathbf{e}$ is reexpressible as

$$\mathbf{y} = \mathbf{X}_1\boldsymbol{\beta}_1 + \mathbf{X}_2\boldsymbol{\beta}_2 + \cdots + \mathbf{X}_L\boldsymbol{\beta}_L + \mathbf{e}. \tag{5.25}$$

Associated with the partitioning (5.24) is a version of the ANOVA (5.23); this version is that obtained by setting $K = L+1$ and taking (for $j = 1, 2, \ldots, L-1$) $\mathbf{W}_j$ to be the model matrix $(\mathbf{X}_1, \mathbf{X}_2, \ldots, \mathbf{X}_j)$ of the submodel identified with the first j terms of expression (5.25). In this version, $\mathbf{y}'\mathbf{P}_{\mathbf{W}_1}\mathbf{y}, \mathbf{y}'\mathbf{P}_{\mathbf{W}_2}\mathbf{y}, \ldots, \mathbf{y}'\mathbf{P}_{\mathbf{W}_{L-1}}\mathbf{y}$ are the model sums of squares for the various submodels. And (for $j = 2, 3, \ldots, L$) the source of the jth sum of squares might be characterized by a phrase such as "$\boldsymbol{\beta}_j$ after $\boldsymbol{\beta}_1, \boldsymbol{\beta}_2, \ldots, \boldsymbol{\beta}_{j-1}$."

When the model matrix $\mathbf{X}$ of the G-M model is partitioned as in expression (5.24), the model might be referred to as a multi-part G-M model. Clearly, the resultant ANOVA depends on how the columns of $\mathbf{X}$ (or the corresponding elements of $\boldsymbol{\beta}$) are "ordered" and on the specifics of how they are partitioned into groups. In practice, there may be relationships among the variables identified with various of the columns of $\mathbf{X}$ that (perhaps in combination with the underlying objectives of the analysis) suggest those variables be "grouped together." And the various terms in the model might be ordered in such a way that the "higher-order" terms come last.

Multiple null hypotheses. An alternative way of arriving at an ANOVA of the form (5.23) is to base the choice of K and of $\mathbf{W}_1, \mathbf{W}_2, \ldots, \mathbf{W}_{K-2}$ on some number, say M, of null hypotheses. Accordingly, suppose that $\boldsymbol{\tau}_j = \boldsymbol{\Lambda}_j'\boldsymbol{\beta}$ $(j = 1, 2, \ldots, M)$ are M vectors of estimable linear combinations of the elements of $\boldsymbol{\beta}$. Then, corresponding to the jth of these vectors is the problem of testing the homogeneous null hypothesis $H_j^{(0)} : \boldsymbol{\tau}_j = \mathbf{0}$ versus the alternative hypothesis $H_j^{(1)} : \boldsymbol{\tau}_j \neq \mathbf{0}$.

For $j = 1, 2, \ldots, M$, let $\tilde{\boldsymbol{\Lambda}}_j = (\boldsymbol{\Lambda}_j, \boldsymbol{\Lambda}_{j+1}, \ldots, \boldsymbol{\Lambda}_M)$ [so that $(\boldsymbol{\tau}_j', \boldsymbol{\tau}_{j+1}', \ldots, \boldsymbol{\tau}_M')' = \tilde{\boldsymbol{\Lambda}}_j'\boldsymbol{\beta}$]. Then, among the versions of the ANOVA (5.23) is that obtained by setting $K = M+2$ and taking

$$\mathbf{W}_j = \mathbf{X}[\mathbf{I} - (\tilde{\boldsymbol{\Lambda}}_j')^{-}\tilde{\boldsymbol{\Lambda}}_j'] \qquad (j = 1, 2, \ldots, K-2) \tag{5.26}$$

—recall that $\mathbf{W}_{K-1} = \mathbf{X}$. That this choice for the $\mathbf{W}_j$'s satisfies condition (5.22) can be readily confirmed by making use of result (1.61).

When $K = M+2$ and $\mathbf{W}_1, \mathbf{W}_2, \ldots, \mathbf{W}_{K-2}$ satisfy condition (5.26) (with $\mathbf{W}_{K-1} = \mathbf{X}$), $\mathbf{y}'(\mathbf{I} - \mathbf{P}_{\mathbf{W}_j})\mathbf{y}$ is the residual sum of squares and $\mathbf{y}'\mathbf{P}_{\mathbf{W}_j}\mathbf{y}$ the model sum of squares for a constrained G-M model with constraint $\tilde{\boldsymbol{\Lambda}}_j'\boldsymbol{\beta} = \mathbf{0}$ (or if $j = K-1$, for an unconstrained G-M model)—for verification, refer to the development in Section 8.4 [including result (4.14)]. Thus, $SS_1 = \mathbf{y}'\mathbf{P}_{\mathbf{W}_1}\mathbf{y}$ is the model sum of squares for a constrained G-M model with constraint $\tilde{\boldsymbol{\Lambda}}_1'\boldsymbol{\beta} = \mathbf{0}$, $SS_j = \mathbf{y}'(\mathbf{P}_{\mathbf{W}_j} - \mathbf{P}_{\mathbf{W}_{j-1}})\mathbf{y}$ is the difference between the model sum of squares for the constrained G-M model with constraint $\tilde{\boldsymbol{\Lambda}}_j'\boldsymbol{\beta} = \mathbf{0}$ and that for the constrained G-M model with constraint $\tilde{\boldsymbol{\Lambda}}_{j-1}'\boldsymbol{\beta} = \mathbf{0}$ (j = 2, 3, …, K-2), and $SS_{K-1} = \mathbf{y}'(\mathbf{P}_{\mathbf{X}} - \mathbf{P}_{\mathbf{W}_{K-2}})\mathbf{y}$ is the difference between the model sum of squares for the (unconstrained) G-M model and that for the constrained G-M model with constraint $\tilde{\boldsymbol{\Lambda}}_{K-2}'\boldsymbol{\beta} = \mathbf{0}$.

The second through $(K-1)$th sums of squares $SS_2, SS_3, \ldots, SS_{K-1}$ can be reexpressed in terms of the constrained or unconstrained least squares estimators of the $\boldsymbol{\tau}_j$'s. Let $\hat{\boldsymbol{\tau}}_M$ represent the (unconstrained) least squares estimator of $\boldsymbol{\tau}_M$; and for $j = 1, 2, \ldots, M-1$, let $\hat{\boldsymbol{\tau}}_j$ represent the (constrained) least squares estimator of $\boldsymbol{\tau}_j$ when $\boldsymbol{\beta}$ is subject to the constraint $\tilde{\boldsymbol{\Lambda}}_{j+1}'\boldsymbol{\beta} = \mathbf{0}$. Further, take $\mathbf{C}$ to be the matrix defined implicitly by the equality $\text{var}(\hat{\boldsymbol{\tau}}_M) = \sigma^2\mathbf{C}$ [or explicitly as $\boldsymbol{\Lambda}_M'(\mathbf{X}'\mathbf{X})^-\boldsymbol{\Lambda}_M]$; and for $j = 1, 2, \ldots, M-1$, take $\mathbf{B}_j$ to be the matrix defined implicitly by the equality $\text{var}(\hat{\boldsymbol{\tau}}_j) = \sigma^2 \mathbf{B}_j$. Then, it follows from the results of Section 8.3 that

$$SS_j = \hat{\boldsymbol{\tau}}_{j-1}'\mathbf{B}_{j-1}^-\hat{\boldsymbol{\tau}}_{j-1} \quad (j = 2, 3, \ldots, K-2) \quad \text{and} \quad SS_{K-1} = \hat{\boldsymbol{\tau}}_M'\mathbf{C}^-\hat{\boldsymbol{\tau}}_M. \tag{5.27}$$

Now, consider the degrees of freedom $r_1, r_2, \ldots, r_{K-1}$ associated with the first $K-1$ terms in the ANOVA. Upon recalling result (1.57), we find (in light of the estimability of the $\boldsymbol{\tau}_j$'s) that (for $j = 1, 2, \ldots, M$)

$$\text{rank}(\mathbf{W}_j) = \text{rank}\begin{pmatrix} \mathbf{X} \\ \tilde{\boldsymbol{\Lambda}}_j' \end{pmatrix} - \text{rank}(\tilde{\boldsymbol{\Lambda}}_j') = \text{rank}(\mathbf{X}) - \text{rank}(\tilde{\boldsymbol{\Lambda}}_j).$$

Thus,

$$r_1 = \text{rank}(\mathbf{X}) - \text{rank}(\tilde{\boldsymbol{\Lambda}}_1),$$

$$r_{K-1} = \text{rank}(\mathbf{X}) - \text{rank}(\mathbf{W}_M) = \text{rank}(\boldsymbol{\Lambda}_M),$$

and (for $j = 2, 3, \ldots, K-2$)

$$r_j = \text{rank}(\mathbf{W}_j) - \text{rank}(\mathbf{W}_{j-1}) = \text{rank}(\tilde{\boldsymbol{\Lambda}}_{j-1}) - \text{rank}(\tilde{\boldsymbol{\Lambda}}_j).$$

Note [in light of results (1.57) and (1.60)] that (for $j = 2, 3, \ldots, K-2$) r_j is reexpressible as

$$\begin{aligned} r_j = \text{rank}\begin{pmatrix} \boldsymbol{\Lambda}_{j-1}' \\ \tilde{\boldsymbol{\Lambda}}_j' \end{pmatrix} - \text{rank}(\tilde{\boldsymbol{\Lambda}}_j') &= \text{rank}\, \boldsymbol{\Lambda}_{j-1}'[\mathbf{I} - (\tilde{\boldsymbol{\Lambda}}_j')^-\tilde{\boldsymbol{\Lambda}}_j'] \\ &= \text{rank}(\boldsymbol{\Lambda}_{j-1}') - \dim[\mathcal{R}(\boldsymbol{\Lambda}_{j-1}') \cap \mathcal{R}(\tilde{\boldsymbol{\Lambda}}_j')]. \end{aligned}$$

It remains to evaluate the expected mean squares. The expected mean square that occupies the $(K-1)$th line of the ANOVA table equals $r_{K-1}^{-1}\, \text{E}(\hat{\boldsymbol{\tau}}_M'\mathbf{C}^-\hat{\boldsymbol{\tau}}_M)$ and (as is evident from the results of Subsection b)

$$r_{K-1}^{-1}\, \text{E}(\hat{\boldsymbol{\tau}}_M'\mathbf{C}^-\hat{\boldsymbol{\tau}}_M) = \sigma^2 + r_{K-1}^{-1}\boldsymbol{\tau}_M'\mathbf{C}^-\boldsymbol{\tau}_M. \tag{5.28}$$

And for $j = 2, 3, \ldots, K$–2, the expected mean square that occupies the jth line of the ANOVA table equals $r_j^{-1}\text{E}(\hat{\boldsymbol{\tau}}_{j-1}'\mathbf{B}_{j-1}^-\hat{\boldsymbol{\tau}}_{j-1})$; and upon proceeding in essentially the same way as in Subsection b in arriving at a result equivalent to result (5.28), we find that

$$\text{E}(\hat{\boldsymbol{\tau}}_{j-1}'\mathbf{B}_{j-1}^-\hat{\boldsymbol{\tau}}_{j-1}) = \sigma^2\, \text{rank}(\mathbf{B}_{j-1}) + [\text{E}(\hat{\boldsymbol{\tau}}_{j-1})]'\mathbf{B}_{j-1}^-\, \text{E}(\hat{\boldsymbol{\tau}}_{j-1}).$$

Moreover, it follows from the results of Section 8.1e—refer specifically to results (1.92) and (1.97)—that

$$\text{rank}(\mathbf{B}_{j-1}) = \text{rank}\begin{pmatrix} \boldsymbol{\Lambda}_{j-1}' \\ \tilde{\boldsymbol{\Lambda}}_j' \end{pmatrix} - \text{rank}(\tilde{\boldsymbol{\Lambda}}_j')$$

and hence that $\text{rank}(\mathbf{B}_{j-1}) = r_j$, leading to the conclusion that (for $j = 2, 3, \ldots, K-2$)

$$r_j^{-1}\text{E}(\hat{\boldsymbol{\tau}}_{j-1}'\mathbf{B}_{j-1}^-\hat{\boldsymbol{\tau}}_{j-1}) = \sigma^2 + r_j^{-1}[\text{E}(\hat{\boldsymbol{\tau}}_{j-1})]'\mathbf{B}_{j-1}^-\, \text{E}(\hat{\boldsymbol{\tau}}_{j-1}). \tag{5.29}$$

In the special case where $M = 1$, the $(M+2)$-term ANOVA presently under consideration simplifies to the 3-term ANOVA considered earlier (in Subsection b) for testing a single null hypothesis. And when the first term of the $(M+2)$-term ANOVA is deleted, the resultant $(M+1)$-term ANOVA simplifies to the 2-term ANOVA considered earlier.

Regardless of the value of M, the degrees of freedom r_{M+1} and sum of squares SS_{M+1} found on the $(M+1)$th line of the $(M+2)$-line ANOVA table are the numerator degrees of freedom and sum of squares for testing the null hypothesis $H_M^{(0)}$ (versus the alternative hypothesis $H_M^{(1)}$) under the (unconstrained) G-M model. However, for $j < M$, the degrees of freedom r_{j+1} and sum of squares SS_{j+1} found on the $(j+1)$th line of the table are not in general the numerator degrees of freedom and sum of squares for testing $H_j^{(0)}$ (versus $H_j^{(1)}$) under the (unconstrained) G-M model. Rather, they are those for testing $H_j^{(0)}$ (versus $H_j^{(1)}$) under a constrained G-M model with constraint $\tilde{\boldsymbol{\Lambda}}_{j+1}'\boldsymbol{\beta} = \mathbf{0}$, with the denominator degrees of freedom and sum of squares for testing $H_j^{(0)}$ (versus $H_j^{(1)}$) under that constrained G-M model being the degrees of freedom $r_{j+2} + r_{j+3} + \cdots + r_{M+2}$ and sum of squares $SS_{j+2} + SS_{j+3} + \cdots + SS_{M+2}$ obtained upon combining the last $M-j+1$ lines of the ANOVA table. Note that the numerator degrees of freedom and sum of squares for testing $H_j^{(0)}$ (versus $H_j^{(1)}$) under the unconstrained G-M model could be obtained from a different ANOVA, one in which what was originally the jth null hypothesis is positioned last relative to whatever other null hypotheses are to be included.

While in general the numerator degrees of freedom and sum of squares for testing $H_j^{(0)}$ (versus $H_j^{(1)}$) under the (unconstrained) G-M model differ from r_{j+1} and SS_{j+1}, $\mathbf{\Lambda}_j$ can be chosen so that they are the same. That can be accomplished by taking into account the value of $\tilde{\mathbf{\Lambda}}_{j+1}$ and imposing the condition that the columns of $\mathbf{\Lambda}_j$ be contained in the column space of the matrix $\mathbf{X}'\mathbf{X}[\mathbf{I}-(\tilde{\mathbf{\Lambda}}'_{j+1})^{-}\tilde{\mathbf{\Lambda}}'_{j+1}]$ or equivalently that

$$\mathbf{\Lambda}_j = \mathbf{X}'\mathbf{X}[\mathbf{I}-(\tilde{\mathbf{\Lambda}}'_{j+1})^{-}\tilde{\mathbf{\Lambda}}'_{j+1}]\mathbf{U}_j \tag{5.30}$$

for some matrix $\mathbf{U}_j$. When $\mathbf{\Lambda}_j$ is of the form (5.30), a solution to the linear system

$$\begin{pmatrix} \mathbf{X}'\mathbf{X} & \tilde{\mathbf{\Lambda}}_{j+1} \\ \tilde{\mathbf{\Lambda}}'_{j+1} & \mathbf{0} \end{pmatrix} \begin{pmatrix} \mathbf{S} \\ \mathbf{T} \end{pmatrix} = \begin{pmatrix} \mathbf{\Lambda}_j \\ \mathbf{0} \end{pmatrix}$$

(in $\mathbf{S}$ and $\mathbf{T}$) can be obtained by taking $\mathbf{S} = \mathbf{S}_j$, where $\mathbf{S}_j = [\mathbf{I}-(\tilde{\mathbf{\Lambda}}'_{j+1})^{-}\tilde{\mathbf{\Lambda}}'_{j+1}]\mathbf{U}_j$, and taking $\mathbf{T} = \mathbf{0}$. Moreover, when $\mathbf{\Lambda}_j$ is of the form (5.30), taking $\mathbf{S} = \mathbf{S}_j$ gives a solution to the ordinary (unconstrained) conjugate normal equations $\mathbf{X}'\mathbf{X}\mathbf{S} = \mathbf{\Lambda}_j$. Thus, when $\mathbf{\Lambda}_j$ is of the form (5.30), the unconstrained least squares estimator of $\boldsymbol{\tau}_j$ is identical to the constrained least squares estimator—both are equal to $\mathbf{S}_j'\mathbf{X}'\mathbf{y}$—so that the numerator degrees of freedom and sum of squares for testing $H_j^{(0)}$ (versus $H_j^{(1)}$) under the unconstrained G-M model are the same as those for testing $H_j^{(0)}$ (versus $H_j^{(1)}$) under the constrained G–M model.

There is an implication that (depending on the value of M) $\mathbf{\Lambda}_1, \mathbf{\Lambda}_2, \ldots, \mathbf{\Lambda}_M$ can be chosen in such a way that the numerator degrees of freedom and sum of squares for each of the M null hypotheses $H_1^{(0)}, H_2^{(0)}, \ldots, H_M^{(0)}$ can be obtained from a single ANOVA. Clearly, this can be accomplished by choosing $\mathbf{\Lambda}_M$ arbitrarily (from among those values for which $\boldsymbol{\tau}_M$ is estimable) and then choosing $\mathbf{\Lambda}_j$ successively (for $j = M-1, M-2, \ldots, 1$) from among those values that are expressible as in condition (5.30) for some matrix $\mathbf{U}_j$. Note that (assuming the degrees of freedom for each of the M null hypotheses is required to be nonzero) the choice for $\mathbf{\Lambda}_1, \mathbf{\Lambda}_2, \ldots, \mathbf{\Lambda}_M$ must be such that

$$0 < \text{rank}(\mathbf{\Lambda}_M) < \text{rank}(\tilde{\mathbf{\Lambda}}_{M-1}) < \text{rank}(\tilde{\mathbf{\Lambda}}_{M-2}) < \cdots < \text{rank}(\tilde{\mathbf{\Lambda}}_1) \leq \text{rank}(\mathbf{X}).$$

Thus, M can be no greater than rank$(\mathbf{X})$; and in choosing $\mathbf{\Lambda}_M, \mathbf{\Lambda}_{M-1}, \ldots, \mathbf{\Lambda}_1$ successively, the choice for $\mathbf{\Lambda}_j$ $(M \geq j \geq 1)$ needs to be one for which

$$\text{rank}(\tilde{\mathbf{\Lambda}}_j) \leq \text{rank}(\mathbf{X}) - (j-1).$$

Accounting for the muliplicity of hypothesis tests: some ANOVA-related discussion. Let us continue the discussion initiated in the preceding part of the present subsection. In the testing of the M null hypotheses $H_j^{(0)}$: $\boldsymbol{\tau}_j = \mathbf{0}$ $(j = 1, 2, \ldots, M)$, we have as a special case that where each of the vectors $\boldsymbol{\tau}_1, \boldsymbol{\tau}_2, \ldots, \boldsymbol{\tau}_M$ has a single element, that is, consists of a single linear combination of the elements of $\boldsymbol{\beta}$. In that special case, a topic that has received a great deal of attention is that of multiple comparisons; in multiple comparisons, the M null hypotheses are tested individually, and the emphasis is on methods that "control" the muliplicity of false rejections. Among such methods are those considered in Sections 7.7a and 7.7b, where the criterion for controlling the multiplicity of false rejections is taken to be the FWER or more generally the k-FWER.

The formulation in Sections 7.7a and 7.7b of the multiple comparison methods for controlling the FWER or k-FWER is in terms of quantities that have t distributions. These methods can be reformulated in terms of quantities that have F distributions [with 1 and $N - \text{rank}(\mathbf{X})$ degrees of freedom] by making use of the relationships described in Part 11 of Section 7.3b. And at least in principle, these methods could be extended for use in controlling the multiplicity of false rejections in the case where some or all of the vectors $\boldsymbol{\tau}_1, \boldsymbol{\tau}_2, \ldots, \boldsymbol{\tau}_M$ have more than one element.

As discussed in Sections 7.7a and 7.7b, whether or not the use of the methods for controlling the FWER or k-FWER is feasible in any particular application depends on the availability or obtainability of various percentage points of the relevant distributions. Typically, the computation of these percentage points requires the use of Monte Carlo methods. The tractability of these computations is considerably enhanced when the circumstances are ones where condition (5.30) is satisfied for all j. When condition (5.30) is satisfied for all j, the procedures for controlling the FWER or k-FWER depend on the value of $\mathbf{y}$ only through the values of the sums of squares from the last $M+1$ lines of a single ANOVA table. And those sums of squares are distributed independently as central or noncentral chi-square random variables.

Matrix decompositions: some ANOVA-related applications. Let us add to the development begun in Part 2 of the present subsection [pertaining to an $(M+2)$-term ANOVA associated with the testing of the M null hypotheses $H_j^{(0)}: \boldsymbol{\tau}_j = \mathbf{0}$ $(j = 1, 2, \ldots, M)$]. Accordingly, take $\mathbf{Q}$ to be a $P \times P$ nonsingular matrix for which

$$\mathbf{Q}'\mathbf{X}'\mathbf{X}\mathbf{Q} = \mathbf{D}$$

for some $(P \times P)$ diagonal matrix $\mathbf{D}$; and denote by $d_1, d_2, \ldots, d_P$ the first through Pth diagonal elements of $\mathbf{D}$ and by $\mathbf{q}_1, \mathbf{q}_2, \ldots, \mathbf{q}_P$ the first through Pth columns of $\mathbf{Q}$. As is evident from Theorem 2.13.20, the matrix $\mathbf{Q}$ can be chosen to be a unit upper triangular matrix [in which case, $\mathbf{X}'\mathbf{X} = (\mathbf{Q}^{-1})'\mathbf{D}\mathbf{Q}^{-1}$ is a decomposition (of $\mathbf{X}'\mathbf{X}$) of a kind known as a U$'$DU decompoosition]. Or as is evident from Theorem 6.7.4, $\mathbf{Q}$ can be chosen to be an orthogonal matrix (in which case, $\mathbf{X}'\mathbf{X} = \mathbf{Q}\mathbf{D}\mathbf{Q}'$ is a decomposition of a kind sometimes referred to as a spectral decomposition). Note that $d_i \geq 0$ for $i = 1, 2, \ldots, P$ and that $\mathbf{Q}$ can be chosen in such a way that all of the nonzero diagonal elements of $\mathbf{D}$ equal 1—if a nonzero diagonal element d_i differs from 1, replace $\mathbf{q}_i$ by $d_i^{-1/2}\mathbf{q}_i$. In particular, $\mathbf{Q}$ can be chosen both to be upper triangular and to be such that all of the nonzero diagonal elements of $\mathbf{D}$ equal 1 [in which case, $\mathbf{X}'\mathbf{X} = (\mathbf{Q}^{-1})'\mathbf{D}\mathbf{Q}^{-1}$ is essentially the Cholesky decomposition of $\mathbf{X}'\mathbf{X}$].

Now, denote by $i_1, i_2, \ldots, i_J$ the subsequence of the sequence $1, 2, \ldots, P$ consisting of those values of the integer i (between 1 and P, inclusive) for which d_i is nonzero; and observe that $J = \operatorname{rank} \mathbf{X}$. Further, for $j = 1, 2, \ldots, J$, let $\boldsymbol{\lambda}_j = \mathbf{X}'\mathbf{X}\mathbf{q}_{i_j}$. And consider the special case of the $(M+2)$-term ANOVA in which $M = J$ and (for $j = 1, 2, \ldots, M$) $\boldsymbol{\Lambda}_j = \boldsymbol{\lambda}_j$ (so that $\boldsymbol{\Lambda}_j$ has a single column and that column equals $\boldsymbol{\lambda}_j$). Then, for $j = 1, 2, \ldots, M$,

$$\tilde{\boldsymbol{\Lambda}}_j = (\boldsymbol{\lambda}_j, \boldsymbol{\lambda}_{j+1}, \ldots, \boldsymbol{\lambda}_M) = \mathbf{X}'\mathbf{X}\mathbf{Q}_j,$$

where $\mathbf{Q}_j = (\mathbf{q}_{i_j}, \mathbf{q}_{i_{j+1}}, \ldots, \mathbf{q}_{i_M})$. And it follows that for $j = M-1, M-2, \ldots, 1$,

$$\tilde{\boldsymbol{\Lambda}}_{j+1}'\mathbf{q}_{i_j} = \mathbf{Q}_{j+1}'\mathbf{X}'\mathbf{X}\mathbf{q}_{i_j} = \mathbf{0}.$$

Thus, for $j = M-1, M-2, \ldots, 1$,

$$\boldsymbol{\Lambda}_j = \boldsymbol{\lambda}_j = \mathbf{X}'\mathbf{X}\mathbf{q}_{i_j} = \mathbf{X}'\mathbf{X}[\mathbf{I} - (\tilde{\boldsymbol{\Lambda}}_{j+1}')^{-}\tilde{\boldsymbol{\Lambda}}_{j+1}']\mathbf{U}_j$$

for some matrix $\mathbf{U}_j$, namely, for $\mathbf{U}_j = \mathbf{q}_{i_j}$, leading to the conclusion that [in the special case where the $(M+2)$-term ANOVA is that for which $M = \operatorname{rank}(\mathbf{X})$ and for which $\boldsymbol{\Lambda}_j = \mathbf{X}'\mathbf{X}\mathbf{q}_{i_j}$ $(j = 1, 2, \ldots, M)$] condition (5.30) is satisfied (by some matrix $\mathbf{U}_j$) for every value of j between $M-1$ and 1, inclusive.

In the present setting, each of the vectors $\boldsymbol{\tau}_j$ $(j = 1, 2, \ldots, M)$ being subjected to testing has a single element. Specifically, $\boldsymbol{\tau}_j$ has a single element, say τ_j, that is expressible as $\tau_j = \boldsymbol{\lambda}_j'\boldsymbol{\beta}$, where (by definition) $\boldsymbol{\lambda}_j = \mathbf{X}'\mathbf{X}\mathbf{q}_{i_j}$. Further, the least squares estimator of τ_j, say $\hat{\tau}_j$, is expressible as $\hat{\tau}_j = \mathbf{q}_{i_j}'\mathbf{X}'\mathbf{y}$ (as is evident upon observing that $\mathbf{q}_{i_j}$ is a solution to the conjugate normal equations). Accordingly, for $j', j = 1, 2, \ldots, M$,

$$\operatorname{cov}(\hat{\tau}_{j'}, \hat{\tau}_j) = \sigma^2\mathbf{q}_{i_{j'}}'\mathbf{X}'\mathbf{X}\mathbf{q}_{i_j}.$$

And it follows that the least squares estimators $\hat{\tau}_1, \hat{\tau}_2, \ldots, \hat{\tau}_M$ are uncorrelated with variances

$$\operatorname{var}(\hat{\tau}_j) = \sigma^2 d_{i_j} \quad (j = 1, 2, \ldots, M).$$

Note that (for any integer between 1 and M, inclusive) the coefficient vector $\boldsymbol{\lambda}_j'$ of the linear combination $\tau_j = \boldsymbol{\lambda}_j'\boldsymbol{\beta}$ is expressible as

$$\boldsymbol{\lambda}_j' = \mathbf{q}_{i_j}'\mathbf{X}'\mathbf{X}\mathbf{Q}\mathbf{Q}^{-1}$$

and that $\mathbf{q}_{i_j}'\mathbf{X}'\mathbf{X}\mathbf{Q}$ is a P-dimensional row vector the i_jth element of which equals d_{i_j} and the other $P-1$ elements of which equal 0. Thus, $\boldsymbol{\lambda}_j'$ is reexpressible as

$$\boldsymbol{\lambda}_j' = d_{i_j} \times \text{ the } i_j\text{th row of the matrix } \mathbf{Q}^{-1}$$

or alternatively as

$$\boldsymbol{\lambda}_j' = \text{ the } i_j\text{th row of the matrix product } \mathbf{D}\mathbf{Q}^{-1}.$$

In the special case where $\mathbf{Q}$ is upper triangular, $\mathbf{D}\mathbf{Q}^{-1}$ is likewise upper triangular (as is evident from the results of Section 2.6b), so that (in that special case) τ_j depends on the vector $\boldsymbol{\beta}$ only through its i_jth, $(i_j + 1)$th, ..., i_Pth elements. And in the special case where $\mathbf{Q}$ is orthogonal, $\mathbf{Q}^{-1} = \mathbf{Q}'$.

Example: application to the ouabain data. The ouabain data can be found in Section 4.2b. They come from a study in which each of 41 cats was injected intravenously with ouabain at one of four rates and the dose that proved lethal was recorded. Interest presumably centered on the relationship between dose and rate.

Let us regard the 41 lethal doses as the realizations of the elements $y_1, y_2, \ldots, y_{41}$ of an observable random vector $\mathbf{y}$ that follows a G-M model. And letting $u_1, u_2, \ldots, u_{41}$ represent the corresponding rates of injection, suppose that (for $i = 1, 2, \ldots, 41$)

$$\text{E}(y_i) = \beta_1 + \beta_2 u_i + \beta_3 u_i^2 + \beta_4 u_i^3.$$

As an example of an ANOVA associated with the partitioning of the model matrix $\mathbf{X}$ into L segments (as in Part 1 of the present subsection), let us form segments consisting of one column each. Then, upon "fitting" the relevant submodels, we obtain an ANOVA table with degreees of freedom and sums of squares as follows:

Source	df	SS
β_1	1	61971.61
β_2 after β_1	1	15700.07
β_3 after β_1 and β_2	1	382.63
β_4 after β_1, β_2, and β_3	1	11.58
Residual	37	5650.11
Total	41	83716

An ANOVA with the same degrees of freedom and sums of squares can be generated from the $M = 4$ null hypotheses $H_j^{(0)} : \beta_j = 0$ $(j = 1, 2, 3, 4)$ via an implementation of the approach described in Part 2 of the present subsection. The degrees of freedom and sum of squares from the fourth line of the ANOVA table are the numerator degrees of freedom and sum of squares for testing the null hypothesis $H_4^{(0)}$. Note, however, that the degrees of freedom and sum of squares found on the third line of the table are not the numerator degrees of freedom and sum of squares for testing the null hypothesis $H_3^{(0)}$, nor are the degrees of freedom and sum of squares found on the second line of the table the numerator degrees of freedom and sum of squares for testing the null hypothesis $H_2^{(0)}$.

Now, let us consider the approach to the generation of an ANOVA described in Part 4 of the present subsection. As a unit upper triangular matrix $\mathbf{Q}$ and a diagonal matrix $\mathbf{D}$ for which

$$\mathbf{Q}'\mathbf{X}'\mathbf{X}\mathbf{Q} = \mathbf{D},$$

we obtain

$$\mathbf{Q} = \begin{pmatrix} 1 & -3.46 & 12.71 & -34.82 \\ 0 & 1 & -9.14 & 42.70 \\ 0 & 0 & 1 & -12.78 \\ 0 & 0 & 0 & 1 \end{pmatrix} \quad \text{and} \quad \mathbf{D} = \begin{pmatrix} 41 & 0 & 0 & 0 \\ 0 & 284.20 & 0 & 0 \\ 0 & 0 & 948.31 & 0 \\ 0 & 0 & 0 & 983.17 \end{pmatrix}.$$

By making use of the matrix $\mathbf{Q}$, the expected values of $y_1, y_2, \ldots, y_{41}$ can be reexpressed in terms of the values of orthogonal polynomials. For $i = 1, 2, \ldots, 41$, we find that

$$\begin{aligned} \mathrm{E}(y_i) &= (1, u_i, u_i^2, u_i^3)\mathbf{Q}\mathbf{Q}^{-1}(\beta_1, \beta_2, \beta_3, \beta_4)' \\ &= \alpha_1 + \alpha_2 r_1(u_i) + \alpha_3 r_2(u_i) + \alpha_4 r_3(u_i), \end{aligned}$$

where

$$\begin{aligned} r_1(u_i) &= u_i - 3.46, \\ r_2(u_i) &= u_i^2 - 9.14u_i + 12.71, \\ r_3(u_i) &= u_i^3 - 12.78u_i^2 + 42.70u_i - 34.82, \\ \alpha_1 &= \beta_1 + 3.46\beta_2 + 18.93\beta_3 + 128.88\beta_4, \\ \alpha_2 &= \beta_2 + 9.14\beta_3 + 74.09\beta_4, \\ \alpha_3 &= \beta_3 + 12.78\beta_4, \quad \text{and} \\ \alpha_4 &= \beta_4. \end{aligned}$$

Here, $r_1(\cdot)$, $r_2(\cdot)$, and $r_3(\cdot)$ are polynomials of degrees one, two, and three, respectively, and as will be established subsequently (in the final part of Section 8.7a) are orthogonal with respect to the inner product that weights each distinct rate of injection in proportion to its frequency of occurrence among the 41 values $u_1, u_2, \ldots, u_{41}$.

The nature of the polynomials $r_1(\cdot)$, $r_2(\cdot)$, and $r_3(\cdot)$ is such that when the model is "parameterized" in terms of $\alpha_1, \alpha_2, \alpha_3$, and α_4 rather than in terms of $\beta_1, \beta_2, \beta_3$, and β_4 and an ANOVA is generated from the $M = 4$ null hypotheses $H_j^{(0)} : \alpha_j = 0$ $(j = 1, 2, 3, 4)$ via an additional implementation of the approach described in Part 2, the resultant degrees of freedom and sums of squares are the same as before. And (also as before) the degrees of freedom and sum of squares from the fourth line of the ANOVA table are the numerator degrees of freedom and sum of squares for testing the null hypothesis $H_4^{(0)}$. However, unlike before, the degrees of freedom and sum of squares from the third line of the table are the numerator degrees of freedom and sum of squares for testing $H_3^{(0)}$ and those from the second line of the table are the numerator degrees of freedom and sum of squares for testing $H_2^{(0)}$.

8.6 Regression Splines

Let us consider further the topic of (univariate) regression. Accordingly, suppose that having observed N values $y_1, y_2, \ldots, y_N$ of a variable y and the corresponding values $u_1, u_2, \ldots, u_N$ of an "explanatory" variable u, we wish to make some inferences about how y varies with u. And suppose further that the inferences are to be based on a statistical model in which (for $i = 1, 2, \ldots, N$) $y_i = s(u_i) + e_i$, where $s(\cdot)$ is a function that is known up to the values of some unknown parameters and where $e_1, e_2, \ldots, e_N$ are the realizations of (unobservable) random variables that have expected values of 0—in what follows, the symbols $e_1, e_2, \ldots, e_N$ and $y_1, y_2, \ldots, y_N$ may sometimes (depending on the context) be used to represent the random variables themselves rather than their realizations.

In the simple case of (univariate) linear regression,

$$s(u) = \beta_1 + \beta_2 u,$$

where β_1 and β_2 are unknown (and unconstrained) parameters—in this simple case, the linearity applies to the dependence on u (for "fixed" β_1 and β_2) as well as to the dependence on β_1 and β_2 (for "fixed" u). Among the simplest (and most common) of the choices for $s(\cdot)$ in which the dependence of $s(u)$ on u may be nonlinear is

$$s(u) = \beta_1 + \beta_2 u + \beta_3 u^2 + \cdots + \beta_{d+1} u^d, \tag{6.1}$$

where d is a positive integer and where $\beta_1, \beta_2, \beta_3, \ldots, \beta_{d+1}$ are unknown (and unconstrained) parameters. A choice of the form (6.1) is that of a polynomial of degree d. It is worth noting that the dependence of such a choice on $\beta_1, \beta_2, \beta_3, \ldots, \beta_{d+1}$ is linear even though the dependence on u is (for $d > 1$) nonlinear. It is also worth noting that the usefulness of these relatively simple kinds of choices can be significantly extended by introducing judiciously chosen transformations of u and/or y and by working with the transformed variable(s) rather than the original(s)—e.g., if u and/or y is inherently positive, it may be advantageous to work with $\log(u)$ and/or $\log(y)$.

Instead of limiting the choice of $s(\cdot)$ to polynomials, we can choose from among a larger class of functions that includes a more flexible kind of function known as a spline function. A spline function (of degree d), say $\tilde{s}(\cdot)$, is formed from some number, say K, of scalars $\lambda_1, \lambda_2, \ldots, \lambda_K$ (where $\lambda_1 < \lambda_2 < \cdots < \lambda_K$) that serve to divide the values of the explanatory variable u into $K+1$ intervals and from $K+1$ polynomials $\tilde{s}_1(\cdot), \tilde{s}_2(\cdot), \ldots, \tilde{s}_{K+1}(\cdot)$ (of degree d) by taking $\tilde{s}(u) = \tilde{s}_k(u)$ for values of u in the kth of the $K+1$ intervals. That is,

$$\tilde{s}(u) = \begin{cases} \tilde{s}_1(u), & \text{for } u \le \lambda_1, \\ \tilde{s}_k(u), & \text{for } \lambda_{k-1} < u \le \lambda_k \ (k = 2, 3, \ldots, K), \\ \tilde{s}_{K+1}(u), & \text{for } u > \lambda_K. \end{cases}$$

For $\tilde{s}(\cdot)$ to qualify as a spline function, there is a requirement that $\tilde{s}_1(\cdot), \tilde{s}_2(\cdot), \ldots, \tilde{s}_{K+1}(\cdot)$ satisfy the conditions

$$\tilde{s}_{k+1}(\lambda_k) = \tilde{s}_k(\lambda_k) \qquad (k = 1, 2, \ldots, K) \tag{6.2}$$

and a further requirement that they satisfy the additional conditions

$$\tilde{s}_{k+1}^{(j)}(\lambda_k) = \tilde{s}_k^{(j)}(\lambda_k) \qquad (k = 1, 2, \ldots, K;\ j = 1, 2, \ldots, d-1), \tag{6.3}$$

where (for $k = 1, 2, \ldots, K+1$) $\tilde{s}_k^{(j)}(\cdot)$ denotes the jth derivative of the polynomial $\tilde{s}_k(\cdot)$. The requirement (6.2) insures that the function $\tilde{s}(\cdot)$ is continuous at each of the points $\lambda_1, \lambda_2, \ldots, \lambda_K$. Accordingly, these points are referred to as join points or knots. And the further requirement (6.3) insures that the derivatives of $\tilde{s}(\cdot)$ of orders 1 through $d-1$ are [like $\tilde{s}(\cdot)$ itself] continuous at each of the points $\lambda_1, \lambda_2, \ldots, \lambda_K$, with the consequence that $\tilde{s}(\cdot)$ is a "smoother" function than might otherwise be the case.

By definition, $\tilde{s}_k(u)$ is expressible in the form

$$\tilde{s}_k(u) = \beta_{k1} + \beta_{k2} u + \beta_{k3} u^2 + \cdots + \beta_{k,d+1} u^d,$$

with coefficients $\beta_{k1}, \beta_{k2}, \beta_{k3}, \ldots, \beta_{k,d+1}$ that (for present purposes) are to be regarded as unknown parameters. And for $j = 1, 2, \ldots, d-1$, $\tilde{s}_k^{(j)}(u)$ is expressible as

$$\tilde{s}_k^{(j)}(u) = 1 \cdots j\beta_{k,j+1} + 2 \cdots (j+1)\beta_{k,j+2} u + \cdots + (d-j+1) \cdots d\beta_{k,d+1} u^{d-j}.$$

a. Application of results on constrained linear models

Let us add to our resuts on the spline function $\tilde{s}(\cdot)$ by introducing some relevant matrix notation and by viewing this function in the context of a constrained linear model. Accordingly, for $k = 1, 2, \ldots, K+1$, let $\boldsymbol{\beta}_k = (\beta'_{k1}, \beta'_{k2}, \ldots, \beta'_{k,d+1})'$, so that $\boldsymbol{\beta}_k$ is the $(d+1)$-dimensional column vector whose elements are the coefficients of the polynomial $\tilde{s}_k(\cdot)$. Further, let $\boldsymbol{\beta}$ represent the $(K+1)(d+1)$-dimensional column vector for which

$$\boldsymbol{\beta}' = (\boldsymbol{\beta}'_1, \boldsymbol{\beta}'_2, \ldots, \boldsymbol{\beta}'_{K+1}),$$

and let

$$\mathbf{y} = (y_1, y_2, \ldots, y_N)'.$$

And observe that when $s(\cdot) = \tilde{s}(\cdot)$,

$$\mathrm{E}(y_i) = (\mathbf{x}'_{i1}, \mathbf{x}'_{i2}, \ldots, \mathbf{x}'_{i,K+1})\boldsymbol{\beta},$$

where $\mathbf{x}_{ik} = (1, u_i, u_i^2, \ldots, u_i^d)'$ if u_i is contained in the kth of the $K+1$ intervals

$$(-\infty, \lambda_1], (\lambda_1, \lambda_2], \ldots, (\lambda_{K-1}, \lambda_K], (\lambda_K, \infty)$$

and $\mathbf{x}_{ik}$ equals a $(d+1)$-dimensional null vector otherwise $(i = 1, 2, \ldots, N)$, and hence that

$$\mathrm{E}(\mathbf{y}) = \mathbf{X}\boldsymbol{\beta},$$

where $\mathbf{X}$ is the $N \times (K+1)(d+1)$ matrix with ith row $(\mathbf{x}'_{i1}, \mathbf{x}'_{i2}, \ldots, \mathbf{x}'_{i,K+1})$.

Now, for purposes of reformulating (in matrix notation) the requirement that $\tilde{s}(\cdot)$ and its derivatives of orders 1 through $d-1$ be continuous, let $\mathbf{C}$ represent a $d \times d$ matrix with jj'th element $c_{jj'}$ defined (for $j, j' = 1, 2, \ldots, d$) as follows:

$$c_{jj'} = \begin{cases} 1 \cdots j, & \text{if } j' = j, \\ (j'-j+1) \cdots j' u^{j'-j}, & \text{if } j' > j, \\ 0, & \text{if } j' < j. \end{cases}$$

Further, let $\mathbf{c} = (u, u^2, \ldots, u^d)'$. And let

$$\mathbf{T} = \begin{pmatrix} 1 & \mathbf{c}' \\ \mathbf{0} & \mathbf{C} \end{pmatrix};$$

let $\mathbf{T}_*$ represent the $d \times (d+1)$ submatrix of $\mathbf{T}$ formed by its first d rows; and (observing that $\mathbf{c}$ and $\mathbf{C}$ depend on $\mathbf{u}$) write $\mathbf{T}_*(\mathbf{u})$ for the value of $\mathbf{T}_*$ at any particular value of $\mathbf{u}$. Then, the requirement that $\tilde{s}(\cdot)$ and its derivatives of orders 1 through $d-1$ be continuous can be reexpressed in the form

$$\mathbf{A}\boldsymbol{\beta} = \mathbf{0}, \tag{6.4}$$

where

$$\mathbf{A} = \begin{pmatrix} \mathbf{T}_*(\lambda_1) & -\mathbf{T}_*(\lambda_1) & \mathbf{0} & \mathbf{0} & \ldots & \mathbf{0} & \mathbf{0} & \mathbf{0} \\ \mathbf{0} & \mathbf{T}_*(\lambda_2) & -\mathbf{T}_*(\lambda_2) & \mathbf{0} & \ldots & \mathbf{0} & \mathbf{0} & \mathbf{0} \\ \mathbf{0} & \mathbf{0} & \mathbf{T}_*(\lambda_3) & -\mathbf{T}_*(\lambda_3) & \ldots & \mathbf{0} & \mathbf{0} & \mathbf{0} \\ \vdots & \vdots & & & & \vdots & & \\ \mathbf{0} & \mathbf{0} & \mathbf{0} & \mathbf{0} & \ldots & \mathbf{0} & \mathbf{T}_*(\lambda_K) & -\mathbf{T}_*(\lambda_K) \end{pmatrix}.$$

Note that the matrix $\mathbf{A}$ has Kd rows [and $(K+1)(d+1)$ columns] and that the rows of $\mathbf{A}$ are linearly independent.

Let us continue to regard the regression coefficients $\boldsymbol{\beta}_{kj}$ $(k = 1, 2, \ldots, K+1; j = 1, 2, \ldots, d+1)$ as unknown parameters, and let us assume (as would often be the case in practice) that the objective is to make inferences about the values of $\tilde{s}(u)$ corresponding to the various values of u. And let us focus on the case where the spline function $\tilde{s}(\cdot)$ is of a kind known as a regression spline. Regression splines differ from so-called smoothing splines in that the number of knots is relatively small (typically K is much smaller than N). Additionally, in the case of regression splines, the extent of the smoothing is determined mostly or entirely by the degree d of the constituent polynomials $\tilde{s}_1(\cdot), \tilde{s}_2(\cdot), \ldots, \tilde{s}_{K+1}(\cdot)$ and by the number and placement of the knots $\lambda_1, \lambda_2, \ldots, \lambda_K$, unlike in the case of smoothing splines where there is extensive reliance on penalty functions as a primary means of controlling the extent of the smoothing.

In the simplest case, d, K, and $\lambda_1, \lambda_2, \ldots, \lambda_K$ are known quantities, as would be the case if their values were predetermined or were determined by making use of the values of N and $u_1, u_2, \ldots, u_N$. Typically, the value of d is predetermined; often, the value of d is taken to be 3. In some cases, the value of K and/or the values of $\lambda_1, \lambda_2, \ldots, \lambda_K$ may likewise be predetermined. Suppose, for example, that u is a measure of time and y a measure of financial activity. Then, conceivably, the knots could be taken to be those points in time that mark significant changes in fiscal policy including ones that may have an effect on consumer interest rates. There are also cases where decisions as to the number and placement of the knots is based in whole or in part on the values of N and $u_1, u_2, \ldots, u_N$. In particular, the knots might be placed so that successive ones of the quantities $\min_i u_i, \lambda_1, \lambda_2, \ldots, \lambda_K, \max_i u_i$ form intervals of prescribed lengths or intervals that contain prescribed numbers of u_i's.

Let $\mathbf{e} = (e_1, e_2, \ldots, e_N)'$, and continue to take $\mathbf{y} = (y_1, y_2, \ldots, y_N)'$. Then, when $s(\cdot) = \tilde{s}(\cdot)$,

$$\mathbf{y} = \mathbf{X}\boldsymbol{\beta} + \mathbf{e}, \tag{6.5}$$

where $\mathbf{X}$ and $\boldsymbol{\beta}$ are as defined earlier in the present subsection and where $\boldsymbol{\beta}$ is subject to constraint (6.4). Moreover, when d, K, and $\lambda_1, \lambda_2, \ldots, \lambda_K$ are known quantities, the elements of $\mathbf{X}$ and the elements of $\mathbf{A}$ [the "coefficient matrix" in constraint (6.4)] are known. Thus, when $s(\cdot) = \tilde{s}(\cdot)$ and when d, K, and $\lambda_1, \lambda_2, \ldots, \lambda_K$ are known quantities, the model for $y_1, y_2, \ldots, y_N$ can (depending on what assumptions are made about the variances and covariances of $e_1, e_2, \ldots, e_N$) be formulated as a constrained G-M, Aitken, or general linear model, and the inferences for the values of $\tilde{s}(u)$ corresponding to the various values of u can be made accordingly. Even when d, K, or $\lambda_1, \lambda_2, \ldots, \lambda_K$ are not known quantities, results on constrained G-M, Aitken, or general linear models could prove useful by serving as a means for reducing the "dimension" of the problem of making inferences for the values of $\tilde{s}(u)$.

b. Cubic splines and natural splines (with an illustration)

Let us specialize the results of Subsection a to the case where $d = 3$, that is, to cubic splines. Cubic splines have been extensively adopted for use in a wide variety of applications.

In the special case where $d = 3$,

$$\tilde{s}(u) = \beta_{k1} + \beta_{k2}u + \beta_{k3}u^2 + \beta_{k4}u^3 = (1, u, u^2, u^3)'\boldsymbol{\beta}_k,$$

with $\boldsymbol{\beta}_k = (\beta_{k1}, \beta_{k2}, \beta_{k3}, \beta_{k4})'$. And the model matrix $\mathbf{X}$ is of dimensions $N \times 4(K+1)$ and the vector $\boldsymbol{\beta}$ of dimensions $4(K+1) \times 1$. Further,

$$\mathbf{T} = \begin{pmatrix} 1 & u & u^2 & u^3 \\ 0 & 1 & 2u & 3u^2 \\ 0 & 0 & 2 & 6u \\ 0 & 0 & 0 & 6 \end{pmatrix},$$

and the coefficient matrix $\mathbf{A}$ of constraint (6.4) is of dimensions $3K \times 4(K+1)$.

Estimators obtained on the basis of the constrained linear model for values of $\tilde{s}(u)$ corresponding to values of u smaller than λ_1 or larger than λ_K tend to be less well-supported by the available data and consequently can be quite variable. That issue is addressed in a variant of the cubic spline called a natural spline. The natural spline is the variant obtained by imposing the additional constraints that the second and third derivatives of $\tilde{s}(u)$ at $u = \lambda_1$ and at $u = \lambda_K$ equal 0 [so that for values of u smaller than λ_1 and for those larger than λ_K, the dependence of $\tilde{s}(u)$ on u is linear]. These (four) additional constraints can be restated in more explicit form as follows:

$$\begin{pmatrix} 0 & 0 & 2 & 6\lambda_1 \\ 0 & 0 & 0 & 6 \end{pmatrix}\boldsymbol{\beta}_1 = \mathbf{0} \quad \text{and} \quad \begin{pmatrix} 0 & 0 & 2 & 6\lambda_K \\ 0 & 0 & 0 & 6 \end{pmatrix}\boldsymbol{\beta}_{K+1} = \mathbf{0}. \tag{6.6}$$

Let (for any value of u) $\mathbf{S}(u)$ represent the 2×4 matrix defined as follows:

$$\mathbf{S}(u) = \begin{pmatrix} 0 & 0 & 2 & 6u \\ 0 & 0 & 0 & 6 \end{pmatrix}.$$

Then, when (in the special case where $d = 3$) the additional constraints are combined with the constraints represented by equality (6.4),we obtain a set of constraints that is expressible in the form of the equality

$$\mathbf{A}_{+}\boldsymbol{\beta} = \mathbf{0}, \tag{6.7}$$

where

$$\mathbf{A}_{+} = \begin{pmatrix} \mathbf{T}_{*}(\boldsymbol{\lambda}_1) & -\mathbf{T}_{*}(\boldsymbol{\lambda}_1) & \mathbf{0} & \mathbf{0} & \dots & \mathbf{0} & \mathbf{0} & \mathbf{0} \\ \mathbf{S}(\lambda_1) & \mathbf{0} & \mathbf{0} & \mathbf{0} & \dots & \mathbf{0} & \mathbf{0} & \mathbf{0} \\ \mathbf{0} & \mathbf{T}_{*}(\boldsymbol{\lambda}_2) & -\mathbf{T}_{*}(\boldsymbol{\lambda}_2) & \mathbf{0} & \dots & \mathbf{0} & \mathbf{0} & \mathbf{0} \\ \mathbf{0} & \mathbf{0} & \mathbf{T}_{*}(\boldsymbol{\lambda}_3) & -\mathbf{T}_{*}(\boldsymbol{\lambda}_3) & \dots & \mathbf{0} & \mathbf{0} & \mathbf{0} \\ \vdots & \vdots & & & & \vdots & & \\ \mathbf{0} & \mathbf{0} & \mathbf{0} & \mathbf{0} & \dots & \mathbf{0} & \mathbf{T}_{*}(\boldsymbol{\lambda}_K) & -\mathbf{T}_{*}(\boldsymbol{\lambda}_K) \\ \mathbf{0} & \mathbf{0} & \mathbf{0} & \mathbf{0} & \dots & \mathbf{0} & \mathbf{0} & \mathbf{S}(\lambda_K) \end{pmatrix}.$$

The matrix $\mathbf{A}_{+}$ has $3K+4$ rows, whereas the matrix $\mathbf{A}$ has (in the special case where $d=3$) $3K$ rows. And the rows of $\mathbf{A}_{+}$ (like those of $\mathbf{A}$) are linearly independent, as can be readily verified.

When the residual effects $e_1, e_2, \dots, e_N$ are taken to be uncorrelated with a common (unknown) variance, the vector $\mathbf{y} = (y_1, y_2, \dots, y_N)'$ follows a constrained G-M model [with constraint (6.4) in the case of a cubic spline and constraint (6.7) in the case of a natural spline]. And inferences for the values of $\tilde{s}(u)$ corresponding to the various values of u can then be made by applying the procedures set forth in Section 8.1 for making inferences about estimable linear combinations of the elements of the vector $\boldsymbol{\beta}$ under such a model.

Recall (from the results of Section 8.1) that under the constrained G-M model, the best (minimum-variance) linear unbiased estimator of an estimable linear combination $\boldsymbol{\ell}'\boldsymbol{\beta}$ is the constrained least squares estimator. Recall further that the constrained least squares estimator is $\boldsymbol{\ell}'\mathbf{b}$, where (when the constraint is $\mathbf{A}\boldsymbol{\beta} = \mathbf{0}$) b is the first (P-dimensional) part of a solution to the linear system

$$\begin{pmatrix} \mathbf{X}'\mathbf{X} & \mathbf{A}' \\ \mathbf{A} & \mathbf{0} \end{pmatrix} \begin{pmatrix} \mathbf{b} \\ \mathbf{r} \end{pmatrix} = \begin{pmatrix} \mathbf{X}'\mathbf{y} \\ \mathbf{0} \end{pmatrix}. \tag{6.8}$$

And in addition recall that $\text{var}(\boldsymbol{\ell}'\mathbf{b}) = \boldsymbol{\ell}'\mathbf{G}_{11}\boldsymbol{\ell}$, where $\mathbf{G}_{11}$ is the $P \times P$ submatrix formed by the first P rows and first P columns of a generalized inverse $\mathbf{G}$ of the coefficient matrix $\begin{pmatrix} \mathbf{X}'\mathbf{X} & \mathbf{A}' \\ \mathbf{A} & \mathbf{0} \end{pmatrix}$ of linear system (6.8). As is evident from Corollary 8.1.2, the rank of the coefficient matrix of linear system (6.8) is expressibe as

$$\text{rank}\begin{pmatrix} \mathbf{X}'\mathbf{X} & \mathbf{A}' \\ \mathbf{A} & \mathbf{0} \end{pmatrix} = \text{rank}\begin{pmatrix} \mathbf{X} \\ \mathbf{A} \end{pmatrix} + \text{rank}(\mathbf{A}').$$

In the present application (i.e., the application to cubic splines and natural splines), the rows of $\mathbf{A}$ and hence the columns of $\mathbf{A}'$ are linearly independent. Thus, in the present application, the matrix $\begin{pmatrix} \mathbf{X}'\mathbf{X} & \mathbf{A}' \\ \mathbf{A} & \mathbf{0} \end{pmatrix}$ is nonsingular if (and only if) the matrix $\begin{pmatrix} \mathbf{X} \\ \mathbf{A} \end{pmatrix}$ is of full column rank, in which case the value of $\tilde{s}(u)$ corresponding to any particular value of u is estimable (as is any other linear combination of the elements of $\boldsymbol{\beta}$). Clearly, a sufficient (but not a necessary) condition for $\begin{pmatrix} \mathbf{X} \\ \mathbf{A} \end{pmatrix}$ to be of full column rank is for $\mathbf{X}$ to be of full column rank, as would be the case if each of the $K+1$ intervals formed by the K knots includes four or more distinct values of u that are among those values represented in the set $\{u_1, u_2, \dots, u_N\}$—refer to Section 5.3d. Note that when the matrix $\begin{pmatrix} \mathbf{X}'\mathbf{X} & \mathbf{A}' \\ \mathbf{A} & \mathbf{0} \end{pmatrix}$ is nonsingular, linear system (6.8) has a unique solution, and the generalized inverse $\mathbf{G}$ is an ordinary inverse.

For purposes of illustration, let us make use of the data presented in Table 4.7. These are data on 101 trees that include a measurement of the height and a measurement of the diameter of each tree (and also include information about the relative locations of the trees). They can be of use in quantifying the extent to which the height of a tree tends to vary with its diameter.

For present purposes, let us suppose that the information on the relative locations of the trees is missing or is simply not to be taken into account. Further, take y to be the log of the height of a tree and u the log of the diameter, take $N = 101$, and take $y_1, y_2, \ldots, y_{101}$ to be the values of y and $u_1, u_2, \ldots, u_{101}$ to be the values of u determined from the 101 heights and diameters in Table 4.7. And regard the problem at hand as one of making inferences for the values of $s(u)$ corresponding to the various values of u based on the model $y_i = s(u_i){+}e_i$ $(i = 1, 2, \ldots, 101)$ (where $e_1, e_2, \ldots, e_{101}$ are uncorrelated with common variance).

Consider first the case where $s(\cdot)$ is the polynomial (6.1) of degree d with coefficients $\beta_1, \beta_2, \ldots, \beta_{d+1}$ and hence where (for any particular value of u) $s(u)$ is a linear combination of $\beta_1, \beta_2, \ldots, \beta_{d+1}$. And assume that this linear combination is estimable (as would be the case if d is smaller than the number of distinct values of u represented among the 101 observed values $u_1, u_2, \ldots, u_{101}$). Then, the ordinary least squares estimator of $s(u)$ is a linear unbiased estimator and in fact (under the assumption that $e_1, e_2, \ldots, e_{101}$ are uncorrelated with common variance) is the best (minimum-variance) linear unbiased estimator. The ordinary least squares estimates of $s(u)$ corresponding to the various values of u were determined for each of the three cases $d = 1$, $d = 2$, and $d = 3$, and the results are displayed in Figure 8.1 (along with the 101 observed values of y).

Alternatively, suppose that instead of taking $s(\cdot)$ to be a polynomial, it is taken to be a cubic spline or a natural spline. Suppose further that the number and placement of the knots is as follows: $K = 5$ and $(\lambda_1, \lambda_2, \lambda_3, \lambda_4, \lambda_5) = (2.37, 2.88, 3.60, 3.93, 4.38)$—these choices are more-or-less in line with the recommendations of Harrell (2015, sec. 2.4.6), And suppose that the inferences for the values of $s(u)$ corresponding to the various values of u are to be made by applying the procedures described earlier in the present subsection [where the values of $s(u)$ are regarded as linear combinations of β_{kj} $(k = 1, 2, \ldots, K;\ j = 1, 2, 3, 4)$ and where β_{kj} $(k = 1, 2, \ldots, K;\ j = 1, 2, 3, 4)$ constitute the elements of the vector $\boldsymbol{\beta}$ that is subject to the constraints]. Moreover, the number and placement of the knots is such that (in the present application) every linear combination of the β_{kj}'s is estimable. Ordinary least squares estimates for the values of $s(u)$ corresponding to the various values of u were obtained from the tree-height data for the case where $s(\cdot)$ is a cubic spline and also for the case where $s(\cdot)$ is a natural spline—under an assumption that $e_1, e_2, \ldots, e_N$ are uncorrelated with a common variance, the ordinary least squares estimators are the minimum-variance linear unbiased estimators— and the results are depicted graphically (along with the data points) in Figure 8.2.

8.7 Reparameterization

Suppose that $\mathbf{y}$ is an $N \times 1$ observable random vector that follows a G–M model, an Aitken model, or a general linear model. As discussed in Section 4.1, each of these models serves to define a set of $N \times 1$ vectors that represents the possible values of $\mathrm{E}(\mathbf{y})$ and a set of $N \times N$ symmetric nonnegative definite matrices that represents the possible values of $\mathrm{var}(\mathbf{y})$—it is implicit in these models that the set of possible pairs of $\mathrm{E}(\mathbf{y})$-values and $\mathrm{var}(\mathbf{y})$-values is the Cartesian product of these two sets. Let us write $\boldsymbol{\mu}$ for $\mathrm{E}(\mathbf{y})$ and recall in particular that (for each of the 3 models) the set of possible values of $\boldsymbol{\mu}$ is the set

$$S_1 = \{\boldsymbol{\mu} : \boldsymbol{\mu} = \mathbf{X}\boldsymbol{\beta},\ \boldsymbol{\beta} \in \mathbb{R}^P\} = \mathcal{C}(\mathbf{X}).$$

Now, consider an alternative version of the model, say

$$\mathbf{y} = \mathbf{W}\boldsymbol{\tau} + \mathbf{e},$$

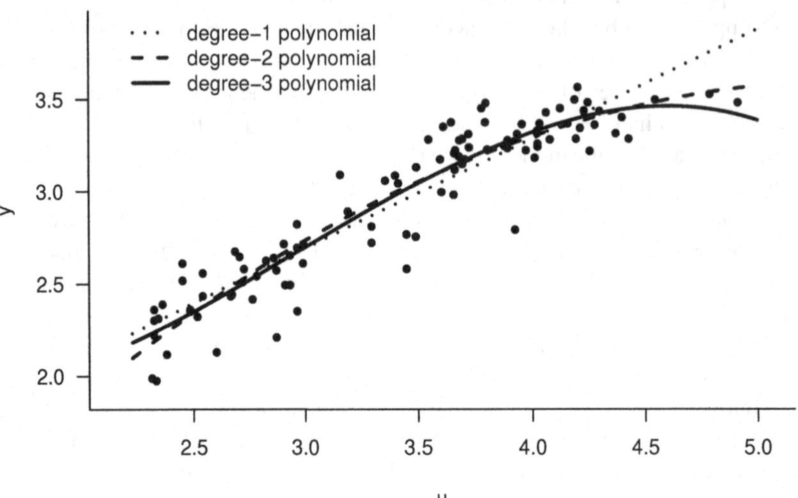

FIGURE 8.1. Observed combinations of values of y = log(tree height) and u = log(tree diameter) together with the least squares estimates of the values of $s(u)$ when $s(\cdot)$ is a polynomial of degree 1, 2, or 3.

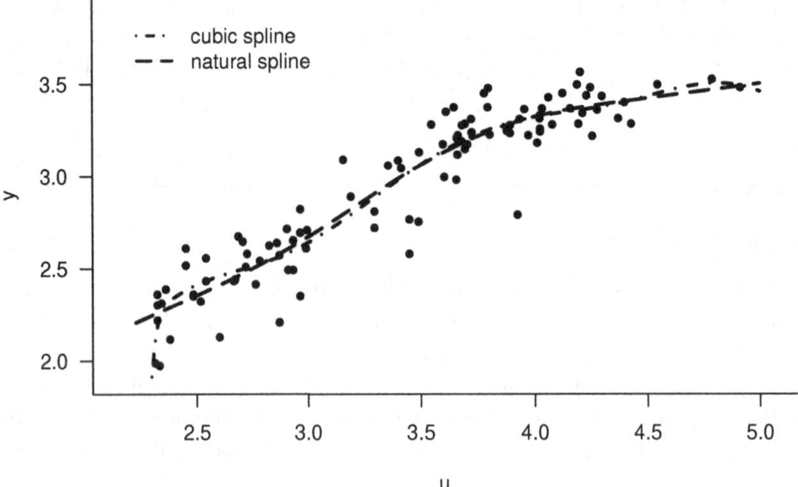

FIGURE 8.2. Observed combinations of values of y = log(tree height) and u = log(tree diameter) together with the least squares estimates of the values of $s(u)$ when $s(\cdot)$ is a cubic spline or a natural spline.

where the model matrix is an $N \times R$ matrix $\mathbf{W}$ (that may differ from the $N \times P$ matrix $\mathbf{X}$) and where the role played by the parameter vector $\boldsymbol{\beta}$ in the original version of the model is played instead by a parametric (column) vector $\boldsymbol{\tau}$ ($\in \mathbb{R}^R$). Under the alternative version of the model, the set of possible values of $\boldsymbol{\mu}$ is the set

$$S_2 = \{\boldsymbol{\mu} : \boldsymbol{\mu} = \mathbf{W}\boldsymbol{\tau},\ \boldsymbol{\tau} \in \mathbb{R}^R\} = \mathcal{C}(\mathbf{W}).$$

It is clear that (at least for purposes of assessing the set of possible values of $\boldsymbol{\mu}$) the alternative version of the model is equivalent to the original if and only if

$$\mathcal{C}(\mathbf{W}) = \mathcal{C}(\mathbf{X}). \tag{7.1}$$

When condition (7.1) is satisfied, the two versions of the model can be regarded as different "parameterizations" of what is essentially the same model. And in that circumstance, it is customary to refer to the alternative version of the model as a reparameterization of the original version. It is worth noting however that the roles of the two versions are essentially interchangeable; when condition (7.1) is satisfied, each of the two versions can be regarded as a reparameterization of the other.

While two parameterizations of what is essentially the same model may (in a sense) be equivalent, there may be reasons for favoring one over the other. In particular, the model matrix of one of them may have orthogonal columns and/or may be of full column rank. Or the parameters of one of them may be more readily amenable to interpretation and "easier to work with."

On occasion there is a need for results on reparameterization that are sufficiently general to cover situations where the original and/or alternative versions of the model are subject to constraints. Sometimes the constraints are an integral part of the underlying model as in the case of regression splines—regression splines were the subject of Section 8.6. Even in cases where the underlying model is an unconstrained model, the results obtained upon carrying out an analysis or analyses based on one or more constrained versions of the model may be of interest and/or relevance. In particular, as discussed in Sections 8.3 and 8.5, such results may be of use in performing an F test of a null hypothesis or in the formation of an ANOVA. Or as discussed in Section 8.2, constraints that are jointly nonestimable may be imposed on an unconstrained model as a matter of "convenience."

The model equation for the constrained (G–M, Aitken, or general linear) model is the same as for the unconstrained model. However, in the case of the constrained model, the value of $\boldsymbol{\beta}$ is confined to those values that satisfy the equality

$$\mathbf{A}\boldsymbol{\beta} = \mathbf{d}, \tag{7.2}$$

whereas in the case of the unconstrained model $\boldsymbol{\beta} \in \mathbb{R}^P$. Analogous to the "original" version of the constrained model (with model equation $\mathbf{y} = \mathbf{X}\boldsymbol{\beta} + \mathbf{e}$ and constraint $\mathbf{A}\boldsymbol{\beta} = \mathbf{d}$) is an "alternative" version in which the model equation is $\mathbf{y} = \mathbf{W}\boldsymbol{\tau} + \mathbf{e}$ and in which the constraint is

$$\mathbf{L}\boldsymbol{\tau} = \mathbf{k}, \tag{7.3}$$

where (for some nonnegative integer S) $\mathbf{L}$ is an $S \times R$ matrix and $\mathbf{k}$ an $S \times 1$ vector of (known) constants [with $\mathbf{k} \in \mathcal{C}(\mathbf{L})$].

Under the original version of the constrained model, the set of possible values of $\boldsymbol{\mu}$ is the set

$$S_1^+ = \{\boldsymbol{\mu} : \boldsymbol{\mu} = \mathbf{X}\boldsymbol{\beta},\ \mathbf{A}\boldsymbol{\beta} = \mathbf{d}\};$$

and under the alternative version of the constrained model, the set of possible values of $\boldsymbol{\mu}$ is the set

$$S_2^+ = \{\boldsymbol{\mu} : \boldsymbol{\mu} = \mathbf{W}\boldsymbol{\tau},\ \mathbf{L}\boldsymbol{\tau} = \mathbf{k}\}.$$

When $S_2^+ = S_1^+$, the two versions of the constrained model can be regarded as different parameterizations of what is essentially the same model. And (when $S_2^+ = S_1^+$) we can refer to the second version as a reparameterization of the first, and vice versa.

The two sets S_1^+ and S_2^+ can be reexpressed in a form that is helpful in ascertaining the conditions under which they are equal to each other and hence in obtaining a generalization of condition (7.1) that is applicable to constrained models. Upon letting $\mathbf{b}$ represent a value of $\boldsymbol{\beta}$ for which $\mathbf{A}\boldsymbol{\beta} = \mathbf{d}$, taking $\mathbf{B}$ to be a matrix whose columns span $\mathcal{N}(\mathbf{A})$, denoting by I the number of columns in $\mathbf{B}$, and observing (in light of Theorem 2.11.6) that $\mathbf{A}\boldsymbol{\beta} = \mathbf{d}$ if and only if $\boldsymbol{\beta} = \mathbf{b}+\mathbf{B}\mathbf{r}$ for some column vector $\mathbf{r}$, we find that

$$S_1^+ = \{\boldsymbol{\mu} : \boldsymbol{\mu} = \mathbf{X}\mathbf{b} + \mathbf{X}\mathbf{B}\mathbf{r},\ \mathbf{r} \in \mathbb{R}^I\}. \tag{7.4}$$

Similarly,

$$S_2^+ = \{\boldsymbol{\mu} : \boldsymbol{\mu} = \mathbf{W}\mathbf{t} + \mathbf{W}\mathbf{T}\mathbf{s},\ \mathbf{s} \in \mathbb{R}^J\}, \tag{7.5}$$

where $\mathbf{t}$ is a value of $\boldsymbol{\tau}$ for which $\mathbf{L}\boldsymbol{\tau} = \mathbf{k}$, where $\mathbf{T}$ is a matrix whose columns span $\mathcal{N}(\mathbf{L})$, and where J represents the number of columns in $\mathbf{T}$.

Upon expressing S_1^+ in the form (7.4) and observing that S_2^+ is expressible in the form (7.5), it is evident that [for any particular choice of the vector $\mathbf{b}$ for which $\mathbf{A}\mathbf{b} = \mathbf{d}$ and any particular choice of the matrix $\mathbf{B}$ whose columns span $\mathcal{N}(\mathbf{A})$] $S_2^+ = S_1^+$ if

$$\mathbf{W}\mathbf{t} = \mathbf{X}\mathbf{b} \text{ for some vector } \mathbf{t} \text{ for which } \mathbf{L}\mathbf{t} = \mathbf{k} \tag{7.6}$$

and

$$\mathcal{C}(\mathbf{W}\mathbf{T}) = \mathcal{C}(\mathbf{X}\mathbf{B}) \text{ for some matrix } \mathbf{T} \text{ whose columns span } \mathcal{N}(\mathbf{L}). \tag{7.7}$$

Conversely, if $S_2^+ = S_1^+$, then conditions (7.6) and (7.7) are satisfied—that condition (7.6) is satisfied follows from the very definition of S_1^+ and S_2^+.

In connection with condition (7.7), note that if $\mathbf{B}_1$ and $\mathbf{B}_2$ are any two choices for the matrix $\mathbf{B}$ [whose columns span $\mathcal{N}(\mathbf{A})$], then $\mathcal{C}(\mathbf{X}\mathbf{B}_1) = \mathcal{C}(\mathbf{X}\mathbf{B}_2)$ (as can be readily verified). And similarly if $\mathbf{T}_1$ and $\mathbf{T}_2$ are any two choices for the matrix $\mathbf{T}$ [whose columns span $\mathcal{N}(\mathbf{L})$], then $\mathcal{C}(\mathbf{W}\mathbf{T}_1) = \mathcal{C}(\mathbf{W}\mathbf{T}_2)$. Among the choices for $\mathbf{B}$ is the $P \times P$ matrix $\mathbf{I} - \mathbf{A}^-\mathbf{A}$, and among the choices for $\mathbf{T}$ is the $R \times R$ matrix $\mathbf{I} - \mathbf{L}^-\mathbf{L}$—that the columns of $\mathbf{I} - \mathbf{A}^-\mathbf{A}$ span $\mathcal{N}(\mathbf{A})$ and the columns of $\mathbf{I} - \mathbf{L}^-\mathbf{L}$ span $\mathcal{N}(\mathbf{L})$ is evident from Corollary 2.11.4.

In regard to condition (7.6), one way of determining whether or not $\mathbf{W}\mathbf{t} = \mathbf{X}\mathbf{b}$ for some vector $\mathbf{t}$ for which $\mathbf{L}\mathbf{t} = \mathbf{k}$ (and of finding such a vector when one exists) is by finding (for values of $\mathbf{t}$ that satisfy the constraint $\mathbf{L}\mathbf{t} = \mathbf{k}$) a value that minimizes the sum of squares $(\mathbf{X}\mathbf{b} - \mathbf{W}\mathbf{t})'(\mathbf{X}\mathbf{b} - \mathbf{W}\mathbf{t})$, that is, by solving a constrained least squares problem in which the role of the data vector is played by the vector $\mathbf{X}\mathbf{b}$ and the role of the model matrix by the matrix $\mathbf{W}$. The constrained normal equations for this constrained least squares problem are

$$\begin{pmatrix} \mathbf{W}'\mathbf{W} & \mathbf{L}' \\ \mathbf{L} & \mathbf{0} \end{pmatrix} \begin{pmatrix} \mathbf{t} \\ \mathbf{r} \end{pmatrix} = \begin{pmatrix} \mathbf{W}'\mathbf{X}\mathbf{b} \\ \mathbf{k} \end{pmatrix}. \tag{7.8}$$

If the residual sum of squares equals 0, condition (7.6) can be satisfied by taking $\mathbf{t}$ to be the first (R-dimensional) part of any solution to the constrained normal equations (7.8); if the residual sum of squares is greater than 0, then a vector $\mathbf{t}$ for which condition (7.6) is satisfied does not exist.

Some simplification of the conditions under which $S_2^+ = S_1^+$ is possible in the special case where

$$\mathbf{d} = \mathbf{0} \text{ or, more generally, } \mathbf{X}\mathbf{b} = \mathbf{0} \text{ for some vector } \mathbf{b} \text{ for which } \mathbf{A}\mathbf{b} = \mathbf{d} \tag{7.9}$$

and where

$$\mathbf{k} = \mathbf{0} \text{ or, more generally, } \mathbf{W}\mathbf{t} = \mathbf{0} \text{ for some vector } \mathbf{t} \text{ for which } \mathbf{L}\mathbf{t} = \mathbf{k}. \tag{7.10}$$

If the setting is one where conditions (7.9) and (7.10) are satisfied, then (assuming that the values of $\mathbf{b}$ and $\mathbf{t}$ are taken to be ones for which $\mathbf{W}\mathbf{t} = \mathbf{X}\mathbf{b} = \mathbf{0}$) condition (7.7) is sufficient as well as necessary. That is, under those conditions (and that assumption),

$$\mathcal{C}(\mathbf{W}\mathbf{T}) = \mathcal{C}(\mathbf{X}\mathbf{B}) \text{ for some matrix } \mathbf{T} \text{ whose columns span } \mathcal{N}(\mathbf{L}) \;\Leftrightarrow\; S_2^+ = S_1^+.$$

Under what conditions is $\mathcal{C}(\mathbf{X}\mathbf{B}) = \mathcal{C}(\mathbf{X})$? And under what conditions is $\mathcal{C}(\mathbf{W}\mathbf{T}) = \mathcal{C}(\mathbf{W})$? Answers to these questions can be obtained by making use of the following result (on the rank of the product of 2 matrices), a derivation of which is obtainable for example from Harville (1997) where this result is among those that form Theorem 17.5.4.

Theorem 8.7.1. Let $\mathbf{A}$ represent an $M \times N$ matrix and $\mathbf{B}$ an $N \times P$ matrix. Then,

$$\operatorname{rank}(\mathbf{A}\mathbf{B}) = \operatorname{rank}(\mathbf{B}) - \dim[\mathcal{C}(\mathbf{B}) \cap \mathcal{N}(\mathbf{A})] \tag{7.11}$$

$$= \operatorname{rank}(\mathbf{A}) - \dim[\mathcal{C}(\mathbf{A}') \cap \mathcal{N}(\mathbf{B}')]. \tag{7.12}$$

To determine the conditions under which $\mathcal{C}(\mathbf{XB}) = \mathcal{C}(\mathbf{X})$ [where $\mathbf{B}$ is a $P \times I$ matrix that spans $\mathcal{N}(\mathbf{A})$ and $\mathbf{A}$ is the coefficient matrix in the constraint $\mathbf{A}\boldsymbol{\beta} = \mathbf{d}$], it suffices (in light of Corollary 2.4.17) to observe that $\text{rank}(\mathbf{XB}) \le \text{rank}(\mathbf{X})$ and to determine the conditions under which equality holds in this inequality. That process is facilitated by observing that

$$\text{rank}(\mathbf{XB}) = \text{rank}[(\mathbf{XB})'] = \text{rank}(\mathbf{B}'\mathbf{X}') \tag{7.13}$$

and (upon applying Theorem 8.7.1) that

$$\text{rank}(\mathbf{B}'\mathbf{X}') = \text{rank}(\mathbf{X}') - \dim[\mathcal{C}(\mathbf{X}') \cap \mathcal{N}(\mathbf{B}')]. \tag{7.14}$$

Moreover, $\mathcal{N}(\mathbf{B}') = \mathcal{C}(\mathbf{A}')$ (as can be readily verified). Thus, upon combining results (7.13) and (7.14) (and recalling Lemma 2.4.6), we find that

$$\begin{aligned} \text{rank}(\mathbf{XB}) &= \text{rank}(\mathbf{X}') - \dim[\mathcal{C}(\mathbf{X}') \cap \mathcal{C}(\mathbf{A}')] \\ &= \text{rank}(\mathbf{X}) - \dim[\mathcal{R}(\mathbf{X}) \cap \mathcal{R}(\mathbf{A})], \end{aligned} \tag{7.15}$$

leading to the conclusion that

$$\begin{aligned} \mathcal{C}(\mathbf{XB}) = \mathcal{C}(\mathbf{X}) &\Leftrightarrow \text{rank}(\mathbf{XB}) = \text{rank}(\mathbf{X}) \\ &\Leftrightarrow \dim[\mathcal{R}(\mathbf{X}) \cap \mathcal{R}(\mathbf{A})] = 0 \\ &\Leftrightarrow \mathcal{R}(\mathbf{X}) \text{ and } \mathcal{R}(\mathbf{A}) \text{ are essentially disjoint.} \end{aligned} \tag{7.16}$$

In similar fashion, it can be shown that

$$\text{rank}(\mathbf{WT}) = \text{rank}(\mathbf{W}) - \dim[\mathcal{R}(\mathbf{W}) \cap \mathcal{R}(\mathbf{L})]. \tag{7.17}$$

And in essentially the same way that result (7.15) leads to conclusion (7.16), result (7.17) leads to the conclusion that

$$\mathcal{C}(\mathbf{WT}) = \mathcal{C}(\mathbf{W}) \Leftrightarrow \mathcal{R}(\mathbf{W}) \text{ and } \mathcal{R}(\mathbf{L}) \text{ are essentially disjoint.} \tag{7.18}$$

Results (7.16) and (7.18) can be restated and reinterpreted in terms of the concept of joint nonestimability, which was introduced and discussed in Section 8.2. Specifically, when $\mathcal{R}(\mathbf{X})$ and $\mathcal{R}(\mathbf{A})$ are essentially disjoint, the elements of the vector $\mathbf{A}\boldsymbol{\beta}$ are jointly nonestimable under the unconstrained model (with model equation $\mathbf{y} = \mathbf{X}\boldsymbol{\beta} + \mathbf{e}$). Similarly, when $\mathcal{R}(\mathbf{W})$ and $\mathcal{R}(\mathbf{L})$ are essentially disjoint, the elements of the vector $\mathbf{L}\boldsymbol{\tau}$ are jointly nonestimable under the unconstrained model (with model equation $\mathbf{y} = \mathbf{W}\boldsymbol{\tau} + \mathbf{e}$).

In what follows, a number of results pertaining to reparameterization are presented and discussed. The initial coverage (in Subsection a) is for the case where one unconstrained version of the model is reparameterized by another unconstrained version. The more general situation where one of the two versions of the model may include constraints is considered in Subsection b, and the still more general situation where either or both versions may include constraints is considered in Subsection c.

a. Some results on the reparameterization of one unconstrained G–M model by another unconstrained G–M model.

Let $\mathbf{y}$ represent an $N \times 1$ observable random vector. Further, consider two different models for $\mathbf{y}$: the original G–M model

$$\mathbf{y} = \mathbf{X}\boldsymbol{\beta} + \mathbf{e} \tag{7.19}$$

(with $N \times P$ model matrix $\mathbf{X}$ and in which $\boldsymbol{\beta}$ is a $P \times 1$ unconstrained parameter vector) and an alternative G–M model

$$\mathbf{y} = \mathbf{W}\boldsymbol{\tau} + \mathbf{e} \tag{7.20}$$

(with $N \times R$ model matrix $\mathbf{W}$ and in which the role played by $\boldsymbol{\beta}$ in the original G–M model is assumed by the $R \times 1$ unconstrained parameter vector $\boldsymbol{\tau}$). And recall (from the introductory part of the present section) that when

$$\mathcal{C}(\mathbf{W}) = \mathcal{C}(\mathbf{X}), \tag{7.21}$$

these two models are said to be reparameterizations of each other.

Note that condition (7.21) is satisfied (for any particular $N \times P$ matrix $\mathbf{X}$ and any particular $N \times R$ matrix $\mathbf{W}$) if and only if there exists a $P \times R$ matrix $\mathbf{F}$ and an $R \times P$ matrix $\mathbf{G}$ for which

$$\mathbf{W} = \mathbf{X}\mathbf{F} \quad \text{and} \quad \mathbf{X} = \mathbf{W}\mathbf{G}. \tag{7.22}$$

Note also that if $\mathbf{F}$ and $\mathbf{G}$ satisfy condition (7.22), then they also satisfy the condition

$$\mathbf{X} = \mathbf{X}\mathbf{F}\mathbf{G}. \tag{7.23}$$

Conversely, if (for any particular $N \times P$ matrix $\mathbf{X}$) $\mathbf{F}$ and $\mathbf{G}$ satisfy condition (7.23), then condition (7.22) and hence condition (7.21) can be satisfied by setting $\mathbf{W} = \mathbf{X}\mathbf{F}$.

Among the choices for a matrix $\mathbf{F}$ that satisfies the condition $\mathbf{W} = \mathbf{X}\mathbf{F}$ (when such a matrix exists) is the matrix

$$\mathbf{F} = (\mathbf{X}'\mathbf{X})^{-}\mathbf{X}'\mathbf{W} \tag{7.24}$$

(as can be readily verified upon recalling from Section 2.12 that $\mathbf{P_X X} = \mathbf{X}$). Similarly, among the choices for a matrix $\mathbf{G}$ that satisfies the condition $\mathbf{X} = \mathbf{W}\mathbf{G}$ (when such a matrix exists) is the matrix

$$\mathbf{G} = (\mathbf{W}'\mathbf{W})^{-}\mathbf{W}'\mathbf{X}. \tag{7.25}$$

Under the original model (7.19), the mean vector $\boldsymbol{\mu}$ of $\mathbf{y}$ is expressible in terms of $\boldsymbol{\beta}$ as $\boldsymbol{\mu} = \mathbf{X}\boldsymbol{\beta}$; and as $\boldsymbol{\beta}$ ranges over $\mathbb{R}^P$, $\boldsymbol{\mu}$ ranges over $\mathcal{C}(\mathbf{X})$. Similarly, under the alternative model (7.20), $\boldsymbol{\mu}$ is expressible in terms of $\boldsymbol{\tau}$ as $\boldsymbol{\mu} = \mathbf{W}\boldsymbol{\tau}$; and as $\boldsymbol{\tau}$ ranges over $\mathbb{R}^R$, $\boldsymbol{\mu}$ ranges over $\mathcal{C}(\mathbf{W})$. When the two models are reparameterizations of each other, $\mathcal{C}(\mathbf{W}) = \mathcal{C}(\mathbf{X})$, so that any particular value, say $\underline{\boldsymbol{\mu}}$, of $\boldsymbol{\mu}$ is expressible as $\underline{\boldsymbol{\mu}} = \mathbf{X}\underline{\boldsymbol{\beta}}$ for some value $\underline{\boldsymbol{\beta}}$ of $\boldsymbol{\beta}$ and is also expressible as $\underline{\boldsymbol{\mu}} = \mathbf{W}\underline{\boldsymbol{\tau}}$ for some value $\underline{\boldsymbol{\tau}}$ of $\boldsymbol{\tau}$. And when the two models are reparameterizations of each other, there exist matrices $\mathbf{F}$ and $\mathbf{G}$ that satisfy condition (7.22), and these two matrices can be used to relate either one of the two values $\underline{\boldsymbol{\beta}}$ and $\underline{\boldsymbol{\tau}}$ to the other. Specifically, if $\underline{\boldsymbol{\mu}} = \mathbf{X}\underline{\boldsymbol{\beta}}$, then $\underline{\boldsymbol{\mu}} = \mathbf{W}\underline{\boldsymbol{\tau}}$ for $\underline{\boldsymbol{\tau}} = \mathbf{G}\underline{\boldsymbol{\beta}}$; and if $\underline{\boldsymbol{\mu}} = \mathbf{W}\underline{\boldsymbol{\tau}}$, then $\underline{\boldsymbol{\mu}} = \mathbf{X}\underline{\boldsymbol{\beta}}$ for $\underline{\boldsymbol{\beta}} = \mathbf{F}\underline{\boldsymbol{\tau}}$.

Interchangeability of models that are reparameterizations of each other. Suppose that the original G–M model (with $N \times P$ model matrix $\mathbf{X}$) and the alternative G–M model (with $N \times R$ model matrix $\mathbf{W}$) are reparameterizations of each other. Then, there exists a $P \times R$ matrix $\mathbf{F}$ and an $R \times P$ matrix $\mathbf{G}$ that satisfy condition (7.22). Let $\mathbf{F}$ and $\mathbf{G}$ represent any such matrices, so that $\mathbf{W}$ is expressible as $\mathbf{W} = \mathbf{X}\mathbf{F}$ and $\mathbf{X}$ as $\mathbf{X} = \mathbf{W}\mathbf{G}$. And consider the implications of the relationship between $\mathbf{W}$ and $\mathbf{X}$ implicit in these two expressions when it comes to making inferences about linear combinations of the elements of $\boldsymbol{\beta}$ or linear combinations of the elements of $\boldsymbol{\tau}$.

The normal equations

$$\mathbf{X}'\mathbf{X}\mathbf{b} = \mathbf{X}'\mathbf{y} \tag{7.26}$$

for the original G–M model are related to the normal equations

$$\mathbf{W}'\mathbf{W}\mathbf{t} = \mathbf{W}'\mathbf{y} \tag{7.27}$$

for the alternative G–M model in the sense that a solution to either one of them can be obtained from a solution to the other. Specifically, if $\tilde{\mathbf{t}}$ is a solution to the equations (7.27), then a solution $\tilde{\mathbf{b}}$ to the equations (7.26) can be obtained by taking

$$\tilde{\mathbf{b}} = \mathbf{F}\tilde{\mathbf{t}}, \tag{7.28}$$

as is evident upon observing that

$$\mathbf{X}'\mathbf{X}\mathbf{F}\tilde{\mathbf{t}} = \mathbf{X}'\mathbf{W}\tilde{\mathbf{t}} = (\mathbf{W}\mathbf{G})'\mathbf{W}\tilde{\mathbf{t}} = \mathbf{G}'\mathbf{W}'\mathbf{W}\tilde{\mathbf{t}} = \mathbf{G}'\mathbf{W}'\mathbf{y} = (\mathbf{W}\mathbf{G})'\mathbf{y} = \mathbf{X}'\mathbf{y}.$$

And similarly, if $\tilde{\mathbf{b}}$ is a solution to the equations (7.26), then a solution $\tilde{\mathbf{t}}$ to the equations (7.27) can be obtained by taking

$$\tilde{\mathbf{t}} = \mathbf{G}\tilde{\mathbf{b}}. \tag{7.29}$$

These results suggest that inferences for a vector $\boldsymbol{\Lambda}'\boldsymbol{\beta}$ of linear combinations of the elements of $\boldsymbol{\beta}$ (under the original G–M model) are closely related to those for the vector $\boldsymbol{\Lambda}'\mathbf{F}\boldsymbol{\tau}$ of linear combinations of the elements of $\boldsymbol{\tau}$ (under the alternative G–M model). And similarly they suggest that inferences for a vector $\boldsymbol{\Delta}'\boldsymbol{\tau}$ (under the alternative G–M model) are closely related to those for the vector $\boldsymbol{\Delta}'\mathbf{G}\boldsymbol{\beta}$ (under the original G–M model).

In that regard, we find that if a linear combination $\boldsymbol{\lambda}'\boldsymbol{\beta}$ is estimable (under the original model), then $\lambda'\mathbf{F}\boldsymbol{\tau}$ is estimable (under the alternative model), as can be readily verified by observing that if $\boldsymbol{\lambda}' = \mathbf{a}'\mathbf{X}$ for some vector $\mathbf{a}$, then $\boldsymbol{\lambda}'\mathbf{F} = \mathbf{a}'\mathbf{X}\mathbf{F} = \mathbf{a}'\mathbf{W}$. Moreover, when $\boldsymbol{\lambda}'\boldsymbol{\beta}$ is estimable, the least squares estimator of $\boldsymbol{\lambda}'\boldsymbol{\beta}$ is identical to the least squares estimator of $\boldsymbol{\lambda}'\mathbf{F}\boldsymbol{\tau}$; both are equal to $\boldsymbol{\lambda}'\mathbf{F}\tilde{\mathbf{t}}$, where $\tilde{\mathbf{t}}$ is any solution to the equations (7.27) [as is evident from result (7.28)]. Similarly, if a linear combination $\boldsymbol{\delta}'\boldsymbol{\tau}$ is estimable, then $\boldsymbol{\delta}'\mathbf{G}\boldsymbol{\beta}$ is estimable, and the least squares estimators of $\boldsymbol{\delta}'\boldsymbol{\tau}$ and $\boldsymbol{\delta}'\mathbf{G}\boldsymbol{\beta}$ are identical, with both estimators being equal to $\boldsymbol{\delta}'\mathbf{G}\tilde{\mathbf{b}}$, where $\tilde{\mathbf{b}}$ is any solution to the equations (7.26).

It is worth noting that in general $\boldsymbol{\lambda}'\mathbf{F}\boldsymbol{\tau}$ can be estimable even if $\boldsymbol{\lambda}'\boldsymbol{\beta}$ is nonestimable, in which case $\boldsymbol{\lambda}'\tilde{\mathbf{b}}$ [where $\tilde{\mathbf{b}}$ is a solution to the equations (7.26)] need not equal $\boldsymbol{\lambda}'\mathbf{F}\tilde{\mathbf{t}}$. And similarly $\boldsymbol{\delta}'\mathbf{G}\boldsymbol{\beta}$ can be estimable even if $\boldsymbol{\delta}'\boldsymbol{\tau}$ is nonestimable. However, if $\boldsymbol{\delta}'\boldsymbol{\tau}$ is estimable, then necessarily $\boldsymbol{\delta}' = \boldsymbol{\lambda}'\mathbf{F}$ for some $\boldsymbol{\lambda}$ for which $\boldsymbol{\lambda}'\boldsymbol{\beta}$ is estimable; and similarly if $\boldsymbol{\lambda}'\boldsymbol{\beta}$ is estimable, then necessarily $\boldsymbol{\lambda}' = \boldsymbol{\delta}'\mathbf{G}$ for some $\boldsymbol{\delta}$ for which $\boldsymbol{\delta}'\boldsymbol{\tau}$ is estimable.

Note that $\text{var}(\mathbf{y}) = \sigma^2\mathbf{I}$ under both the original and alternative G–M models. Thus, the variance-covariance matrices of the least squares estimators of $\boldsymbol{\Lambda}'\boldsymbol{\beta}$ and $\boldsymbol{\Lambda}'\mathbf{F}\boldsymbol{\tau}$ or of $\boldsymbol{\Delta}'\boldsymbol{\tau}$ and $\boldsymbol{\Delta}'\mathbf{G}\boldsymbol{\beta}$ (like the estimators themselves) are identical. Note also that the least squares estimator of $\boldsymbol{\mu}$ under the alternative G–M model is identical to the least squares estimator of $\boldsymbol{\mu}$ under the original G–M model [as is evident from result (7.28) or result (7.29)], implying in particular that the residual sum of squares is the same for both models. Accordingly, we are led to the conclusion that the original and alternative G-M models can be used more or less interchangeably.

Full-rank reparameterizations and orthogonal reparameterizations. Suppose that the alternative model is a reparameterization of the original model. Then, necessarily, the model matrix $\mathbf{W}$ of the alternative model is of the same rank as the model matrix $\mathbf{X}$ of the original model. And it follows that the number R of columns of $\mathbf{W}$ is greater than or equal to $\text{rank}(\mathbf{X})$. When $\mathbf{W}$ is of full column rank or equivalently when $R = \text{rank}(\mathbf{X})$, the reparameterization is said to be a *full-rank reparameterization*—that would be the case even if the reparameterization were a reparameterization of an Aitken model or a general linear model rather than of a G–M model.

In the special case where the alternative model is a full-rank reparameterization, the coefficient matrix $\mathbf{W}'\mathbf{W}$ of the normal equations (7.27) for the alternative model is nonsingular (and, in fact, positive definite). In the further special case where $\mathbf{W}'\mathbf{W}$ is not only nonsingular but is diagonal (with diagonal elements that are strictly positive), the alternative model is referred to as an *orthogonal reparameterization*. When the alternative model is a full-rank reparameterization, the individual elements of the vector $\boldsymbol{\tau}$ are estimable. And when the alternative model is an orthogonal reparameterization, the least squares estimators of the individual elements of $\boldsymbol{\tau}$ are uncorrelated with each other.

A full-rank reparameterization always exists. In fact, a model matrix $\mathbf{W}$ for a full-rank reparameterization can be formed from any columns of $\mathbf{X}$ that are linearly independent and equal in number to $\text{rank}(\mathbf{X})$. A set of such columns can sometimes be identified by "inspection," in which case $\mathbf{W}$ can be formed from the columns so identified and matrices $\mathbf{F}$ and $\mathbf{G}$ that satisfy condition (7.22) can be readily determined.

In more opaque situations or in situations for which an orthogonal reparameterization is sought, resort can be made to numerical methods. In particular, such methods can be used to generate a $P\times P$ nonsingular matrix $\mathbf{U}$ for which

$$\mathbf{X}\mathbf{U} = (\mathbf{Z},\ \mathbf{0}) \tag{7.30}$$

for some $N\times(\text{rank}\,\mathbf{X})$ matrix $\mathbf{Z}$. Then, a full-rank reparameterization can be created by setting $\mathbf{W} = \mathbf{Z}$, in which case condition (7.22) can be satisfied by taking $\mathbf{F} = \mathbf{U}_1$ and $\mathbf{G} = \mathbf{C}_1$, where $\mathbf{U} = (\mathbf{U}_1,\ \mathbf{U}_2)$ and $\mathbf{U}^{-1} = \begin{pmatrix}\mathbf{C}_1\\ \mathbf{C}_2\end{pmatrix}$. In fact, the results of Section 5.4e (on the QR decomposition) can be used to generate the matrix $\mathbf{U}$ in such a way that the columns of the matrix $\mathbf{Z}$ are orthonormal, in which case the full-rank reparameterization created by setting $\mathbf{W} = \mathbf{Z}$ is an orthogonal reparameterization (one with $\mathbf{W}'\mathbf{W} = \mathbf{I}$).

As an alternative to generating a $P \times P$ nonsingular matrix $\mathbf{U}$ [for which $\mathbf{XU}$ is of the form (7.30)] by operating directly on the $N \times P$ model matrix $\mathbf{X}$, one can be generated by operating on the $P \times P$ matrix $\mathbf{X'X}$ [which is the coefficient matrix of the normal equations (7.26)]. It suffices to take $\mathbf{U}$ to be such that

$$\mathbf{U'X'XU} = \begin{pmatrix} \mathbf{D} & \mathbf{0} \\ \mathbf{0} & \mathbf{0} \end{pmatrix}, \tag{7.31}$$

where $\mathbf{D}$ is a $(\operatorname{rank}\mathbf{X}) \times (\operatorname{rank}\mathbf{X})$ nonsingular matrix—upon partitioning $\mathbf{U}$ as $\mathbf{U} = (\mathbf{U}_1, \mathbf{U}_2)$ [where $\mathbf{U}_1$ is of dimensions $P \times (\operatorname{rank}\mathbf{X})$], we find that $(\mathbf{XU}_2)'\mathbf{XU}_2 = \mathbf{U}_2'\mathbf{X}'\mathbf{XU}_2 = \mathbf{0}$, implying that $\mathbf{XU}_2 = \mathbf{0}$ and hence that $\mathbf{XU} = (\mathbf{XU}_1, \mathbf{0})$. Moreover, by taking the rows of $\mathbf{U}$ to be the rows of a suitably chosen upper triangular matrix (formed via a process related to that undertaken in forming a U′DU or Cholesky decomposition of a symmetric nonnegative definite matrix) or by taking the columns of $\mathbf{U}$ to be orthogonal eigenvectors of $\mathbf{X'X}$, we can obtain a $\mathbf{U}$ for which the matrix $\mathbf{D}$ of expression (7.31) is diagonal (in which case the full-rank reparameterization created upon setting $\mathbf{W} = \mathbf{XU}_1$ is an orthogonal reparameterization).

There are two opposing points of view concerning the utility of full-rank reparameterization. Inferences for a vector $\boldsymbol{\Lambda}'\boldsymbol{\beta}$ of linear combinations of the elements of $\boldsymbol{\beta}$ under the original (possibly less than full rank) model are more-or-less the same as those for the vector $\boldsymbol{\Lambda}'\mathbf{F}\boldsymbol{\tau}$ under the alternative reparameterized full-rank model. Thus, it can be argued that the inclusion (in the coverage of linear-model methodology) of the possibility that the model matrix may be of less than full column rank represents a needless complication. A counterargument is that the computations required to devise and utilize a full-rank reparameterization are much the same as those encountered in basing the inferences directly on the original (possibly less than full rank) model, leading to the conclusion that the coverage of linear-model methodology might as well be sufficiently general to allow one to deal "directly" with the less-than-full-rank case.

Orthogonal polynomials. Let us consider orthogonal reparameterization in the special case where the expected values of the N elements, say $y_1, y_2, \ldots, y_N$, of the observable random vector $\mathbf{y}$ (under the original model) are of the form

$$\mathrm{E}(y_i) = \delta(u_i) \quad (i = 1, 2, \ldots, N), \tag{7.32}$$

where

$$\delta(u) = \beta_1 + \beta_2 u + \beta_3 u^2 + \cdots + \beta_P u^{P-1}$$

is a polynomial (of degree $P-1$) in a single explanatory variable u and where $u_1, u_2, \ldots, u_N$ are N values of u (that include at least P distinct values). In this special case, the model matrix $\mathbf{X}$ (of the original model $\mathbf{y} = \mathbf{X}\boldsymbol{\beta} + \mathbf{e}$) is that with ith row $(1, u_i, u_i^2, \ldots, u_i^{P-1})$ $(i = 1, 2, \ldots, N)$ and is of full column rank P, and the column vector $\boldsymbol{\beta}$ represents the vector with elements $\beta_1, \beta_2, \ldots, \beta_P$ [consisting of the coefficients of the polynomial $\delta(\cdot)$].

Now, let $\mathbf{Q}$ represent a $P \times P$ nonsingular upper triangular matrix—a triangular matrix is nonsingular if and only if all its diagonal elements are nonzero—for which the columns of $\mathbf{XQ}$ are orthogonal and hence for which

$$(\mathbf{XQ})'\mathbf{XQ} = \mathbf{D} \tag{7.33}$$

for some diagonal matrix $\mathbf{D}$ (with nonzero diagonal elements). The existence of such a matrix was established (and processes for generating such a matrix were discussed) in Part 2 of the present subsection. Further, take the alternative model $\mathbf{y} = \mathbf{W}\boldsymbol{\tau} + \mathbf{e}$ to be that obtained upon setting $\mathbf{W} = \mathbf{XQ}$, in which case the alternative model is an orthogonal reparameterization of the original model.

The expressions (7.32) for the expected values of $y_1, y_2, \ldots, y_N$ under the original model can be converted into expressions for the expected values under the reparameterized model by "reexpressing" the function $\delta(u)$ as follows:

$$\delta(u) = [\mathbf{r}(u)]'\boldsymbol{\tau}, \tag{7.34}$$

where $\mathbf{r}(u)$ is the $P \times 1$ vector of functions of u whose transpose is

$$[\mathbf{r}(u)]' = (1, u, u^2, u^3, ..., u^{P-1})\mathbf{Q}.$$

Note that expression (7.34) can be rewritten in the form

$$\delta(u) = \sum_{j=1}^{P} r_{j-1}(u)\tau_j,$$

where (for $j = 1, 2, \ldots, P$) $r_{j-1}(u)$ is the jth element of $\mathbf{r}(u)$ and τ_j the jth element of $\boldsymbol{\tau}$. Note also that $r_{j-1}(u)$ is a polynomial of degree $j-1$, as becomes evident upon letting $\mathbf{q}_j$ represent the jth column of the upper triangular matrix $\mathbf{Q}$ and observing that $r_{j-1}(u) = (1, u, u^2, u^3, ..., u^{P-1})\mathbf{q}_j$. Moreover, for $i = 1, 2, \ldots, N$, the row vector $[\mathbf{r}(u_i)]' = [r_0(u_i), r_1(u_i), \ldots, r_{P-1}(u_i)]$ is the ith row of the model matrix $\mathbf{W}$ of the reparameterized model.

The polynomials $r_0(\cdot), r_1(\cdot), \ldots, r_{P-1}(\cdot)$ are orthogonal. More specifically, they are orthogonal when the inner product of any two of them, say r_j and $r_{j'}$, is taken to be

$$\sum_{k=1}^{K} w(u_k^*) r_j(u_k^*) r_{j'}(u_k^*), \tag{7.35}$$

where $u_1^*, u_2^*, \ldots, u_K^*$ are the distinct values of u represented among the N values $u_1, u_2, \ldots, u_N$ and where (for $k = 1, 2, \ldots, K$) $w(u_k^*)$ represents the proportion of the N values $u_1, u_2, \ldots, u_N$ that equal u_k^*. To verify that expression (7.35) equals 0 for $j' \neq j = 0, 1, \ldots, P-1$ and hence to verify that the polynomials $r_0(\cdot), r_1(\cdot), \ldots, r_{P-1}(\cdot)$ are orthogonal with respect to the inner product (7.35), it suffices to observe that

$$\begin{aligned}\sum_{k=1}^{K} w(u_k^*)\mathbf{r}(u_k^*)[\mathbf{r}(u_k^*)]' = (1/N)\sum_{i=1}^{N} \mathbf{r}(u_i)[\mathbf{r}(u_i)]' &= (1/N)\mathbf{W}'\mathbf{W}\\ &= (1/N)(\mathbf{XQ})'\mathbf{XQ} = (1/N)\mathbf{D}.\end{aligned}$$

Note that (since $\mathbf{W} = \mathbf{XQ}$ and since $\mathbf{Q}$ is nonsingular) $\mathbf{X} = \mathbf{W}\mathbf{Q}^{-1}$. And let $\boldsymbol{\alpha} = \mathbf{Q}^{-1}\boldsymbol{\beta}$. Then, as is evident from the results of Part 1 of the present subsection, inferences for the vector $\boldsymbol{\alpha}$ (under the original model) are essentially the same as those for the vector $\boldsymbol{\tau}$ (under the alternative model).

As an example of orthogonal reparameterization (one for which the underlying model is a polynomial), consider an application to the ouabain data. Do so by revisiting Section 8.5e, where these data were subjected to an analysis of variance. The results presented therein are suitable for the additional purpose of illustrating orthogonal reparameterization (and can serve also to demonstrate the close relationship between the computational procedures employed in devising a reparameterization and those that might be employed in solving some normal equations). In fact, the results obtained earlier for the ouabain data include an upper triangular matrix $\mathbf{Q}$ (one that is unit upper triangular) and the diagonal matrix $\mathbf{D}$ and expressions for the quantities $r_j(u_i)$ $(j = 1, 2, \ldots, P-1)$ and for the elements $\alpha_1, \alpha_2, \ldots, \alpha_P$ of the vector $\boldsymbol{\alpha}$.

b. Some results on the reparameterization of a constrained G–M model by an unconstrained G–M model.

Let $\mathbf{y}$ represent an $N \times 1$ observable random vector. Further, consider two different models for $\mathbf{y}$: the (original) constrained G–M model

$$\mathbf{y} = \mathbf{X}\boldsymbol{\beta} + \mathbf{e}, \tag{7.36}$$

with $N \times P$ model matrix $\mathbf{X}$ and in which $\boldsymbol{\beta}$ is a $P \times 1$ parameter vector that is subject to the constraint $\mathbf{A}\boldsymbol{\beta} = \mathbf{d}$ [where $\mathbf{d} \in \mathcal{C}(\mathbf{A})$], and an unconstained G–M model

$$\mathbf{y} = \mathbf{W}\boldsymbol{\tau} + \mathbf{e}, \tag{7.37}$$

with $N \times R$ model matrix $\mathbf{W}$ and in which $\boldsymbol{\tau}$ is an $R \times 1$ unconstrained parameter vector.

Note that the second (unconstrained) model can be regarded as a special case of a constrained G–M model with the same model equation $\mathbf{y} = \mathbf{W}\boldsymbol{\tau} + \mathbf{e}$ and same model matrix $\mathbf{W}$ but in which $\boldsymbol{\tau}$ is an $R \times 1$ parameter vector that is subject to the constraint $\mathbf{L}\boldsymbol{\tau} = \mathbf{k}$; it is the special case where $\mathbf{L} = \mathbf{0}$ and $\mathbf{k} = \mathbf{0}$. In that special case, $\mathcal{N}(\mathbf{L})$ is spanned by the columns of $\mathbf{I}_R$, and $\mathbf{Lt} = \mathbf{k}$ for every $\mathbf{t} \in \mathcal{R}^R$. And conditions that are necessary and sufficient for the unconstrained G–M model (7.37) to reparameterize the constrained G–M model (7.36) are provided by conditions (7.6) and (7.7), which in the special case at hand can be restated in the following simplified form:

$$\mathbf{Wt} = \mathbf{Xb} \text{ for some } \mathbf{t} \in \mathcal{R}^R \tag{7.38}$$

(where $\mathbf{b}$ is a vector for which $\mathbf{Ab} = \mathbf{d}$) and

$$\mathcal{C}(\mathbf{W}) = \mathcal{C}(\mathbf{XB}) \tag{7.39}$$

[where $\mathbf{B}$ is a matrix whose columns span $\mathcal{N}(\mathbf{A})$].

Now, suppose that $\mathbf{d} = \mathbf{0}$ or, more generally, that condition (7.9) is satisfied. And take $\mathbf{b}$ to be any vector that satisfies the constraint $\mathbf{Ab} = \mathbf{d}$ and for which $\mathbf{Xb} = \mathbf{0}$—the existence of such a vector is insured by the supposition. Further, take $\mathbf{t} = \mathbf{0}$, in which case $\mathbf{Wt} = \mathbf{0}$ and condition (7.38) is satisfied. Then, for the unconstrained G–M model (7.37) to reparameterize the constrained G–M model (7.36), it is necessary and sufficient that $\mathbf{W}$ satisfy condition (7.39).

Let $\mathbf{U} = \mathbf{XB}$. And observe that condition (7.39) can be satisfied by setting $\mathbf{W} = \mathbf{U}$ and hence that the constrained G–M model (7.36) is reparameterized in particular by the unconstrained G–M model

$$\mathbf{y} = \mathbf{U}\boldsymbol{\alpha} + \mathbf{e}, \tag{7.40}$$

with model matrix $\mathbf{U}$ and in which $\boldsymbol{\alpha}$ is an $I \times 1$ (unconstrained) parameter vector (where I represents the number of columns in the matrix $\mathbf{B}$ and hence in the matrix $\mathbf{U}$). Observe also that for any choice of $\mathbf{W}$ that satisfies condition (7.39), the unconstrained G–M model (7.37) (with model matrix $\mathbf{W}$) is a reparameterization of the unconstrained G–M model (7.40).

How [for an arbitrary choice of the matrix $\mathbf{W}$ that satisfies condition (7.39)] are inferences made under the unconstrained G–M model (7.37) related to inferences made under the constrained G–M model (7.36)? In answering this question, it is convenient to proceed in two steps by first determining how inferences made under the unconstrained G–M model (7.40) are related to inferences made under the constrained G–M model and then determining how inferences made under the unconstrained G–M model (7.37) are related to those made under the unconstrained G–M model (7.40). This two-step approach is advantageous because the first step is facilitated by the results of Section 8.1f and the second is facilitated by the results of Subsection a of the present section.

Specifically, it follows from the results of Section 8.1f that if $\boldsymbol{\lambda}'\boldsymbol{\beta}$ is estimable [under the constrained G–M model (7.36)], then $\boldsymbol{\lambda}'\mathbf{B}\boldsymbol{\alpha}$ is estimable [under the unconstrained G–M model (7.40)]. And if $\boldsymbol{\lambda}'\boldsymbol{\beta}$ is estimable, then the constrained least squares estimator of $\boldsymbol{\lambda}'\boldsymbol{\beta}$ equals $\boldsymbol{\lambda}'\mathbf{b}$ (a constant) plus the unconstrained least squares estimator of $\boldsymbol{\lambda}'\mathbf{B}\boldsymbol{\alpha}$—when $\mathbf{d} = \mathbf{0}$, $\boldsymbol{\lambda}'\mathbf{b} = 0$. Further, the residual vector, the residual sum of squares, and the ordinary unbiased estimator of σ^2 obtained on the basis of the constrained G–M model (7.36) are identical to those obtained on the basis of the unconstrained G–M model (7.40).

Turning to the question of how inferences made under the unconstrained G–M model (7.37) are related to those made under the unconstrained G–M model (7.40), let $\mathbf{F}$ represent a matrix for which $\mathbf{W} = \mathbf{UF}$ (or equivalently for which $\mathbf{W} = \mathbf{XBF}$). And observe (in light of the results of Subsection a) that if $\boldsymbol{\lambda}'\mathbf{B}\boldsymbol{\alpha}$ is estimable [under the unconstrained G–M model (7.40)], then $\boldsymbol{\lambda}'\mathbf{BF}\boldsymbol{\tau}$ is estimable [under the unconstrained G–M model (7.37)] and the least squares estimator of $\boldsymbol{\lambda}'\mathbf{B}\boldsymbol{\alpha}$ is identical to the least squares estimator of $\boldsymbol{\lambda}'\mathbf{BF}\boldsymbol{\tau}$. Further, the residual vector, the residual sum of squares, and the ordinary unbiased estimator of σ^2 obtained on the basis of the unconstrained G–M model (7.40) are identical to those obtained on the basis of the unconstrained G–M model (7.37).

By combining the various results, we are able to establish a number of fundamental ways in which inferences made under the unconstrained G–M model (7.37) are related to those made under the constrained G–M model (7.36), leading to the conclusion that these two models can be used more or less interchangeably. Specifically, if $\boldsymbol{\lambda}'\boldsymbol{\beta}$ is estimable [under the constrained G–M model (7.36)], then $\boldsymbol{\lambda}'\mathbf{BF}\boldsymbol{\tau}$ is estimable [under the unconstrained G–M model (7.37)]. And if $\boldsymbol{\lambda}'\boldsymbol{\beta}$ is estimable, then the constrained least squares estimator of $\boldsymbol{\lambda}'\boldsymbol{\beta}$ equals a (possibly nonzero) constant $\boldsymbol{\lambda}'\mathbf{b}$ (where $\mathbf{b}$ is such that $\mathbf{Ab} = \mathbf{d}$ and $\mathbf{Xb} = \mathbf{0}$) plus the unconstrained least squares estimator of $\boldsymbol{\lambda}'\mathbf{BF}\boldsymbol{\tau}$. Further, the residual vector, the residual sum of squares, and the ordinary unbiased estimator of σ^2 obtained on the basis of the constrained G–M model (7.36) are identical to those obtained on the basis of the unconstrained G–M model (7.37).

c. Some results on the reparameterization of one constrained G–M model by another constrained G–M model.

Let $\mathbf{y}$ represent an $N \times 1$ observable random vector. And consider two different models for $\mathbf{y}$: the (original) constrained G–M model

$$\mathbf{y} = \mathbf{X}\boldsymbol{\beta} + \mathbf{e}, \tag{7.41}$$

with $N{\times}P$ model matrix $\mathbf{X}$ and in which $\boldsymbol{\beta}$ is a $P{\times}1$ parameter vector that is subject to the constraint $\mathbf{A}\boldsymbol{\beta} = \mathbf{d}$ [where $\mathbf{d} \in \mathcal{C}(\mathbf{A})$], and an alternative constrained G–M model

$$\mathbf{y} = \mathbf{W}\boldsymbol{\tau} + \mathbf{e}, \tag{7.42}$$

with $N \times R$ model matrix $\mathbf{W}$ and in which $\boldsymbol{\tau}$ is an $R{\times}1$ parameter vector that is subject to the constraint $\mathbf{L}\boldsymbol{\tau} = \mathbf{k}$ [where $\mathbf{k} \in \mathcal{C}(\mathbf{W})$]. Further, take $\mathbf{B}$ to be a $P \times I$ matrix whose columns span $\mathcal{N}(\mathbf{A})$ and $\mathbf{b}$ to be a $P \times 1$ vector for which $\mathbf{A}\mathbf{b} = \mathbf{d}$; and similarly take $\mathbf{T}$ to be an $R{\times}J$ matrix whose columns span $\mathcal{N}(\mathbf{W})$ and $\mathbf{t}$ to be an $R{\times}1$ vector for which $\mathbf{L}\mathbf{t} = \mathbf{k}$. As noted earlier (in the first part of Section 8.7), the alternative constrained G–M model qualifies as a reparameterization of the original constrained G–M model when the matrix $\mathbf{T}$ and the vector $\mathbf{t}$ are such that

$$\mathcal{C}(\mathbf{WT}) = \mathcal{C}(\mathbf{XB}) \quad \text{and} \quad \mathbf{Wt} = \mathbf{Xb}. \tag{7.43}$$

Let us consider how inferences made under the alternative constrained G–M model are related to those made under the original constrained G–M model when condition (7.43) is satisfied. In doing so, let us focus on the case where $\mathbf{k} \neq \mathbf{0}$. The motivation for focusing on this case comes from the observation that when $\mathbf{k} = \mathbf{0}$, $\mathbf{t} \in \mathcal{N}(\mathbf{L})$ and hence $\mathbf{t} = \mathbf{Tr}$ for some vector $\mathbf{r}$, so that when $\mathbf{k} = \mathbf{0}$, we find that

$$\mathbf{Xb} = \mathbf{Wt} = \mathbf{WTr} = \mathbf{XBs} \ \text{ for some vector } \mathbf{s}$$

and hence that

$$\mathbf{X}(\mathbf{b} - \mathbf{Bs}) = \mathbf{0} \ \text{ for some vector } \mathbf{s}.$$

Moreover,

$$\mathbf{A}(\mathbf{b} - \mathbf{Bs}) = \mathbf{Ab} = \mathbf{d} \quad \text{and} \quad \mathbf{W0} = \mathbf{X}(\mathbf{b} - \mathbf{Bs}).$$

There is an implication that if (in the special case where $\mathbf{k} = \mathbf{0}$) the model matrix $\mathbf{W}$ and the coefficient matrix $\mathbf{L}$ of the alternative constrained G–M model are such that condition (7.43) can be satisfied, then condition (7.43) can be satisfied in particular when a choice for $\mathbf{b}$ is made for which $\mathbf{Xb} = \mathbf{0}$. The significance of this observation is that the case where $\mathbf{Xb} = \mathbf{0}$ was covered in Subsection b, where it was shown in effect that when $\mathbf{Xb} = \mathbf{0}$ the constrained G–M model (7.41) can be reparameterized by an unconstrained version of model (7.42) and where the relationship between inferences made on the basis of the unconstrained version of model (7.42) and those made on the basis of the constrained model (7.41) was determined.

Now, suppose that condition (7.43) is satisfied. And starting with that supposition, let us undertake a series of steps, culminating in the determination of the relationship between inferences made on the basis of the alternative (reparameterized) constrained G–M model and those made on the basis of the original constrained G–M model. This approach bears some resemblance to that adopted in Subsection b (in dealing with a different case) and is advantageous for the same reason; namely, it makes use of the results of Sections 8.1f and 8.7a. More specifically, Steps 1 through 3 consist of relating inferences made under each of a succession of four models to those made under its predecessor; the first and last of these models are the constrained models (7.41) and (7.42), and the second and third are unconstrained models. The fourth and final step consists of assimilating the results from the first three steps.

Step 1. Letting $\mathbf{z} = \mathbf{y} - \mathbf{Xb}$ and $\mathbf{U} = \mathbf{XB}$, consider the following unconstrained G–M model (as applied to $\mathbf{z}$ rather than to $\mathbf{y}$):

$$\mathbf{z} = \mathbf{U}\boldsymbol{\alpha} + \mathbf{e}, \tag{7.44}$$

where $\mathbf{U}$ is the model matrix and $\boldsymbol{\alpha}$ is an (unconstrained) parameter vector. It follows from the results of Section 8.1f that if $\boldsymbol{\lambda}'\boldsymbol{\beta}$ is estimable under the constrained G–M model (7.41), then $\boldsymbol{\lambda}'\mathbf{B}\boldsymbol{\alpha}$ is estimable under the unconstrained G–M model (7.44). And if $\boldsymbol{\lambda}'\boldsymbol{\beta}$ is estimable, then the constrained least squares estimator of $\boldsymbol{\lambda}'\boldsymbol{\beta}$ equals $\boldsymbol{\lambda}'\mathbf{b}$ (a constant) plus the unconstrained least squares estimator of $\boldsymbol{\lambda}'\mathbf{B}\boldsymbol{\alpha}$ [as determined from $\mathbf{z}$ on the basis of model (7.44)]. Further, the residual vector, the residual sum of squares, and the ordinary unbiased estimator of σ^2 obtained on the basis of the constrained G–M model (7.41) are identical to those obtained on the basis of the unconstrained G–M model (7.44).

Step 2. Taking $\mathbf{z} = \mathbf{y} - \mathbf{X}\mathbf{b}$ (as in Step 1), noting that $\mathbf{z}$ is reexpressible as $\mathbf{z} = \mathbf{y} - \mathbf{W}\mathbf{t}$ [as is evident from condition (7.43)], and letting $\mathbf{Q} = \mathbf{W}\mathbf{T}$, consider the alternative (for the transformed vector $\mathbf{z}$) unconstrained G–M model

$$\mathbf{z} = \mathbf{Q}\boldsymbol{\nu} + \mathbf{e}, \tag{7.45}$$

where $\mathbf{Q}$ is the model matrix and $\boldsymbol{\nu}$ is an (unconstrained) parameter vector. Observe [in light of condition (7.43)] that the unconstrained G–M model (7.45) is a reparameterization of the unconstrained G–M model (7.44) and that there exists a matrix $\mathbf{F}$ for which $\mathbf{Q} = \mathbf{U}\mathbf{F}$. Accordingly, it follows from the results of Subsection a that if $\boldsymbol{\lambda}'\mathbf{B}\boldsymbol{\alpha}$ is estimable under the unconstrained G–M model (7.44), then $\boldsymbol{\lambda}'\mathbf{B}\mathbf{F}\boldsymbol{\nu}$ is estimable under the alternative unconstrained G–M model (7.45), and the unconstrained least squares estimator of $\boldsymbol{\lambda}'\mathbf{B}\boldsymbol{\alpha}$ is identical to the unconstrained least squares estimator of $\boldsymbol{\lambda}'\mathbf{B}\mathbf{F}\boldsymbol{\nu}$. Moreover, the residual vector, the residual sum of squares, and the ordinary unbiased estimator of σ^2 obtained on the basis of either of these two unconstrained models are the same as those obtained on the basis of the other.

Step 3. The coefficient vector of an estimable linear combination $\boldsymbol{\lambda}'\boldsymbol{\beta}$ of the elements of the vector $\boldsymbol{\beta}$ in the (original) constrained G–M model (7.41) is expressible in the form

$$\boldsymbol{\lambda}' = \mathbf{a}'\mathbf{X} + \mathbf{c}'\mathbf{A}$$

for some vectors $\mathbf{a}$ and $\mathbf{c}$. And by making use of this expression, we are able to reexpress the quantity $\boldsymbol{\lambda}'\mathbf{B}\mathbf{F}$ (from Step 2) in the form

$$\boldsymbol{\lambda}'\mathbf{B}\mathbf{F} = \mathbf{a}'\mathbf{X}\mathbf{B}\mathbf{F} + \mathbf{c}'\mathbf{A}\mathbf{B}\mathbf{F} = \mathbf{a}'\mathbf{W}\mathbf{T} + \mathbf{0} = \boldsymbol{\delta}'\mathbf{T},$$

where (for any vector $\mathbf{h}$)

$$\boldsymbol{\delta}' = \mathbf{a}'\mathbf{W} + \mathbf{h}'\mathbf{L}.$$

Thus, $\boldsymbol{\lambda}'\mathbf{B}\mathbf{F}\boldsymbol{\nu}$ is reexpressible as

$$\boldsymbol{\lambda}'\mathbf{B}\mathbf{F}\boldsymbol{\nu} = \boldsymbol{\delta}'\mathbf{T}\boldsymbol{\nu}.$$

Upon making further use of the results of Section 8.1f, we find that $\boldsymbol{\delta}'\boldsymbol{\tau}$ is estimable under the alternative constrained G–M model (7.42) if and only if $\boldsymbol{\delta}'\mathbf{T}\boldsymbol{\nu}$ is estimable under the unconstrained G–M model (7.45), in which case the constrained least squares estimator of $\boldsymbol{\delta}'\boldsymbol{\tau}$ equals $\boldsymbol{\delta}'\mathbf{t}$ plus the unconstrained least squares estimator of $\boldsymbol{\delta}'\mathbf{T}\boldsymbol{\nu}$. Moreover, the residual vector, the residual sum of squares, and the ordinary unbiased estimator of σ^2 obtained on the basis of the alternative constrained G–M model (7.42) are identical to those obtained on the basis of the unconstrained G–M model (7.45).

Fourth (and final) step. Assume that the vector $\mathbf{h}$ (in Step 3) is chosen so that $\mathbf{h}'\mathbf{k} = \mathbf{c}'\mathbf{d}$, as would be possible if (as discussed earlier in the present subsection) $\mathbf{k} \neq \mathbf{0}$. Then,

$$\boldsymbol{\delta}'\mathbf{t} = \mathbf{a}'\mathbf{W}\mathbf{t} + \mathbf{h}'\mathbf{L}\mathbf{t} = \mathbf{a}'\mathbf{W}\mathbf{t} + \mathbf{h}'\mathbf{k} = \mathbf{a}'\mathbf{X}\mathbf{b} + \mathbf{c}'\mathbf{d} = \mathbf{a}'\mathbf{X}\mathbf{b} + \mathbf{c}'\mathbf{A}\mathbf{b} = \boldsymbol{\lambda}'\mathbf{b}. \tag{7.46}$$

And upon combining the results obtained in Steps 1, 2, and 3, we find that if $\boldsymbol{\lambda}'\boldsymbol{\beta}$ is estimable under the original constrained G–M model (7.41), then $\boldsymbol{\delta}'\boldsymbol{\tau}$ is estimable under the alternative constrained G–M model (7.42), and [in light of result (7.46)] the constrained least squares estimator of $\boldsymbol{\delta}'\boldsymbol{\tau}$ is the same as that of $\boldsymbol{\lambda}'\boldsymbol{\beta}$. Moreover, the residual vector, the residual sum of squares, and the ordinary unbiased estimator of σ^2 obtained on the basis of the alternative constrained G–M model are identical to those obtained on the basis of the original constrained G–M model. Thus, the alternative constrained G–M model can be used more or less interchangeably with the original constrained G–M model.

Exercises

Exercise 1. Using the technique described in Section 8.1a (or otherwise), extend the coverage of the following results in Section 5.3 [about estimability under an unconstrained G–M, Aitken, or general linear model (with model equation $\mathbf{y} = \mathbf{X}\boldsymbol{\beta} = \mathbf{e}$)] to estimability under a constrained G–M, Aitken, or general linear model (with model equation $\mathbf{y} = \mathbf{X}\boldsymbol{\beta} + \mathbf{e}$ and constraint $\mathbf{A}\boldsymbol{\beta} = \mathbf{d}$):

(a) the results pertaining to the issue of "how many essentially different estimable functions there are";

(b) the result that condition (5.3.6) is necessary and sufficient for the estimability of $\boldsymbol{\lambda}'\boldsymbol{\beta}$; and

(c) the result that condition (5.3.9) is necessary and sufficient for the estimability of $\boldsymbol{\lambda}'\boldsymbol{\beta}$.

Exercise 2. Suppose that $\mathbf{y}$ is an $N \times 1$ observable random vector that follows a constrained G–M (or Aitken or general linear) model, with model equation $\mathbf{y} = \mathbf{X}\boldsymbol{\beta} + \mathbf{e}$ and constraint $\mathbf{A}\boldsymbol{\beta} = \mathbf{d}$. Show that for any $P \times 1$ vector $\mathbf{r}$ (of constants), $\mathbf{r}'\mathbf{X}'\mathbf{y}$ is the constrained least squares estimator of its expected value (i.e., of $\mathbf{r}'\mathbf{X}'\mathbf{X}\boldsymbol{\beta}$) if and only if

$$\mathbf{X}\mathbf{r} \in \mathcal{C}[\mathbf{X}(\mathbf{I}-\mathbf{A}^{-}\mathbf{A})]. \tag{E.1}$$

Exercise 3. Let $\mathbf{B}$ represent a partitioned matrix of the form $\mathbf{B} = \begin{pmatrix} \mathbf{X}'\mathbf{X} & \mathbf{A}' \\ \mathbf{A} & \mathbf{0} \end{pmatrix}$, and let $\mathbf{G}$ represent a generalized inverse (of $\mathbf{B}$) that has been partitioned as $\mathbf{G} = \begin{pmatrix} \mathbf{G}_{11} & \mathbf{G}_{12} \\ \mathbf{G}_{21} & \mathbf{G}_{22} \end{pmatrix}$, conformally to the partitioning of $\mathbf{B}$ (so that the dimensions of $\mathbf{G}_{11}$ are the same as those of $\mathbf{X}'\mathbf{X}$). Add to the results of Corollary 8.1.4 by using Lemma 8.1.3 (or other means) to show (1) that

$$\mathbf{A}\mathbf{G}_{11}\mathbf{A}' = \mathbf{0}, \quad \mathbf{X}'\mathbf{X}\mathbf{G}_{11}\mathbf{A}' = \mathbf{0}, \text{ and } \mathbf{A}\mathbf{G}_{11}\mathbf{X}'\mathbf{X} = \mathbf{0},$$

(2) that

$$\mathbf{X}'\mathbf{X}\mathbf{G}_{12}\mathbf{A} = \mathbf{A}'\mathbf{G}_{21}\mathbf{X}'\mathbf{X} = -\mathbf{A}'\mathbf{G}_{22}\mathbf{A},$$

and (3) that

$$\mathbf{X}'\mathbf{X} = \mathbf{X}'\mathbf{X}\mathbf{G}_{11}\mathbf{X}'\mathbf{X} - \mathbf{A}'\mathbf{G}_{22}\mathbf{A}.$$

Exercise 4. Suppose that $\mathbf{y}$ is an $N \times 1$ observable random vector that follows a constrained G–M model, with model equation $\mathbf{y} = \mathbf{X}\boldsymbol{\beta} + \mathbf{e}$ and constraint $\mathbf{A}\boldsymbol{\beta} = \mathbf{d}$. Further, let $\hat{\boldsymbol{\alpha}}$ represent a $P \times 1$ vector whose elements are the constrained least squares estimators of the corresponding elements of the vector $\mathbf{X}'\mathbf{X}\boldsymbol{\beta}$, and let $\tilde{\boldsymbol{\alpha}}$ represent a $P \times 1$ vector whose elements are the unconstrained least squares estimators. And take $\mathbf{G}$ to be a generalized inverse of the partitioned matrix $\begin{pmatrix} \mathbf{X}'\mathbf{X} & \mathbf{A}' \\ \mathbf{A} & \mathbf{0} \end{pmatrix}$, and partition $\mathbf{G}$ conformally as $\mathbf{G} = \begin{pmatrix} \mathbf{G}_{11} & \mathbf{G}_{12} \\ \mathbf{G}_{21} & \mathbf{G}_{22} \end{pmatrix}$.

(a) Making use of the results of Exercise 3 (or otherwise), show that

$$\operatorname{var}(\tilde{\boldsymbol{\alpha}}) = \operatorname{var}(\hat{\boldsymbol{\alpha}}) + \sigma^2(-\mathbf{A}'\mathbf{G}_{22}\mathbf{A}).$$

(b) Show that the result of Part (a) implies that the matrix $-\mathbf{A}'\mathbf{G}_{22}\mathbf{A}$ is symmetric and nonnegative definite.

Exercise 5. Suppose that $\mathbf{y}$ is an $N \times 1$ observable random vector that follows a G–M model with model equation $\mathbf{y} = \mathbf{X}\boldsymbol{\beta} + \mathbf{e}$ (where $\boldsymbol{\beta}$ is a $P \times 1$ vector of unconstrained parameters), assume that the distribution of $\mathbf{e}$ is MVN (with mean vector $\mathbf{0}$ and variance-covariance matrix $\sigma^2\mathbf{I}$), let $\boldsymbol{\tau} = \boldsymbol{\Lambda}'\boldsymbol{\beta}$ represent an $M \times 1$ vector of linearly independent estimable linear combinations of the elements of

$\boldsymbol{\beta}$, and take F to be the F statistic for testing the null hypothesis $\boldsymbol{\tau} = \boldsymbol{\tau}^{(0)}$ (versus the alternative $\boldsymbol{\tau} \neq \boldsymbol{\tau}^{(0)}$). Further, let $\boldsymbol{\tau}_1 = \boldsymbol{\Lambda}'_1\boldsymbol{\beta}$ represent an M_1-dimensional subvector of $\boldsymbol{\tau}$, and take F_1 to be the test statistic that would be applicable for testing the null hypothesis $\boldsymbol{\tau}_1 = \boldsymbol{\tau}_1^{(0)}$ (versus $\boldsymbol{\tau}_1 \neq \boldsymbol{\tau}_1^{(0)}$) if $\boldsymbol{\beta}$ were subject to the constraint $\mathbf{A}\boldsymbol{\beta} = \mathbf{d}$, where $\boldsymbol{\Lambda}' = \begin{pmatrix} \boldsymbol{\Lambda}'_1 \\ \mathbf{A} \end{pmatrix}$ (so that the elements of $\boldsymbol{\tau}_1$ are the first M_1 elements of $\boldsymbol{\tau}$) and where $\boldsymbol{\tau}^{(0)} = \begin{pmatrix} \boldsymbol{\tau}_1^{(0)} \\ \mathbf{d} \end{pmatrix}$. Recalling that F and F_1 have noncentral F distributions, letting $\boldsymbol{\delta} = (1/\sigma)(\boldsymbol{\tau} - \boldsymbol{\tau}^{(0)})$ and $\boldsymbol{\delta}_1 = (1/\sigma)(\boldsymbol{\tau}_1 - \boldsymbol{\tau}_1^{(0)})$, and observing that the noncentrality parameter of the noncentral F distribution is expressible as a function, say $\lambda(\boldsymbol{\delta})$, of $\boldsymbol{\delta}$ in the case of the distribution of F and as a function, say $\lambda_1(\boldsymbol{\delta}_1)$, of $\boldsymbol{\delta}_1$ in the case of the distribution of F_1, (1) compare the numerator and denominator degrees of freedom of the distribution of F_1 with those of the distribution of F, and (2) show that when $\boldsymbol{\delta} = \begin{pmatrix} \boldsymbol{\delta}_1 \\ \mathbf{0} \end{pmatrix}$, $\lambda(\boldsymbol{\delta}) = \lambda_1(\boldsymbol{\delta}_1)$.

Exercise 6. Take the setting to be that of the final part of Section 8.1f, and adopt the notation and terminology employed therein. Suppose that the transformed vector $\mathbf{z}$ were taken to be $\mathbf{z} = \mathbf{y} - \mathbf{X}\mathbf{O}_1\underline{\mathbf{h}}$, where $\underline{\mathbf{h}}$ is the unique solution to the linear system $\mathbf{U}_1'\mathbf{h} = \mathbf{d}$ (in $\mathbf{h}$), rather than $\mathbf{z} = \mathbf{y} - \mathbf{X}\tilde{\mathbf{b}}$. Show that if (under the constrained G–M model with model equation $\mathbf{y} = \mathbf{X}\boldsymbol{\beta} + \mathbf{e}$ and constraint $\mathbf{A}\boldsymbol{\beta} = \mathbf{d}$) $\boldsymbol{\lambda}'\boldsymbol{\beta}$ is estimable, then for any solution $\begin{pmatrix} \underline{\mathbf{b}} \\ \underline{\mathbf{r}} \end{pmatrix}$ to the constrained normal equations $\begin{pmatrix} \mathbf{X}'\mathbf{X} & \mathbf{A}' \\ \mathbf{A} & \mathbf{0} \end{pmatrix} \begin{pmatrix} \mathbf{b} \\ \mathbf{r} \end{pmatrix} = \begin{pmatrix} \mathbf{X}'\mathbf{y} \\ \mathbf{d} \end{pmatrix}$ and any solution $\underline{\mathbf{t}}$ to the unconstrained normal equations $\mathbf{W}'\mathbf{W}\mathbf{t} = \mathbf{W}'\mathbf{z}$ (where $\mathbf{W} = \mathbf{X}\mathbf{O}_2$),

$$\boldsymbol{\lambda}'\underline{\mathbf{b}} = (\mathbf{O}_1'\boldsymbol{\lambda})'\underline{\mathbf{h}} + (\mathbf{O}_2'\boldsymbol{\lambda})'\underline{\mathbf{t}}. \tag{E.2}$$

Exercise 7. Let $\mathbf{y}$ represent an $N \times 1$ observable random vector and $\mathbf{d}$ a $Q \times 1$ observable random vector, and suppose that the combined $(N+Q)$-dimensional vector $\begin{pmatrix} \mathbf{y} \\ \mathbf{d} \end{pmatrix}$ follows a G–M model with model equation $\begin{pmatrix} \mathbf{y} \\ \mathbf{d} \end{pmatrix} = \begin{pmatrix} \mathbf{X} \\ \mathbf{A} \end{pmatrix} \boldsymbol{\beta} + \mathbf{e}$, where $\boldsymbol{\beta}$ is a $P \times 1$ vector of unknown (and unconstrained) parameters (and where $\mathbf{X}$ is of dimensions $N \times P$ and the residual vector $\mathbf{e}$ is of dimension $N+Q$). Further, take $\boldsymbol{\lambda}$ to be a $P \times 1$ vector of constants such that $\boldsymbol{\lambda}' \in \mathcal{R}(\mathbf{X})$ (so that $\boldsymbol{\lambda}'\boldsymbol{\beta}$ is estimable from $\mathbf{y}$ alone), and suppose that $\mathcal{R}(\mathbf{X}) \cap \mathcal{R}(\mathbf{A}) = \{\mathbf{0}\}$. Show that the least squares estimator of $\boldsymbol{\lambda}'\boldsymbol{\beta}$ is the same (when both $\mathbf{y}$ and $\mathbf{d}$ are observable) as when only $\mathbf{y}$ is observable. And show that for $\mathbf{d} = \mathbf{0}$ or more generally for $\mathbf{d} \in \mathcal{C}(\mathbf{A})$, the residual sum of squares is also the same.

Exercise 8. Suppose that $\mathbf{y}$ is an $N \times 1$ observable random vector that follows a G–M model with model equation $\mathbf{y} = \mathbf{X}\boldsymbol{\beta} + \mathbf{e}$, where $\boldsymbol{\beta}$ is a $P \times 1$ vector of unknown parameters and where $\mathrm{E}(\mathbf{e}) = \mathbf{0}$ and $\mathrm{var}(\mathbf{e}) = \sigma^2\mathbf{I}$. Suppose further—refer to Section 5.9d—that $\mathbf{e} \sim \sigma\mathbf{u}$, where $\mathbf{u}$ is a random vector that has an absolutely continuous spherical distribution with variance-covariance matrix $\mathbf{I}$ and with a pdf $h(\cdot)$ that is expressible in the form $h(\mathbf{u}) = c^{-1}g(\mathbf{u}'\mathbf{u})$ for some strictly positive constant c and some nonnegative strictly decreasing function $g(\cdot)$. And let $L(\boldsymbol{\beta}, \sigma)$ represent the likelihood function.

(a) Show that (regardless of the value of σ) $L(\boldsymbol{\beta}, \sigma)$ attains its maximum value with respect to $\boldsymbol{\beta}$ at any solution to the normal equations $\mathbf{X}'\mathbf{X}\mathbf{b} = \mathbf{X}'\mathbf{y}$ and that if $\boldsymbol{\beta}$ were subject to the constraint $\mathbf{A}\boldsymbol{\beta} = \mathbf{d}$ [where $\mathbf{d} \in \mathcal{C}(\mathbf{A})$] $L(\boldsymbol{\beta}, \sigma)$ would attain its maximum value with respect to $\boldsymbol{\beta}$ at the first ($\mathbf{b}$-part) of any solution to the constrained normal equations $\begin{pmatrix} \mathbf{X}'\mathbf{X} & \mathbf{A}' \\ \mathbf{A} & \mathbf{0} \end{pmatrix} \begin{pmatrix} \mathbf{b} \\ \mathbf{r} \end{pmatrix} = \begin{pmatrix} \mathbf{X}'\mathbf{y} \\ \mathbf{d} \end{pmatrix}$.

(b) Let $k(s)$ represent the function of s defined (for $s > 0$) as follows: $k(s) = \dfrac{d \log g(s)}{ds}$. Further, let $\underline{\boldsymbol{\beta}}$ represent any particular value of $\boldsymbol{\beta}$ for which $q(\boldsymbol{\beta}) > 0$. And take a to be the strictly positive scalar defined implicitly (in terms of σ) by the equality $\sigma^2 = aq(\underline{\beta})/N$. Show that $L(\underline{\beta}, \sigma)$ attains its maximum value (with respect to σ) when the value of a is such that $a = -2k(N/a)$.

(c) (1) Show that among the possibilities for the random vector $\mathbf{u}$ (having an absolutely continuous spherical distribution with variance-covariance matrix $\mathbf{I}$) is that obtained by taking

$$\mathbf{u} \sim \sqrt{(M-2)/M}\,\mathbf{t}, \tag{E.3}$$

where (for $M > 2$) $\mathbf{t} \sim MVt(M,\, \mathbf{I}_N)$; and (2) show that in the special case (E.3), the solution for the strictly positive scalar a in Part (b) is $a = M/(M-2)$.

(d) Let $\boldsymbol{\tau} = \boldsymbol{\Lambda}'\boldsymbol{\beta}$ represent an $M \times 1$ vector of linear combinations of the elements of $\boldsymbol{\beta}$, and consider the test of the null hypothesis $H_0 : \boldsymbol{\tau} = \boldsymbol{\tau}^{(0)}$ [where $\boldsymbol{\tau}^{(0)} \in \mathcal{C}(\boldsymbol{\Lambda}')$] versus the alternative hypothesis $H_1 : \boldsymbol{\tau} \neq \boldsymbol{\tau}^{(0)}$. Show that the $\dot{\gamma}$-level F test is the equivalent of a $\dot{\gamma}$-level likelihood ratio test.

Exercise 9. Suppose that $\mathbf{y}$ is an $N \times 1$ observable random vector that follows a constrained G–M model with model equation $\mathbf{y} = \mathbf{X}\boldsymbol{\beta} + \mathbf{e}$, where $\boldsymbol{\beta}$ is a $P \times 1$ vector of unknown parameters that [for some $Q \times P$ matrix $\mathbf{A}$ and $Q \times 1$ vector $\mathbf{d} \in \mathcal{C}(\mathbf{A})$] is confined to the set $\{\boldsymbol{\beta} : \mathbf{A}\boldsymbol{\beta} = \mathbf{d}\}$. Suppose further that $\mathbf{k}'\mathbf{d} \neq 0$ for some $Q \times 1$ vector $\mathbf{k}$ for which $\mathbf{k}'\mathbf{A} \in \mathcal{R}(\mathbf{X})$. Show that when $\mathbf{y} = \mathbf{0}$, $ResSS > 0$ (thereby verifying an assertion made in the introductory part of Section 8.4).

Exercise 10. Suppose that $\mathbf{y}$ is an $N \times 1$ observable random vector that follows a G–M model with model equation $\mathbf{y} = \mathbf{X}\boldsymbol{\beta} + \mathbf{e}$, where $\boldsymbol{\beta}$ is a $P \times 1$ vector of unknown parameters. Further, denote the coefficient of determination by R^2 and adopt the general definition (5.12). And let $ResSS = \mathbf{y}'(\mathbf{I} - \mathbf{P}_\mathbf{X})\mathbf{y}$ represent the residual SS; and extend the definition of the regression SS to this setting by taking $RegSS = \mathbf{y}'\mathbf{P}_\mathbf{X}\mathbf{y} - N\bar{y}^2$ (where $\bar{y} = N^{-1}\mathbf{1}'\mathbf{y}$) or $RegSS = \mathbf{y}'\mathbf{P}_\mathbf{X}\mathbf{y}$ depending on whether or not $\mathbf{1} \in \mathcal{C}(\mathbf{X})$. Show that

$$\frac{RegSS}{ResSS} = \frac{R^2}{1-R^2}.$$

Exercise 11. Let y represent a random variable and $\mathbf{x}$ a $K \times 1$ random vector. Further, let $\sigma_y = \sqrt{\mathrm{var}(y)}$, $\boldsymbol{\sigma}_{\mathbf{x}y} = \mathrm{cov}(\mathbf{x}, y)$, and $\boldsymbol{\Sigma}_\mathbf{x} = \mathrm{var}(\mathbf{x})$, so that $\mathrm{var}\begin{pmatrix} y \\ \mathbf{x} \end{pmatrix} = \begin{pmatrix} \sigma_y^2 & \boldsymbol{\sigma}'_{\mathbf{x}y} \\ \boldsymbol{\sigma}_{\mathbf{x}y} & \boldsymbol{\Sigma}_\mathbf{x} \end{pmatrix}$ and there exists a matrix $\boldsymbol{\Gamma} = (\boldsymbol{\gamma}_y, \boldsymbol{\Gamma}_\mathbf{x})$ (where $\boldsymbol{\gamma}_y$ is a column vector) such that $\mathrm{var}\begin{pmatrix} y \\ \mathbf{x} \end{pmatrix} = \boldsymbol{\Gamma}'\boldsymbol{\Gamma}$ and hence such that $\sigma_y^2 = \boldsymbol{\gamma}_y'\boldsymbol{\gamma}_y$, $\boldsymbol{\sigma}_{\mathbf{x}y} = \boldsymbol{\Gamma}_\mathbf{x}'\boldsymbol{\gamma}_y$, and $\boldsymbol{\Sigma}_\mathbf{x} = \boldsymbol{\Gamma}_\mathbf{x}'\boldsymbol{\Gamma}_\mathbf{x}$. And let $\boldsymbol{\beta}$ represent any $K \times 1$ vector such that $\boldsymbol{\Sigma}_\mathbf{x}\boldsymbol{\beta} = \boldsymbol{\sigma}_{\mathbf{x}y}$.

(a) Show that for any constant a and any $K \times 1$ vector of constants $\mathbf{b}$,

$$\mathrm{var}(a + \mathbf{b}'\mathbf{x}) = \mathbf{b}'\boldsymbol{\Sigma}_\mathbf{x}\mathbf{b} \quad \text{and} \quad \mathrm{cov}(y,\, a + \mathbf{b}'\mathbf{x}) = \boldsymbol{\beta}'\boldsymbol{\Sigma}_\mathbf{x}\mathbf{b}. \tag{E.4}$$

(b) Using the Cauchy-Schwarz inequality (or other means), show that

$$|\boldsymbol{\beta}'\boldsymbol{\Sigma}_\mathbf{x}\mathbf{b}| \leq (\boldsymbol{\beta}'\boldsymbol{\Sigma}_\mathbf{x}\boldsymbol{\beta})^{1/2}(\mathbf{b}'\boldsymbol{\Sigma}_\mathbf{x}\mathbf{b})^{1/2}. \tag{E.5}$$

(c) Show that (for any constant a and any $K \times 1$ vector of constants $\mathbf{b}$)

$$|\mathrm{corr}(y,\, a + \mathbf{b}'\mathbf{x})| \leq (\boldsymbol{\beta}'\boldsymbol{\Sigma}_\mathbf{x}\boldsymbol{\beta})^{1/2}/\sigma_y = \mathrm{corr}(y,\, a + \boldsymbol{\beta}'\mathbf{x}),$$

so that the maximum value (with respect to $\mathbf{b}$) of $|\mathrm{corr}(y,\, a + \mathbf{b}'\mathbf{x})|$ is $(\boldsymbol{\beta}'\boldsymbol{\Sigma}_\mathbf{x}\boldsymbol{\beta})^{1/2}/\sigma_y$ and that value is attained at $\mathbf{b} = \boldsymbol{\beta}$ [a value of $\mathbf{b}$ for which $\mathrm{corr}(y,\, a + \mathbf{b}'\mathbf{x})$ is nonnegative].

(d) Let $\alpha = \mu_y - \boldsymbol{\beta}'\boldsymbol{\mu}_\mathbf{x}$, where $\mu_y = \mathrm{E}(y)$ and $\boldsymbol{\mu}_\mathbf{x} = \mathrm{E}(\mathbf{x})$. And take $\eta(\mathbf{x})$ to be a function of $\mathbf{x}$ defined as follows: $\eta(\mathbf{x}) = \alpha + \boldsymbol{\beta}'\mathbf{x}$. If α and $\boldsymbol{\beta}$ were known, $\eta(\mathbf{x})$ could serve as a point predictor (for predicting the value of y from that of $\mathbf{x}$). What can be said about the MSE (mean squared error) of that predictor (relative to the MSE of other point predictors)? Provide justification for your answer.

(e) The correlation $\operatorname{corr}(y, \alpha+\boldsymbol{\beta}'\mathbf{x}) = (\boldsymbol{\beta}'\boldsymbol{\Sigma}_{\mathbf{x}}\boldsymbol{\beta})^{1/2}/\sigma_y$ between y and $\alpha+\boldsymbol{\beta}'\mathbf{x}$ [where α is as defined in Part (d)] is referred to as the multiple correlation coefficient. Let $\mathbf{x}_1, \mathbf{x}_2, \ldots, \mathbf{x}_N$ represent N values of $\mathbf{x}$ and $y_1, y_2, \ldots, y_N$ the corresponding values of y. Show that in the special case where the joint distribution of y and $\mathbf{x}$ is the empirical distribution that assigns probability $1/N$ to each of the vectors $\begin{pmatrix} y_1 \\ \mathbf{x}_1 \end{pmatrix}, \begin{pmatrix} y_2 \\ \mathbf{x}_2 \end{pmatrix}, \ldots, \begin{pmatrix} y_N \\ \mathbf{x}_N \end{pmatrix}$,

$$\frac{\boldsymbol{\beta}'\boldsymbol{\Sigma}_{\mathbf{x}}\boldsymbol{\beta}}{\sigma_y^2} = \frac{\mathbf{y}'\mathbf{P}_{\mathbf{X}}\mathbf{y} - N\bar{y}^2}{\mathbf{y}'\mathbf{y} - N\bar{y}^2},$$

where $\mathbf{X} = (\mathbf{1}, \mathbf{X}_*)$ with $\mathbf{X}_* = (\mathbf{x}_1, \mathbf{x}_2, \ldots, \mathbf{x}_N)'$, $\mathbf{y} = (y_1, y_2, \ldots, y_N)'$, and $\bar{y} = N^{-1}\sum_i y_i$, so that (in this special case) the square of the multiple correlation coefficient equals the coefficient of determination [that for an observable random vector $\mathbf{y}$ that follows a G–M model with a model matrix $\mathbf{X} = (\mathbf{1}, \mathbf{X}_*)$ for which $\mathbf{1} \in \mathcal{C}(\mathbf{X})$ and whose observed value is $\mathbf{y} = (y_1, y_2, \ldots, y_N)'$].

Exercise 12. Take the setting to be that in Part 2 of Section 8.5e, and adopt the terminology and notation employed therein. Further, for $j = 1, 2, \ldots, M$, denote by $\bar{\boldsymbol{\tau}}_j$ the ordinary (unconstrained) least squares estimator of $\boldsymbol{\tau}_j$. Show that each of the following two conditions is equivalent to the condition that [as in Condition (5.30)]

$$\boldsymbol{\Lambda}_j = \mathbf{X}'\mathbf{X}[\mathbf{I} - (\tilde{\boldsymbol{\Lambda}}_{j+1}')^{-}\tilde{\boldsymbol{\Lambda}}_{j+1}]\mathbf{U}_j$$

for some matrix $\mathbf{U}_j$ (where $M-1 \geq j \geq 1$):

(1) $\boldsymbol{\Lambda}_j = \mathbf{X}'\mathbf{X}\mathbf{T}_j$ for some matrix $\mathbf{T}_j$ such that $\tilde{\boldsymbol{\Lambda}}_{j+1}'\mathbf{T}_j = \mathbf{0}$ [i.e., some matrix $\mathbf{T}_j$ whose columns are contained in the null space of $\tilde{\boldsymbol{\Lambda}}_{j+1}'$];

(2) $\operatorname{cov}(\bar{\boldsymbol{\tau}}_j, \bar{\boldsymbol{\tau}}_k) = \mathbf{0}$ for $k = j+1, j+2, \ldots, M$.

Exercise 13. Consider further the application to the ouabain data of the results of Section 8.5e [on K or $(K-1)$-line ANOVA tables]. Suppose that instead of taking the "independent variable" in this application to be the rate of injection (as in the approach to the application taken in Part 5 of Section 8.5e), it were taken to be the log to the base 2 of the rate. Determine the effects of this change.

Exercise 14. Take the setting to be that of Section 8.6, and adopt the notation and terminology employed therein.

(a) Devise a $(K+1)(d+1)$-dimensional column vector $\tilde{\mathbf{x}}(u)$ (that is functionally dependent on u) for which
$$\tilde{s}(u) = \tilde{\mathbf{x}}'(u)\boldsymbol{\beta} \quad \text{(for all } u\text{)},$$
where $\tilde{\mathbf{x}}'(u) = [\tilde{\mathbf{x}}(u)]'$.

(b) Let $\tilde{s}^{(1)}(\cdot), \tilde{s}^{(2)}(\cdot), \ldots, \tilde{s}^{(M)}(\cdot)$ represent any splines of the same degree d and with the same knots $\lambda_1, \lambda_2, \ldots, \lambda_K$, so that they are completely characterized by their $(K+1)(d+1) \times 1$ vectors of coefficients, say $\boldsymbol{\beta}^{(1)}, \boldsymbol{\beta}^{(2)}, \ldots, \boldsymbol{\beta}^{(M)}$. Making use of your solution to Part (a) (or other means), show that for any constants $a_1, a_2, \ldots, a_M$, the linear combination $\sum_{i=1}^{M} a_i\tilde{s}^{(i)}(\cdot)$ is a spline function of degree d with knots $\lambda_1, \lambda_2, \ldots, \lambda_K$ and with vector of coefficients $\sum_{i=1}^{M} a_i\boldsymbol{\beta}^{(i)}$. Note that there is an implication that the set consisting of spline functions of the same degree and with the same number and placement of knots constitutes what is known as a vector space (or linear space).

(c) Taking the matrices $\mathbf{X}$ and $\mathbf{A}$ to be as defined in Subsection a of Section 8.6 and taking the functionally dependent row vector $\tilde{\mathbf{x}}'(\cdot)$ to be the same as in Part (a), letting $R = \dim[\mathcal{N}(\mathbf{A})]$, letting $\boldsymbol{\beta}^{(1)}, \boldsymbol{\beta}^{(2)}, \ldots, \boldsymbol{\beta}^{(M)}$ represent any $(K+1)(d+1)$-dimensional column vectors that span $\mathcal{N}(\mathbf{A})$, letting $\mathbf{L} = (\boldsymbol{\beta}^{(1)}, \boldsymbol{\beta}^{(2)}, \ldots, \boldsymbol{\beta}^{(M)})$, and letting $\mathbf{W} = \mathbf{X}\mathbf{L}$, show (1) that the ith row of $\mathbf{X}$ equals $\tilde{\mathbf{x}}'(u_i)$ $(i = 1, 2, \ldots, N)$, (2) that
$$R = K + d + 1,$$

(3) that the constrained least squares estimator of the value of $\tilde{s}(u)$ corresponding to any value of u for which $\tilde{s}(u)$ is estimable equals $\tilde{\mathbf{x}}'(u)\mathbf{L}\mathbf{t}$, where $\mathbf{t}$ is any solution to the linear system $\mathbf{W}'\mathbf{W}\mathbf{t} = \mathbf{W}'\mathbf{y}$, and that the constrained least squares estimator is reexpressible in the form $\sum_{j=1}^{M} t_j \tilde{s}(u; j)$, where (for $j = 1, 2, \ldots, M$) t_j is the jth element of $\mathbf{t}$ and $\tilde{s}(u; j)$ equals $\tilde{\mathbf{x}}'(u)\beta^{(j)}$, (4) that if the values of $\tilde{s}(u)$ corresponding to any $d+1$ distinct values of u contained in the kth of the intervals defined by the knots $\lambda_1, \lambda_2, \ldots, \lambda_K$ are estimable, then $\beta_{k1}, \beta_{k2}, \ldots, \beta_{k,d+1}$ are all estimable [which implies that the value of $\tilde{s}(u)$ corresponding to every value of u in the kth of these intervals is estimable], (5) that

$$\text{rank}(\mathbf{W}) \leq K + d + 1,$$

with equality holding if and only if all $(K+1)(d+1)$ of the parameters β_{kj} $(k = 1, 2, \ldots, K+1;\ j = 1, 2, \ldots, d+1)$ are estimable, and (6) that if $M = R$ [so that $\boldsymbol{\beta}^{(1)}, \boldsymbol{\beta}^{(2)}, \ldots, \boldsymbol{\beta}^{(M)}$ form a basis for $\mathcal{N}(\mathbf{A})$] and if all $(K+1)(d+1)$ of the parameters β_{kj} $(k = 1, 2, \ldots, K+1;\ j = 1, 2, \ldots, d+1)$ are estimable, then the linear system $\mathbf{W}'\mathbf{W}\mathbf{t} = \mathbf{W}'\mathbf{y}$ has a unique solution.

(d) What modifications need to be made to the results of Part (c) to make them applicable to natural splines [for which $d = 3$ and for which the constraint is $\mathbf{A}_{+}\boldsymbol{\beta} = \mathbf{0}$ rather than $\mathbf{A}\boldsymbol{\beta} = \mathbf{0}$].

(e) Let us add to the development in Part (d) pertaining to natural splines by taking $M = K$ and considering the effect of imposing on $\mathbf{L}$ [and hence on $\boldsymbol{\beta}^{(1)}, \boldsymbol{\beta}^{(2)}, \ldots, \boldsymbol{\beta}^{(K)}$] the additional constraint $\mathbf{U}\mathbf{L} = \mathbf{I}$,where $\mathbf{U}$ is the $K \times 4(K+1)$ matrix with kth row $\tilde{\mathbf{x}}'(\boldsymbol{\lambda}_k)$. Show that there is a unique matrix $\mathbf{L}$ that satisfies this constraint as well as the constraint $\mathbf{A}_{+}\mathbf{L} = \mathbf{0}$ and that (for this choice of $\mathbf{L}$) the kth element of the solution $\mathbf{t}$ to the linear system $\mathbf{W}'\mathbf{W}\mathbf{t} = \mathbf{W}'\mathbf{y}$ is the least squares estimator of $\tilde{\mathbf{x}}'(\boldsymbol{\lambda}_k)\boldsymbol{\beta}$—the K splines $\tilde{s}(u; k) = \tilde{\mathbf{x}}'(u)\boldsymbol{\beta}^{(k)}$ $(k = 1, 2, \ldots, K)$ are known as cardinal splines.

Exercise 15. Use the tree-height data of Section 4.4h to further illustrate the results of Section 8.5e [on K- or $(K-1)$-line ANOVA tables]. For purposes of doing so, let $y = \log(\text{tree height})$ and $u = \log(\text{tree diameter})$. Further, regard the 101 values of y as the observed values of random variables $y_1, y_2, \ldots, y_{101}$ that follow a G–M model with

$$\text{E}(y_t) = \beta_1 + \beta_2 u_t + \beta_3 u_t^2 + \beta_4 u_t^3 + \beta_5 u_t^4 + \beta_6 u_t^5 \qquad (t = 1, 2, \ldots, 101),$$

where $u_1, u_2, \ldots, u_{101}$ are the corresponding values of u—assume (for purposes of this exercise) that (as in the use of these data in Section 8.6b) the information on the relative locations of the trees is to be disregarded. Subject these data to the same kinds of procedures that were applied to the ouabain data in Part 5 of Section 8.5e (and in Exercise 13).

Exercise 16. Consider the two "multiple regression" models defined implicitly by expressions (4.2.10) and (4.2.11). Show that these two models can be regarded as reparameterizations of each other. Do so by regarding one of these models as the original model with model matrix $\mathbf{X}$ and the other as the alternative model with model matrix $\mathbf{W}$ and by finding matrices $\mathbf{F}$ and $\mathbf{G}$ that satisfy condition (7.22).

Exercise 17. Take the setting to be that of Section 8.7a, and adopt the notation and terminology employed therein. Suppose that the alternative model (with model equation $\mathbf{y} = \mathbf{W}\boldsymbol{\tau} + \mathbf{e}$ and $N \times R$ model matrix $\mathbf{W}$) is a reparameterization of the original model (with model equation $\mathbf{y} = \mathbf{X}\boldsymbol{\beta} + \mathbf{e}$ and $N \times P$ model matrix $\mathbf{X}$). And take the matrix $\mathbf{G}$ to be as defined by condition (7.22), that is, to be an $R \times P$ matrix for which $\mathbf{X} = \mathbf{W}\mathbf{G}$.

(a) Show that $\mathcal{R}(\mathbf{X}) \subset \mathcal{R}(\mathbf{G})$.

(b) Show that in the special case where the alternative model is a full-rank reparameterization of the original model, (1) $\mathcal{R}(\mathbf{X}) = \mathcal{R}(\mathbf{G})$ and (2) a linear combination $\boldsymbol{\lambda}'\boldsymbol{\beta}$ of the elements of $\boldsymbol{\beta}$ is estimable if and only if $\boldsymbol{\lambda}' \in \mathcal{R}(\mathbf{G})$.

Exercise 18. Take the setting to be that of Section 8.7a, and adopt the notation and terminology employed therein. Further, let $\boldsymbol{\Lambda}'\boldsymbol{\beta}$ represent a vector of linearly independent linear combinations of the elements of the vector $\boldsymbol{\beta}$ that are estimable and that are equal in number to rank($\mathbf{X}$). And suppose that the model matrix $\mathbf{W}$ of the alternative model is that for which $\mathbf{W} = \mathbf{X}\boldsymbol{\Lambda}(\boldsymbol{\Lambda}'\boldsymbol{\Lambda})^{-1}$.

(a) Show that the alternative model is a full-rank reparameterization of the original model and find matrices $\mathbf{F}$ and $\mathbf{G}$ that satisfy condition (7.22).

(b) Show that the least squares estimator of $\boldsymbol{\tau}$ equals the least squares estimator of $\boldsymbol{\Lambda}'\boldsymbol{\beta}$.

Exercise 19. Suppose that $\mathbf{y}$ is an $N \times 1$ observable random vector. And suppose that $\mathbf{y}$ follows a G–M model for which the $N \times P$ model matrix $\mathbf{X}$ is expressible in the form $\mathbf{X} = (\mathbf{1}, \mathbf{X}_1, \mathbf{X}_2)$, where for some number a $(0 < a < 1)$ and some integer C $(2 \le C \le P-2)$ $\mathbf{X}_1$ has C columns that sum to $a\mathbf{1}$ and $\mathbf{X}_2$ $P-C-1$ columns that sum to $(1-a)\mathbf{1}$. Suppose also that rank($\mathbf{X}$) $= P-2$. A G–M model with these characteristics might be encountered in the analysis of mixture data of the kind considered by Cornell (2002) in his Section 4.14.

(a) Show that for a linear combination $\sum_{j=1}^{P} \lambda_j \beta_j$ of the elements $\beta_1, \beta_2, \ldots, \beta_P$ of $\boldsymbol{\beta}$ to be estimable, it is necessary and sufficient that (1) $\lambda_1 = \sum_{j=2}^{P} \lambda_j$ and (2) $a \sum_{j=C+2}^{P} \lambda_j = (1-a) \sum_{j=2}^{C+1} \lambda_j$.

(b) Show that each of the two quantities, β_1 and $(a/C) \sum_{j=2}^{C+1} \beta_j + [(1-a)/(P-C-1)] \sum_{j=C+2}^{P} \beta_j$, is nonestimable but that the two of them are not jointly nonestimable.

(c) Show that the two quantities, $\sum_{j=2}^{C+1} \beta_j$ and $\sum_{j=C+2}^{P} \beta_j$, are jointly nonestimable.

(d) Let $\mathbf{B}_1$ represent a $(C-1)\times C$ matrix whose rows are the last $C-1$ rows of a $C\times C$ orthogonal matrix (such as a $C\times C$ Helmert matrix) whose first row is $C^{-1/2}\mathbf{1}_C'$ and $\mathbf{B}_2$ a $(P-C-2)\times(P-C-1)$ matrix whose rows are the last $P-C-2$ rows of a $(P-C-1)\times(P-C-1)$ orthogonal matrix whose first row is $(P-C-1)^{-1/2}\mathbf{1}_{P-C-1}'$. And let $\mathbf{U} = \mathbf{XB}$, where $\mathbf{B} = \begin{pmatrix} 1 & \mathbf{0} & \mathbf{0} \\ \mathbf{0} & \mathbf{B}_1' & \mathbf{0} \\ \mathbf{0} & \mathbf{0} & \mathbf{B}_2' \end{pmatrix}$. Show that the unconstrained G–M model
$$\mathbf{y} = \mathbf{U}\boldsymbol{\alpha} + \mathbf{e}$$
[with model matrix $\mathbf{U}$ and in which $\boldsymbol{\alpha}$ is an (unconstrained) parameter vector] is a reparameterization of the constrained G–M model with model matrix $\mathbf{X}$ and with constraints $\sum_{j=2}^{C+1} \beta_j = 0$ and $\sum_{j=C+2}^{P} \beta_j = 0$ and is also a reparameterization of the unconstrained G–M model with model matrix $\mathbf{X}$.

(e) Letting $\mathbf{b}_j$ represent the jth column of the $P\times(P-2)$ matrix $\mathbf{B}$ [defined in Part (d)], show that for $j = 2, 3, \ldots, P-2$ or for $j = 1, 2, \ldots, P-2$ (depending on whether the model is the unconstrained G–M model with model matrix $\mathbf{X}$ or the constrained G–M model with model matrix $\mathbf{X}$ and constraints $\sum_{j=2}^{C+1} \beta_j = 0$ and $\sum_{j=C+2}^{P} \beta_j = 0$), $\mathbf{b}_j'\boldsymbol{\beta}$ is estimable and its unconstrained or constrained least squares estimator equals the (unconstrained) least squares estimator of α_j (as determined under the unconstrained G–M model $\mathbf{y} = \mathbf{U}\boldsymbol{\alpha} + \mathbf{e}$), where α_j is the jth element of $\boldsymbol{\alpha}$.

Bibliographic and Supplementary Notes

§1d and Exercise 3. The results of Corollary 8.1.4 and Exercise 3 on generalized inverses of partitioned matrices of the form $\begin{pmatrix} \mathbf{X}'\mathbf{X} & \mathbf{A}' \\ \mathbf{A} & \mathbf{0} \end{pmatrix}$ are among those covered by Harville (1997, theorem 19.4.2). However, the approach taken herein in the derivation of those results (that taken in the derivation of Corollary 8.1.4 and suggested for use in the derivation of the results of Exercise 3) differs somewhat from that adopted in the earlier source.

Exercise 5. Inspiration for Exercise 5 came from Griffiths and Hill's (2022) article.

§1f and Exercise 6. Exercise 6 and the results of the final part of Section 8.1f are closely related to the results presented by Golub and Van Loan (2013) in their Section 6.2.3.

§6 and Exercise 14. The approach taken herein to regression splines is based on the observation that (for regression splines of the same degree and with the same number and placement of knots) the underlying model can be regarded as a constrained linear model. That approach differs from the conventional approach, which is based on the observation that regression splines (of the same degree and with the same number and placement of knots) form a vector (linear) space. An overview of the latter approach is provided by Perperoglou, Sauerbrei, Abrahamowicz, and Schmid (2019). Various connections between the two approaches are addressed in Exercise 14.

Index

Part II

Exercises and Solutions

2

Matrix Algebra: a Primer

EXERCISE 1. Let $\mathbf{A}$ represent an $M \times N$ matrix and $\mathbf{B}$ an $N \times M$ matrix. Can the value of $\mathbf{A} + \mathbf{B}'$ be determined from the value of $\mathbf{A}' + \mathbf{B}$ (in the absence of any other information about $\mathbf{A}$ and $\mathbf{B}$)? Describe your reasoning.

Solution. Yes! In light of results (1.12) and (1.10), we have that

$$(\mathbf{A}' + \mathbf{B})' = (\mathbf{A}')' + \mathbf{B}' = \mathbf{A} + \mathbf{B}'.$$

Thus, to obtain the value of $\mathbf{A} + \mathbf{B}'$, it suffices to transpose the value of $\mathbf{A}' + \mathbf{B}$.

EXERCISE 2. Show that for any $M \times N$ matrix $\mathbf{A} = \{a_{ij}\}$ and $N \times P$ matrix $\mathbf{B} = \{b_{ij}\}$, $(\mathbf{AB})' = \mathbf{B}'\mathbf{A}'$ [thereby verifying result (1.13)].

Solution. Since (for $k = 1, 2, \ldots, N$) the ikth element of the $P \times N$ matrix $\mathbf{B}'$ is the kith element of $\mathbf{B}$ and the kjth element of the $N \times M$ matrix $\mathbf{A}'$ is the jkth element of $\mathbf{A}$, it follows from the very definition of a matrix product that the ijth element of the $P \times M$ matrix $\mathbf{B}'\mathbf{A}'$ is $\sum_{k=1}^{N} b_{ki} a_{jk}$ $(i = 1, 2, \ldots, P;\ j = 1, 2, \ldots, M)$. And upon rewriting $\sum_{k=1}^{N} b_{ki} a_{jk}$ as $\sum_{k=1}^{N} a_{jk} b_{ki}$, it is evident that the ijth element of $\mathbf{B}'\mathbf{A}'$ equals the jith element of $\mathbf{AB}$. Moreover, the jith element of $\mathbf{AB}$ is the ijth element of $(\mathbf{AB})'$. Thus, each element of $\mathbf{B}'\mathbf{A}'$ equals the corresponding element of $(\mathbf{AB})'$, and we conclude that $(\mathbf{AB})' = \mathbf{B}'\mathbf{A}'$.

EXERCISE 3. Let $\mathbf{A} = \{a_{ij}\}$ and $\mathbf{B} = \{b_{ij}\}$ represent $N \times N$ symmetric matrices.

(a) Show that in the special case where $N = 2$, $\mathbf{AB}$ is symmetric if and only if $b_{12}(a_{11} - a_{22}) = a_{12}(b_{11} - b_{22})$.

(b) Give a numerical example where $\mathbf{AB}$ is nonsymmetric.

(c) Show that $\mathbf{A}$ and $\mathbf{B}$ commute if and only if $\mathbf{AB}$ is symmetric.

Solution. (a) Suppose that $N = 2$. Then,

$$\mathbf{AB} = \begin{pmatrix} a_{11} & a_{12} \\ a_{12} & a_{22} \end{pmatrix} \begin{pmatrix} b_{11} & b_{12} \\ b_{12} & b_{22} \end{pmatrix} = \begin{pmatrix} a_{11}b_{11} + a_{12}b_{12} & a_{11}b_{12} + a_{12}b_{22} \\ a_{12}b_{11} + a_{22}b_{12} & a_{12}b_{12} + a_{22}b_{22} \end{pmatrix}.$$

Thus, $\mathbf{AB}$ is symmetric if and only if

$$a_{11}b_{12} + a_{12}b_{22} = a_{12}b_{11} + a_{22}b_{12}$$

or, equivalently, if and only if

$$b_{12}(a_{11} - a_{22}) = a_{12}(b_{11} - b_{22}).$$

(b) Take

$$\begin{pmatrix} a_{11} & a_{12} \\ a_{21} & a_{22} \end{pmatrix} = \begin{pmatrix} 2 & 1 \\ 1 & 1 \end{pmatrix} \quad \text{and} \quad \begin{pmatrix} b_{11} & b_{12} \\ b_{21} & b_{22} \end{pmatrix} = \begin{pmatrix} 2 & 0 \\ 0 & 1 \end{pmatrix}.$$

And, for $i, j = 3, 4, \ldots, N$, take $a_{ij} = b_{ij} = 0$. Then,

$$\mathbf{AB} = \begin{pmatrix} 4 & 1 & 0 & \ldots & 0 \\ 2 & 1 & 0 & \ldots & 0 \\ 0 & 0 & 0 & \ldots & 0 \\ \vdots & \vdots & \vdots & \ddots & \vdots \\ 0 & 0 & 0 & \ldots & 0 \end{pmatrix}.$$

(c) Since $\mathbf{A}$ and $\mathbf{B}$ are symmetric,

$$(\mathbf{AB})' = \mathbf{B}'\mathbf{A}' = \mathbf{BA}.$$

Thus, $\mathbf{BA} = \mathbf{AB}$ if and only if $(\mathbf{AB})' = \mathbf{AB}$, that is, if and only if $\mathbf{AB}$ is symmetric.

EXERCISE 4. Let $\mathbf{A}$ represent an $M \times N$ partitioned matrix comprising R rows and U columns of blocks, the ijth of which is an $M_i \times N_j$ matrix $\mathbf{A}_{ij}$ that (for some scalar c_{ij}) is expressible as $\mathbf{A}_{ij} = c_{ij}\mathbf{1}_{M_i}\mathbf{1}'_{N_j}$ (a scalar multiple of an $M_i \times N_j$ matrix of 1's). Similarly, let $\mathbf{B}$ represent an $N \times Q$ partitioned matrix comprising U rows and V columns of blocks, the ijth of which is an $N_i \times Q_j$ matrix $\mathbf{B}_{ij}$ that (for some scalar d_{ij}) is expressible as $\mathbf{B}_{ij} = d_{ij}\mathbf{1}_{N_i}\mathbf{1}'_{Q_j}$. Obtain (in as simple form as possible) the conditions that must be satisfied by the scalars c_{ij} $(i = 1, 2, \ldots, R;\ j = 1, 2, \ldots, U)$ and d_{ij} $(i = 1, 2, \ldots, U;\ j = 1, 2, \ldots, V)$ in order for $\mathbf{AB}$ to equal a null matrix.

Solution. According to result (2.5), $\mathbf{AB}$ is expressible as a partitioned matrix comprising R rows and V columns of blocks, the ijth of which is

$$\mathbf{F}_{ij} = \sum_{k=1}^{U} \mathbf{A}_{ik}\mathbf{B}_{kj}.$$

Moreover,

$$\begin{aligned} \mathbf{A}_{ik}\mathbf{B}_{kj} &= \big(c_{ik}\mathbf{1}_{M_i}\mathbf{1}'_{N_k}\big)\big(d_{kj}\mathbf{1}_{N_k}\mathbf{1}'_{Q_j}\big) \\ &= c_{ik}d_{kj}\mathbf{1}_{M_i}\mathbf{1}'_{N_k}\mathbf{1}_{N_k}\mathbf{1}'_{Q_j} \\ &= c_{ik}d_{kj}\mathbf{1}_{M_i}(N_k)\mathbf{1}'_{Q_j} \\ &= N_k c_{ik} d_{kj}\mathbf{1}_{M_i}\mathbf{1}'_{Q_j}. \end{aligned}$$

Thus, $\mathbf{AB} = \mathbf{0}$ if and only if for $i = 1, \ldots, R$ and $j = 1, \ldots, V$, $\sum_{k=1}^{U} N_k c_{ik} d_{kj} = 0$.

EXERCISE 5. Show that for any $M \times N$ matrix $\mathbf{A}$ and $N \times M$ matrix $\mathbf{B}$,

$$\operatorname{tr}(\mathbf{AB}) = \operatorname{tr}(\mathbf{A}'\mathbf{B}').$$

Solution. Making use of results (3.3), (1.13), and (3.7), we find that

$$\operatorname{tr}(\mathbf{AB}) = \operatorname{tr}[(\mathbf{AB})'] = \operatorname{tr}(\mathbf{B}'\mathbf{A}') = \operatorname{tr}(\mathbf{A}'\mathbf{B}').$$

EXERCISE 6. Show that for any $M \times N$ matrix $\mathbf{A}$, $N \times P$ matrix $\mathbf{B}$, and $P \times M$ matrix $\mathbf{C}$,

$$\operatorname{tr}(\mathbf{ABC}) = \operatorname{tr}(\mathbf{CAB}) = \operatorname{tr}(\mathbf{BCA})$$

(i.e., the cyclic permutation of the 3 matrices in the product $\mathbf{ABC}$ does not affect the trace of the product).

Solution. Making use of Lemma 2.3.1, we find that

$$\operatorname{tr}(\mathbf{ABC}) = \operatorname{tr}[(\mathbf{AB})\mathbf{C}] = \operatorname{tr}[\mathbf{C}(\mathbf{AB})] = \operatorname{tr}(\mathbf{CAB})$$

and similarly that

$$\operatorname{tr}(\mathbf{CAB}) = \operatorname{tr}[(\mathbf{CA})\mathbf{B}] = \operatorname{tr}[\mathbf{B}(\mathbf{CA})] = \operatorname{tr}(\mathbf{BCA}).$$

EXERCISE 7. Let $\mathbf{A}$, $\mathbf{B}$, and $\mathbf{C}$ represent square matrices of order N.

(a) Using the result of Exercise 5 (or otherwise), show that if $\mathbf{A}$, $\mathbf{B}$, and $\mathbf{C}$ are symmetric, then $\mathrm{tr}(\mathbf{ABC}) = \mathrm{tr}(\mathbf{BAC})$.

(b) Show that [aside from special cases like that considered in Part (a)] $\mathrm{tr}(\mathbf{BAC})$ is not necessarily equal to $\mathrm{tr}(\mathbf{ABC})$.

Solution. (a) Making use of the result of Exercise 5 and of result (1.13), we find that

$$\mathrm{tr}(\mathbf{ABC}) = \mathrm{tr}[(\mathbf{AB})'\mathbf{C}'] = \mathrm{tr}(\mathbf{B}'\mathbf{A}'\mathbf{C}').$$

Now, suppose that $\mathbf{A}$, $\mathbf{B}$, and $\mathbf{C}$ are symmetric. Then, $\mathbf{B}'\mathbf{A}'\mathbf{C}' = \mathbf{BAC}$, and we conclude that $\mathrm{tr}(\mathbf{ABC}) = \mathrm{tr}(\mathbf{BAC})$.

(b) Let $\mathbf{A} = \mathrm{diag}(\mathbf{A}_*, \mathbf{0})$, $\mathbf{B} = \mathrm{diag}(\mathbf{B}_*, \mathbf{0})$, and $\mathbf{C} = \mathrm{diag}(\mathbf{C}_*, \mathbf{0})$, where

$$\mathbf{A}_* = \begin{pmatrix} 1 & 1 \\ 0 & 0 \end{pmatrix}, \quad \mathbf{B}_* = \begin{pmatrix} 1 & 0 \\ -1 & 0 \end{pmatrix}, \quad \text{and} \quad \mathbf{C}_* = \begin{pmatrix} 1 & 0 \\ 0 & -1 \end{pmatrix}.$$

Then, $\mathbf{A}_*\mathbf{B}_*\mathbf{C}_* = \mathbf{0}$ and $\mathbf{B}_*\mathbf{A}_*\mathbf{C}_* = \begin{pmatrix} 1 & -1 \\ -1 & 1 \end{pmatrix}$, and, observing that $\mathbf{ABC} = \mathrm{diag}(\mathbf{A}_*\mathbf{B}_*\mathbf{C}_*, \mathbf{0})$ and $\mathbf{BAC} = \mathrm{diag}(\mathbf{B}_*\mathbf{A}_*\mathbf{C}_*, \mathbf{0})$ and making use of result (2.5), we find that

$$\mathrm{tr}(\mathbf{BAC}) = \mathrm{tr}(\mathbf{B}_*\mathbf{A}_*\mathbf{C}_*) = 2 \neq 0 = \mathrm{tr}(\mathbf{A}_*\mathbf{B}_*\mathbf{C}_*) = \mathrm{tr}(\mathbf{ABC}).$$

EXERCISE 8. Which of the following sets are linear spaces: (1) the set of all $N \times N$ diagonal matrices, (2) the set of all $N \times N$ upper triangular matrices, and (3) the set of all $N \times N$ nonsymmetric matrices?

Solution. Clearly, the sum of two $N \times N$ diagonal matrices is a diagonal matrix, and the sum of two $N \times N$ upper triangular matrices is upper triangular. Likewise, a scalar multiple of an $N \times N$ diagonal matrix is a diagonal matrix, and a scalar multiple of an $N \times N$ upper triangular matrix is upper triangular. However, the sum of two $N \times N$ nonsymmetric matrices is not necessarily nonsymmetric. For example, if $\mathbf{A}$ is an $N \times N$ nonsymmetic matrix, then $-\mathbf{A}$ and $\mathbf{A}'$ are also $N \times N$ nonsymmetric matrices, yet the sums $\mathbf{A} + (-\mathbf{A}) = \mathbf{0}$ and $\mathbf{A} + \mathbf{A}'$ are symmetric. Also, the product of the scalar 0 and any $N \times N$ nonsymmetric matrix is the $N \times N$ null matrix, which is symmetric. Thus, the set of all $N \times N$ diagonal matrices and the set of all $N \times N$ upper triangular matrices are linear spaces, but the set of all $N \times N$ nonsymmetric matrices is not a linear space.

EXERCISE 9. Define

$$\mathbf{A} = \begin{pmatrix} 1 & 2 & -1 & 0 \\ 2 & 1 & 1 & 1 \\ 1 & -1 & 2 & 1 \end{pmatrix},$$

and (for $i = 1, 2, 3$) let $\mathbf{a}_i'$ represent the ith row of $\mathbf{A}$.

(a) Show that the set $\{\mathbf{a}_1', \mathbf{a}_2'\}$ is a basis for $\mathcal{R}(\mathbf{A})$.

(b) Find $\mathrm{rank}(\mathbf{A})$.

(c) Making use of the answer to Part (b) (or otherwise), find a basis for $\mathcal{C}(\mathbf{A})$.

Solution. (a) For any scalars k_1 and k_2 such that $k_1\mathbf{a}_1' + k_2\mathbf{a}_2' = \mathbf{0}$, we have that $k_1 + 2k_2 = 0$, $2k_1 + k_2 = 0$, $k_2 - k_1 = 0$, and $k_2 = 0$, implying that $k_1 = k_2 = 0$. Thus, the set $\{\mathbf{a}_1', \mathbf{a}_2'\}$ is linearly independent. Moreover, $\mathbf{a}_3' = \mathbf{a}_2' - \mathbf{a}_1'$, so that every row of $\mathbf{A}$ belongs to $\mathrm{sp}(\mathbf{a}_1', \mathbf{a}_2')$, implying (in light of Lemma 2.4.2) that $\mathcal{R}(\mathbf{A}) \subset \mathrm{sp}(\mathbf{a}_1', \mathbf{a}_2')$ and hence that $\{\mathbf{a}_1', \mathbf{a}_2'\}$ spans $\mathcal{R}(\mathbf{A})$. And we conclude that $\{\mathbf{a}_1', \mathbf{a}_2'\}$ is a basis for $\mathcal{R}(\mathbf{A})$.

(b) $\mathrm{Rank}(\mathbf{A}) = 2$, as is evident from the result of Part (a) upon observing that $\mathrm{rank}(\mathbf{A}) = \dim[\mathcal{R}(\mathbf{A})]$.

(c) Any two of the four columns of **A** are linearly independent (as is easily verified). Moreover, $\dim[\mathcal{C}(\mathbf{A})] = \text{rank}(\mathbf{A}) = 2$. And we conclude, on the basis of Theorem 2.4.11, that the set comprising any two of the columns of **A** is a basis for $\mathcal{C}(\mathbf{A})$; in particular, the set comprising the first two columns of **A** is a basis for $\mathcal{C}(\mathbf{A})$.

EXERCISE 10. Let $\mathbf{A}_1, \mathbf{A}_2, \ldots, \mathbf{A}_K$ represent matrices in a linear space $\mathcal{V}$, and let $\mathcal{U}$ represent a subspace of $\mathcal{V}$. Show that $\text{sp}(\mathbf{A}_1, \mathbf{A}_2, \ldots, \mathbf{A}_K) \subset \mathcal{U}$ if and only if $\mathbf{A}_1, \mathbf{A}_2, \ldots, \mathbf{A}_K$ are contained in $\mathcal{U}$ (thereby establishing what is essentially a generalization of Lemma 2.4.2).

Solution. If $\text{sp}(\mathbf{A}_1, \mathbf{A}_2, \ldots, \mathbf{A}_K) \subset \mathcal{U}$, then, obviously, $\mathbf{A}_1, \mathbf{A}_2, \ldots, \mathbf{A}_K$ are contained in $\mathcal{U}$.

Now (for purposes of establishing the converse), suppose that $\mathbf{A}_1, \mathbf{A}_2, \ldots, \mathbf{A}_K$ are contained in $\mathcal{U}$. And let **A** represent an arbitrary matrix in $\mathcal{V}$. If $\mathbf{A} \in \text{sp}(\mathbf{A}_1, \mathbf{A}_2, \ldots, \mathbf{A}_K)$, then **A** is expressible as a linear combination of $\mathbf{A}_1, \mathbf{A}_2, \ldots, \mathbf{A}_K$, implying (since $\mathcal{U}$ is a linear space) that $\mathbf{A} \in \mathcal{U}$. Thus, $\text{sp}(\mathbf{A}_1, \mathbf{A}_2, \ldots, \mathbf{A}_K) \subset \mathcal{U}$.

EXERCISE 11. Let $\mathcal{V}$ represent a K-dimensional linear space of $M \times N$ matrices (where $K \geq 1$). Further, let $\{\mathbf{A}_1, \mathbf{A}_2, \ldots, \mathbf{A}_K\}$ represent a basis for $\mathcal{V}$, and, for arbitrary scalars $x_1, x_2, \ldots, x_K$ and $y_1, y_2, \ldots, y_K$, define $\mathbf{A} = \sum_{i=1}^K x_i \mathbf{A}_i$ and $\mathbf{B} = \sum_{j=1}^K y_j \mathbf{A}_j$. Show that

$$\mathbf{A} \bullet \mathbf{B} = \sum_{i=1}^{K} x_i y_i$$

for all choices of $x_1, x_2, \ldots, x_K$ and $y_1, y_2, \ldots, y_K$ if and only if the basis $\{\mathbf{A}_1, \mathbf{A}_2, \ldots, \mathbf{A}_K\}$ is orthonormal.

Solution. Making use of result (2.8) (and of basic properties of an inner product), we find that

$$\begin{aligned}
\mathbf{A} \bullet \mathbf{B} &= \sum_{i=1}^{K} x_i \mathbf{A}_i \sum_{j=1}^{K} y_j \mathbf{A}_j \\
&= \sum_{i=1}^{K} x_i \Big(\mathbf{A}_i \bullet \sum_{j=1}^{K} y_j \mathbf{A}_j\Big) \\
&= \sum_{i=1}^{K} x_i \Big[\Big(\sum_{j=1}^{K} y_j \mathbf{A}_j\Big) \bullet \mathbf{A}_i\Big] \\
&= \sum_{i=1}^{K} x_i \sum_{j=1}^{K} y_j (\mathbf{A}_j \bullet \mathbf{A}_i).
\end{aligned}$$

Now, suppose that the basis $\{\mathbf{A}_1, \mathbf{A}_2, \ldots, \mathbf{A}_K\}$ is orthonormal. Then, $\mathbf{A}_i \bullet \mathbf{A}_i = 1$ and, for $j \neq i$, $\mathbf{A}_j \mathbf{A}_i = 0$, implying that $\sum_{j=1}^K y_j (\mathbf{A}_j \bullet \mathbf{A}_i) = y_i$. Thus, $\mathbf{A} \bullet \mathbf{B} = \sum_{i=1}^K x_i y_i$.

Conversely, suppose that $\mathbf{A} \bullet \mathbf{B} = \sum_{i=1}^K x_i y_i$ (for all choices of $x_1, x_2, \ldots, x_K$ and $y_1, y_2, \ldots, y_K$). Upon setting $x_s = 1$ and $y_t = 1$ (where s and t are integers between 1 and K, inclusive) and setting $x_i = 0$ for $i \neq s$ and $y_j = 0$ for $j \neq t$, we find that $\mathbf{A} \bullet \mathbf{B} = \mathbf{A}_s \bullet \mathbf{A}_t$ and that $\sum_{i=1}^K x_i y_i$ equals 1 if $t = s$ and equals 0 if $t \neq s$. Thus, $\mathbf{A}_s \bullet \mathbf{A}_s = 1$ and, for $t \neq s$, $\mathbf{A}_s \bullet \mathbf{A}_t = 0$ ($s = 1, 2, \ldots, K$), so that the basis $\{\mathbf{A}_1, \mathbf{A}_2, \ldots, \mathbf{A}_K\}$ is orthonormal.

EXERCISE 12. An $N \times N$ matrix **A** is said to be *involutory* if $\mathbf{A}^2 = \mathbf{I}$, that is, if **A** is invertible and is its own inverse.

(a) Show that an $N \times N$ matrix **A** is involutory if and only if $(\mathbf{I} - \mathbf{A})(\mathbf{I} + \mathbf{A}) = \mathbf{0}$.

(b) Show that a 2×2 matrix $\mathbf{A} = \begin{pmatrix} a & b \\ c & d \end{pmatrix}$ is involutory if and only if (1) $a^2 + bc = 1$ and $d = -a$ or (2) $b = c = 0$ and $d = a = \pm 1$.

Solution. (a) Clearly,

$$(\mathbf{I}-\mathbf{A})(\mathbf{I}+\mathbf{A}) = \mathbf{I}-\mathbf{A}+(\mathbf{I}-\mathbf{A})\mathbf{A} = \mathbf{I}-\mathbf{A}+\mathbf{A}-\mathbf{A}^2 = \mathbf{I}-\mathbf{A}^2.$$

Thus,

$$(\mathbf{I}-\mathbf{A})(\mathbf{I}+\mathbf{A}) = \mathbf{0} \;\Leftrightarrow\; \mathbf{I}-\mathbf{A}^2 = \mathbf{0} \;\Leftrightarrow\; \mathbf{A}^2 = \mathbf{I}.$$

(b) Clearly,

$$\mathbf{A}^2 = \begin{pmatrix} a^2+bc & ab+bd \\ ac+cd & bc+d^2 \end{pmatrix}.$$

And if Condition (1) or (2) is satisfied, it is easy to see that $\mathbf{A}$ is involutory.

Conversely, suppose that $\mathbf{A}$ is involutory. Then, $ab = -db$ and $ac = -dc$, implying that $d = -a$ or $b = c = 0$. Moreover, $a^2+bc = 1$ and $d^2+bc = 1$. Consequently, if $d = -a$, Condition (1) is satisfied. Alternatively, if $b = c = 0$, then $d^2 = a^2 = 1$, implying that $d = a = \pm 1$ [in which case Condition (2) is satisfied] or that $d = -a = \pm 1$ [in which case Condition (1) is satisfied].

EXERCISE 13. Let $\mathbf{A} = \{a_{ij}\}$ represent an $M \times N$ matrix of full row rank.

(a) Show that in the special case $M = 1$ (i.e., in the special case where $\mathbf{A}$ is an N-dimensional row vector), there exists an N-dimensional column vector $\mathbf{b}$, $N-1$ elements of which are 0, that is a right inverse of $\mathbf{A}$.

(b) Generalize from Part (a) (to an arbitrary value of M) by showing that there exists an $N \times M$ matrix $\mathbf{B}$, $N-M$ rows of which are null vectors, that is a right inverse of $\mathbf{A}$.

Solution. (a) Since $\mathbf{A}$ is of full row rank, it contains at least one nonzero element. Take k to be any integer between 1 and N, inclusive, such that $a_{1k} \neq 0$. And take $\mathbf{b} = \{b_j\}$ to be the N-dimensional column vector whose kth element is $1/a_{1k}$ and whose other $N-1$ elements equal 0. Then, $\mathbf{Ab}$ is the 1×1 matrix (c), where

$$c = \sum_{j=1}^{N} a_{1j} b_j = a_{1k}(1/a_{1k}) = 1.$$

Thus, $\mathbf{b}$ is a right inverse of $\mathbf{A}$.

(b) Since $\mathbf{A}$ is of full row rank (i.e., of rank M), it contains M linearly independent columns, say columns $j_1, j_2, \ldots, j_M$ ($j_1 < j_2 < \cdots < j_M$). Denote by $\mathbf{A}_*$ the $M \times M$ submatrix of $\mathbf{A}$ obtained by striking out all of the columns of $\mathbf{A}$ except the $j_1, j_2, \ldots, j_M$th columns (and observe that $\mathbf{A}_*$ is nonsingular). Now, take $\mathbf{B}$ to be the $N \times M$ matrix whose $j_1, j_2, \ldots, j_M$th rows are respectively the first through Mth rows of $\mathbf{A}_*^{-1}$ and whose other $N-M$ rows are null vectors. Then, letting $\mathbf{a}_j$ represent the jth column of $\mathbf{A}$ and $\mathbf{b}'_j$ the jth row of $\mathbf{B}$ ($j = 1, 2, \ldots, N$), we find that

$$\begin{aligned} \mathbf{AB} &= (\mathbf{a}_1, \mathbf{a}_2, \ldots, \mathbf{a}_N)\begin{pmatrix} \mathbf{b}'_1 \\ \mathbf{b}'_2 \\ \vdots \\ \mathbf{b}'_N \end{pmatrix} \\ &= \sum_{j=1}^{N} \mathbf{a}_j \mathbf{b}'_j = \sum_{s=1}^{M} \mathbf{a}_{j_s} \mathbf{b}'_{j_s} = (\mathbf{a}_{j_1}, \mathbf{a}_{j_2}, \ldots, \mathbf{a}_{j_M})\begin{pmatrix} \mathbf{b}'_{j_1} \\ \mathbf{b}'_{j_2} \\ \vdots \\ \mathbf{b}'_{j_M} \end{pmatrix} = \mathbf{A}_*\mathbf{A}_*^{-1} = \mathbf{I}. \end{aligned}$$

Thus, $\mathbf{B}$ is a right inverse of $\mathbf{A}$.

EXERCISE 14. Provide an alternative verification of equality (6.10) by premultiplying or postmultiplying the right side of the equality by $\begin{pmatrix} \mathbf{T} & \mathbf{U} \\ \mathbf{V} & \mathbf{W} \end{pmatrix}$ and by confirming that the resultant product equals $\mathbf{I}_{M+N}$.

Solution. Let

$$\mathbf{B} = \begin{pmatrix} \mathbf{T} & \mathbf{U} \\ \mathbf{V} & \mathbf{W} \end{pmatrix} \begin{pmatrix} \mathbf{T}^{-1} + \mathbf{T}^{-1}\mathbf{U}\mathbf{Q}^{-1}\mathbf{V}\mathbf{T}^{-1} & -\mathbf{T}^{-1}\mathbf{U}\mathbf{Q}^{-1} \\ -\mathbf{Q}^{-1}\mathbf{V}\mathbf{T}^{-1} & \mathbf{Q}^{-1} \end{pmatrix}.$$

Then, $\mathbf{B} = \begin{pmatrix} \mathbf{B}_{11} & \mathbf{B}_{12} \\ \mathbf{B}_{21} & \mathbf{B}_{22} \end{pmatrix}$, where

$$\begin{aligned} \mathbf{B}_{11} &= \mathbf{T}\,(\mathbf{T}^{-1} + \mathbf{T}^{-1}\mathbf{U}\mathbf{Q}^{-1}\mathbf{V}\mathbf{T}^{-1}) + \mathbf{U}\,(-\mathbf{Q}^{-1}\mathbf{V}\mathbf{T}^{-1}) \\ &= \mathbf{I} + \mathbf{U}\mathbf{Q}^{-1}\mathbf{V}\mathbf{T}^{-1} - \mathbf{U}\mathbf{Q}^{-1}\mathbf{V}\mathbf{T}^{-1} \\ &= \mathbf{I}_M, \end{aligned}$$

$$\mathbf{B}_{12} = \mathbf{T}\,(-\mathbf{T}^{-1}\mathbf{U}\mathbf{Q}^{-1}) + \mathbf{U}\mathbf{Q}^{-1} = -\mathbf{U}\mathbf{Q}^{-1} + \mathbf{U}\mathbf{Q}^{-1} = \mathbf{0},$$

$$\begin{aligned} \mathbf{B}_{21} &= \mathbf{V}\,(\mathbf{T}^{-1} + \mathbf{T}^{-1}\mathbf{U}\mathbf{Q}^{-1}\mathbf{V}\mathbf{T}^{-1}) + \mathbf{W}\,(-\mathbf{Q}^{-1}\mathbf{V}\mathbf{T}^{-1}) \\ &= \mathbf{V}\mathbf{T}^{-1} + \mathbf{V}\mathbf{T}^{-1}\mathbf{U}\mathbf{Q}^{-1}\mathbf{V}\mathbf{T}^{-1} - \mathbf{W}\mathbf{Q}^{-1}\mathbf{V}\mathbf{T}^{-1} \\ &= \mathbf{V}\mathbf{T}^{-1} - \mathbf{Q}\mathbf{Q}^{-1}\mathbf{V}\mathbf{T}^{-1} \\ &= \mathbf{V}\mathbf{T}^{-1} - \mathbf{V}\mathbf{T}^{-1} \\ &= \mathbf{0}, \qquad \text{and} \end{aligned}$$

$$\begin{aligned} \mathbf{B}_{22} &= \mathbf{V}\,(-\mathbf{T}^{-1}\mathbf{U}\mathbf{Q}^{-1}) + \mathbf{W}\mathbf{Q}^{-1} \\ &= -\mathbf{V}\mathbf{T}^{-1}\mathbf{U}\mathbf{Q}^{-1} + \mathbf{W}\mathbf{Q}^{-1} \\ &= \mathbf{Q}\mathbf{Q}^{-1} \\ &= \mathbf{I}_N. \end{aligned}$$

Thus, $\mathbf{B} = \mathbf{I}_{M+N}$.

EXERCISE 15. Let $\mathbf{A} = \begin{pmatrix} 2 & 0 & 4 \\ 3 & 5 & 6 \\ 4 & 2 & 12 \end{pmatrix}$. Use the results of Section 2.6 to show that $\mathbf{A}$ is nonsingular and to obtain $\mathbf{A}^{-1}$. (*Hint*. Partition $\mathbf{A}$ as $\mathbf{A} = \begin{pmatrix} \mathbf{T} & \mathbf{U} \\ \mathbf{V} & \mathbf{W} \end{pmatrix}$, where $\mathbf{T}$ is a square matrix of order 2.)

Solution. Partition $\mathbf{A}$ as $\mathbf{A} = \begin{pmatrix} \mathbf{T} & \mathbf{U} \\ \mathbf{V} & \mathbf{W} \end{pmatrix}$, where $\mathbf{T}$ is a square matrix of order 2. Upon observing that $\mathbf{T}$ is lower triangular and that its diagonal elements are nonzero, it follows from Part (1) of Lemma 2.6.4 that $\mathbf{T}$ is nonsingular and that

$$\mathbf{T}^{-1} = \begin{pmatrix} 1/2 & 0 \\ -(1/5)3(1/2) & 1/5 \end{pmatrix} = \begin{pmatrix} 0.5 & 0 \\ -0.3 & 0.2 \end{pmatrix}.$$

Now, defining $\mathbf{Q}$ as in Theorem 2.6.6, that is, as $\mathbf{Q} = \mathbf{W} - \mathbf{V}\mathbf{T}^{-1}\mathbf{U}$, we find that

$$\mathbf{Q} = \left[12 - (4, 2) \begin{pmatrix} 0.5 & 0 \\ -0.3 & 0.2 \end{pmatrix} \begin{pmatrix} 4 \\ 6 \end{pmatrix} \right] = (12 - 8) = (4).$$

And, based on Part (1) of Theorem 2.6.6, we conclude that $\mathbf{A}$ is nonsingular, and, upon observing that

$$\mathbf{Q}^{-1} = (4)^{-1} = (0.25),$$

$$\mathbf{Q}^{-1}\mathbf{V}\mathbf{T}^{-1} = (0.25)(4,2)\begin{pmatrix} 0.5 & 0 \\ -0.3 & 0.2 \end{pmatrix} = (0.35, 0.1),$$

$$\mathbf{T}^{-1}\mathbf{U}\mathbf{Q}^{-1} = \begin{pmatrix} 0.5 & 0 \\ -0.3 & 0.2 \end{pmatrix}\begin{pmatrix} 4 \\ 6 \end{pmatrix}(0.25) = \begin{pmatrix} 0.5 \\ 0 \end{pmatrix}, \quad \text{and}$$

$$\begin{aligned}\mathbf{T}^{-1} + \mathbf{T}^{-1}\mathbf{U}\mathbf{Q}^{-1}\mathbf{V}\mathbf{T}^{-1} &= \mathbf{T}^{-1} + (\mathbf{T}^{-1}\mathbf{U}\mathbf{Q}^{-1})\,\mathbf{Q}\,(\mathbf{Q}^{-1}\mathbf{V}\mathbf{T}^{-1}) \\ &= \begin{pmatrix} 0.5 & 0 \\ -0.3 & 0.2 \end{pmatrix} + \begin{pmatrix} 0.5 \\ 0 \end{pmatrix}(4)(0.35, 0.1) \\ &= \begin{pmatrix} 0.5 & 0 \\ -0.3 & 0.2 \end{pmatrix} + \begin{pmatrix} 0.7 & 0.2 \\ 0 & 0 \end{pmatrix} \\ &= \begin{pmatrix} 1.2 & 0.2 \\ -0.3 & 0.2 \end{pmatrix},\end{aligned}$$

we further conclude that

$$\mathbf{A}^{-1} = \begin{pmatrix} 1.2 & 0.2 & -0.5 \\ -0.3 & 0.2 & 0 \\ -0.35 & -0.1 & 0.25 \end{pmatrix}.$$

EXERCISE 16. Let $\mathbf{T} = \{t_{ij}\}$ represent an $N \times N$ triangular matrix. Show that if $\mathbf{T}$ is orthogonal, then $\mathbf{T}$ is diagonal. If $\mathbf{T}$ is orthogonal, what can be inferred about the values of the diagonal elements $t_{11}, t_{22}, \ldots, t_{NN}$ of $\mathbf{T}$?

Solution. Suppose that $\mathbf{T}$ is orthogonal, and consider the case where $\mathbf{T}$ is upper triangular (i.e., the case where $t_{ij} = 0$ for $i > j = 1, 2, \ldots, N$). For $j = 1, 2, \ldots, N$, let $\mathbf{t}_j = (t_{1j}, t_{2j}, \ldots, t_{Nj})'$ represent the jth column of $\mathbf{T}$. Then, the ijth element of $\mathbf{T}'\mathbf{T}$ equals $\mathbf{t}_i'\mathbf{t}_j$. And, for $i \le j = 1, 2, \ldots, N$,

$$\mathbf{t}_i'\mathbf{t}_j = \sum_{k=1}^{N} t_{ki}t_{kj} = \sum_{k=1}^{i} t_{ki}t_{kj}.$$

In particular, for $j = 1, 2, \ldots, N$, we have that

$$\mathbf{t}_1'\mathbf{t}_j = t_{11}t_{1j}.$$

Thus, $t_{11}^2 = 1$ and (for $j > 1$) $t_{11}t_{1j} = 0$, implying that $t_{11} = \pm 1$ and that (for $j > 1$) $t_{1j} = 0$.
We have established that, for $i = 1$,

$$t_{ii} = \pm 1 \quad \text{and} \quad t_{i,i+1} = t_{i,i+2} = \cdots = t_{iN} = 0. \tag{S.1}$$

Let us now show that result (S.1) holds for all i. To do so, let us proceed by mathematical induction. Suppose that result (S.1) holds for all i less than or equal to an arbitrary integer I (between 1 and $N-1$, inclusive). Then, for $j = I+1, I+2, \ldots, N$,

$$\mathbf{t}_{I+1}'\mathbf{t}_j = \sum_{k=1}^{I+1} t_{k,I+1}t_{kj} = t_{I+1,I+1}t_{I+1,j}.$$

Thus, $t_{I+1,I+1}^2 = 1$ and (for $j > I+1$) $t_{I+1,I+1}t_{I+1,j} = 0$, implying that $t_{I+1,I+1} = \pm 1$ and that (for $j > I+1$) $t_{I+1,j} = 0$ and hence that result (S.1) holds for $i = 1, \ldots, I, I+1$ (which completes the mathematical induction argument). And we conclude that $\mathbf{T}$ is diagonal and that its diagonal elements equal ± 1—this is also the case when $\mathbf{T}$ is lower triangular, as can be established via a similar argument.

EXERCISE 17. Let $\mathbf{A}$ represent an $N \times N$ matrix. Show that for any $N \times N$ nonsingular matrix $\mathbf{B}$, $\mathbf{B}^{-1}\mathbf{A}\mathbf{B}$ is idempotent if and only if $\mathbf{A}$ is idempotent.

Solution. If $\mathbf{A}$ is idempotent, then

$$(\mathbf{B}^{-1}\mathbf{A}\mathbf{B})^2 = \mathbf{B}^{-1}\mathbf{A}\mathbf{B}\mathbf{B}^{-1}\mathbf{A}\mathbf{B} = \mathbf{B}^{-1}\mathbf{A}^2\mathbf{B} = \mathbf{B}^{-1}\mathbf{A}\mathbf{B}$$

and hence $\mathbf{B}^{-1}\mathbf{A}\mathbf{B}$ is idempotent. Conversely, if $\mathbf{B}^{-1}\mathbf{A}\mathbf{B}$ is idempotent, then

$$\mathbf{A}^2 = \mathbf{A}\mathbf{A} = \mathbf{I}_N\mathbf{A}\mathbf{I}_N\mathbf{A}\mathbf{I}_N = \mathbf{B}\mathbf{B}^{-1}\mathbf{A}\mathbf{B}\mathbf{B}^{-1}\mathbf{A}\mathbf{B}\mathbf{B}^{-1} = \mathbf{B}\mathbf{B}^{-1}\mathbf{A}\mathbf{B}\mathbf{B}^{-1} = \mathbf{A},$$

so that $\mathbf{A}$ is idempotent.

EXERCISE 18. Let $\mathbf{x} = \{x_i\}$ and $\mathbf{y} = \{y_i\}$ represent nonnull N-dimensional column vectors. Show that $\mathbf{x}\mathbf{y}'$ is a scalar multiple of an idempotent matrix (i.e., that $\mathbf{x}\mathbf{y}' = c\mathbf{A}$ for some scalar c and some idempotent matrix $\mathbf{A}$) if and only if $\sum_{i=1}^N x_i y_i \neq 0$ (i.e., if and only if $\mathbf{x}$ and $\mathbf{y}$ are not orthogonal with respect to the usual inner product).

Solution. Let $k = \sum_{i=1}^N x_i y_i$. And observe that

$$(\mathbf{x}\mathbf{y}')^2 = \mathbf{x}\mathbf{y}'\mathbf{x}\mathbf{y}' = \mathbf{x}(\mathbf{y}'\mathbf{x})\mathbf{y}' = k\mathbf{x}\mathbf{y}'. \tag{S.2}$$

Now, suppose that $\mathbf{x}\mathbf{y}' = c\mathbf{A}$ for some scalar c and some idempotent matrix $\mathbf{A}$. Then, $c \neq 0$ and $\mathbf{A} \neq \mathbf{0}$, as is evident upon observing that the ijth element of $\mathbf{x}\mathbf{y}'$ equals $x_i y_j$ $(i, j = 1, 2, \ldots, N)$ and hence that $\mathbf{x}\mathbf{y}'$ (like $\mathbf{x}$ and $\mathbf{y}$) is nonnull. Moreover,

$$(\mathbf{x}\mathbf{y}')^2 = c\mathbf{A}(c\mathbf{A}) = c^2\mathbf{A}^2 = c^2\mathbf{A}.$$

Thus, $(\mathbf{x}\mathbf{y}')^2 \neq \mathbf{0}$, and we conclude [on the basis of result (S.2)] that $k \neq 0$.

Conversely, suppose that $k \neq 0$. And take $c = k$ and $\mathbf{A} = (1/k)\mathbf{x}\mathbf{y}'$. Then, $\mathbf{x}\mathbf{y}' = c\mathbf{A}$. Moreover,

$$\mathbf{A}^2 = (1/k)\mathbf{x}\mathbf{y}'[(1/k)\mathbf{x}\mathbf{y}'] = (1/k^2)\mathbf{x}(\mathbf{y}'\mathbf{x})\mathbf{y}' = (1/k^2)k\mathbf{x}\mathbf{y}' = (1/k)\mathbf{x}\mathbf{y}' = \mathbf{A},$$

and hence $\mathbf{A}$ is idempotent.

EXERCISE 19. Let $\mathbf{A}$ represent a $4 \times N$ matrix of rank 2, and take $\mathbf{b} = \{b_i\}$ to be a 4-dimensional column vector. Suppose that $b_1 = 1$ and $b_2 = 0$ and that two of the N columns of $\mathbf{A}$ are the vectors $\mathbf{a}_1 = (5, 4, 3, 1)'$ and $\mathbf{a}_2 = (1, 2, 0, -1)'$. Determine for which values of b_3 and b_4 the linear system $\mathbf{A}\mathbf{x} = \mathbf{b}$ (in $\mathbf{x}$) is consistent.

Solution. According to Theorem 2.9.2, $\mathbf{A}\mathbf{x} = \mathbf{b}$ is consistent if and only if $\mathbf{b} \in \mathcal{C}(\mathbf{A})$. Moreover, $\mathcal{C}(\mathbf{A}) = \text{sp}(\mathbf{a}_1, \mathbf{a}_2)$, as is evident from Theorem 2.4.11 upon observing that neither $\mathbf{a}_1$ nor $\mathbf{a}_2$ is a scalar multiple of the other and hence that $\mathbf{a}_1$ and $\mathbf{a}_2$ are linearly independent. Thus, $\mathbf{A}\mathbf{x} = \mathbf{b}$ is consistent if and only if $\mathbf{b} \in \text{sp}(\mathbf{a}_1, \mathbf{a}_2)$ or, equivalently, if and only if there exist scalars k_1 and k_2 such that

$$\mathbf{b} = k_1\mathbf{a}_1 + k_2\mathbf{a}_2.$$

The first and second elements of the vector $k_1\mathbf{a}_1 + k_2\mathbf{a}_2$ equal b_1 $(= 1)$ and b_2 $(= 0)$, respectively, if and only if $k_1 = 1/3$ and $k_2 = -2/3$, as is easily verified. And, for $k_1 = 1/3$ and $k_2 = -2/3$, both the third and fourth elements of $k_1\mathbf{a}_1 + k_2\mathbf{a}_2$ equal 1. Accordingly, there exist scalars k_1 and k_2 such that $\mathbf{b} = k_1\mathbf{a}_1 + k_2\mathbf{a}_2$ if and only if $b_3 = b_4 = 1$, and hence $\mathbf{A}\mathbf{x} = \mathbf{b}$ is consistent if and only if $b_3 = b_4 = 1$.

EXERCISE 20. Let $\mathbf{A}$ represent an $M \times N$ matrix. Show that for any generalized inverses $\mathbf{G}_1$ and $\mathbf{G}_2$ of $\mathbf{A}$ and for any scalars w_1 and w_2 such that $w_1 + w_2 = 1$, the linear combination $w_1\mathbf{G}_1 + w_2\mathbf{G}_2$ is a generalized inverse of $\mathbf{A}$.

Solution. We find that

$$\mathbf{A}(w_1\mathbf{G}_1 + w_2\mathbf{G}_2)\mathbf{A} = w_1\mathbf{A}\mathbf{G}_1\mathbf{A} + w_2\mathbf{A}\mathbf{G}_2\mathbf{A} = w_1\mathbf{A} + w_2\mathbf{A} = (w_1 + w_2)\mathbf{A} = \mathbf{A}.$$

Thus, $w_1\mathbf{G}_1 + w_2\mathbf{G}_2$ is a generalized inverse of $\mathbf{A}$.

EXERCISE 21. Let $\mathbf{A}$ represent an $N \times N$ matrix.

(a) Using the result of Exercise 20 in combination with Corollary 2.10.11 (or otherwise), show that if $\mathbf{A}$ is symmetric, then $\mathbf{A}$ has a symmetric generalized inverse.

(b) Show that if $\mathbf{A}$ is singular (i.e., of rank less than N) and if $N > 1$, then (even if $\mathbf{A}$ is symmetric) $\mathbf{A}$ has a nonsymmetric generalized inverse. (*Hint.* Make use of the second part of Theorem 2.10.7.)

Solution. (a) Suppose that $\mathbf{A}$ is symmetric. Then, according to Corollary 2.10.11, $(\mathbf{A}^-)'$ is a generalized inverse of $\mathbf{A}$, and hence it follows from the result of Exercise 20 that $(1/2)\mathbf{A}^- + (1/2)(\mathbf{A}^-)'$ is a generalized inverse of $\mathbf{A}$. Moreover,

$$\begin{aligned}[(1/2)\mathbf{A}^- + (1/2)(\mathbf{A}^-)']' &= (1/2)(\mathbf{A}^-)' + (1/2)[(\mathbf{A}^-)']' \\ &= (1/2)(\mathbf{A}^-)' + (1/2)\mathbf{A}^- = (1/2)\mathbf{A}^- + (1/2)(\mathbf{A}^-)'.\end{aligned}$$

Thus, $(1/2)\mathbf{A}^- + (1/2)(\mathbf{A}^-)'$ is a symmetric generalized inverse of $\mathbf{A}$.

(b) Suppose that $\mathbf{A}$ is singular and that $N > 1$. And suppose further (for purposes of establishing a contradiction) that every generalized inverse of $\mathbf{A}$ is symmetric.

Let $\mathbf{G}$ represent any generalized inverse of $\mathbf{A}$. Since $\text{rank}(\mathbf{A}) < N$, $\mathbf{G}$ is not a left inverse of $\mathbf{A}$ (as is evident from Lemma 2.5.1). Thus, $\mathbf{I} - \mathbf{G}\mathbf{A} \neq \mathbf{0}$, and hence, among the elements of $\mathbf{I} - \mathbf{G}\mathbf{A}$, is an element, say the ikth element, that differs from 0. Now, take j to be any of the first N positive integers $1, 2, \ldots, N$ other than i, and take $\mathbf{T}$ to be an $N \times N$ matrix whose kjth element is nonzero and whose remaining $N^2 - 1$ elements equal 0. Then, the ijth element of $(\mathbf{I} - \mathbf{G}\mathbf{A})\mathbf{T}$ equals the product of the ikth element of $\mathbf{I} - \mathbf{G}\mathbf{A}$ and the kjth element of $\mathbf{T}$ and consequently is nonzero, while the jith element of $(\mathbf{I} - \mathbf{G}\mathbf{A})\mathbf{T}$ equals 0. This implies that the ijth element of $\mathbf{G} + (\mathbf{I} - \mathbf{G}\mathbf{A})\mathbf{T}$ differs from the jith element (since, in keeping with our supposition, $\mathbf{G}$ is symmetric), so that $\mathbf{G} + (\mathbf{I} - \mathbf{G}\mathbf{A})\mathbf{T}$ is nonsymmetric. Moreover, $\mathbf{G} + (\mathbf{I} - \mathbf{G}\mathbf{A})\mathbf{T}$ is a generalized inverse of $\mathbf{A}$, as is evident from the second part of Theorem 2.10.7. Accordingly, we have arrived at the sought-after contradiction (of the supposition that every generalized inverse of $\mathbf{A}$ is symmetric). And we conclude that $\mathbf{A}$ has a nonsymmetric generalized inverse.

EXERCISE 22. Let $\mathbf{A}$ represent an $M \times N$ matrix of rank $N - 1$. And let $\mathbf{x}$ represent any nonnull vector in $\mathcal{N}(\mathbf{A})$, that is, any N-dimensional nonnull column vector such that $\mathbf{A}\mathbf{x} = \mathbf{0}$. Show that a matrix $\mathbf{Z}^*$ is a solution to the homogeneous linear system $\mathbf{A}\mathbf{Z} = \mathbf{0}$ (in an $N \times P$ matrix $\mathbf{Z}$) if and only if $\mathbf{Z}^* = \mathbf{x}\mathbf{k}'$ for some P-dimensional row vector $\mathbf{k}'$.

Solution. Suppose that $\mathbf{Z}^* = \mathbf{x}\mathbf{k}'$ for some P-dimensional row vector $\mathbf{k}'$. Then,

$$\mathbf{A}\mathbf{Z}^* = (\mathbf{A}\mathbf{x})\mathbf{k}' = \mathbf{0}\mathbf{k}' = \mathbf{0}.$$

Conversely, suppose that $\mathbf{A}\mathbf{Z}^* = \mathbf{0}$. And denote the columns of $\mathbf{Z}^*$ by $\mathbf{z}_1^*, \mathbf{z}_2^*, \ldots, \mathbf{z}_P^*$, respectively. Then, clearly, $\mathbf{A}\mathbf{z}_j^* = \mathbf{0}$ and hence $\mathbf{z}_j^* \in \mathcal{N}(\mathbf{A})$ $(j = 1, 2, \ldots, P)$. Now, upon observing (in light of Lemma 2.11.5) that $\dim[\mathcal{N}(\mathbf{A})] = N - (N - 1) = 1$, it follows from Theorem 2.4.11 that $\{\mathbf{x}\}$ is a basis for $\mathcal{N}(\mathbf{A})$. Thus, $\mathbf{z}_j^* = k_j^*\mathbf{x}$ for some scalar k_j^* $(j = 1, 2, \ldots, P)$. And we conclude that $\mathbf{Z}^* = \mathbf{x}\mathbf{k}'$ for $\mathbf{k}' = (k_1^*, k_2^*, \ldots, k_P^*)$.

EXERCISE 23. Suppose that $\mathbf{AX} = \mathbf{B}$ is a consistent linear system (in an $N \times P$ matrix $\mathbf{X}$).

(a) Show that if $\text{rank}(\mathbf{A}) = N$ or $\text{rank}(\mathbf{B}) = P$, then, corresponding to any solution $\mathbf{X}^*$ to $\mathbf{AX} = \mathbf{B}$, there is a generalized inverse $\mathbf{G}$ of $\mathbf{A}$ such that $\mathbf{X}^* = \mathbf{GB}$.

(b) Show that if $\text{rank}(\mathbf{A}) < N$ and $\text{rank}(\mathbf{B}) < P$, then there exists a solution $\mathbf{X}^*$ to $\mathbf{AX} = \mathbf{B}$ such that there is no generalized inverse $\mathbf{G}$ of $\mathbf{A}$ for which $\mathbf{X}^* = \mathbf{GB}$.

Solution. (a) Assume that $\text{rank}(\mathbf{A}) = N$. Then, according to Lemma 2.10.3, every generalized inverse of $\mathbf{A}$ is a left inverse. Thus, for every generalized inverse $\mathbf{G}$ of $\mathbf{A}$, we have that

$$\mathbf{X}^* = \mathbf{I}_N\mathbf{X}^* = \mathbf{GAX}^* = \mathbf{GB}.$$

Alternatively, assume that $\text{rank}(\mathbf{B}) = P$. Then, according to Lemma 2.5.1, $\mathbf{B}$ has a left inverse, say $\mathbf{L}$. Moreover, according to Theorem 2.11.7,

$$\mathbf{X}^* = \mathbf{A}^-\mathbf{B} + (\mathbf{I} - \mathbf{A}^-\mathbf{A})\mathbf{Y}$$

for some matrix $\mathbf{Y}$. Thus,

$$\mathbf{X}^* = \mathbf{A}^-\mathbf{B} + (\mathbf{I} - \mathbf{A}^-\mathbf{A})\mathbf{Y}\mathbf{I}_P = \mathbf{A}^-\mathbf{B} + (\mathbf{I} - \mathbf{A}^-\mathbf{A})\mathbf{YLB} = [\mathbf{A}^- + (\mathbf{I} - \mathbf{A}^-\mathbf{A})\mathbf{YL}]\mathbf{B}.$$

And $\mathbf{A}^- + (\mathbf{I} - \mathbf{A}^-\mathbf{A})\mathbf{YL}$ is a generalized inverse of $\mathbf{A}$, as is evident from the second part of Theorem 2.10.7.

(b) Suppose that $\text{rank}(\mathbf{A}) < N$ and $\text{rank}(\mathbf{B}) < P$. Then, there exist a nonnull vector, say $\mathbf{r}$, in $\mathcal{N}(\mathbf{A})$ and a nonnull vector, say $\mathbf{s}$, in $\mathcal{N}(\mathbf{B})$.

For any generalized inverse $\mathbf{G}$ of $\mathbf{A}$, we have that

$$\mathbf{GBs} = \mathbf{G}(\mathbf{Bs}) = \mathbf{G0} = \mathbf{0}. \tag{S.3}$$

Now, taking $\mathbf{G}_0$ to be any particular generalized inverse of $\mathbf{A}$, define

$$\mathbf{X}^* = \mathbf{X}_0 + \mathbf{Z}^*,$$

where $\mathbf{X}_0 = \mathbf{G}_0\mathbf{B}$ and $\mathbf{Z}^* = \mathbf{rs}'$. Then,

$$\mathbf{AX}^* = \mathbf{AX}_0 + \mathbf{AZ}^* = \mathbf{B} + (\mathbf{Ar})\mathbf{s}' = \mathbf{B} + \mathbf{0s}' = \mathbf{B},$$

so that $\mathbf{X}^*$ is a solution to $\mathbf{AX} = \mathbf{B}$. Moreover,

$$\mathbf{X}^*\mathbf{s} = \mathbf{X}_0\mathbf{s} + \mathbf{Z}^*\mathbf{s} = \mathbf{G}_0\mathbf{Bs} + \mathbf{rs}'\mathbf{s} = \mathbf{0} + \mathbf{rs}'\mathbf{s} = (\mathbf{s}'\mathbf{s})\mathbf{r},$$

implying (since $\mathbf{s}'\mathbf{s} \neq 0$) that $\mathbf{X}^*\mathbf{s} \neq \mathbf{0}$. And, based on result (S.3), we conclude that there is no generalized inverse $\mathbf{G}$ of $\mathbf{A}$ for which $\mathbf{X}^* = \mathbf{GA}$.

EXERCISE 24. Show that a matrix $\mathbf{A}$ is symmetric and idempotent if and only if there exists a matrix $\mathbf{X}$ such that $\mathbf{A} = \mathbf{P_X}$.

Solution. That $\mathbf{A}$ is symmetric and idempotent if there exists a matrix $\mathbf{X}$ such that $\mathbf{A} = \mathbf{P_X}$ is an immediate consequence of Parts (4) and (5) of Theorem 2.12.2.

Conversely, suppose that $\mathbf{A}$ is symmetric and idempotent. Then, upon observing that $\mathbf{A}' = \mathbf{A}$ and $\mathbf{A}'\mathbf{A} = \mathbf{A}^2 = \mathbf{A}$, we find that

$$\mathbf{P_A} = \mathbf{A}(\mathbf{A}'\mathbf{A})^-\mathbf{A}' = \mathbf{AA}^-\mathbf{A} = \mathbf{A}.$$

Thus, $\mathbf{A} = \mathbf{P_X}$ for $\mathbf{X} = \mathbf{A}$.

EXERCISE 25. Show that corresponding to any quadratic form $\mathbf{x}'\mathbf{A}\mathbf{x}$ (in an N-dimensional vector $\mathbf{x}$), there exists a unique lower triangular matrix $\mathbf{B}$ such that $\mathbf{x}'\mathbf{A}\mathbf{x}$ and $\mathbf{x}'\mathbf{B}\mathbf{x}$ are identically equal, and express the elements of $\mathbf{B}$ in terms of the elements of $\mathbf{A}$.

Solution. Let $\mathbf{A} = \{a_{ij}\}$ represent the ijth element of $\mathbf{A}$ $(i, j = 1, 2, \ldots, N)$. For an $N \times N$ matrix $\mathbf{B} = \{b_{ij}\}$ that is lower triangular (i.e., having $b_{ij} = 0$ for $j > i = 1, 2, \ldots, N$), the conditions $a_{ii} = b_{ii}$ and $a_{ij} + a_{ji} = b_{ij} + b_{ji}$ $(j \neq i = 1, 2, \ldots, N)$ of Lemma 2.13.1 are equivalent to the conditions $a_{ii} = b_{ii}$ and $a_{ij} + a_{ji} = b_{ij}$ $(j < i = 1, 2, \ldots, N)$. Thus, it follows from the lemma that there exists a unique lower triangular matrix $\mathbf{B}$ such that $\mathbf{x}'\mathbf{A}\mathbf{x}$ and $\mathbf{x}'\mathbf{B}\mathbf{x}$ are identically equal, namely, the lower triangular matrix $\mathbf{B} = \{b_{ij}\}$, where $b_{ii} = a_{ii}$ and $b_{ij} = a_{ij} + a_{ji}$ $(j < i = 1, 2, \ldots, N)$.

EXERCISE 26. Show, via an example, that the sum of two positive semidefinite matrices can be positive definite.

Solution. Consider the two $N \times N$ matrices $\begin{pmatrix} 1 & \mathbf{0} \\ \mathbf{0} & \mathbf{0} \end{pmatrix}$ and $\begin{pmatrix} 0 & \mathbf{0} \\ \mathbf{0} & \mathbf{I}_{N-1} \end{pmatrix}$. Clearly, both of these two matrices are positive semidefinite; however, their sum is the $N \times N$ matrix $\mathbf{I}_N$, which is positive definite.

EXERCISE 27. Let $\mathbf{A}$ represent an $N \times N$ symmetric nonnegative definite matrix (where $N \geq 2$). Define $\mathbf{A}_0 = \mathbf{A}$, and, for $k = 1, 2, \ldots, N-1$, take $\mathbf{Q}_k$ to be an $(N-k+1) \times (N-k+1)$ unit upper triangular matrix, $\mathbf{A}_k$ an $(N-k) \times (N-k)$ matrix, and d_k a scalar that satisfy the recursive relationship

$$\mathbf{Q}_k'\mathbf{A}_{k-1}\mathbf{Q}_k = \operatorname{diag}(d_k,\ \mathbf{A}_k) \tag{E.1}$$

—$\mathbf{Q}_k$, $\mathbf{A}_k$, and d_k can be constructed by making use of Lemma 2.13.19 and by proceeding as in the proof of Theorem 2.13.20.

(a) Indicate how $\mathbf{Q}_1, \mathbf{Q}_2, \ldots, \mathbf{Q}_{N-1}, \mathbf{A}_1, \mathbf{A}_2, \ldots, \mathbf{A}_{N-1}$, and $d_1, d_2, \ldots, d_{N-1}$ could be used to form an $N \times N$ unit upper triangular matrix $\mathbf{Q}$ and a diagonal matrix $\mathbf{D}$ such that $\mathbf{Q}'\mathbf{A}\mathbf{Q} = \mathbf{D}$.

(b) Taking $\mathbf{A} = \begin{pmatrix} 2 & 0 & 0 & 0 \\ 0 & 4 & -2 & -4 \\ 0 & -2 & 1 & 2 \\ 0 & -4 & 2 & 7 \end{pmatrix}$ (which is a symmetric nonnegative definite matrix), determine unit upper triangular matrices $\mathbf{Q}_1$, $\mathbf{Q}_2$, and $\mathbf{Q}_3$, matrices $\mathbf{A}_1$, $\mathbf{A}_2$, and $\mathbf{A}_3$, and scalars d_1, d_2, and d_3 that satisfy the recursive relationship (E.1), and illustrate the procedure devised in response to Part (a) by using it to find a 4×4 unit upper triangular matrix $\mathbf{Q}$ and a diagonal matrix $\mathbf{D}$ such that $\mathbf{Q}'\mathbf{A}\mathbf{Q} = \mathbf{D}$.

Solution. (a) Take $\mathbf{D}$ to be the diagonal matrix $\mathbf{D} = \operatorname{diag}(d_1, d_2, \ldots, d_{N-1}, \mathbf{A}_{N-1})$. And take $\mathbf{Q}$ to be the $N \times N$ unit upper triangular matrix $\mathbf{Q} = \mathbf{P}_1\mathbf{P}_2 \cdots \mathbf{P}_{N-1}$, where $\mathbf{P}_1 = \mathbf{Q}_1$ and where, for $k = 2, 3, \ldots, N-1$, $\mathbf{P}_k = \operatorname{diag}(\mathbf{I}_{k-1},\ \mathbf{Q}_k)$—that $\mathbf{P}_1\mathbf{P}_2 \cdots \mathbf{P}_{N-1}$ is unit upper triangular is evident upon observing that a product of unit upper triangular matrices is itself unit upper triangular. Then,

$$\mathbf{Q}'\mathbf{A}\mathbf{Q} = \mathbf{D}.$$

To see this, define, for $k = 1, 2, \ldots, N-1$,

$$\mathbf{B}_k = \operatorname{diag}(\mathbf{D}_k,\ \mathbf{A}_k),$$

where $\mathbf{D}_k = \operatorname{diag}(d_1, d_2, \ldots, d_k)$. And observe that $\mathbf{B}_1 = \mathbf{P}_1'\mathbf{A}\mathbf{P}_1$ and that, for $k = 2, 3, \ldots, N-1$,

$$\mathbf{B}_k = \begin{pmatrix} \mathbf{D}_{k-1} & \mathbf{0} & \mathbf{0} \\ \mathbf{0} & d_k & \mathbf{0} \\ \mathbf{0} & \mathbf{0} & \mathbf{A}_k \end{pmatrix} = \begin{pmatrix} \mathbf{D}_{k-1} & \mathbf{0} \\ \mathbf{0} & \mathbf{Q}_k'\mathbf{A}_{k-1}\mathbf{Q}_k \end{pmatrix} = \mathbf{P}_k'\mathbf{B}_{k-1}\mathbf{P}_k.$$

Repeated substitution gives

$$\mathbf{B}_{N-1} = \mathbf{P}'_{N-1} \cdots \mathbf{P}'_2\mathbf{P}'_1\mathbf{A}\mathbf{P}_1\mathbf{P}_2 \cdots \mathbf{P}_{N-1} = \mathbf{Q}'\mathbf{A}\mathbf{Q}.$$

It remains only to observe that $\mathbf{B}_{N-1} = \mathbf{D}$.

(b) Recursive relationship (E.1) can be satisfied for $k = 1$ simply by taking $\mathbf{Q}_1 = \mathbf{I}_4$, $d_1 = 2$, and

$$\mathbf{A}_1 = \begin{pmatrix} 4 & -2 & -4 \\ -2 & 1 & 2 \\ -4 & 2 & 7 \end{pmatrix}.$$

Relationship (E.1) can be satisfied for $k = 2$ by proceeding as in Case (1) of the proof of Theorem 2.13.20. Specifically, take

$$\mathbf{Q}_2 = \begin{pmatrix} 1 & 1/2 & 1 \\ 0 & 1 & 0 \\ 0 & 0 & 1 \end{pmatrix},$$

so that

$$\mathbf{Q}'_2\mathbf{A}_1\mathbf{Q}_2 = \begin{pmatrix} 4 & 0 & 0 \\ 0 & 0 & 0 \\ 0 & 0 & 3 \end{pmatrix}.$$

And take $d_2 = 4$ and

$$\mathbf{A}_2 = \begin{pmatrix} 0 & 0 \\ 0 & 3 \end{pmatrix}.$$

Finally, relationship (E.1) can be satisfied for $k = 3$ by proceeding as in Case (2) of the proof of Theorem 2.13.20. Accordingly, take

$$\mathbf{Q}_3 = \begin{pmatrix} 1 & 0 \\ 0 & 1 \end{pmatrix},$$

in which case

$$\mathbf{Q}'_3\mathbf{A}_3\mathbf{Q}_3 = \text{diag}(0, 3).$$

And take $d_3 = 0$ and $\mathbf{A}_3 = (3)$.

Now, making use of the solution to Part (a), we find that

$$\mathbf{Q}'\mathbf{A}\mathbf{Q} = \mathbf{D},$$

with

$$\mathbf{Q} = \mathbf{Q}_1\begin{pmatrix} 1 & \mathbf{0} \\ \mathbf{0} & \mathbf{Q}_2 \end{pmatrix}\begin{pmatrix} \mathbf{I}_2 & \mathbf{0} \\ \mathbf{0} & \mathbf{Q}_3 \end{pmatrix} = \begin{pmatrix} 1 & 0 & 0 & 0 \\ 0 & 1 & 1/2 & 1 \\ 0 & 0 & 1 & 0 \\ 0 & 0 & 0 & 1 \end{pmatrix}$$

and

$$\mathbf{D} = \text{diag}(d_1, d_2, d_3, \mathbf{A}_3) = \text{diag}(2, 4, 0, 3).$$

EXERCISE 28. Let $\mathbf{A} = \{a_{ij}\}$ represent an $N \times N$ symmetric positive definite matrix, and let $\mathbf{B} = \{b_{ij}\} = \mathbf{A}^{-1}$. Show that, for $i = 1, 2, \ldots, N$,

$$b_{ii} \geq 1/a_{ii},$$

with equality holding if and only if $a_{ij} = 0$ for all $j \neq i$.

Solution. Let $\mathbf{U} = (\mathbf{u}_1, \mathbf{U}_2)$, where $\mathbf{u}_1$ is the ith column of $\mathbf{I}_N$ and $\mathbf{U}_2$ is the submatrix of $\mathbf{I}_N$ obtained by striking out the ith column, and observe that $\mathbf{U}$ is orthogonal (so that $\mathbf{U}$ is nonsingular and $\mathbf{U}^{-1} = \mathbf{U}'$). Define $\mathbf{R} = \mathbf{U}'\mathbf{A}\mathbf{U}$ and $\mathbf{S} = \mathbf{U}'\mathbf{B}\mathbf{U}$. And partition $\mathbf{R}$ and $\mathbf{S}$ as

$$\mathbf{R} = \begin{pmatrix} r_{11} & \mathbf{r}' \\ \mathbf{r} & \mathbf{R}_* \end{pmatrix} \quad \text{and} \quad \mathbf{S} = \begin{pmatrix} s_{11} & \mathbf{s}' \\ \mathbf{s} & \mathbf{S}_* \end{pmatrix}$$

[where the dimensions of both $\mathbf{R}_*$ and $\mathbf{S}_*$ are $(N-1) \times (N-1)$]. Then,

$$r_{11} = \mathbf{u}_1'\mathbf{A}\mathbf{u}_1 = a_{ii}, \tag{S.4}$$

$$\mathbf{r}' = \mathbf{u}_1'\mathbf{A}\mathbf{U}_2 = (a_{i1}, a_{i2}, \ldots, a_{i,i-1}, a_{i,i+1}, \ldots, a_{i,N-1}, a_{iN}), \tag{S.5}$$

and

$$s_{11} = \mathbf{u}_1'\mathbf{B}\mathbf{u}_1 = b_{ii}. \tag{S.6}$$

It follows from Corollary 2.13.11 that $\mathbf{R}$ is positive definite, implying (in light of Corollary 2.13.13) that $\mathbf{R}_*$ is positive definite and hence (in light of Corollary 2.13.12) that $\mathbf{R}_*$ is invertible and $\mathbf{R}_*^{-1}$ is positive definite. Also,

$$\mathbf{R}^{-1} = \mathbf{U}^{-1}\mathbf{A}^{-1}(\mathbf{U}^{-1})' = \mathbf{U}'\mathbf{B}\mathbf{U} = \mathbf{S}.$$

Thus, in light of results (S.4) and (S.6), it follows from Theorem 2.13.32 and result (6.11) that $a_{ii} - \mathbf{r}'\mathbf{R}_*^{-1}\mathbf{r} > 0$ and that

$$b_{ii} = (a_{ii} - \mathbf{r}'\mathbf{R}_*^{-1}\mathbf{r})^{-1}.$$

Moreover, $\mathbf{r}'\mathbf{R}_*^{-1}\mathbf{r} \geq 0$, with equality holding if and only if $\mathbf{r} = \mathbf{0}$. Accordingly, we conclude that $b_{ii} \geq 1/a_{ii}$, with equality holding if and only if $\mathbf{r} = \mathbf{0}$ or equivalently [in light of result (S.5)] if and only if $a_{ij} = 0$ for $j \neq i$.

EXERCISE 29. Let

$$\mathbf{A} = \begin{pmatrix} a_{11} & a_{12} & a_{13} & \boxed{a_{14}} \\ a_{21} & a_{22} & \boxed{a_{23}} & a_{24} \\ \boxed{a_{31}} & a_{32} & a_{33} & a_{34} \\ a_{41} & \boxed{a_{42}} & a_{43} & a_{44} \end{pmatrix}.$$

(a) Write out all of the pairs that can be formed from the four " boxed" elements of $\mathbf{A}$.

(b) Indicate which of the pairs from Part (a) are positive and which are negative.

(c) Use formula (14.1) to compute the number of pairs from Part (a) that are negative, and check that the result of this computation is consistent with your answer to Part (b).

Solution. (a) and (b)

Pair	"Sign"
$a_{14},\ a_{23}$	$-$
$a_{14},\ a_{31}$	$-$
$a_{14},\ a_{42}$	$-$
$a_{23},\ a_{31}$	$-$
$a_{23},\ a_{42}$	$-$
$a_{31},\ a_{42}$	$+$

(c) $\phi_4(4, 3, 1, 2) = 3 + 2 + 0 = 5$ [or alternatively $\phi_4(3, 4, 2, 1) = 2 + 2 + 1 = 5$].

EXERCISE 30. Obtain (in as simple form as possible) an expression for the determinant of each of the following two matrices: (1) an $N \times N$ matrix $\mathbf{A} = \{a_{ij}\}$ of the general form

$$\mathbf{A} = \begin{pmatrix} 0 & \ldots & 0 & 0 & a_{1N} \\ 0 & \ldots & 0 & a_{2,N-1} & a_{2N} \\ 0 & & a_{3,N-2} & a_{3,N-1} & a_{3N} \\ & & \vdots & \vdots & \vdots \\ a_{N1} & \ldots & a_{N,N-2} & a_{N,N-1} & a_{NN} \end{pmatrix}$$

(where $a_{ij} = 0$ for $j = 1, 2, \ldots, N-i$; $i = 1, 2, \ldots, N-1$); (2) an $N \times N$ matrix $\mathbf{B} = \{b_{ij}\}$ of the general form

$$\mathbf{B} = \begin{pmatrix} 0 & 1 & 0 & \ldots & 0 \\ 0 & 0 & 1 & & 0 \\ \vdots & \vdots & & \ddots & \\ 0 & 0 & 0 & & 1 \\ -k_0 & -k_1 & -k_2 & \ldots & -k_{N-1} \end{pmatrix}$$

—a matrix of this general form is called a *companion matrix.*

Solution. (1) Consider the expression for the determinant given by the sum (14.2) or (14.2'). For a matrix $\mathbf{A}$ of the specified form, there is only one term of this sum that can be nonzero, namely, the term corresponding to $j_1 = N,\ j_2 = N-1,\ \ldots,\ j_N = 1$. (To verify formally that this is the only term that can be nonzero, let $j_1, j_2, \ldots, j_N$ represent an arbitrary permutation of the first N positive integers, and suppose that $a_{1j_1}a_{2j_2}\cdots a_{Nj_N} \neq 0$ or, equivalently, that $a_{ij_i} \neq 0$ for $i = 1, 2, \ldots, N$. Then, it is clear that $j_1 = N$ and that $j_1 = N \Rightarrow j_2 = N-1$ and more generally $j_{i-1} = N+1-(i-1) \Rightarrow j_i = N+1-i$. We conclude, on the basis of mathematical induction, that $j_1 = N, j_2 = N-1, \ldots, j_N = 1$.) Now, upon observing that

$$\phi_N(N, N-1, \ldots, 1) = (N-1) + (N-2) + \cdots + 1$$

and "recalling" that the sum of the first $N-1$ positive integers equals $N(N-1)/2$, we find that

$$|\mathbf{A}| = (-1)^{N(N-1)/2}\, a_{1N}a_{2,N-1}\cdots a_{N1}.$$

(2) Consider the expression for the determinant given by the sum (14.2) or (14.2'). For a matrix $\mathbf{B}$ of the specified form, there is only one term of this sum that can be nonzero, namely, the term corresponding to $j_1 = 2,\ j_2 = 3, \ldots,\ j_{N-1} = N,\ j_N = 1$. Upon observing that

$$\phi_N(2, 3, \ldots, N, 1) = 1 + 1 + \cdots + 1 = N-1,$$

we conclude that

$$|\mathbf{B}| = (-1)^{N-1}\, b_{12}b_{23}\cdots b_{N-1,N}b_{N1} = (-1)^{N-1}1^{N-1}(-k_0) = (-1)^N k_0.$$

Alternatively, this expression could be derived by making use of result (14.14). Partitioning $\mathbf{B}$ as $\mathbf{B} = \begin{pmatrix} \mathbf{B}_* \\ \mathbf{c}' \end{pmatrix}$, where $\mathbf{c}'$ is the last row of $\mathbf{B}$, it follows from result (14.14) and from formula (14.7) (for the determinant of a triangular matrix) that

$$|\mathbf{A}| = \begin{vmatrix} \mathbf{B}_* \\ \mathbf{c}' \end{vmatrix} = (-1)^{(N-1)1} \begin{vmatrix} \mathbf{c}' \\ \mathbf{B}_* \end{vmatrix} = (-1)^{N-1}(-k_0)\,1^{N-1} = (-1)^N k_0.$$

Still another possibility is to apply formula (14.22). Taking $\mathbf{T} = \mathbf{I}_{N-1}$, $\mathbf{W} = (-k_0)$, and $\mathbf{V} = -(k_1, k_2, \ldots, k_{N-1})$, we find that

$$|\mathbf{B}| = \begin{vmatrix} \mathbf{0} & \mathbf{T} \\ \mathbf{W} & \mathbf{V} \end{vmatrix} = (-1)^{(N-1)1}|\mathbf{T}||\mathbf{W}| = (-1)^{N-1}1(-k_0) = (-1)^N k_0.$$

EXERCISE 31. Verify the part of result (14.16) that pertains to the postmultiplication of a matrix by a unit upper triangular matrix by showing that for any $N \times N$ matrix $\mathbf{A}$ and any $N \times N$ unit upper triangular matrix $\mathbf{T}$, $|\mathbf{AT}| = |\mathbf{A}|$.

Solution. Define $\mathbf{T}_i$ to be a matrix formed from $\mathbf{I}_N$ by replacing the ith column of $\mathbf{I}_N$ with the ith column of $\mathbf{T}$ $(i = 1, 2, \ldots, N)$. Then,

$$\mathbf{T} = \mathbf{T}_N\mathbf{T}_{N-1}\cdots\mathbf{T}_1. \tag{S.7}$$

[That $\mathbf{T}$ is expressible in this form can be established via a mathematical induction argument: let $\mathbf{t}_i$ represent the ith column of $\mathbf{T}$ and $\mathbf{u}_i$ the ith column of $\mathbf{I}_N$ $(i = 1, 2, \ldots, N)$; and observe that $\mathbf{T}_N = (\mathbf{u}_1, \mathbf{u}_2, \ldots, \mathbf{u}_{N-1}, \mathbf{t}_N)$ and that if $\mathbf{T}_N\mathbf{T}_{N-1}\cdots\mathbf{T}_{i+1} = (\mathbf{u}_1, \mathbf{u}_2, \ldots, \mathbf{u}_i, \mathbf{t}_{i+1}, \ldots, \mathbf{t}_{N-1}, \mathbf{t}_N)$, then $\mathbf{T}_N\mathbf{T}_{N-1}\cdots\mathbf{T}_i = (\mathbf{T}_N\mathbf{T}_{N-1}\cdots\mathbf{T}_{i+1})\mathbf{T}_i = (\mathbf{u}_1, \mathbf{u}_2, \ldots, \mathbf{u}_{i-1}, \mathbf{t}_i, \mathbf{t}_{i+1}, \ldots, \mathbf{t}_{N-1}, \mathbf{t}_N)$.] In light of result (S.7), $\mathbf{AT}$ is expressible as

$$\mathbf{AT} = \mathbf{AT}_N\mathbf{T}_{N-1}\cdots\mathbf{T}_1.$$

Now, define $\mathbf{B}_{N+1} = \mathbf{A}$, and $\mathbf{B}_i = \mathbf{AT}_N\mathbf{T}_{N-1}\cdots\mathbf{T}_i$ $(i = N, N-1, \ldots, 2)$. Clearly, to show that $|\mathbf{AT}| = |\mathbf{A}|$, it suffices to show that, for $i = N, N-1, \ldots, 1$, the postmultiplication of $\mathbf{B}_{i+1}$ by $\mathbf{T}_i$ does not alter the determinant of $\mathbf{B}_{i+1}$. Observe that the columns of $\mathbf{B}_{i+1}\mathbf{T}_i$ are the same as those of $\mathbf{B}_{i+1}$, except for the ith column of $\mathbf{B}_{i+1}\mathbf{T}_i$, which consists of the ith column of $\mathbf{B}_{i+1}$ plus scalar multiples of the first through $(i-1)$th columns of $\mathbf{B}_{i+1}$. Accordingly, it follows from Theorem 2.14.12 that $|\mathbf{B}_{i+1}\mathbf{T}_i| = |\mathbf{B}_{i+1}|$. We conclude that $|\mathbf{AT}| = |\mathbf{A}|$.

EXERCISE 32. Show that for any $N \times N$ matrix $\mathbf{A}$ and any $N \times N$ nonsingular matrix $\mathbf{C}$,

$$|\mathbf{C}^{-1}\mathbf{AC}| = |\mathbf{A}|.$$

Solution. Making use of results (14.25) and (14.28), we find that

$$|\mathbf{C}^{-1}\mathbf{AC}| = |\mathbf{C}^{-1}||\mathbf{A}||\mathbf{C}| = (1/|\mathbf{C}|)|\mathbf{A}||\mathbf{C}| = |\mathbf{A}|.$$

EXERCISE 33. Let $\mathbf{A} = \begin{pmatrix} a & b \\ c & d \end{pmatrix}$, where a, b, c, and d are scalars.

(a) Show that in the special case where $\mathbf{A}$ is symmetric (i.e., where $c = b$), $\mathbf{A}$ is nonnegative definite if and only if $a \geq 0$, $d \geq 0$, and $|b| \leq \sqrt{ad}$ and is positive definite if and only if $a > 0$, $d > 0$, and $|b| < \sqrt{ad}$.

(b) Extend the result of Part (a) by showing that in the general case where $\mathbf{A}$ is not necessarily symmetric (i.e., where possibly $c \neq b$), $\mathbf{A}$ is nonnegative definite if and only if $a \geq 0$, $d \geq 0$, and $|b+c|/2 \leq \sqrt{ad}$ and is positive definite if and only if $a > 0$, $d > 0$, and $|b+c|/2 < \sqrt{ad}$. [*Hint*. Take advantage of the result of Part (a).]

Solution. (a) Suppose that $\mathbf{A}$ is symmetric (i.e., that $c = b$). And, based on Corollary 2.13.14 and Lemma 2.14.21, observe that if $\mathbf{A}$ is nonnegative definite, then $a \geq 0$, $d \geq 0$, and $|\mathbf{A}| \geq 0$, and that if $\mathbf{A}$ is positive definite, then $a > 0$, $d > 0$, and $|\mathbf{A}| > 0$. Moreover,

$$|\mathbf{A}| = ad - b^2$$

[as is evident from result (14.4)], so that

$$|\mathbf{A}| \geq 0 \;\Leftrightarrow\; b^2 \leq ad \qquad \text{and} \qquad |\mathbf{A}| > 0 \;\Leftrightarrow\; b^2 < ad.$$

Thus, if $\mathbf{A}$ is nonnegative definite, then $a \geq 0$, $d \geq 0$, and $|b| \leq \sqrt{ad}$, and if $\mathbf{A}$ is positive definite, then $a > 0$, $d > 0$, and $|b| < \sqrt{ad}$.

Conversely, if $a > 0$, then

$$\mathbf{A} = \mathbf{U}'\mathbf{DU},$$

where

$$\mathbf{U} = \begin{pmatrix} 1 & b/a \\ 0 & 1 \end{pmatrix} \quad \text{and} \quad \mathbf{D} = \begin{pmatrix} a & 0 \\ 0 & d - (b^2/a) \end{pmatrix}.$$

And based on Corollary 2.13.16, we conclude that if $a > 0$, $d \geq 0$, and $|b| \leq \sqrt{ad}$ [in which case $d - (b^2/a) \geq 0$], then $\mathbf{A}$ is nonnegative definite and that if $a > 0$, $d > 0$, and $|b| < \sqrt{ad}$ [in which case $d - (b^2/a) > 0$], then $\mathbf{A}$ is positive definite. It remains only to observe that if $a = 0$, $d \geq 0$, and $|b| \leq \sqrt{ad}$ (in which case $b = 0$), then $\mathbf{A} = \text{diag}(0, d)$ and $\text{diag}(0, d)$ is positive semidefinite.

(b) Suppose that $\mathbf{A}$ is not necessarily symmetric (i.e., that possibly $c \neq b$). Then, according to Corollary 2.13.2, there is a unique symmetric matrix $\mathbf{B}$ such that the quadratic forms $\mathbf{x}'\mathbf{A}\mathbf{x}$ and $\mathbf{x}'\mathbf{B}\mathbf{x}$ (in $\mathbf{x}$) are identically equal, namely, the matrix

$$\mathbf{B} = (1/2)(\mathbf{A} + \mathbf{A}') = \begin{pmatrix} a & (b+c)/2 \\ (b+c)/2 & d \end{pmatrix}.$$

Accordingly, $\mathbf{A}$ is nonnegative definite if and only if $\mathbf{B}$ is nonnegative definite and is positive definite if and only if $\mathbf{B}$ is positive definite. And upon applying the result of Part (a) to the matrix $\mathbf{B}$, we find that $\mathbf{A}$ is nonnegative definite if and only if $a \geq 0$, $d \geq 0$, and $|b + c|/2 \leq \sqrt{ad}$ and is positive definite if and only if $a > 0$, $d > 0$, and $|b + c|/2 < \sqrt{ad}$.

EXERCISE 34. Let $\mathbf{A} = \{a_{ij}\}$ represent an $N \times N$ symmetric matrix. And suppose that $\mathbf{A}$ is nonnegative definite (in which case its diagonal elements are nonnegative). By, for example, making use of the result of Part (a) of Exercise 33, show that, for $j \neq i = 1, 2, \ldots, N$,

$$|a_{ij}| \leq \sqrt{a_{ii}a_{jj}} \leq \max(a_{ii}, a_{jj}),$$

with $|a_{ij}| < \sqrt{a_{ii}a_{jj}}$ if $\mathbf{A}$ is positive definite.

Solution. Observe that the 2×2 matrix $\begin{pmatrix} a_{ii} & a_{ij} \\ a_{ji} & a_{jj} \end{pmatrix}$ is a principal submatrix of $\mathbf{A}$ and, accordingly, that it is symmetric and (in light of Corollary 2.13.13) nonnegative definite. And it is positive definite if $\mathbf{A}$ is positive definite. Thus, it follows from Part (a) of Exercise 33 that $|a_{ij}| \leq \sqrt{a_{ii}a_{jj}}$, with $|a_{ij}| < \sqrt{a_{ii}a_{jj}}$ if $\mathbf{A}$ is positive definite.

Moreover, if $a_{ii} \geq a_{jj}$, then

$$\sqrt{a_{ii}a_{jj}} \leq \sqrt{a_{ii}a_{ii}} = a_{ii} = \max(a_{ii}, a_{jj});$$

and similarly if $a_{ii} \leq a_{jj}$, then

$$\sqrt{a_{ii}a_{jj}} \leq \sqrt{a_{jj}a_{jj}} = a_{jj} = \max(a_{ii}, a_{jj}).$$

EXERCISE 35. Let $\mathbf{A} = \{a_{ij}\}$ represent an $N \times N$ symmetric positive definite matrix. Show that

$$\det \mathbf{A} \leq \prod_{i=1}^{N} a_{ii},$$

with equality holding if and only if $\mathbf{A}$ is diagonal.

Solution. That $\det \mathbf{A} = \prod_{i=1}^{N} a_{ii}$ if $\mathbf{A}$ is diagonal is an immediate consequence of Corollary 2.14.2. Thus, it suffices to show that if $\mathbf{A}$ is not diagonal, then $\det \mathbf{A} < \prod_{i=1}^{N} a_{ii}$. This is accomplished by mathematical induction.

Consider a 2×2 symmetric matrix

$$\mathbf{A} = \begin{pmatrix} a_{11} & a_{12} \\ a_{12} & a_{22} \end{pmatrix}$$

that is not diagonal. (Every 1×1 matrix is diagonal.) Upon recalling formula (14.4), we find that

$$|\mathbf{A}| = a_{11}a_{22} - a_{12}^2 < a_{11}a_{22}.$$

Suppose now that, for every $(N-1) \times (N-1)$ symmetric positive definite matrix that is not diagonal, the determinant of the matrix is (strictly) less than the product of its diagonal elements, and consider the determinant of an $N \times N$ symmetric positive definite matrix $\mathbf{A} = \{a_{ij}\}$ that is not diagonal (where $N \geq 3$). Partition $\mathbf{A}$ as

$$\mathbf{A} = \begin{pmatrix} \mathbf{A}_* & \mathbf{a} \\ \mathbf{a}' & a_{NN} \end{pmatrix}$$

[where $\mathbf{A}_*$ is of dimensions $(N-1) \times (N-1)$]. And observe (in light of Corollary 2.13.13) that $\mathbf{A}_*$ is (symmetric) positive definite. Upon recalling (from Lemma 2.13.9) that a positive definite matrix is nonsingular, it follows from Theorem 2.14.22 that

$$|\mathbf{A}| = |\mathbf{A}_*|(a_{NN} - \mathbf{a}'\mathbf{A}_*^{-1}\mathbf{a}). \tag{S.8}$$

Moreover, it follows from Lemma 2.14.21 that $|\mathbf{A}_*| > 0$ and from Theorem 2.13.32 that $a_{NN} - \mathbf{a}'\mathbf{A}_*^{-1}\mathbf{a} > 0$.

In the case where $\mathbf{A}_*$ is diagonal, we have (since $\mathbf{A}$ is not diagonal) that $\mathbf{a} \neq \mathbf{0}$, implying (since, in light of Corollary 2.13.12, $\mathbf{A}_*^{-1}$ is positive definite) that $\mathbf{a}'\mathbf{A}_*^{-1}\mathbf{a} > 0$ and hence that $a_{NN} > a_{NN} - \mathbf{a}'\mathbf{A}_*^{-1}\mathbf{a}$, so that [in light of result (S.8)]

$$|\mathbf{A}| < a_{NN}|\mathbf{A}_*| = a_{NN} \prod_{i=1}^{N-1} a_{ii} = \prod_{i=1}^{N} a_{ii}.$$

In the alternative case where $\mathbf{A}_*$ is not diagonal, we have that $\mathbf{a}'\mathbf{A}_*^{-1}\mathbf{a} \geq 0$, implying that $a_{NN} \geq a_{NN} - \mathbf{a}'\mathbf{A}_*^{-1}\mathbf{a}$, and we have (by supposition) that $|\mathbf{A}_*| < \prod_{i=1}^{N-1} a_{ii}$, so that

$$|\mathbf{A}| < (a_{NN} - \mathbf{a}'\mathbf{A}_*^{-1}\mathbf{a}) \prod_{i=1}^{N-1} a_{ii} \leq a_{NN} \prod_{i=1}^{N-1} a_{ii} = \prod_{i=1}^{N} a_{ii}.$$

Thus, in either case, $|\mathbf{A}| < \prod_{i=1}^{N} a_{ii}$.

3

Random Vectors and Matrices

EXERCISE 1. Provide detailed verifications for (1) equality (1.7), (2) equality (1.8), (3) equality (1.10), and (4) equality (1.9).

Solution. (1) Let c_i represent the ith element of $\mathbf{c}$ and $\mathbf{a}_i'$ the ith row of $\mathbf{A}$. Then, the ith element of $\mathrm{E}(\mathbf{c}+\mathbf{Ax})$ equals $\mathrm{E}(c_i+\mathbf{a}_i'\mathbf{x})$, and the ith element of $\mathbf{c}+\mathbf{A}\mathrm{E}(\mathbf{x})$ equals $c_i+\mathbf{a}_i'\mathrm{E}(\mathbf{x})$. Accordingly, it follows from equality (1.6) that the ith element of $\mathrm{E}(\mathbf{c}+\mathbf{Ax})$ equals the ith element of $\mathbf{c}+\mathbf{A}\mathrm{E}(\mathbf{x})$. Thus, $\mathrm{E}(\mathbf{c}+\mathbf{Ax}) = \mathbf{c}+\mathbf{A}\mathrm{E}(\mathbf{x})$.

(2) Let c_i represent the ith element of $\mathbf{c}$, and, for $j=1,2,\ldots,N$, let x_{ij} represent the ith element of $\mathbf{x}_j$ $(i=1,2,\ldots,M)$. Then, the ith element of $\mathrm{E}(\mathbf{c}+\sum_{j=1}^N a_j\mathbf{x}_j)$ equals $\mathrm{E}(c_i+\sum_{j=1}^N a_j x_{ij})$, and the ith element of $\mathbf{c}+\sum_{j=1}^N a_j\mathrm{E}(\mathbf{x}_j)$ equals $c_i+\sum_{j=1}^N a_j\mathrm{E}(x_{ij})$. Accordingly, it follows from equality (1.5) that the ith element of $\mathrm{E}(\mathbf{c}+\sum_{j=1}^N a_j\mathbf{x}_j)$ equals the ith element of $\mathbf{c}+\sum_{j=1}^N a_j\mathrm{E}(\mathbf{x}_j)$. Thus, $\mathrm{E}(\mathbf{c}+\sum_{j=1}^N a_j\mathbf{x}_j) = \mathbf{c}+\sum_{j=1}^N a_j\mathrm{E}(\mathbf{x}_j)$.

(3) Let $\mathbf{c}_p$ represent the pth column of $\mathbf{C}$, and for $j=1,2,\ldots,N$, let $\mathbf{x}_p^{(j)}$ represent the pth column of $\mathbf{X}_j$ $(p=1,2,\ldots,P)$. Then, the pth column of $\mathrm{E}(\mathbf{C}+\sum_{j=1}^N a_j\mathbf{X}_j)$ equals $\mathrm{E}(\mathbf{c}_p+\sum_{j=1}^N a_j\mathbf{x}_p^{(j)})$, and the pth column of $\mathbf{C}+\sum_{j=1}^N a_j\mathrm{E}(\mathbf{X}_j)$ equals $\mathbf{c}_p+\sum_{j=1}^N a_j\mathrm{E}(\mathbf{x}_p^{(j)})$. Accordingly, it follows from equality (1.8) that the pth column of $\mathrm{E}(\mathbf{C}+\sum_{j=1}^N a_j\mathbf{X}_j)$ equals the pth column of $\mathbf{C}+\sum_{j=1}^N a_j\mathrm{E}(\mathbf{X}_j)$. Thus, $\mathrm{E}(\mathbf{C}+\sum_{j=1}^N a_j\mathbf{X}_j) = \mathbf{C}+\sum_{j=1}^N a_j\mathrm{E}(\mathbf{X}_j)$.

(4) Let $\mathbf{c}_q$ represent the qth column of $\mathbf{C}$ and $\mathbf{k}_q$ the qth column of $\mathbf{K}$, and define $\mathbf{y}_q=\mathbf{Xk}_q$ $(q=1,2,\ldots,Q)$. To verify equality (1.9), it suffices to show that each column of $\mathrm{E}(\mathbf{C}+\mathbf{AXK})$ equals the corresponding column of $\mathbf{C}+\mathbf{A}\mathrm{E}(\mathbf{X})\mathbf{K}$. The qth column of $\mathrm{E}(\mathbf{C}+\mathbf{AXK})$ equals $\mathrm{E}(\mathbf{c}_q+\mathbf{Ay}_q)$. And it follows from equality (1.7) that

$$\mathrm{E}(\mathbf{c}_q+\mathbf{Ay}_q) = \mathbf{c}_q+\mathbf{A}\mathrm{E}(\mathbf{y}_q). \tag{S.1}$$

Moreover, letting k_{pq} represent the pqth element of $\mathbf{K}$ (or, equivalently, the pth element of $\mathbf{k}_q$), letting $\mathbf{x}_p$ represent the pth column of $\mathbf{X}$, and making use of result (1.8), we find that

$$\mathrm{E}(\mathbf{y}_q) = \mathrm{E}(\mathbf{Xk}_q) = \mathrm{E}\Bigl(\sum_{p=1}^P k_{pq}\mathbf{x}_p\Bigr) = \sum_{p=1}^P k_{pq}\,\mathrm{E}(\mathbf{x}_p) = \mathrm{E}(\mathbf{X})\mathbf{k}_q.$$

Upon substituting this expression into expression (S.1), we obtain the equality

$$\mathrm{E}(\mathbf{c}_q+\mathbf{Ay}_q) = \mathbf{c}_q+\mathbf{A}\mathrm{E}(\mathbf{X})\mathbf{k}_q,$$

the right side of which equals the qth column of $\mathbf{C}+\mathbf{A}\mathrm{E}(\mathbf{X})\mathbf{K}$. Thus, each column of $\mathrm{E}(\mathbf{C}+\mathbf{AXK})$ equals the corresponding column of $\mathbf{C}+\mathbf{A}\mathrm{E}(\mathbf{X})\mathbf{K}$.

EXERCISE 2.

(a) Let w and z represent random variables [such that $\mathrm{E}(w^2)<\infty$ and $\mathrm{E}(z^2)<\infty$]. Show that

$$|\mathrm{E}(wz)| \le \mathrm{E}(|wz|) \le \sqrt{\mathrm{E}(w^2)}\,\sqrt{\mathrm{E}(z^2)}; \tag{E.1}$$

and determine the conditions under which the first inequality holds as an equality, the conditions under which the second inequality holds as an equality, and the conditions under which both inequalities hold as equalities.

(b) Let x and y represent random variables [such that $\mathrm{E}(x^2) < \infty$ and $\mathrm{E}(y^2) < \infty$]. Using Part (a) (or otherwise), show that

$$|\mathrm{cov}(x, y)| \leq \mathrm{E}[|x - \mathrm{E}(x)||y - \mathrm{E}(y)|] \leq \sqrt{\mathrm{var}(x)}\sqrt{\mathrm{var}(y)}; \qquad \text{(E.2)}$$

and determine the conditions under which the first inequality holds as an equality, the conditions under which the second inequality holds as an equality, and the conditions under which both inequalities hold as equalities.

Solution. (a) When $\mathrm{E}(w^2) = 0$ or $\mathrm{E}(z^2) = 0$, $w = 0$ with probability 1 or $z = 0$ with probability 1, implying that $wz = 0$ with probability 1 and hence that $\mathrm{E}(wz) = 0$ and $\mathrm{E}(|wz|) = 0$, so that both inequalities in result (E.1) hold as equalities.

Now, suppose that $\mathrm{E}(w^2) > 0$ and $\mathrm{E}(z^2) > 0$. Clearly, $-|wz| \leq wz \leq |wz|$, and it follows that $-\mathrm{E}(|wz|) \leq \mathrm{E}(wz) \leq \mathrm{E}(|wz|)$ and hence that $|\mathrm{E}(wz)| \leq \mathrm{E}(|wz|)$. And to establish the inequality $\mathrm{E}(|wz|) \leq \sqrt{\mathrm{E}(w^2)}\sqrt{\mathrm{E}(z^2)}$, it suffices to observe that this inequality is equivalent to the inequality $|\mathrm{E}(|w||z|)| \leq \sqrt{\mathrm{E}(w^2)}\sqrt{\mathrm{E}(z^2)}$, obtained from inequality (2.13) by applying inequality (2.13) with $|w|$ and $|z|$ in place of w and z, respectively.

Further, the inequality $|\mathrm{E}(wz)| \leq \mathrm{E}(|wz|)$ holds as the equality $\mathrm{E}(wz) = \mathrm{E}(|wz|)$ if and only if $\mathrm{E}(wz - |wz|) = 0$ and hence if and only if $wz \geq 0$ with probability 1. Similarly, the inequality $|\mathrm{E}(wz)| \leq \mathrm{E}(|wz|)$ holds as the equality $\mathrm{E}(wz) = -\mathrm{E}(|wz|)$ if and only if $\mathrm{E}(wz + |wz|) = 0$ and hence if and only if $wz \leq 0$ with probability 1. And equality holds in the inequality $\mathrm{E}(|wz|) \leq \sqrt{\mathrm{E}(w^2)}\sqrt{\mathrm{E}(z^2)}$ if and only if $\mathrm{E}(|w||z|) = \sqrt{\mathrm{E}(w^2)}\sqrt{\mathrm{E}(z^2)}$ and hence if and only if $|z|/\sqrt{\mathrm{E}(z^2)} = |w|/\sqrt{\mathrm{E}(w^2)}$ with probability 1 [as is evident from the conditions under which equality is attained in inequality (2.13) when that inequality is applied with $|w|$ and $|z|$ in place of w and z]. Finally, equality holds in both of the inequalities $|\mathrm{E}(wz)| \leq \mathrm{E}(|wz|)$ and $\mathrm{E}(|wz|) \leq \sqrt{\mathrm{E}(w^2)}\sqrt{\mathrm{E}(z^2)}$ if and only if $|\mathrm{E}(wz)| = \sqrt{\mathrm{E}(w^2)}\sqrt{\mathrm{E}(z^2)}$ and hence if and only if $z/\sqrt{\mathrm{E}(z^2)} = w/\sqrt{\mathrm{E}(w^2)}$ with probability 1 or $z/\sqrt{\mathrm{E}(z^2)} = -w/\sqrt{\mathrm{E}(w^2)}$ with probability 1.

(b) Upon setting $w = x - \mathrm{E}(x)$ and $z = y - \mathrm{E}(y)$ in result (E.1), we obtain result (E.2). Moreover, upon setting $w = x - \mathrm{E}(x)$ and $z = y - \mathrm{E}(y)$ in the conditions under which equality is attained in the first and/or second of the two inequalities in result (E.1), we obtain the conditions under which equality is attained in the corresponding inequality or inequalities in result (E.2). Accordingly, when $\mathrm{var}(x) = 0$ or $\mathrm{var}(y) = 0$, both inequalities in result (E.2) hold as equalities.

Now, suppose that $\mathrm{var}(x) > 0$ and $\mathrm{var}(y) > 0$. Then, the inequality $|\mathrm{cov}(x, y)| \leq \mathrm{E}[|x - \mathrm{E}(x)||y - \mathrm{E}(y)|]$ holds as the equality $\mathrm{cov}(x, y) = \mathrm{E}[|x - \mathrm{E}(x)||y - \mathrm{E}(y)|]$ if and only if $[x - \mathrm{E}(x)][y - \mathrm{E}(y)] \geq 0$ with probability 1 and holds as the equality $\mathrm{cov}(x, y) = -\mathrm{E}[|x - \mathrm{E}(x)||y - \mathrm{E}(y)|]$ if and only if $[x - \mathrm{E}(x)][y - \mathrm{E}(y)] \leq 0$ with probability 1. Further, the inequality $\mathrm{E}[|x - \mathrm{E}(x)||y - \mathrm{E}(y)|] \leq \sqrt{\mathrm{var}(x)}\sqrt{\mathrm{var}(y)}$ holds as an equality if and only if $|y - \mathrm{E}(y)|/\sqrt{\mathrm{var}(y)} = |x - \mathrm{E}(x)|/\sqrt{\mathrm{var}(x)}$ with probability 1. And equality is attained in both of the inequalities in result (E.2) if and only if $[y - \mathrm{E}(y)]/\sqrt{\mathrm{var}(y)} = [x - \mathrm{E}(x)]/\sqrt{\mathrm{var}(x)}$ with probability 1 or $[y - \mathrm{E}(y)]/\sqrt{\mathrm{var}(y)} = -[x - \mathrm{E}(x)]/\sqrt{\mathrm{var}(x)}$ with probability 1.

EXERCISE 3. Let $\mathbf{x}$ represent an N-dimensional random column vector and $\mathbf{y}$ a T-dimensional random column vector. And define $\mathbf{x}_*$ to be an R-dimensional subvector of $\mathbf{x}$ and $\mathbf{y}_*$ an S-dimensional subvector of $\mathbf{y}$ (where $1 \leq R \leq N$ and $1 \leq S \leq T$). Relate $\mathrm{E}(\mathbf{x}_*)$ to $\mathrm{E}(\mathbf{x})$, $\mathrm{var}(\mathbf{x}_*)$ to $\mathrm{var}(\mathbf{x})$, and $\mathrm{cov}(\mathbf{x}_*, \mathbf{y}_*)$ and $\mathrm{cov}(\mathbf{y}_*, \mathbf{x}_*)$ to $\mathrm{cov}(\mathbf{x}, \mathbf{y})$.

Solution. The subvector $\mathbf{x}_*$ is obtained from $\mathbf{x}$ by striking out all except R elements (of $\mathbf{x}$), say the $i_1, i_2, \ldots, i_R$th elements. Similarly, the subvector $\mathbf{y}_*$ is obtained from $\mathbf{y}$ by striking out all except S elements (of $\mathbf{y}$), say the $j_1, j_2, \ldots, j_S$th elements.

Accordingly, $\mathrm{E}(\mathbf{x}_*)$ is the R-dimensional subvector of $\mathrm{E}(\mathbf{x})$ obtained by striking out all of the elements of $\mathrm{E}(\mathbf{x})$ except the $i_1, i_2, \ldots, i_R$th elements. And $\mathrm{var}(\mathbf{x}_*)$ is the $R \times R$ submatrix of $\mathrm{var}(\mathbf{x})$ obtained by striking out all of the rows and columns of $\mathrm{var}(\mathbf{x})$ except the $i_1, i_2, \ldots, i_R$th rows and columns. Further, $\mathrm{cov}(\mathbf{x}_*, \mathbf{y}_*)$ is the $R \times S$ submatrix of $\mathrm{cov}(\mathbf{x}, \mathbf{y})$ obtained by striking out all of the rows and columns of $\mathrm{cov}(\mathbf{x}, \mathbf{y})$ except the $i_1, i_2, \ldots, i_R$th rows and the $j_1, j_2, \ldots, j_S$th columns; and $\mathrm{cov}(\mathbf{y}_*, \mathbf{x}_*)$ is the $S \times R$ submatrix of $[\mathrm{cov}(\mathbf{x}, \mathbf{y})]'$ obtained by striking out all of the rows and columns of $[\mathrm{cov}(\mathbf{x}, \mathbf{y})]'$ except the $j_1, j_2, \ldots, j_S$th rows and $i_1, i_2, \ldots, i_R$th columns.

EXERCISE 4. Let x represent a random variable that is distributed symmetrically about 0 (so that $-x \sim x$); and suppose that the distribution of x is "nondegenerate" in the sense that there exists a nonnegative constant c such that $0 < \Pr(x > c) < \frac{1}{2}$ [and assume that $\mathrm{E}(x^2) < \infty$]. Further, define $y = |x|$.

(a) Show that $\mathrm{cov}(x, y) = 0$.

(b) Are x and y statistically independent? Why or why not?

Solution. (a) Since $-x \sim x$, we have that $\mathrm{cov}(x, y) = \mathrm{cov}(-x, |-x|)$. And upon observing that

$$\mathrm{cov}(-x, |-x|) = \mathrm{cov}(-x, y) = -\mathrm{cov}(x, y),$$

it follows that $\mathrm{cov}(x, y) = -\mathrm{cov}(x, y)$ and hence that $\mathrm{cov}(x, y) = 0$.

(b) The random variables x and y are statistically dependent. To formally verify that conclusion, let c represent a nonnegative constant such that $0 < \Pr(x > c) < \frac{1}{2}$. And observe (in light of the symmetry about 0 of the distribution of x) that $\Pr(x < -c) = \Pr(-x < -c) = \Pr(x > c)$ and hence that

$$\Pr(y \le c) = 1 - \Pr(x < -c) - \Pr(x > c) = 1 - 2\Pr(x > c) > 0.$$

Thus,

$$\Pr(x > c, y \le c) = 0 < \Pr(x > c)\Pr(y \le c),$$

which confirms that x and y are statistically dependent.

EXERCISE 5. Provide detailed verifications for (1) equality (2.39), (2) equality (2.45), and (3) equality (2.48).

Solution. (1) By definition,

$$\begin{aligned} \mathrm{cov}\Big(c + \sum_{i=1}^{N} a_i x_i,\, k + \sum_{j=1}^{T} b_j y_j\Big) &= \mathrm{E}\Big\{\Big[c + \sum_{i=1}^{N} a_i x_i - \mathrm{E}\Big(c + \sum_{i=1}^{N} a_i x_i\Big)\Big] \\ &\qquad \times \Big[k + \sum_{j=1}^{T} b_j y_j - \mathrm{E}\Big(k + \sum_{j=1}^{T} b_j y_j\Big)\Big]\Big\}. \end{aligned} \tag{S.2}$$

In light of equality (1.5), we have that

$$\begin{aligned} c + \sum_{i=1}^{N} a_i x_i - \mathrm{E}\Big(c + \sum_{i=1}^{N} a_i x_i\Big) &= c + \sum_{i=1}^{N} a_i x_i - \Big[c + \sum_{i=1}^{N} a_i \mathrm{E}(x_i)\Big] \\ &= \sum_{i=1}^{N} a_i [x_i - \mathrm{E}(x_i)] \end{aligned} \tag{S.3}$$

and similarly that

$$k + \sum_{j=1}^{T} b_j y_j - \mathrm{E}\Big(k + \sum_{j=1}^{T} b_j y_j\Big) = \sum_{j=1}^{T} b_j [y_j - \mathrm{E}(y_j)]. \tag{S.4}$$

And upon substituting expressions (S.3) and (S.4) into expression (S.2) and making further use of equality (1.5), we find that

$$\begin{aligned}\mathrm{cov}\Big(c + \sum_{i=1}^{N} a_i x_i,\ k + \sum_{j=1}^{T} b_j y_j\Big) &= \mathrm{E}\Big\{\sum_{i=1}^{N} a_i [x_i - \mathrm{E}(x_i)] \sum_{j=1}^{T} b_j [y_j - \mathrm{E}(y_j)]\Big\} \\ &= \mathrm{E}\Big\{\sum_{i=1}^{N}\sum_{j=1}^{T} a_i b_j [x_i - \mathrm{E}(x_i)][y_j - \mathrm{E}(y_j)]\Big\} \\ &= \sum_{i=1}^{N}\sum_{j=1}^{T} a_i b_j\, \mathrm{E}\{[x_i - \mathrm{E}(x_i)][y_j - \mathrm{E}(y_j)]\} \\ &= \sum_{i=1}^{N}\sum_{j=1}^{T} a_i b_j\, \mathrm{cov}(x_i, y_j).\end{aligned}$$

(2) Let c_i represent the ith element of $\mathbf{c}$ and $\mathbf{a}_i'$ the ith row of $\mathbf{A}$ $(i = 1, 2, \ldots, M)$. Similarly, let k_j represent the jth element of $\mathbf{k}$, and $\mathbf{b}_j'$ the jth row of $\mathbf{B}$ $(j = 1, 2, \ldots, S)$. Then, the ith element of $\mathbf{c} + \mathbf{Ax}$ is $c_i + \mathbf{a}_i'\mathbf{x}$, and the jth element of $\mathbf{k} + \mathbf{By}$ is $k_j + \mathbf{b}_j'\mathbf{y}$. Thus, the ijth element of the $M \times S$ matrix $\mathrm{cov}(\mathbf{c} + \mathbf{Ax},\ \mathbf{k} + \mathbf{By})$ is $\mathrm{cov}(c_i + \mathbf{a}_i'\mathbf{x},\ k_j + \mathbf{b}_j'\mathbf{y})$. And it follows from result (2.42) that $\mathrm{cov}(c_i + \mathbf{a}_i'\mathbf{x},\ k_j + \mathbf{b}_j'\mathbf{y})$ equals $\mathbf{a}_i'\mathrm{cov}(\mathbf{x}, \mathbf{y})\mathbf{b}_j$. Also, $\mathbf{a}_i'\mathrm{cov}(\mathbf{x}, \mathbf{y})\mathbf{b}_j$ is the ijth element of the $M \times S$ matrix $\mathbf{A}\mathrm{cov}(\mathbf{x}, \mathbf{y})\mathbf{B}'$, as is evident from result (2.2.12) upon observing that $\mathbf{b}_j$ is the jth column of $\mathbf{B}'$. We conclude that each element of $\mathrm{cov}(\mathbf{c} + \mathbf{Ax},\ \mathbf{k} + \mathbf{By})$ equals the corresponding element of $\mathbf{A}\mathrm{cov}(\mathbf{x}, \mathbf{y})\mathbf{B}'$ and hence that

$$\mathrm{cov}(\mathbf{c} + \mathbf{Ax},\ \mathbf{k} + \mathbf{By}) = \mathbf{A}\mathrm{cov}(\mathbf{x}, \mathbf{y})\mathbf{B}'.$$

(3) Let c_p represent the pth element of $\mathbf{c}$, and, for $i = 1, 2, \ldots, N$, let x_{pi} represent the pth element of $\mathbf{x}_i$ $(p = 1, 2, \ldots, M)$. Similarly, let k_q represent the qth element of $\mathbf{k}$, and, for $j = 1, 2, \ldots, T$, let y_{qj} represent the qth element of $\mathbf{y}_j$ $(q = 1, 2, \ldots, S)$. Then, the pth element of $\mathbf{c} + \sum_{i=1}^{N} a_i \mathbf{x}_i$ is $c_p + \sum_{i=1}^{N} a_i x_{pi}$, and the qth element of $\mathbf{k} + \sum_{j=1}^{T} b_j \mathbf{y}_j$ is $k_q + \sum_{j=1}^{T} b_j y_{qj}$. Thus, the pqth element of the $M \times S$ matrix $\mathrm{cov}(\mathbf{c} + \sum_{i=1}^{N} a_i \mathbf{x}_i,\ \mathbf{k} + \sum_{j=1}^{T} b_j \mathbf{y}_j)$ is $\mathrm{cov}(c_p + \sum_{i=1}^{N} a_i x_{pi},\ k_q + \sum_{j=1}^{T} b_j y_{qj})$. And it follows from result (2.39) that $\mathrm{cov}(c_p + \sum_{i=1}^{N} a_i x_{pi},\ k_q + \sum_{j=1}^{T} b_j y_{qj})$ equals $\sum_{i=1}^{N}\sum_{j=1}^{T} a_i b_j\, \mathrm{cov}(x_{pi}, y_{qj})$, which is the pqth element of the $M \times S$ matrix $\sum_{i=1}^{N}\sum_{j=1}^{T} a_i b_j\, \mathrm{cov}(\mathbf{x}_i, \mathbf{y}_j)$. We conclude that each element of $\mathrm{cov}(\mathbf{c} + \sum_{i=1}^{N} a_i \mathbf{x}_i,\ \mathbf{k} + \sum_{j=1}^{T} b_j \mathbf{y}_j)$ equals the corresponding element of $\sum_{i=1}^{N}\sum_{j=1}^{T} a_i b_j\, \mathrm{cov}(\mathbf{x}_i, \mathbf{y}_j)$ and hence that

$$\mathrm{cov}\Big(\mathbf{c} + \sum_{i=1}^{N} a_i \mathbf{x}_i,\ \mathbf{k} + \sum_{j=1}^{T} b_j \mathbf{y}_j\Big) = \sum_{i=1}^{N}\sum_{j=1}^{T} a_i b_j\, \mathrm{cov}(\mathbf{x}_i, \mathbf{y}_j).$$

EXERCISE 6.

(a) Let x and y represent random variables. Show that $\mathrm{cov}(x, y)$ can be determined from knowledge of $\mathrm{var}(x)$, $\mathrm{var}(y)$, and $\mathrm{var}(x + y)$, and give a formula for doing so.

(b) Let $\mathbf{x} = (x_1, x_2)'$ and $\mathbf{y} = (y_1, y_2)'$ represent 2-dimensional random column vectors. Can $\text{cov}(\mathbf{x}, \mathbf{y})$ be determined from knowledge of $\text{var}(\mathbf{x})$, $\text{var}(\mathbf{y})$, and $\text{var}(\mathbf{x}+\mathbf{y})$? Why or why not?

Solution. (a) As a special case of result (2.41), we have that

$$\text{var}(x + y) = \text{var}(x) + \text{var}(y) + 2\,\text{cov}(x, y).$$

Thus,

$$\text{cov}(x, y) = [\text{var}(x + y) - \text{var}(x) - \text{var}(y)]/2.$$

(b) Clearly, $\text{cov}(\mathbf{x}, \mathbf{y})$ can be determined from knowledge of $\text{var}(\mathbf{x})$, $\text{var}(\mathbf{y})$, and $\text{var}(\mathbf{x}+\mathbf{y})$ if and only if it can be determined from knowledge of $\text{var}(\mathbf{x})$, $\text{var}(\mathbf{y})$, and $\text{var}(\mathbf{x}+\mathbf{y})-\text{var}(\mathbf{x})-\text{var}(\mathbf{y})$. And, as a special case of result (2.50), we have that

$$\text{var}(\mathbf{x}+\mathbf{y}) = \text{var}(\mathbf{x}) + \text{var}(\mathbf{y}) + \text{cov}(\mathbf{x}, \mathbf{y}) + \text{cov}(\mathbf{y}, \mathbf{x})$$

and hence that

$$\text{var}(\mathbf{x}+\mathbf{y}) - \text{var}(\mathbf{x}) - \text{var}(\mathbf{y}) = \text{cov}(\mathbf{x}, \mathbf{y}) + \text{cov}(\mathbf{y}, \mathbf{x}).$$

Further,

$$\text{cov}(\mathbf{x}, \mathbf{y}) + \text{cov}(\mathbf{y}, \mathbf{x}) = \begin{pmatrix} 2\,\text{cov}(x_1, y_1) & \text{cov}(x_1, y_2) + \text{cov}(x_2, y_1) \\ \text{cov}(x_1, y_2) + \text{cov}(x_2, y_1) & 2\,\text{cov}(x_2, y_2) \end{pmatrix}.$$

Now, observe that

$$\text{cov}(\mathbf{x}, \mathbf{y}) = \begin{pmatrix} \text{cov}(x_1, y_1) & \text{cov}(x_1, y_2) \\ \text{cov}(x_2, y_1) & \text{cov}(x_2, y_2) \end{pmatrix}.$$

Observe also that

$$\text{var}(\mathbf{x}) = \begin{pmatrix} \text{var}(x_1) & \text{cov}(x_1, x_2) \\ \text{cov}(x_2, x_1) & \text{var}(x_2) \end{pmatrix} \text{ and } \text{var}(\mathbf{y}) = \begin{pmatrix} \text{var}(y_1) & \text{cov}(y_1, y_2) \\ \text{cov}(y_2, y_1) & \text{var}(y_2) \end{pmatrix}.$$

Accordingly, the diagonal elements of $\text{cov}(\mathbf{x}, \mathbf{y})$ and the average of the two off-diagonal elements of $\text{cov}(\mathbf{x}, \mathbf{y})$ can be determined from knowledge of $\text{var}(\mathbf{x})$, $\text{var}(\mathbf{y})$, and $\text{var}(\mathbf{x}+\mathbf{y})$—they are respectively equal to the diagonal elements and either off-diagonal element of $(1/2)[\text{var}(\mathbf{x}+\mathbf{y})-\text{var}(\mathbf{x})-\text{var}(\mathbf{y})]$. But the knowledge provided by $\text{var}(\mathbf{x})$, $\text{var}(\mathbf{y})$, and $\text{var}(\mathbf{x}+\mathbf{y})$ is insufficient to "separate" one off-diagonal element of $\text{cov}(\mathbf{x}, \mathbf{y})$ from the other, leading to the conclusion that $\text{cov}(\mathbf{x}, \mathbf{y})$ cannot be completely determined from knowledge of $\text{var}(\mathbf{x})$, $\text{var}(\mathbf{y})$, and $\text{var}(\mathbf{x}+\mathbf{y})$.

EXERCISE 7. Let $\mathbf{x}$ represent an M-dimensional random column vector with mean vector $\boldsymbol{\mu}$ and variance-covariance matrix $\boldsymbol{\Sigma}$. Show that there exist $M(M+1)/2$ linear combinations of the M elements of $\mathbf{x}$ such that $\boldsymbol{\mu}$ can be determined from knowledge of the expected values of M of these linear combinations and $\boldsymbol{\Sigma}$ can be determined from knowledge of the $[M(M+1)/2]$ variances of these linear combinations.

Solution. Denote by x_i the ith element of $\mathbf{x}$ and by μ_i the ith element of $\boldsymbol{\mu}$, and let σ_{ij} represent the ijth element of $\boldsymbol{\Sigma}$. Further, for $i = 1, 2, \ldots, M$, define $w_i = x_i$ and, for $j > i = 1, 2, \ldots, M$, define $y_{ij} = x_i + x_j$. Then, clearly,

$$\text{E}(w_i) = \mu_i \quad \text{and} \quad \text{var}(w_i) = \sigma_{ii} \qquad (i = 1, 2, \ldots, M),$$

and

$$\text{var}(y_{ij}) = \sigma_{ii} + \sigma_{jj} + 2\sigma_{ij} \qquad (j > i = 1, 2, \ldots, M).$$

And it follows that

$$\mu_i = \text{E}(w_i) \quad \text{and} \quad \sigma_{ii} = \text{var}(w_i) \qquad (i = 1, 2, \ldots, M)$$

and

$$\sigma_{ji} = \sigma_{ij} = [\text{var}(y_{ij}) - \text{var}(w_i) - \text{var}(w_j)]/2 \qquad (j > i = 1, 2, \ldots, M).$$

Thus, $\boldsymbol{\mu}$ can be determined from knowledge of the expected values of the M linear combinations $w_1, w_2, \ldots, w_M$, and $\boldsymbol{\Sigma}$ can be determined from knowledge of the variances of $w_1, w_2, \ldots, w_M$ and the variances of the $M(M-1)/2$ additional linear combinations y_{ij} $(j > i = 1, 2, \ldots, M)$—note that $M + [M(M-1)/2] = M(M+1)/2$.

EXERCISE 8. Let x and y represent random variables (whose expected values and variances exist), let $\mathbf{V}$ represent the variance-covariance matrix of the random vector (x, y), and suppose that $\mathrm{var}(x) > 0$ and $\mathrm{var}(y) > 0$.

(a) Show that if $|\mathbf{V}| = 0$, then, for scalars a and b,

$$(a,\, b)' \in \mathcal{N}(\mathbf{V}) \;\Leftrightarrow\; b\sqrt{\mathrm{var}\, y} = ca\sqrt{\mathrm{var}\, x},$$

where $c = \begin{cases} +1, & \text{when } \mathrm{cov}(x, y) < 0, \\ -1, & \text{when } \mathrm{cov}(x, y) > 0. \end{cases}$

(b) Use the result of Part (a) and the results of Section 3.2e to devise an alternative proof of the result (established in Section 3.2a) that $\mathrm{cov}(x, y) = \sqrt{\mathrm{var}\, x}\sqrt{\mathrm{var}\, y}$ if and only if $[y - \mathrm{E}(y)]/\sqrt{\mathrm{var}\, y} = [x - \mathrm{E}(x)]/\sqrt{\mathrm{var}\, x}$ with probability 1 and that $\mathrm{cov}(x, y) = -\sqrt{\mathrm{var}\, x}\sqrt{\mathrm{var}\, y}$ if and only if $[y - \mathrm{E}(y)]/\sqrt{\mathrm{var}\, y} = -[x - \mathrm{E}(x)]/\sqrt{\mathrm{var}\, x}$ with probability 1.

Solution. (a) Suppose that $|\mathbf{V}| = 0$. Then, in light of result (2.54) [or result (2.14.4)],

$$[\mathrm{cov}(x, y)]^2 = \mathrm{var}(x)\,\mathrm{var}(y),$$

or, equivalently,

$$\mathrm{cov}(x, y) = \pm\sqrt{\mathrm{var}\, x}\sqrt{\mathrm{var}\, y},$$

so that

$$\mathbf{V} = \begin{pmatrix} \sqrt{\mathrm{var}\, x} & 0 \\ 0 & \sqrt{\mathrm{var}\, y} \end{pmatrix} \begin{pmatrix} 1 & -c \\ -c & 1 \end{pmatrix} \begin{pmatrix} \sqrt{\mathrm{var}\, x} & 0 \\ 0 & \sqrt{\mathrm{var}\, y} \end{pmatrix}.$$

Thus, $(a,\, b)' \in \mathcal{N}(\mathbf{V})$ if and only if

$$\begin{pmatrix} \sqrt{\mathrm{var}\, x} & 0 \\ 0 & \sqrt{\mathrm{var}\, y} \end{pmatrix} \begin{pmatrix} 1 & -c \\ -c & 1 \end{pmatrix} \begin{pmatrix} a\sqrt{\mathrm{var}\, x} \\ b\sqrt{\mathrm{var}\, y} \end{pmatrix} = \begin{pmatrix} 0 \\ 0 \end{pmatrix}$$

and hence $\left[\text{since } a\sqrt{\mathrm{var}\, x} - cb\sqrt{\mathrm{var}\, y} = -c\,(b\sqrt{\mathrm{var}\, y} - ca\sqrt{\mathrm{var}\, x})\right]$ if and only if

$$b\sqrt{\mathrm{var}\, y} - ca\sqrt{\mathrm{var}\, x} = 0$$

or, equivalently, if and only if

$$b\sqrt{\mathrm{var}\, y} = ca\sqrt{\mathrm{var}\, x}.$$

(b) If $[y - \mathrm{E}(y)]/\sqrt{\mathrm{var}\, y} = [x - \mathrm{E}(x)]/\sqrt{\mathrm{var}\, x}$ with probability 1, then

$$\begin{aligned} \mathrm{cov}(x, y) &= \mathrm{E}\{[x - \mathrm{E}(x)][y - \mathrm{E}(y)]\} \\ &= \mathrm{E}\{[x - \mathrm{E}(x)]^2 \sqrt{\mathrm{var}\,(y)}/\sqrt{\mathrm{var}\,(x)}\} \\ &= \sqrt{\mathrm{var}\,(x)}\sqrt{\mathrm{var}\,(y)}; \end{aligned}$$

and, similarly, if $[y - \mathrm{E}(y)]/\sqrt{\mathrm{var}\, y} = -[x - \mathrm{E}(x)]/\sqrt{\mathrm{var}\, x}$ with probability 1, then

$$\begin{aligned} \mathrm{cov}(x, y) &= \mathrm{E}\{[x - \mathrm{E}(x)][y - \mathrm{E}(y)]\} \\ &= \mathrm{E}\{-[x - \mathrm{E}(x)]^2 \sqrt{\mathrm{var}\,(y)}/\sqrt{\mathrm{var}\,(x)}\} \\ &= -\sqrt{\mathrm{var}\,(x)}\sqrt{\mathrm{var}\,(y)}. \end{aligned}$$

Conversely, suppose that $\operatorname{cov}(x, y) = \sqrt{\operatorname{var} x}\sqrt{\operatorname{var} y}$. Then, $|\mathbf{V}| = 0$ [as is evident from result (2.54) or result (2.14.4)]. Now, take a to be an arbitrary nonzero (nonrandom) scalar, and take $b = -a\sqrt{\operatorname{var} x}/\sqrt{\operatorname{var} y}$. Then, it follows from the result of Part (a) that $(a, b)' \in \mathcal{N}(\mathbf{V})$, implying [in light of result (2.51)] that $\operatorname{var}(ax + by) = 0$. Moreover,

$$\begin{aligned}\operatorname{var}(ax + by) &= \operatorname{var}\Big[-a\sqrt{\operatorname{var} x}\Big(\frac{y}{\sqrt{\operatorname{var} y}} - \frac{x}{\sqrt{\operatorname{var} x}}\Big)\Big] \\ &= a^2 \operatorname{var}(x) \operatorname{var}\Big(\frac{y}{\sqrt{\operatorname{var} y}} - \frac{x}{\sqrt{\operatorname{var} x}}\Big).\end{aligned}$$

Thus,

$$\operatorname{var}\Big(\frac{y}{\sqrt{\operatorname{var} y}} - \frac{x}{\sqrt{\operatorname{var} x}}\Big) = 0,$$

and, consequently,

$$\frac{y}{\sqrt{\operatorname{var} y}} - \frac{x}{\sqrt{\operatorname{var} x}} = \mathrm{E}\Big(\frac{y}{\sqrt{\operatorname{var} y}} - \frac{x}{\sqrt{\operatorname{var} x}}\Big) \quad \text{with probability 1}$$

or, equivalently,

$$\frac{y - \mathrm{E}(y)}{\sqrt{\operatorname{var} y}} = \frac{x - \mathrm{E}(x)}{\sqrt{\operatorname{var} x}} \quad \text{with probability 1.}$$

That $\operatorname{cov}(x, y) = -\sqrt{\operatorname{var} x}\sqrt{\operatorname{var} y}$ implies that $[y - \mathrm{E}(y)]/\sqrt{\operatorname{var} y} = -[x - \mathrm{E}(x)]/\sqrt{\operatorname{var} x}$ with probability 1 can be established via a similar argument.

EXERCISE 9. Let $\mathbf{x}$ represent an N-dimensional random column vector (with elements whose expected values and variances exist). Show that (regardless of the rank of $\operatorname{var} \mathbf{x}$) there exist a nonrandom column vector $\mathbf{c}$ and an $N \times N$ nonsingular nonrandom matrix $\mathbf{A}$ such that the random vector $\mathbf{w}$, defined implicitly by $\mathbf{x} = \mathbf{c} + \mathbf{A}'\mathbf{w}$, has mean $\mathbf{0}$ and a variance-covariance matrix of the form $\operatorname{diag}(\mathbf{I}, \mathbf{0})$ [where $\operatorname{diag}(\mathbf{I}, \mathbf{0})$ is to be regarded as including $\mathbf{0}$ and $\mathbf{I}$ as special cases].

Solution. Let $\mathbf{V} = \operatorname{var}(\mathbf{x})$ and $K = \operatorname{rank}(\mathbf{V})$, and observe that

$$\mathbf{x} = \mathbf{c} + \mathbf{A}'\mathbf{w} \quad \Leftrightarrow \quad \mathbf{w} = (\mathbf{A}^{-1})'(\mathbf{x} - \mathbf{c}).$$

Take $\mathbf{c} = \mathrm{E}(\mathbf{x})$. Then, clearly, $\mathrm{E}(\mathbf{w}) = \mathbf{0}$.

Now, consider the condition $\operatorname{var}(\mathbf{w}) = \operatorname{diag}(\mathbf{I}, \mathbf{0})$. Observe that

$$\operatorname{var}(\mathbf{w}) = (\mathbf{A}^{-1})'\mathbf{V}\mathbf{A}^{-1},$$

and suppose that $\mathbf{V}$ is nonnull (in which case $K \geq 1$)—in the degenerate special case where $\mathbf{V} = \mathbf{0}$, take $\mathbf{A}$ to be $\mathbf{I}_N$ (or any other $N \times N$ nonsingular matrix). Then, proceeding as in Section 3.3b, take $\mathbf{T}$ to be any $K \times N$ nonrandom matrix such that $\mathbf{V} = \mathbf{T}'\mathbf{T}$, and observe that $\operatorname{rank}(\mathbf{T}) = K$—that such a matrix exists and is of rank K follows from Corollary 2.13.23.

If $K = N$, take $\mathbf{A} = \mathbf{T}$, with the result that

$$\operatorname{var}(\mathbf{w}) = (\mathbf{T}^{-1})'\mathbf{V}\mathbf{T}^{-1} = (\mathbf{T}\mathbf{T}^{-1})'\mathbf{T}\mathbf{T}^{-1} = \mathbf{I}'\mathbf{I} = \mathbf{I}.$$

Alternatively, suppose that $K < N$, and consider the matrix $(\mathbf{R}, \mathbf{S})$, where $\mathbf{R}$ is any right inverse of $\mathbf{T}$ and $\mathbf{S}$ is any $N \times (N - K)$ matrix whose columns form a basis for $\mathcal{N}(\mathbf{V})$—Lemma 2.11.5 implies that $\dim[\mathcal{N}(\mathbf{V})] = N - K$. The matrix $(\mathbf{R}, \mathbf{S})$ is nonsingular—to see this, take $\mathbf{L}$ to be a left inverse of $\mathbf{S}$, and observe that $\mathbf{T}\mathbf{S} = \mathbf{0}$ (as is evident from Corollary 2.3.4 upon observing that $\mathbf{T}'\mathbf{T}\mathbf{S} = \mathbf{V}\mathbf{S} = \mathbf{0}$), so that

$$\begin{pmatrix}\mathbf{T}\\ \mathbf{L}\end{pmatrix}(\mathbf{R}, \mathbf{S}) = \begin{pmatrix}\mathbf{T}\mathbf{R} & \mathbf{T}\mathbf{S}\\ \mathbf{L}\mathbf{R} & \mathbf{L}\mathbf{S}\end{pmatrix} = \begin{pmatrix}\mathbf{I}_K & \mathbf{0}\\ \mathbf{L}\mathbf{R} & \mathbf{I}_{N-K}\end{pmatrix},$$

implying (in light of Corollary 2.4.17 and Lemma 2.6.2) that

$$\text{rank}\,(\mathbf{R},\ \mathbf{S}) \geq \text{rank}\begin{pmatrix} \mathbf{I}_K & \mathbf{0} \\ \mathbf{LR} & \mathbf{I}_{N-K} \end{pmatrix} = N.$$

Further, since $\mathbf{R}'\mathbf{VR} = (\mathbf{TR})'\mathbf{TR} = \mathbf{I}'\mathbf{I} = \mathbf{I}$, we have that

$$(\mathbf{R},\ \mathbf{S})'\mathbf{V}(\mathbf{R},\ \mathbf{S}) = \begin{pmatrix} \mathbf{R}'\mathbf{VR} & \mathbf{R}'\mathbf{VS} \\ (\mathbf{VS})'\mathbf{R} & \mathbf{S}'\mathbf{VS} \end{pmatrix} = \begin{pmatrix} \mathbf{I} & \mathbf{0} \\ \mathbf{0} & \mathbf{0} \end{pmatrix}.$$

Thus, to satisfy the condition $\text{var}(\mathbf{w}) = \text{diag}(\mathbf{I},\ \mathbf{0})$, it suffices to take $\mathbf{A} = (\mathbf{R},\ \mathbf{S})^{-1}$ [in which case $\mathbf{A}^{-1} = (\mathbf{R},\ \mathbf{S})$].

EXERCISE 10. Establish the validity of result (5.11).

Solution. For $n = 1, 2, 3, \ldots,$

$$\begin{aligned} \frac{(2n)!\sqrt{\pi}}{4^n\, n!} &= \frac{2n(2n-1)(2n-2)\cdots 7\cdot 6\cdot 5\cdot 4\cdot 3\cdot 2\cdot 1}{2n(2n-2)\cdots 6\cdot 4\cdot 2}\,\frac{\sqrt{\pi}}{2^n} \\ &= \frac{1\cdot 3\cdot 5\cdot 7\cdots(2n-1)}{2^n}\sqrt{\pi}. \end{aligned}$$

Thus, to establish the validity of result (5.11), it suffices to establish the validity (for $n = 0, 1, 2, \ldots$) of the formula

$$\Gamma\big(n + \tfrac{1}{2}\big) = \frac{(2n)!\sqrt{\pi}}{4^n\, n!}. \tag{S.5}$$

Let us proceed by mathematical induction. Formula (S.5) is valid for $n = 0$, as is evident from result (5.10). Now, let r represent an arbitrary one of the integers $1, 2, 3, \ldots,$ and suppose that formula (S.5) is valid for $n = r - 1$, in which case

$$\Gamma\big(r - \tfrac{1}{2}\big) = \Gamma\big(r - 1 + \tfrac{1}{2}\big) = \frac{[2(r-1)]!\sqrt{\pi}}{4^{r-1}\,(r-1)!}.$$

Then, making use of result (5.6), we find that

$$\begin{aligned} \Gamma\big(r + \tfrac{1}{2}\big) &= \Gamma\big(r - \tfrac{1}{2} + 1\big) \\ &= \big(r - \tfrac{1}{2}\big)\,\Gamma\big(r - \tfrac{1}{2}\big) \\ &= \frac{(2r-1)(2r)[2(r-1)]!\sqrt{\pi}}{2(2r)\,4^{r-1}\,(r-1)!} \\ &= \frac{(2r)!\sqrt{\pi}}{4^r\, r!}, \end{aligned}$$

which establishes that formula (S.5) is valid for $n = r$ and thereby completes the mathematical-induction argument.

EXERCISE 11. Let $w = |z|$, where z is a standard normal random variable.

(a) Find a probability density function for the distribution of w.

(b) Use the expression obtained in Part (a) (for the probability density function of the distribution of w) to derive formula (5.15) for $\text{E}(w^r)$—in Section 3.5c, this formula is derived from the probability density function of the distribution of z.

(c) Find $\text{E}(w)$ and $\text{var}(w)$.

Solution. (a) Let $G(\cdot)$ represent the (cumulative) distribution function of w, let $f(\cdot)$ represent the probabiltiy density function of the standard-normal distribution, and denote by t an arbitrary scalar. For $t > 0$, we find that

$$\begin{aligned} G(t) &= \Pr(w \le t) \\ &= \Pr(-t \le z \le t) \\ &= \int_0^t f(s)\,ds + \int_{-t}^0 f(s)\,ds, \end{aligned}$$

and because $\int_{-t}^0 f(s)\,ds = \int_0^t f(s)\,ds$ [as can be formally verified by making the change of variable $u = -s$ and recalling result (5.1)], it follows that

$$G(t) = 2\int_0^t f(s)\,ds = \int_0^t \sqrt{\frac{2}{\pi}}\,e^{-s^2/2}\,ds.$$

Moreover, $G(t) = 0$ for $t < 0$. And we conclude that the function $g(\cdot)$, defined by

$$g(s) = \begin{cases} \sqrt{\dfrac{2}{\pi}}\,e^{-s^2/2} & \text{for } s > 0, \\ 0 & \text{for } s \le 0, \end{cases}$$

is a probability density function for the distribution of w.

(b) Making use of result (5.9) [and defining $g(\cdot)$ as in Part (a)], we find that

$$\begin{aligned} \mathrm{E}(w^r) &= \int_0^\infty s^r g(s)\,ds \\ &= \sqrt{\frac{2}{\pi}}\int_0^\infty s^r e^{-s^2/2}\,ds \\ &= \sqrt{\frac{2^r}{\pi}}\,\Gamma\!\left(\frac{r+1}{2}\right). \end{aligned}$$

(c) Making use of result (5.16), we find that

$$\mathrm{E}(w) = \sqrt{\frac{2}{\pi}}$$

and

$$\mathrm{var}(w) = \mathrm{E}(w^2) - [\mathrm{E}(w)]^2 = 1 - \frac{2}{\pi}.$$

EXERCISE 12. Let x represent a random variable having mean μ and variance σ^2. Then, $\mathrm{E}(x^2) = \mu^2 + \sigma^2$ [as is evident from result (2.3)]. Thus, the second moment of x depends on the distribution of x only through μ and σ^2. If the distribution of x is normal, then the third and higher moments of x also depend only on μ and σ^2. Taking the distribution of x to be normal, obtain explicit expressions for $\mathrm{E}(x^3)$, $\mathrm{E}(x^4)$, and, more generally, $\mathrm{E}(x^r)$ (where r is an arbitrary positive integer).

Solution. Take z to be a standard normal random variable. Then,

$$\mathrm{E}(x^r) = \mathrm{E}[(\mu + \sigma z)^r].$$

And, based on the binomial theorem (e.g., Casella and Berger 2002, sec 3.2), we have that

$$(\mu + \sigma z)^r = \sum_{k=0}^{r} \binom{r}{k} \mu^{r-k} \sigma^k z^k$$

—interpret 0^0 as 1. Thus, letting r_* represent the largest even integer that does not exceed r (so that $r_* = r$ if r is even, and $r_* = r - 1$ if r is odd) and making use of results (5.17) and (5.18), we find that

$$\begin{aligned} \mathrm{E}(x^r) &= \sum_{k=0}^{r} \binom{r}{k} \mu^{r-k} \sigma^k \, \mathrm{E}(z^k) \\ &= \sum_{s=0}^{r_*/2} \binom{r}{2s} \mu^{r-2s} \sigma^{2s} E(z^{2s}) \\ &= \mu^r + \sum_{s=1}^{r_*/2} \frac{r(r-1)(r-2)\cdots(r-2s+1)}{2s(2s-2)(2s-4)\cdots 6\cdot 4\cdot 2} \mu^{r-2s} \sigma^{2s}. \end{aligned} \tag{S.6}$$

As special cases of result (S.6), we have that

$$\mathrm{E}(x^3) = \mu^3 + \frac{3\cdot 2}{2} \mu^{3-2} \sigma^2 = \mu^3 + 3\mu\sigma^2$$

and

$$\begin{aligned} \mathrm{E}(x^4) &= \mu^4 + \frac{4\cdot 3}{2} \mu^{4-2} \sigma^2 + \frac{4\cdot 3\cdot 2\cdot 1}{4\cdot 2} \mu^{4-4} \sigma^4 \\ &= \mu^4 + 6\mu^2\sigma^2 + 3\sigma^4. \end{aligned}$$

EXERCISE 13. Let $\mathbf{x}$ represent an N-dimensional random column vector whose distribution is $N(\boldsymbol{\mu}, \boldsymbol{\Sigma})$. Further, let $R = \mathrm{rank}(\boldsymbol{\Sigma})$, and assume that $\boldsymbol{\Sigma}$ is nonnull (so that $R \geq 1$). Show that there exist an R-dimensional nonrandom column vector $\mathbf{c}$ and an $R \times N$ nonrandom matrix $\mathbf{A}$ such that $\mathbf{c} + \mathbf{A}\mathbf{x} \sim N(\mathbf{0}, \mathbf{I})$ (i.e., such that the distribution of $\mathbf{c} + \mathbf{A}\mathbf{x}$ is R-variate standard normal).

Solution. Take $\boldsymbol{\Gamma}$ to be an $R \times N$ nonrandom matrix such that $\boldsymbol{\Sigma} = \boldsymbol{\Gamma}'\boldsymbol{\Gamma}$, and observe that $\mathrm{rank}(\boldsymbol{\Gamma}) = R$—that such a matrix exists and is of rank R follows from Corollary 2.13.23. Further, take $\boldsymbol{\Lambda}$ to be a right inverse of $\boldsymbol{\Gamma}$—the existence of a right inverse follows from Lemma 2.5.1. Then, upon taking $\mathbf{A} = \boldsymbol{\Lambda}'$ and $\mathbf{c} = -\boldsymbol{\Lambda}'\boldsymbol{\mu}$, we find that

$$\mathbf{A}\boldsymbol{\Sigma}\mathbf{A}' = \boldsymbol{\Lambda}'\boldsymbol{\Gamma}'\boldsymbol{\Gamma}(\boldsymbol{\Lambda}')' = (\boldsymbol{\Gamma}\boldsymbol{\Lambda})'\boldsymbol{\Gamma}\boldsymbol{\Lambda} = \mathbf{I}'\mathbf{I} = \mathbf{I}$$

and

$$\mathbf{c} + \mathbf{A}\boldsymbol{\mu} = -\boldsymbol{\Lambda}'\boldsymbol{\mu} + \boldsymbol{\Lambda}'\boldsymbol{\mu} = \mathbf{0}.$$

And we conclude (on the basis of Theorem 3.5.1) that, for $\mathbf{A} = \boldsymbol{\Lambda}'$ and $\mathbf{c} = -\boldsymbol{\Lambda}'\boldsymbol{\mu}$, the distribution of $\mathbf{c} + \mathbf{A}\mathbf{x}$ is $N(\mathbf{0}, \mathbf{I})$.

EXERCISE 14. Let x and y represent random variables, and suppose that $x + y$ and $x - y$ are independently and normally distributed and have the same mean, say γ, and the same variance, say τ^2. Show that x and y are statistically independent, and determine the distribution of x and the distribution of y.

Solution. Let $u = x + y$ and $w = x - y$. And observe (in light of Theorem 3.5.7) that the joint distribution of u and w is MVN. Observe also that

$$x = \tfrac{1}{2}(u + w) \quad \text{and} \quad y = \tfrac{1}{2}(u - w)$$

and that

$$\begin{aligned} \mathrm{E}(x) &= \tfrac{1}{2}\,\mathrm{E}(u) + \tfrac{1}{2}\,\mathrm{E}(w) = \gamma, \\ \mathrm{E}(y) &= \tfrac{1}{2}\,\mathrm{E}(u) - \tfrac{1}{2}\,\mathrm{E}(w) = 0, \\ \mathrm{var}(x) &= \tfrac{1}{4}\,\mathrm{var}(u) + \tfrac{1}{4}\,\mathrm{var}(w) + \tfrac{1}{2}\,\mathrm{cov}(u, w) = \tfrac{1}{2}\tau^2, \\ \mathrm{var}(y) &= \tfrac{1}{4}\,\mathrm{var}(u) + \tfrac{1}{4}\,\mathrm{var}(w) - \tfrac{1}{2}\,\mathrm{cov}(u, w) = \tfrac{1}{2}\tau^2, \end{aligned}$$

and
$$\text{cov}(x, y) = \tfrac{1}{4}\,\text{var}(u) - \tfrac{1}{4}\,\text{cov}(u, w) + \tfrac{1}{4}\,\text{cov}(w, u) - \tfrac{1}{4}\,\text{var}(w) = 0.$$

Accordingly, it follows from Corollary 3.5.6 that x and y are statistically independent and from Theorem 3.5.1 that $x \sim N(\gamma, \tfrac{1}{2}\tau^2)$ and $y \sim N(0, \tfrac{1}{2}\tau^2)$.

EXERCISE 15. Suppose that two or more random column vectors $\mathbf{x}_1, \mathbf{x}_2, \ldots, \mathbf{x}_P$ are pairwise independent (i.e., $\mathbf{x}_i$ and $\mathbf{x}_j$ are statistically independent for $j > i = 1, 2, \ldots, P$) and that the joint distribution of $\mathbf{x}_1, \mathbf{x}_2, \ldots, \mathbf{x}_P$ is MVN. Is it necessarily the case that $\mathbf{x}_1, \mathbf{x}_2, \ldots, \mathbf{x}_P$ are mutually independent? Why or why not?

Solution. For $j > i = 1, 2, \ldots, P$, $\text{cov}(\mathbf{x}_i, \mathbf{x}_j) = \mathbf{0}$, as is evident from Lemma 3.2.1 (upon observing that the independence of $\mathbf{x}_i$ and $\mathbf{x}_j$ implies the independence of each element of $\mathbf{x}_i$ and each element of $\mathbf{x}_j$) and as is also evident from Corollary 3.5.5. Thus, it follows from Theorem 3.5.4 that $\mathbf{x}_1, \mathbf{x}_2, \ldots, \mathbf{x}_P$ are mutually independent.

EXERCISE 16. Let x represent a random variable whose distribution is $N(0, 1)$, and define $y = ux$, where u is a discrete random variable that is distributed independently of x with $\Pr(u = 1) = \Pr(u = -1) = \tfrac{1}{2}$.

(a) Show that $y \sim N(0, 1)$.

(b) Show that $\text{cov}(x, y) = 0$.

(c) Show that x and y are statistically dependent.

(d) Is the joint distribution of x and y bivariate normal? Why or why not?

Solution. (a) We find that (for an arbitrary constant c)

$$\begin{aligned}
\Pr(y \le c) &= \Pr(y \le c,\ u = 1) + \Pr(y \le c,\ u = -1) \\
&= \Pr(x \le c,\ u = 1) + \Pr(-x \le c,\ u = -1) \\
&= \Pr(x \le c)\Pr(u = 1) + \Pr(-x \le c)\Pr(u = -1) \\
&= \tfrac{1}{2}\Pr(x \le c) + \tfrac{1}{2}\Pr(-x \le c).
\end{aligned}$$

Moreover, upon observing that $-x \sim N(0, 1)$ and hence that $-x \sim x$, it is clear that $\Pr(-x \le c) = \Pr(x \le c)$. Thus, $\Pr(y \le c) = \Pr(x \le c)$, implying that $y \sim x$ and hence that $y \sim N(0, 1)$.

(b) Making use of the statistical independence of u and x, we find that

$$\text{cov}(x, y) = \text{E}(xy) = \text{E}(ux^2) = \text{E}(u)\,\text{E}(x^2) = 0(1) = 0.$$

(c) Clearly, $|y| = |x|$. This suggests that x and y are statistically dependent. To formally verify their statistical dependence, let c represent a (strictly) positive constant, and observe that $\Pr(|x| \le c) > 0$ and that $\Pr(|y| \le c) = \Pr(|x| \le c) < 1$. Then,

$$\Pr(|y| \le c,\ |x| \le c) = \Pr(|x| \le c) > \Pr(|x| \le c)\Pr(|y| \le c),$$

implying that $|x|$ and $|y|$ are statistically dependent. If x and y were statistically independent, then "any" function of x would be statistically independent of "any" function of y (e.g., Casella and Berger 2002, sec. 4.3; Bickel and Doksum 2001, app. A) and, in particular, $|x|$ would be distributed independently of $|y|$. Thus, x and y are not statistically independent; they are statistically dependent.

(d) The joint distribution of x and y is not bivariate normal. If the joint distribution of x and y were bivariate normal, then [in light of the result of Part (b)] it would follow from Corollary 3.5.5 that x and y would be statistically independent, in contradiction to what was established in Part (c).

EXERCISE 17. Let $\mathbf{x}_1, \mathbf{x}_2, \ldots, \mathbf{x}_K$ represent N-dimensional random column vectors, and suppose that $\mathbf{x}_1, \mathbf{x}_2, \ldots, \mathbf{x}_K$ are mutually independent and that (for $i = 1, 2, \ldots, K$) $\mathbf{x}_i \sim N(\boldsymbol{\mu}_i, \boldsymbol{\Sigma}_i)$. Derive (for arbitrary scalars $a_1, a_2, \ldots, a_K$) the distribution of the linear combination $\sum_{i=1}^K a_i \mathbf{x}_i$.

Solution. Take $\mathbf{x}$ to be the (KN-dimensional) random column vector defined by $\mathbf{x}' = (\mathbf{x}_1', \mathbf{x}_2', \ldots, \mathbf{x}_K')$, take $\boldsymbol{\mu}$ to be the (KN-dimensional) column vector defined by $\boldsymbol{\mu}' = (\boldsymbol{\mu}_1', \boldsymbol{\mu}_2', \ldots, \boldsymbol{\mu}_K')$, and define $\boldsymbol{\Sigma} = \text{diag}(\boldsymbol{\Sigma}_1, \boldsymbol{\Sigma}_2, \ldots, \boldsymbol{\Sigma}_K)$. Then, according to Theorem 3.5.7, $\mathbf{x} \sim N(\boldsymbol{\mu}, \boldsymbol{\Sigma})$. Moreover, $\sum_{i=1}^K a_i \mathbf{x}_i = \mathbf{A}\mathbf{x}$, where $\mathbf{A} = (a_1 \mathbf{I}_N, a_2 \mathbf{I}_N, \ldots, a_K \mathbf{I}_N)$. And upon applying Theorem 3.5.1, it follows that

$$\sum_{i=1}^K a_i \mathbf{x}_i \sim N(\mathbf{A}\boldsymbol{\mu},\ \mathbf{A}\boldsymbol{\Sigma}\mathbf{A}').$$

Clearly, $\mathbf{A}\boldsymbol{\mu} = \sum_{i=1}^K a_i \boldsymbol{\mu}_i$ and $\mathbf{A}\boldsymbol{\Sigma}\mathbf{A}' = \sum_{i=1}^K a_i^2 \boldsymbol{\Sigma}_i$. Thus,

$$\sum_{i=1}^K a_i \mathbf{x}_i \sim N\Bigl(\sum_{i=1}^K a_i \boldsymbol{\mu}_i,\ \sum_{i=1}^K a_i^2 \boldsymbol{\Sigma}_i\Bigr).$$

EXERCISE 18. Let x and y represent random variables whose joint distribution is bivariate normal. Further, let $\mu_1 = \text{E}(x)$, $\mu_2 = \text{E}(y)$, $\sigma_1^2 = \text{var}\, x$, $\sigma_2^2 = \text{var}\, y$, and $\rho = \text{corr}(x, y)$ (where $\sigma_1 \geq 0$ and $\sigma_2 \geq 0$). Assuming that $\sigma_1 > 0$, $\sigma_2 > 0$, and $-1 < \rho < 1$, show that the conditional distribution of y given x is $N[\mu_2 + \rho\sigma_2(x - \mu_1)/\sigma_1,\ \sigma_2^2(1 - \rho^2)]$.

Solution. The assumption that $\sigma_1 > 0$, $\sigma_2 > 0$, and $-1 < \rho < 1$ implies [in light, e.g., of result (5.33)] that the determinant of the variance-covariance matrix of the vector $(x,\ y)$ is nonzero (and, in fact, strictly positive) and hence (recalling Lemma 2.14.21) that this matrix is positive definite. Thus, it follows from Theorem 3.5.8 that the conditional distribution of y given x is normal with mean

$$\mu_2 + \rho\sigma_2\sigma_1(\sigma_1^2)^{-1}(x - \mu_1) = \mu_2 + \rho\sigma_2(x - \mu_1)/\sigma_1$$

and variance

$$\sigma_2^2 - \rho\sigma_2\sigma_1(\sigma_1^2)^{-1}\rho\sigma_1\sigma_2 = \sigma_2^2(1 - \rho^2).$$

EXERCISE 19. Let x and y represent random variables whose joint distribution is bivariate normal. Further, let $\sigma_1^2 = \text{var}\, x$, $\sigma_2^2 = \text{var}\, y$, and $\sigma_{12} = \text{cov}(x, y)$ (where $\sigma_1 \geq 0$ and $\sigma_2 \geq 0$). Describe (in as simple terms as possible) the marginal distributions of x and y and the conditional distributions of y given x and of x given y. Do so for each of the following two "degenerate" cases: (1) $\sigma_1^2 = 0$; and (2) $\sigma_1^2 > 0$, $\sigma_2^2 > 0$, and $|\sigma_{12}| = \sigma_1\sigma_2$.

Solution. Let $\mu_1 = \text{E}(x)$ and $\mu_2 = \text{E}(y)$.

(1) Suppose that $\sigma_1^2 = 0$. Then, according to result (2.12), $\sigma_{12} = 0$, implying (in light of Corollary 3.5.5) that x and y are statistically independent. Further, $x = \mu_1$ with probability 1 (both unconditionally and conditionally on y). And $y \sim N(\mu_2,\ \sigma_2^2)$ or, in the special case where $\sigma_2^2 = 0$, $y = \mu_2$ with probability 1 (both unconditionally and conditionally on x).

(2) Suppose that $\sigma_1^2 > 0$, $\sigma_2^2 > 0$, and $|\sigma_{12}| = \sigma_1\sigma_2$, and let $\rho = \text{corr}(x, y)$. (And observe that $\rho = \pm 1$.) The marginal distribution of x is $N(\mu_1,\ \sigma_1^2)$. Further, the conditional distribution of y given x is normal with mean

$$\mu_2 + \rho\sigma_2\sigma_1(\sigma_1^2)^{-1}(x - \mu_1) = \mu_2 + \rho\sigma_2(x - \mu_1)/\sigma_1$$

and variance

$$\sigma_2^2 - \rho\sigma_2\sigma_1(\sigma_1^2)^{-1}\rho\sigma_1\sigma_2 = \sigma_2^2(1 - \rho^2) = 0,$$

so that, conditional on x,

$$y = \mu_2 + \rho\sigma_2(x - \mu_1)/\sigma_1 \text{ with probability 1}$$

or, equivalently,

$$\frac{y - \mu_2}{\sigma_2} = \rho\frac{x - \mu_1}{\sigma_1} \text{ with probability 1.}$$

Similarly, the marginal distribution of y is $N(\mu_2, \sigma_2^2)$, and the conditional distribution of x given y is $N[\mu_1 + \rho\sigma_1(y - \mu_2)/\sigma_2,\ 0]$, with the implication that, conditional on y,

$$x = \mu_1 + \rho\sigma_1(y - \mu_2)/\sigma_2 \text{ with probability 1}$$

or, equivalently,

$$\frac{x - \mu_1}{\sigma_1} = \rho\frac{y - \mu_2}{\sigma_2} \text{ with probability 1.}$$

EXERCISE 20. Let $\mathbf{x}$ represent an N-dimensional random column vector, and take $\mathbf{y}$ to be the M-dimensional random column vector defined by $\mathbf{y} = \mathbf{c} + \mathbf{A}\mathbf{x}$, where $\mathbf{c}$ is an M-dimensional nonrandom column vector and $\mathbf{A}$ an $M \times N$ nonrandom matrix.

(a) Express the moment generating function of the distribution of $\mathbf{y}$ in terms of the moment generating function of the distribution of $\mathbf{x}$.

(b) Use the result of Part (a) to show that if the distribution of $\mathbf{x}$ is $N(\boldsymbol{\mu}, \boldsymbol{\Sigma})$, then the moment generating function of the distribution of $\mathbf{y}$ is the same as that of the $N(\mathbf{c} + \mathbf{A}\boldsymbol{\mu},\ \mathbf{A}\boldsymbol{\Sigma}\mathbf{A}')$ distribution, thereby (since distributions having the same moment generating function are identical) providing an alternative way of arriving at Theorem 3.5.1.

Solution. Let $m(\cdot)$ represent the moment generating function of the distribution of $\mathbf{y}$.

(a) For an arbitrary M-dimensional nonrandom column vector $\mathbf{t}$,

$$\begin{aligned} m(\mathbf{t}) &= \mathrm{E}[\exp(\mathbf{t}'\mathbf{y})] \\ &= \mathrm{E}[\exp(\mathbf{t}'\mathbf{c} + \mathbf{t}'\mathbf{A}\mathbf{x})] \\ &= \mathrm{E}\{\exp(\mathbf{t}'\mathbf{c})\exp[(\mathbf{A}'\mathbf{t})'\mathbf{x}]\} \\ &= \exp(\mathbf{t}'\mathbf{c})\,\mathrm{E}\{\exp[(\mathbf{A}'\mathbf{t})'\mathbf{x}]\} \\ &= \exp(\mathbf{t}'\mathbf{c})\,s(\mathbf{A}'\mathbf{t}), \end{aligned}$$

where $s(\cdot)$ is the moment generating function of the distribution of $\mathbf{x}$.

(b) Suppose that $\mathbf{x} \sim N(\boldsymbol{\mu}, \boldsymbol{\Sigma})$. Then, starting with the result of Part (a) and making use of formula (5.47), we find that (for an arbitrary $\mathbf{t}$)

$$\begin{aligned} m(\mathbf{t}) &= \exp(\mathbf{t}'\mathbf{c})\exp[(\mathbf{A}'\mathbf{t})'\boldsymbol{\mu} + \tfrac{1}{2}(\mathbf{A}'\mathbf{t})'\boldsymbol{\Sigma}\mathbf{A}'\mathbf{t}] \\ &= \exp[\mathbf{t}'(\mathbf{c} + \mathbf{A}\boldsymbol{\mu}) + \tfrac{1}{2}\mathbf{t}'\mathbf{A}\boldsymbol{\Sigma}\mathbf{A}'\mathbf{t}]. \end{aligned}$$

And upon substituting $\mathbf{c} + \mathbf{A}\boldsymbol{\mu}$ for $\boldsymbol{\mu}$ and $\mathbf{A}\boldsymbol{\Sigma}\mathbf{A}'$ for $\boldsymbol{\Sigma}$ in formula (5.47), it is clear that $m(\cdot)$ is the same as the moment generating function of the $N(\mathbf{c} + \mathbf{A}\boldsymbol{\mu},\ \mathbf{A}\boldsymbol{\Sigma}\mathbf{A}')$ distribution.

4

The General Linear Model

EXERCISE 1. Verify formula (2.3).

Solution. We have that

$$\delta(u) = \beta_1 + \sum_{j=2}^{P} \beta_j \delta_j(u),$$

where (for $j = 2, 3, \ldots, P$)

$$\delta_j(u) = (u - a)^{j-1}.$$

And making use of basic results on differentiation (and employing a standard notation), we find that

$$\frac{d^{k-1}\delta_j(u)}{du^{k-1}} = \begin{cases} (j-1)(j-2)\cdots(j-k+1)(u-a)^{j-k} & \text{for } j \geq k, \\ 0 & \text{for } j < k. \end{cases}$$

Thus,

$$\begin{aligned} \delta^{(k-1)}(u) &= \sum_{j=2}^{P} \beta_j \frac{d^{k-1}\delta_j(u)}{du^{k-1}} \\ &= \sum_{j=k}^{P} (j-1)(j-2)\cdots(j-k+1)(u-a)^{j-k}\beta_j \\ &= (k-1)!\beta_k + \sum_{j=k+1}^{P} (j-1)(j-2)\cdots(j-k+1)(u-a)^{j-k}\beta_j. \end{aligned}$$

EXERCISE 2. Write out the elements of the vector $\boldsymbol{\beta}$, of the observed value of the vector $\mathbf{y}$, and of the matrix $\mathbf{X}$ (in the model equation $\mathbf{y} = \mathbf{X}\boldsymbol{\beta} + \mathbf{e}$) in an application of the G–M model to the cement data of Section 4.2d. In doing so, regard the measurements of the heat that evolves during hardening as the data points, take $C = 4$, take u_1, u_2, u_3, and u_4 to be the respective amounts of tricalcium aluminate, tricalcium silicate, tetracalcium aluminoferrite, and β-dicalcium silicate, and take $\delta(\mathbf{u})$ to be of the form (2.11).

Solution. The parameter vector $\boldsymbol{\beta}$ is the $(P = 5)$-dimensional vector whose transpose is $\boldsymbol{\beta}' = (\beta_1, \beta_2, \beta_3, \beta_4, \beta_5)$. And, assuming that the $N = 13$ data points are numbered 1 through 13 in the order in which they are listed in Table 4.2, the observed value of the vector $\mathbf{y}$ and the model matrix

$\mathbf{X}$ are

$$\mathbf{y} = \begin{pmatrix} 78.5 \\ 74.3 \\ 104.3 \\ 87.6 \\ 95.9 \\ 109.2 \\ 102.7 \\ 72.5 \\ 93.1 \\ 115.9 \\ 83.8 \\ 113.3 \\ 109.4 \end{pmatrix} \quad \text{and} \quad \mathbf{X} = \begin{pmatrix} 1 & 7 & 26 & 6 & 60 \\ 1 & 1 & 29 & 15 & 52 \\ 1 & 11 & 56 & 8 & 20 \\ 1 & 11 & 31 & 8 & 47 \\ 1 & 7 & 52 & 6 & 33 \\ 1 & 11 & 55 & 9 & 22 \\ 1 & 3 & 71 & 17 & 6 \\ 1 & 1 & 31 & 22 & 44 \\ 1 & 2 & 54 & 18 & 22 \\ 1 & 21 & 47 & 4 & 26 \\ 1 & 1 & 40 & 23 & 34 \\ 1 & 11 & 66 & 9 & 12 \\ 1 & 10 & 68 & 8 & 12 \end{pmatrix}.$$

EXERCISE 3. Write out the elements of the vector $\boldsymbol{\beta}$, of the observed value of the vector $\mathbf{y}$, and of the matrix $\mathbf{X}$ (in the model equation $\mathbf{y} = \mathbf{X}\boldsymbol{\beta} + \mathbf{e}$) in an application of the G–M model to the lettuce data of Section 4.2e. In doing so, regard the yields of lettuce as the data points, take $C = 3$, take u_1, u_2, and u_3 to be the transformed amounts of Cu, Mo, and Fe, respectively, and take $\delta(\mathbf{u})$ to be of the form (2.14).

Solution. The parameter vector $\boldsymbol{\beta}$ is the $(P = 10)$-dimensional vector whose transpose is

$$\boldsymbol{\beta}' = (\beta_1, \beta_2, \beta_3, \beta_4, \beta_{11}, \beta_{12}, \beta_{13}, \beta_{22}, \beta_{23}, \beta_{33}).$$

And, assuming that the $N = 20$ data points are numbered 1 through 20 in the order in which they are listed in Table 4.3, the observed value of the vector $\mathbf{y}$ and the model matrix $\mathbf{X}$ are

$$\mathbf{y} = \begin{pmatrix} 21.42 \\ 15.92 \\ 22.81 \\ 14.90 \\ 14.95 \\ 7.83 \\ 19.90 \\ 4.68 \\ 0.20 \\ 17.65 \\ 18.16 \\ 25.39 \\ 11.99 \\ 7.37 \\ 22.22 \\ 19.49 \\ 22.76 \\ 24.27 \\ 27.88 \\ 27.53 \end{pmatrix} \quad \text{and } \mathbf{X} = \begin{pmatrix} 1 & -1 & -1 & -0.5 & 1 & 1 & 0.5 & 1 & 0.5 & 0.25 \\ 1 & 1 & -1 & -0.5 & 1 & -1 & -0.5 & 1 & 0.5 & 0.25 \\ 1 & -1 & 1 & -0.5 & 1 & -1 & 0.5 & 1 & -0.5 & 0.25 \\ 1 & -1 & -1 & 1 & 1 & 1 & -1 & 1 & -1 & 1 \\ 1 & 1 & 1 & -0.5 & 1 & 1 & -0.5 & 1 & -0.5 & 0.25 \\ 1 & 1 & -1 & 1 & 1 & -1 & 1 & 1 & -1 & 1 \\ 1 & -1 & 1 & 1 & 1 & -1 & -1 & 1 & 1 & 1 \\ 1 & 1 & 1 & 1 & 1 & 1 & 1 & 1 & 1 & 1 \\ 1 & \sqrt[4]{8} & 0 & 0 & \sqrt{8} & 0 & 0 & 0 & 0 & 0 \\ 1 & -\sqrt[4]{8} & 0 & 0 & \sqrt{8} & 0 & 0 & 0 & 0 & 0 \\ 1 & 0 & \sqrt[4]{8} & 0 & 0 & 0 & 0 & \sqrt{8} & 0 & 0 \\ 1 & 0 & -\sqrt[4]{8} & 0 & 0 & 0 & 0 & \sqrt{8} & 0 & 0 \\ 1 & 0 & 0 & -\sqrt[4]{8} & 0 & 0 & 0 & 0 & 0 & \sqrt{8} \\ 1 & 0 & 0 & \sqrt[4]{8} & 0 & 0 & 0 & 0 & 0 & \sqrt{8} \\ 1 & 0 & 0 & 0 & 0 & 0 & 0 & 0 & 0 & 0 \\ 1 & 0 & 0 & 0 & 0 & 0 & 0 & 0 & 0 & 0 \\ 1 & 0 & 0 & 0 & 0 & 0 & 0 & 0 & 0 & 0 \\ 1 & 0 & 0 & 0 & 0 & 0 & 0 & 0 & 0 & 0 \\ 1 & 0 & 0 & 0 & 0 & 0 & 0 & 0 & 0 & 0 \\ 1 & 0 & 0 & 0 & 0 & 0 & 0 & 0 & 0 & 0 \end{pmatrix}.$$

EXERCISE 4. Let y represent a random variable and $\mathbf{u}$ a C-dimensional random column vector such that the joint distribution of y and $\mathbf{u}$ is MVN (with a nonsingular variance-covariance matrix).

And take $\mathbf{z} = \{z_j\}$ to be a transformation (of $\mathbf{u}$) of the form

$$\mathbf{z} = \mathbf{R}'[\mathbf{u} - \mathrm{E}(\mathbf{u})],$$

where $\mathbf{R}$ is a nonsingular (nonrandom) matrix such that $\mathrm{var}(\mathbf{z}) = \mathbf{I}$—the existence of such a matrix follows from the results of Section 3.3b. Show that

$$\frac{\mathrm{E}(y \mid \mathbf{u}) - \mathrm{E}(y)}{\sqrt{\mathrm{var}\, y}} = \sum_{j=1}^{C} \mathrm{corr}(y, z_j)\, z_j.$$

Solution. According to result (3.2) (or Theorem 3.5.8),

$$\mathrm{E}(y \mid \mathbf{u}) = \mathrm{E}(y) + \mathrm{cov}(y, \mathbf{u})(\mathrm{var}\, \mathbf{u})^{-1}[\mathbf{u} - \mathrm{E}(\mathbf{u})].$$

Now, let $\mathbf{T} = \mathbf{R}^{-1}$, in which case $(\mathbf{R}')^{-1} = (\mathbf{R}^{-1})' = \mathbf{T}'$. Then,

$$\mathbf{u} = \mathrm{E}(\mathbf{u}) + \mathbf{T}'\mathbf{z},$$

implying that

$$\mathrm{var}\, \mathbf{u} = \mathbf{T}'(\mathrm{var}\, \mathbf{z})\mathbf{T} = \mathbf{T}'\mathbf{T}$$

and hence (since $\mathbf{T}^{-1} = \mathbf{R}$) that

$$(\mathrm{var}\, \mathbf{u})^{-1} = \mathbf{T}^{-1}(\mathbf{T}')^{-1} = \mathbf{T}^{-1}(\mathbf{T}^{-1})' = \mathbf{R}\mathbf{R}'.$$

Thus, in light of formula (3.2.45), we have that

$$\mathrm{cov}(y, \mathbf{u})(\mathrm{var}\, \mathbf{u})^{-1}[\mathbf{u} - \mathrm{E}(\mathbf{u})] = \mathrm{cov}(y, \mathbf{u})\mathbf{R}\mathbf{R}'[\mathbf{u} - \mathrm{E}(\mathbf{u})] = \mathrm{cov}(y, \mathbf{z})\, \mathbf{z}.$$

And since $\mathrm{var}(z_j) = 1$ (for $j = 1, 2, \ldots, C$), we conclude that

$$\begin{aligned} \frac{\mathrm{E}(y \mid \mathbf{u}) - \mathrm{E}(y)}{\sqrt{\mathrm{var}\, y}} &= \frac{1}{\sqrt{\mathrm{var}\, y}} \mathrm{cov}(y, \mathbf{u})(\mathrm{var}\, \mathbf{u})^{-1}[\mathbf{u} - \mathrm{E}(\mathbf{u})] \\ &= \frac{1}{\sqrt{\mathrm{var}\, y}} \mathrm{cov}(y, \mathbf{z})\, \mathbf{z} \\ &= \sum_{j=1}^{C} \mathrm{corr}(y, z_j)\, z_j. \end{aligned}$$

EXERCISE 5. Let y represent a random variable and $\mathbf{u} = (u_1, u_2, \ldots, u_C)'$ a C-dimensional random column vector, assume that $\mathrm{var}(\mathbf{u})$ is nonsingular, and suppose that $\mathrm{E}(y \mid \mathbf{u})$ is expressible in the form

$$\mathrm{E}(y \mid \mathbf{u}) = \beta_1 + \beta_2(u_1 - a_1) + \beta_3(u_2 - a_2) + \cdots + \beta_{C+1}(u_C - a_C),$$

where $a_1, a_2, \ldots, a_C$ and $\beta_1, \beta_2, \beta_3, \ldots, \beta_{C+1}$ are nonrandom scalars.

(a) Using the results of Section 3.4 (or otherwise), show that

$$(\beta_2, \beta_3, \ldots, \beta_{C+1}) = \mathrm{cov}(y, \mathbf{u})(\mathrm{var}\, \mathbf{u})^{-1},$$

and that

$$\beta_1 = \mathrm{E}(y) + \sum_{j=2}^{C+1} \beta_j[a_{j-1} - \mathrm{E}(u_{j-1})],$$

in agreement with the results obtained in Section 4.3 (under the assumption that the joint distribution of y and $\mathbf{u}$ is MVN).

(b) Show that

$$\mathrm{E}[\mathrm{var}(y \mid \mathbf{u})] = \mathrm{var}(y) - \mathrm{cov}(y, \mathbf{u})(\mathrm{var}\,\mathbf{u})^{-1}\mathrm{cov}(\mathbf{u}, y).$$

Solution. (a) Using result (3.4.1), we find that

$$\begin{aligned}
\mathrm{E}(y) &= \mathrm{E}[\mathrm{E}(y \mid \mathbf{u})] \\
&= \mathrm{E}[\beta_1 + \beta_2(u_1 - a_1) + \beta_3(u_2 - a_2) + \cdots + \beta_{C+1}(u_C - a_C)] \\
&= \beta_1 + \sum_{j=2}^{C+1} \beta_j[\mathrm{E}(u_{j-1}) - a_{j-1}]
\end{aligned}$$

and hence that

$$\beta_1 = \mathrm{E}(y) - \sum_{j=2}^{C+1} \beta_j[\mathrm{E}(u_{j-1}) - a_{j-1}] = \mathrm{E}(y) + \sum_{j=2}^{C+1} \beta_j[a_{j-1} - \mathrm{E}(u_{j-1})].$$

Further, appylying result (3.4.3), observing that $\mathrm{cov}(y, \mathbf{u} \mid \mathbf{u}) = \mathbf{0}$ and $\mathrm{E}(\mathbf{u} \mid \mathbf{u}) = \mathbf{u}$ (with probability 1), reexpressing $\mathrm{E}(y \mid \mathbf{u})$ in the form

$$\mathrm{E}(y \mid \mathbf{u}) = \beta_1 - \sum_{j=2}^{C+1} \beta_j a_{j-1} + (\beta_2, \beta_3, \ldots, \beta_{C+1})\mathbf{u},$$

and recalling formula (3.2.46), we determine that

$$\begin{aligned}
\mathrm{cov}(y, \mathbf{u}) &= \mathrm{E}[\mathrm{cov}(y, \mathbf{u} \mid \mathbf{u})] + \mathrm{cov}[\mathrm{E}(y \mid \mathbf{u}),\ \mathrm{E}(\mathbf{u} \mid \mathbf{u})] \\
&= \mathrm{E}(\mathbf{0}) + \mathrm{cov}\Big[\beta_1 - \sum_{j=2}^{C+1} \beta_j a_{j-1} + (\beta_2, \beta_3, \ldots, \beta_{C+1})\mathbf{u},\ \mathbf{u}\Big] \\
&= (\beta_2, \beta_3, \ldots, \beta_{C+1})(\mathrm{var}\,\mathbf{u}).
\end{aligned}$$

And it follows that

$$(\beta_2, \beta_3, \ldots, \beta_{C+1}) = \mathrm{cov}(y, \mathbf{u})(\mathrm{var}\,\mathbf{u})^{-1}.$$

(b) Using result (3.4.2) and formula (3.2.43) and applying the result of Part (a), we find that

$$\begin{aligned}
\mathrm{var}(y) &= \mathrm{E}[\mathrm{var}(y \mid \mathbf{u})] + \mathrm{var}[\mathrm{E}(y \mid \mathbf{u})] \\
&= \mathrm{E}[\mathrm{var}(y \mid \mathbf{u})] + \mathrm{var}\Big[\beta_1 - \sum_{j=2}^{C+1} \beta_j a_{j-1} + (\beta_2, \beta_3, \ldots, \beta_{C+1})\mathbf{u}\Big] \\
&= \mathrm{E}[\mathrm{var}(y \mid \mathbf{u})] + (\beta_2, \beta_3, \ldots, \beta_{C+1})(\mathrm{var}\,\mathbf{u})(\beta_2, \beta_3, \ldots, \beta_{C+1})' \\
&= \mathrm{E}[\mathrm{var}(y \mid \mathbf{u})] + \mathrm{cov}(y, \mathbf{u})(\mathrm{var}\,\mathbf{u})^{-1}(\mathrm{var}\,\mathbf{u})(\mathrm{var}\,\mathbf{u})^{-1}[\mathrm{cov}(y, \mathbf{u})]' \\
&= \mathrm{E}[\mathrm{var}(y \mid \mathbf{u})] + \mathrm{cov}(y, \mathbf{u})(\mathrm{var}\,\mathbf{u})^{-1}\mathrm{cov}(\mathbf{u}, y),
\end{aligned}$$

implying that

$$\mathrm{E}[\mathrm{var}(y \mid \mathbf{u})] = \mathrm{var}(y) - \mathrm{cov}(y, \mathbf{u})(\mathrm{var}\,\mathbf{u})^{-1}\mathrm{cov}(\mathbf{u}, y).$$

EXERCISE 6. Suppose that (in conformance with the development in Section 4.4b) the residual effects in the general linear model have been partitioned into K mutually exclusive and exhaustive subsets or classes numbered $1, 2, \ldots, K$. And for $k = 1, 2, \ldots, K$, write $e_{k1}, e_{k2}, \ldots, e_{kN_k}$ for the residual effects in the kth class. Take a_k^* $(k = 1, 2, \ldots, K)$ and r_{ks}^* $(k = 1, 2, \ldots, K;\ s = 1, 2, \ldots, N_k)$ to be uncorrelated random variables, each with mean 0, such that $\mathrm{var}(a_k^*) = \tau_k^{*2}$ for some nonnegative scalar τ_k^* and $\mathrm{var}(r_{ks}^*) = \phi_k^{*2}$ $(s = 1, 2, \ldots, N_k)$ for some strictly positive scalar ϕ_k^*. Consider the effect of taking the residual effects to be of the form

$$e_{ks} = a_k^* + r_{ks}^*, \qquad \text{(E.1)}$$

rather than of the form (4.26). Are there values of τ_k^{*2} and ϕ_k^{*2} for which the value of var(**e**) is the same when the residual effects are taken to be of the form (E.1) as when they are taken to be of the form (4.26)? If so, what are those values; if not, why not?

Solution. When the residual effects are taken to be of the form (E.1), we find that (for $k = 1, 2, \ldots, K$ and $s = 1, 2, \ldots, N_k$)

$$\text{var}(e_{ks}) = \phi_k^{*2} + \tau_k^{*2},$$

that (for $k = 1, 2, \ldots, K$ and $t \neq s = 1, 2, \ldots, N_k$)

$$\text{cov}(e_{ks}, e_{kt}) = \tau_k^{*2},$$

and that (for $k' \neq k = 1, 2, \ldots, K$, $s = 1, 2, \ldots, N_k$, and $t = 1, 2, \ldots, N_{k'}$)

$$\text{cov}(e_{ks}, e_{k't}) = 0.$$

For the value of var(**e**) to be the same when the residual effects are taken to be of the form (E.1) as when they are taken to be of the form (4.26), it would [in light of results (4.30) and (4.31)] be necessary and sufficient that

$$\phi_k^{*2} + \tau_k^{*2} = \phi_k^2 + \tau_k^2 + \omega_k^2$$

and

$$\tau_k^{*2} = \tau_k^2 - \frac{1}{N_k^* - 1}\omega_k^2$$

or, equivalently, that

$$\phi_k^{*2} = \phi_k^2 + \frac{N_k^*}{N_k^* - 1}\omega_k^2$$

and

$$\tau_k^{*2} = \tau_k^2 - \frac{1}{N_k^* - 1}\omega_k^2. \qquad \text{(S.1)}$$

These conditions can be satisfied if $\tau_k^2 \geq \frac{1}{N_k^*-1}\omega_k^2$, but not otherwise—if $\tau_k^2 < \frac{1}{N_k^*-1}\omega_k^2$, then condition (S.1) could only be satisfied by taking τ_k^{*2} (which is inherently nonnegative) to be negative.

EXERCISE 7. Develop a correlation structure for the residual effects in the general linear model that, in the application of the model to the shear-strength data (of Section 4.4d), would allow for the possibility that steel aliquots chosen at random from those on hand on different dates may tend to be more alike when the intervening time is short than when it is long. Do so by making use of the results (in Section 4.4e) on stationary first-order autoregressive processes.

Solution. For $k = 1, 2, \ldots, K$ and $s = 1, 2, \ldots, N_k$, let e_{ks} represent the sth of the residual effects associated with the kth class, and suppose that

$$e_{ks} = a_k + r_{ks}.$$

Here, a_k ($k = 1, 2, \ldots, K$) and r_{ks} ($k = 1, 2, \ldots, K$; $s = 1, 2, \ldots, N_k$) are taken to be random variables, each with mean 0. Further, the a_k's are taken to have a common variance τ^2 and the r_{ks}'s a common variance ϕ^2, and the r_{ks}'s are assumed to be uncorrelated with the a_k's and with each other. However, instead of taking the a_k's to be uncorrelated (with each other), they are assumed to be such that (for $k' \neq k = 1, 2, \ldots, K$)

$$\text{corr}(a_k, a_{k'}) = \xi^{|k'-k|}$$

for some nonrandom scalar ξ in the interval $0 \leq \xi < 1$ (as would be the case if $a_1, a_2, \ldots, a_K$ were members of a stationary first-order autoregressive process).

The setting is such that (for $k = 1, 2, \ldots, K$ and $s = 1, 2, \ldots, N_k$)

$$\text{var}(e_{ks}) = \sigma^2, \tag{S.2}$$

where $\sigma^2 = \phi^2 + \tau^2$. Further, for $k = 1, 2, \ldots, K$ and $t \neq s = 1, 2, \ldots, N_k$,

$$\text{cov}(e_{ks}, e_{kt}) = \text{var}(a_k) = \tau^2 = \sigma^2 \rho, \tag{S.3}$$

where $\rho = \tau^2/(\phi^2 + \tau^2)$. And for $k' \neq k = 1, 2, \ldots, K$, $s = 1, 2, \ldots, N_k$, and $t = 1, 2, \ldots, N_{k'}$,

$$\text{cov}(e_{ks}, e_{k't}) = \text{cov}(a_k, a_{k'}) = \tau^2 \xi^{|k'-k|} = \sigma^2 \rho\, \xi^{|k'-k|}. \tag{S.4}$$

The vector $\mathbf{e}$ is expressible as $\mathbf{e} = (\mathbf{e}_1', \mathbf{e}_2', \ldots, \mathbf{e}_K')'$, where (for $k = 1, 2, \ldots, K$) $\mathbf{e}_k = (e_{k1}, e_{k2}, \ldots, e_{kN_k})'$. Accordingly,

$$\text{var}(\mathbf{e}) = \sigma^2 \begin{pmatrix} \mathbf{R}_{11} & \mathbf{R}_{12} & \ldots & \mathbf{R}_{1K} \\ \mathbf{R}_{21} & \mathbf{R}_{22} & \ldots & \mathbf{R}_{2K} \\ \vdots & \vdots & \ddots & \vdots \\ \mathbf{R}_{K1} & \mathbf{R}_{K2} & \ldots & \mathbf{R}_{KK} \end{pmatrix}, \tag{S.5}$$

where (for $k, k' = 1, 2, \ldots, K$) $\mathbf{R}_{kk'}$ is the $N_k \times N_{k'}$ matrix whose stth element is $\text{corr}(e_{ks}, e_{k't})$. And in light of results (S.2), (S.3), and (S.4), we have (for $k = 1, 2, \ldots, K$) that

$$\mathbf{R}_{kk} = (1 - \rho)\mathbf{I}_{N_k} + \rho \mathbf{1}_{N_k} \mathbf{1}_{N_k}'$$

and (for $k' \neq k = 1, 2, \ldots, K$) that

$$\mathbf{R}_{kk'} = \rho\, \xi^{|k'-k|} \mathbf{1}_{N_k} \mathbf{1}_{N_{k'}}'.$$

In the case of the shear-strength data, $K = 12$ and

$$N_k = \begin{cases} 9 & \text{if } k = 2, 3, 5, 6, 7, 8, 10, \text{ or } 12, \\ 11 & \text{if } k = 9 \text{ or } 11, \\ 12 & \text{if } k = 1 \text{ or } 4. \end{cases}$$

Taking $\text{var}(\mathbf{e})$ to be of the form (S.5) may (or may not) be preferable to taking $\text{var}(\mathbf{e})$ to be of the more-restrictive (block-diagonal) form (4.34).

EXERCISE 8. Suppose (as in Section 4.4g) that the residual effects $e_1, e_2, \ldots, e_N$ in the general linear model correspond to locations in D-dimensional space, that these locations are represented by D-dimensional column vectors $\mathbf{s}_1, \mathbf{s}_2, \ldots, \mathbf{s}_N$ of coordinates, and that S is a finite or infinite set of D-dimensional column vectors that includes $\mathbf{s}_1, \mathbf{s}_2, \ldots, \mathbf{s}_N$. Suppose further that $e_1, e_2, \ldots, e_N$ are expressible in the form (4.54) and that conditions (4.55) and (4.57) are applicable. And take $\psi(\cdot)$ to be the function defined on the set $H = \{\mathbf{h} \in \mathcal{R}^D : \mathbf{h} = \mathbf{s} - \mathbf{t} \text{ for } \mathbf{s}, \mathbf{t} \in S\}$ by

$$\psi(\mathbf{h}) = \phi^2 + \tau^2[1 - K(\mathbf{h})].$$

(a) Show that, for $j \neq i = 1, 2, \ldots, N$,

$$\tfrac{1}{2}\,\text{var}(e_i - e_j) = \psi(\mathbf{s}_i - \mathbf{s}_j)$$

—this result serves to establish the function $\psi^*(\cdot)$, defined by

$$\psi^*(\mathbf{h}) = \begin{cases} \psi(\mathbf{h}) & \text{if } \mathbf{h} \neq \mathbf{0}, \\ 0 & \text{if } \mathbf{h} = \mathbf{0}, \end{cases}$$

as what in spatial statistics is known as a semivariogram (e.g., Cressie 1993).

(b) Show that (1) $\psi(\mathbf{0}) = \phi^2$; that (2) $\psi(-\mathbf{h}) = \psi(\mathbf{h})$ for $\mathbf{h} \in H$; and that (3) $\sum_{i=1}^{M}\sum_{j=1}^{M} x_i x_j \psi(\mathbf{t}_i - \mathbf{t}_j) \le 0$ for every positive integer M, for all not-necessarily-distinct vectors $\mathbf{t}_1, \mathbf{t}_2, \dots, \mathbf{t}_M$ in S, and for all scalars $x_1, x_2, \dots, x_M$ such that $\sum_{i=1}^{M} x_i = 0$.

Solution. (a) Making use of results (4.56) and (4.58), we find that for $j \neq i = 1, 2, \dots, N$,

$$\begin{aligned}
\tfrac{1}{2}\operatorname{var}(e_i - e_j) &= \tfrac{1}{2}[\operatorname{var}(e_i) - \operatorname{var}(e_j) - 2\operatorname{cov}(e_i, e_j)] \\
&= \tfrac{1}{2}[\tau^2 + \phi^2 + \tau^2 + \phi^2 - 2\tau^2 K(\mathbf{s}_i - \mathbf{s}_j)] \\
&= \tfrac{1}{2}\{2\phi^2 + 2\tau^2[1 - K(\mathbf{s}_i - \mathbf{s}_j)]\} \\
&= \phi^2 + \tau^2[1 - K(\mathbf{s}_i - \mathbf{s}_j)] \\
&= \psi(\mathbf{s}_i - \mathbf{s}_j)
\end{aligned}$$

(b) In light of the three requisite properties of the function $K(\cdot)$, we have that (1)

$$\psi(\mathbf{0}) = \phi^2 + \tau^2[1 - K(\mathbf{0})] = \phi^2 + \tau^2(1 - 1) = \phi^2;$$

that (2) for $\mathbf{h} \in H$,

$$\phi(-\mathbf{h}) = \phi^2 + \tau^2[1 - K(-\mathbf{h})] = \phi^2 + \tau^2[1 - K(\mathbf{h})] = \psi(\mathbf{h});$$

and that (3) for every positive integer M, for any M not-necessarily-distinct vectors $\mathbf{t}_1, \mathbf{t}_2, \dots, \mathbf{t}_M$ in S, and for any M scalars $x_1, x_2, \dots, x_M$ such that $\sum_{i=1}^{M} x_i = 0$,

$$\begin{aligned}
\sum_{i=1}^{M}\sum_{j=1}^{M} x_i x_j \psi(\mathbf{t}_i - \mathbf{t}_j) &= \sum_{i=1}^{M}\sum_{j=1}^{M} x_i x_j \{\phi^2 + \tau^2[1 - K(\mathbf{t}_i - \mathbf{t}_j)]\} \\
&= (\phi^2 + \tau^2)\sum_{i=1}^{M} x_i \sum_{j=1}^{M} x_j - \tau^2 \sum_{i=1}^{M}\sum_{j=1}^{M} x_i x_j K(\mathbf{t}_i - \mathbf{t}_j) \\
&= 0 - \tau^2 \sum_{i=1}^{M}\sum_{j=1}^{M} x_i x_j K(\mathbf{t}_i - \mathbf{t}_j) \\
&\le 0.
\end{aligned}$$

EXERCISE 9. Suppose that the general linear model is applied to the example of Section 4.5b (in the way described in Section 4.5b). What is the form of the function $\delta(\mathbf{u})$?

Solution. Take $C = 3$, and define u_1, u_2, and u_3 as in Section 4.5b. Then, $\delta(\mathbf{u})$ is such that for $u_3 = s$ $(= 1, 2, 3, 4)$,

$$\delta(\mathbf{u}) = \beta_{s1} + \beta_{s2}u_1 + \beta_{s3}u_2 + \beta_{s4}u_1^2 + \beta_{s5}u_2^2 + \beta_{s6}u_1u_2,$$

where $\beta_{s1}, \beta_{s2}, \dots, \beta_{s6}$ are unknown parameters.

5

Estimation and Prediction: Classical Approach

EXERCISE 1. Take the context to be that of estimating parametric functions of the form $\boldsymbol{\lambda}'\boldsymbol{\beta}$ from an $N \times 1$ observable random vector $\mathbf{y}$ that follows a G–M, Aitken, or general linear model. Verify (1) that linear combinations of estimable functions are estimable and (2) that linear combinations of nonestimable functions are not necessarily nonestimable.

Solution. (1) Let $\sum_{j=1}^{K} c_j(\boldsymbol{\lambda}_j'\boldsymbol{\beta})$ represent an arbitrary linear combination of an arbitrary number K of estimable linear functions $\boldsymbol{\lambda}_1'\boldsymbol{\beta}, \boldsymbol{\lambda}_2'\boldsymbol{\beta}, \ldots, \boldsymbol{\lambda}_K'\boldsymbol{\beta}$. Then, for $j = 1, 2, \ldots, K$, there exists an $N \times 1$ vector of constants $\mathbf{a}_j$ such that $\mathrm{E}(\mathbf{a}_j'\mathbf{y}) = \boldsymbol{\lambda}_j'\boldsymbol{\beta}$. And letting $\mathbf{a} = \sum_{j=1}^{K} c_j \mathbf{a}_j$, we find that

$$\mathrm{E}(\mathbf{a}'\mathbf{y}) = \mathrm{E}\Big[\sum_{j=1}^{K} c_j(\mathbf{a}_j'\mathbf{y})\Big] = \sum_{j=1}^{K} c_j \mathrm{E}(\mathbf{a}_j'\mathbf{y}) = \sum_{j=1}^{K} c_j(\boldsymbol{\lambda}_j'\boldsymbol{\beta}).$$

Thus, upon observing that $\sum_{j=1}^{K} c_j(\boldsymbol{\lambda}_j'\boldsymbol{\beta}) = \big(\sum_{j=1}^{K} c_j\boldsymbol{\lambda}_j\big)'\boldsymbol{\beta}$, we conclude that $\sum_{j=1}^{K} c_j(\boldsymbol{\lambda}_j'\boldsymbol{\beta})$ is an estimable function.

(2) Take $\boldsymbol{\lambda}_1'\boldsymbol{\beta}$ to be any estimable function and $\boldsymbol{\lambda}_2'\boldsymbol{\beta}$ to be any nonestimable function. And define $\boldsymbol{\lambda}_3 = \boldsymbol{\lambda}_1 + \boldsymbol{\lambda}_2$. Then, $\boldsymbol{\lambda}_3'\boldsymbol{\beta}$ is a nonestimable function—as is evident upon observing that $\boldsymbol{\lambda}_2'\boldsymbol{\beta} = \boldsymbol{\lambda}_3'\boldsymbol{\beta} - \boldsymbol{\lambda}_1'\boldsymbol{\beta}$ and hence that if $\boldsymbol{\lambda}_3'\boldsymbol{\beta}$ were estimable, then [in light of the result of Part (1)] $\boldsymbol{\lambda}_2'\boldsymbol{\beta}$ would be estimable. Moreover, $\boldsymbol{\lambda}_3'\boldsymbol{\beta} - \boldsymbol{\lambda}_2'\boldsymbol{\beta} = \boldsymbol{\lambda}_1'\boldsymbol{\beta}$. Thus, $\boldsymbol{\lambda}_3'\boldsymbol{\beta} - \boldsymbol{\lambda}_2'\boldsymbol{\beta}$, which is a linear combination of 2 nonestimable functions, is estimable.

EXERCISE 2. Take the context to be that of estimating parametric functions of the form $\boldsymbol{\lambda}'\boldsymbol{\beta}$ from an $N \times 1$ observable random vector $\mathbf{y}$ that follows a G–M, Aitken, or general linear model. And let $R = \mathrm{rank}(\mathbf{X})$.

(a) Verify (1) that there exists a set of R linearly independent estimable functions; (2) that no set of estimable functions contains more than R linearly independent estimable functions; and (3) that if the model is not of full rank (i.e., if $R < P$), then at least one and, in fact, at least $P - R$ of the individual parameters $\beta_1, \beta_2, \ldots, \beta_P$ are nonestimable.

(b) Show that the jth of the individual parameters $\beta_1, \beta_2, \ldots, \beta_P$ is estimable if and only if the jth element of every vector in $\mathcal{N}(\mathbf{X})$ equals 0 ($j = 1, 2, \ldots, P$).

Solution. (a) (1) Take $\boldsymbol{\lambda}_1, \boldsymbol{\lambda}_2, \ldots, \boldsymbol{\lambda}_R$ to be any R P-dimensional column vectors whose transposes $\boldsymbol{\lambda}_1', \boldsymbol{\lambda}_2', \ldots, \boldsymbol{\lambda}_R'$ are linearly independent rows of the model matrix $\mathbf{X}$—that $\mathbf{X}$ contains R linearly independent rows follows from Theorem 2.4.19. Then, obviously, $\boldsymbol{\lambda}_i' \in \mathcal{R}(\mathbf{X})$ ($i = 1, 2, \ldots, R$). And in light of the results of Section 5.3, it follows that $\boldsymbol{\lambda}_1'\boldsymbol{\beta}, \boldsymbol{\lambda}_2'\boldsymbol{\beta}, \ldots, \boldsymbol{\lambda}_R'\boldsymbol{\beta}$ are R linearly independent estimable functions.

(2) Let $\boldsymbol{\lambda}_1'\boldsymbol{\beta}, \boldsymbol{\lambda}_2'\boldsymbol{\beta}, \ldots, \boldsymbol{\lambda}_K'\boldsymbol{\beta}$ represent any K linearly independent estimable functions (where K is an arbitrary positive integer). Then, in light of the results of Section 5.3, $\boldsymbol{\lambda}_1', \boldsymbol{\lambda}_2', \ldots, \boldsymbol{\lambda}_K'$ are linearly independent vectors in $\mathcal{R}(\mathbf{X})$. And upon observing that $\dim[\mathcal{R}(\mathbf{X})] = R$, it follows from Therorem 2.4.9 that $K \le R$.

(3) Suppose that $R < P$. Then, upon observing that the individual parameters $\beta_1, \beta_2, \ldots, \beta_P$ can be regarded as P linearly independent combinations of the elements of $\boldsymbol{\beta}$, it follows from the

result of Part (2) that no more than R of the individual parameters are estimable (in which case, at least $P - R$ of them are nonestimable).

(b) Clearly, $\beta_j = \boldsymbol{\lambda}_j'\boldsymbol{\beta}$, where $\boldsymbol{\lambda}_j$ is the jth column of $\mathbf{I}_P$. Accordingly, it follows from the results of Section 5.3b that β_j is estimable if and only if

$$\mathbf{k}'\boldsymbol{\lambda}_j = 0 \tag{S.1}$$

for every $P \times 1$ vector $\mathbf{k}$ in $\mathcal{N}(\mathbf{X})$.

Now, let k_j represent the jth element of $\mathbf{k}$. Then, upon observing that

$$\mathbf{k}'\boldsymbol{\lambda}_j = k_j,$$

condition (S.1) is reexpressible as

$$k_j = 0.$$

Thus, β_j is estimable if and only if the jth element of every vector in $\mathcal{N}(\mathbf{X})$ equals 0.

EXERCISE 3. Show that for a parametric function of the form $\boldsymbol{\lambda}'\boldsymbol{\beta}$ to be estimable from an $N \times 1$ observable random vector $\mathbf{y}$ that follows a G–M, Aitken, or general linear model, it is necessary and sufficient that

$$\text{rank}(\mathbf{X}', \boldsymbol{\lambda}) = \text{rank}(\mathbf{X}).$$

Solution. It follows from the results of Section 5.3 that for $\boldsymbol{\lambda}'\boldsymbol{\beta}$ to be estimable, it is necessary and sufficient that the linear system $\mathbf{X}'\mathbf{a} = \boldsymbol{\lambda}$ (in $\mathbf{a}$) be consistent. And upon observing (in light of Theorem 2.9.2) that the linear system $\mathbf{X}'\mathbf{a} = \boldsymbol{\lambda}$ is consistent if and only if $\text{rank}(\mathbf{X}', \boldsymbol{\lambda}) = \text{rank}(\mathbf{X}')$ and observing [in light of equality (2.4.1)] that $\text{rank}(\mathbf{X}') = \text{rank}(\mathbf{X})$, it is clear that for $\boldsymbol{\lambda}'\boldsymbol{\beta}$ to be estimable, it is necessary and sufficient that $\text{rank}(\mathbf{X}', \boldsymbol{\lambda}) = \text{rank}(\mathbf{X})$.

EXERCISE 4. Suppose that $\mathbf{y}$ is an $N \times 1$ observable random vector that follows a G–M, Aitken, or general linear model. Further, take $\underline{\mathbf{y}}$ to be any value of $\mathbf{y}$, and consider the quantity $\boldsymbol{\lambda}'\tilde{\mathbf{b}}$, where $\boldsymbol{\lambda}$ is an arbitrary $P \times 1$ vector of constants and $\tilde{\mathbf{b}}$ is any solution to the linear system $\mathbf{X}'\mathbf{X}\mathbf{b} = \mathbf{X}'\underline{\mathbf{y}}$ (in the $P \times 1$ vector $\mathbf{b}$). Show that if $\boldsymbol{\lambda}'\tilde{\mathbf{b}}$ is invariant to the choice of the solution $\tilde{\mathbf{b}}$, then $\boldsymbol{\lambda}'\boldsymbol{\beta}$ is an estimable function. And discuss the implications of this result.

Solution. Theorem 2.11.7 implies that a $P \times 1$ vector is a solution to $\mathbf{X}'\mathbf{X}\mathbf{b} = \mathbf{X}'\underline{\mathbf{y}}$ if and only if it is expressible in the form

$$(\mathbf{X}'\mathbf{X})^-\mathbf{X}'\underline{\mathbf{y}} + [\mathbf{I} - (\mathbf{X}'\mathbf{X})^-\mathbf{X}'\mathbf{X}]\mathbf{z}$$

for some $P \times 1$ vector $\mathbf{z}$. Thus, $\boldsymbol{\lambda}'\tilde{\mathbf{b}}$ is invariant to the choice of the solution $\tilde{\mathbf{b}}$ if and only if

$$\boldsymbol{\lambda}'(\mathbf{X}'\mathbf{X})^-\mathbf{X}'\underline{\mathbf{y}} + \boldsymbol{\lambda}'[\mathbf{I} - (\mathbf{X}'\mathbf{X})^-\mathbf{X}'\mathbf{X}]\mathbf{z}$$

is invariant to the choice of $\mathbf{z}$, or equivalently (since one choice for $\mathbf{z}$ is $\mathbf{z} = \mathbf{0}$) if and only if

$$\boldsymbol{\lambda}'[\mathbf{I} - (\mathbf{X}'\mathbf{X})^-\mathbf{X}'\mathbf{X}]\mathbf{z} = 0$$

for every $P \times 1$ vector $\mathbf{z}$. Moreover, in light of Lemma 2.2.2,

$$\begin{aligned}
\boldsymbol{\lambda}'[\mathbf{I} - (\mathbf{X}'\mathbf{X})^-\mathbf{X}'\mathbf{X}]\mathbf{z} = 0 \ \text{for every } \mathbf{z} \\
&\Rightarrow \ \boldsymbol{\lambda}'[\mathbf{I} - (\mathbf{X}'\mathbf{X})^-\mathbf{X}'\mathbf{X}] = \mathbf{0} \\
&\Rightarrow \ \boldsymbol{\lambda}' = \boldsymbol{\lambda}'(\mathbf{X}'\mathbf{X})^-\mathbf{X}'\mathbf{X} \\
&\Rightarrow \ \boldsymbol{\lambda}' = \mathbf{a}'\mathbf{X} \ \text{for some } N \times 1 \text{ vector } \mathbf{a} \\
&\Rightarrow \ \boldsymbol{\lambda}'\boldsymbol{\beta} \ \text{is an estimable function.}
\end{aligned}$$

The primary implication of this result is that it is not feasible to extend the definition of a least squares estimator of an estimable parametric function of the form $\boldsymbol{\lambda}'\boldsymbol{\beta}$ to a nonestimable parametric function of the form $\boldsymbol{\lambda}'\boldsymbol{\beta}$. Such an extension would result in a function of $\mathbf{y}$ that is not uniquely defined. The least squares estimator of an estimable parametric function $\boldsymbol{\lambda}'\boldsymbol{\beta}$ is (by definition) the function of $\mathbf{y}$ whose value at $\mathbf{y} = \underline{\mathbf{y}}$ is $\boldsymbol{\lambda}'\tilde{\mathbf{b}}$ (where $\tilde{\mathbf{b}}$ is an any solution to the linear system $\mathbf{X}'\mathbf{X}\mathbf{b} = \mathbf{X}'\underline{\mathbf{y}}$). If $\boldsymbol{\lambda}'\boldsymbol{\beta}$ is an estimable function, $\boldsymbol{\lambda}'\tilde{\mathbf{b}}$ is invariant to the choice of the solution $\tilde{\mathbf{b}}$. If $\boldsymbol{\lambda}'\boldsymbol{\beta}$ were nonestimable, $\boldsymbol{\lambda}'\tilde{\mathbf{b}}$ would not be invariant to the choice of $\tilde{\mathbf{b}}$.

EXERCISE 5. Suppose that $\mathbf{y}$ is an $N \times 1$ observable random vector that follows a G–M, Aitken, or general linear model. And let $\mathbf{a}$ represent an arbitrary $N \times 1$ vector of constants. Show that $\mathbf{a}'\mathbf{y}$ is the least squares estimator of its expected value $\mathrm{E}(\mathbf{a}'\mathbf{y})$ (i.e., of the parametric function $\mathbf{a}'\mathbf{X}\boldsymbol{\beta}$) if and only if $\mathbf{a} \in \mathcal{C}(\mathbf{X})$.

Solution. Let $\ell(\mathbf{y})$ represent the least squares estimator of $\mathbf{a}'\mathbf{X}\boldsymbol{\beta}$. And observe [in light of result (4.24)] that

$$\ell(\mathbf{y}) = \mathbf{a}'\mathbf{X}(\mathbf{X}'\mathbf{X})^{-}\mathbf{X}'\mathbf{y}$$

and [in light of result (4.35)] that

$$\ell(\mathbf{y}) = \tilde{\mathbf{r}}'\mathbf{X}'\mathbf{y},$$

where $\tilde{\mathbf{r}}$ is any solution to the conjugate normal equations $\mathbf{X}'\mathbf{X}\mathbf{r} = \mathbf{X}'\mathbf{a}$. Either of these two representations can be used to establish that $\mathbf{a}'\mathbf{y}$ is the least squares estimator of its expected value if and only if $\mathbf{a} \in \mathcal{C}(\mathbf{X})$.

Now, suppose that $\mathbf{a} \in \mathcal{C}(\mathbf{X})$. Then, $\mathbf{a} = \mathbf{X}\mathbf{k}$ for some $P \times 1$ vector $\mathbf{k}$. And in light of Part (2) of Theorem 2.12.2, we have that

$$\ell(\mathbf{y}) = \mathbf{a}'\mathbf{X}(\mathbf{X}'\mathbf{X})^{-}\mathbf{X}'\mathbf{y} = \mathbf{k}'\mathbf{X}'\mathbf{X}(\mathbf{X}'\mathbf{X})^{-}\mathbf{X}'\mathbf{y} = \mathbf{k}'\mathbf{X}'\mathbf{y} = \mathbf{a}'\mathbf{y}.$$

Or, alternatively, upon observing that $\mathbf{X}'\mathbf{a} = \mathbf{X}'\mathbf{X}\mathbf{k}$ and hence that $\mathbf{k}$ is a solution to the conjugate normal equations, we have that

$$\ell(\mathbf{y}) = \mathbf{k}'\mathbf{X}'\mathbf{y} = \mathbf{a}'\mathbf{y}.$$

Conversely, suppose that $\ell(\mathbf{y}) = \mathbf{a}'\mathbf{y}$ (for every value of $\mathbf{y}$). Then, $\mathbf{a}'\mathbf{X}(\mathbf{X}'\mathbf{X})^{-}\mathbf{X}' = \mathbf{a}'$ and hence $\mathbf{a} = \mathbf{X}[(\mathbf{X}'\mathbf{X})^{-}]'\mathbf{X}'\mathbf{a} \in \mathcal{C}(\mathbf{X})$. Or, alternatively, $\tilde{\mathbf{r}}'\mathbf{X}' = \mathbf{a}'$ and hence $\mathbf{a} = \mathbf{X}\tilde{\mathbf{r}} \in \mathcal{C}(\mathbf{X})$.

EXERCISE 6. Let $\mathcal{U}$ represent a subspace of the linear space $\mathcal{R}^M$ of all M-dimensional column vectors. Verify that the set $\mathcal{U}^{\perp}$ (comprising all M-dimensional column vectors that are orthogonal to $\mathcal{U}$) is a linear space.

Solution. The $M \times 1$ null vector $\mathbf{0}$ is a member of $\mathcal{U}^{\perp}$, and consequently $\mathcal{U}^{\perp}$ is nonempty. Moreover, for any scalar k and for any M-dimensional column vector $\mathbf{x}$ in $\mathcal{U}^{\perp}$, we have that $k\mathbf{x} \in \mathcal{U}^{\perp}$, as is evident upon observing that for every M-dimensional column vector $\mathbf{w}$ in $\mathcal{U}$,

$$(k\mathbf{x}) \bullet \mathbf{w} = k(\mathbf{x} \bullet \mathbf{w}) = 0.$$

Similarly, for any M-dimensional column vectors $\mathbf{x}$ and $\mathbf{y}$ in $\mathcal{U}^{\perp}$, we have that $\mathbf{x}+\mathbf{y} \in \mathcal{U}^{\perp}$, as is evident upon observing that for every M-dimensional column vector $\mathbf{w}$ in $\mathcal{U}$,

$$(\mathbf{x} + \mathbf{y}) \bullet \mathbf{w} = (\mathbf{x} \bullet \mathbf{w}) + (\mathbf{y} \bullet \mathbf{w}) = 0 + 0 = 0.$$

And it follows that $\mathcal{U}^{\perp}$ is a linear space.

EXERCISE 7. Let $\mathbf{X}$ represent an $N \times P$ matrix. A $P \times N$ matrix $\mathbf{G}$ is said to be a *least squares generalized inverse* of $\mathbf{X}$ if it is a generalized inverse of $\mathbf{X}$ (i.e., if $\mathbf{X}\mathbf{G}\mathbf{X} = \mathbf{X}$) and if, in addition, $(\mathbf{X}\mathbf{G})' = \mathbf{X}\mathbf{G}$ (i.e., $\mathbf{X}\mathbf{G}$ is symmetric).

(a) Show that $\mathbf{G}$ is a least squares generalized inverse of $\mathbf{X}$ if and only if $\mathbf{X}'\mathbf{X}\mathbf{G} = \mathbf{X}'$.

(b) Using Part (a) (or otherwise), establish the existence of a least squares generalized inverse of $\mathbf{X}$.

(c) Show that if $\mathbf{G}$ is a least squares generalized inverse of $\mathbf{X}$, then, for any $N \times Q$ matrix $\mathbf{Y}$, the matrix $\mathbf{GY}$ is a solution to the linear system $\mathbf{X'XB} = \mathbf{X'Y}$ (in the $P \times Q$ matrix $\mathbf{B}$).

Solution. (a) Suppose that $\mathbf{X'XG} = \mathbf{X'}$. Then, $\mathbf{X'XGX} = \mathbf{X'X}$, implying (in light of Corollary 2.3.4) that $\mathbf{XGX} = \mathbf{X}$. Moreover,

$$\mathbf{XG} = (\mathbf{X'})'\mathbf{G} = (\mathbf{X'XG})'\mathbf{G} = (\mathbf{XG})'\mathbf{XG},$$

so that $\mathbf{XG}$ is symmetric. Thus, $\mathbf{G}$ is a least squares generalized inverse of $\mathbf{X}$.

Conversely, if $\mathbf{G}$ is a least squares generalized inverse of $\mathbf{X}$, then

$$\mathbf{X'XG} = \mathbf{X'}(\mathbf{XG})' = (\mathbf{XGX})' = \mathbf{X'}.$$

(b) In light of Part (a), it suffices to establish the existence of a matrix $\mathbf{G}$ such that $\mathbf{X'XG} = \mathbf{X'}$. Moreover, according to Part (2) of Theorem 2.12.2, $\mathbf{X'XG} = \mathbf{X'}$ for $\mathbf{G} = (\mathbf{X'X})^{-}\mathbf{X'}$.

(c) Suppose that $\mathbf{G}$ is a least squares generalized inverse of $\mathbf{X}$. Then, in light of Part (a),

$$\mathbf{X'XGY} = \mathbf{X'Y},$$

so that $\mathbf{GY}$ is a solution to the linear system $\mathbf{X'XB} = \mathbf{X'Y}$.

EXERCISE 8. Let $\mathbf{A}$ represent an $M \times N$ matrix. An $N \times M$ matrix $\mathbf{H}$ is said to be a *minimum norm generalized inverse* of $\mathbf{A}$ if it is a generalized inverse of $\mathbf{A}$ (i.e., if $\mathbf{AHA} = \mathbf{A}$) and if, in addition, $(\mathbf{HA})' = \mathbf{HA}$ (i.e., $\mathbf{HA}$ is symmetric).

(a) Show that $\mathbf{H}$ is a minimum norm generalized inverse of $\mathbf{A}$ if and only if $\mathbf{H'}$ is a least squares generalized inverse of $\mathbf{A'}$ (where least squares generalized inverse is as defined in Exercise 7).

(b) Using the results of Exercise 7 (or otherwise), establish the existence of a minimum norm generalized inverse of $\mathbf{A}$.

(c) Show that if $\mathbf{H}$ is a minimum norm generalized inverse of $\mathbf{A}$, then, for any vector $\mathbf{b} \in \mathcal{C}(\mathbf{A})$, $\|\mathbf{x}\|$ attains its minimum value over the set $\{\mathbf{x} : \mathbf{Ax} = \mathbf{b}\}$ [comprising all solutions to the linear system $\mathbf{Ax} = \mathbf{b}$ (in $\mathbf{x}$)] uniquely at $\mathbf{x} = \mathbf{Hb}$ (where $\|\cdot\|$ denotes the usual norm).

Solution. (a) Suppose that $\mathbf{H'}$ is a least squares generalized inverse of $\mathbf{A'}$. Then, $\mathbf{A'H'A'} = \mathbf{A'}$ and $(\mathbf{A'H'})' = \mathbf{A'H'}$, implying that

$$\mathbf{AHA} = (\mathbf{A'H'A'})' = (\mathbf{A'})' = \mathbf{A}$$

and that

$$(\mathbf{HA})' = \mathbf{A'H'} = (\mathbf{A'H'})' = \mathbf{HA}.$$

Thus, $\mathbf{H}$ is a minimum norm generalized inverse of $\mathbf{A}$.

Conversely, suppose that $\mathbf{H}$ is a minimum norm generalized inverse of $\mathbf{A}$. Then,

$$\mathbf{A'H'A'} = (\mathbf{AHA})' = \mathbf{A'},$$

and

$$(\mathbf{A'H'})' = [(\mathbf{HA})']' = (\mathbf{HA})' = \mathbf{A'H'}.$$

And it follows that $\mathbf{H'}$ is a least squares generalized inverse of $\mathbf{A'}$.

(b) It follows from Part (a) that the transpose of any least squares generalized inverse of $\mathbf{A'}$ is a minimum norm generalized inverse of $\mathbf{A}$. Thus, the existence of a minimum norm generalized inverse of $\mathbf{A}$ follows from the existence of a least squares generalized inverse of $\mathbf{A'}$; the existence of a least squares generalized inverse of $\mathbf{A'}$ is a consequence of the result established in Part (b) of Exercise 7.

(c) Suppose that $\mathbf{H}$ is a minimum norm generalized inverse of $\mathbf{A}$. And let $\mathbf{x}$ represent an arbitrary solution to $\mathbf{Ax} = \mathbf{b}$ [where $\mathbf{b} \in \mathcal{C}(\mathbf{A})$]. Then, according to Theorem 2.11.7,

$$\mathbf{x} = \mathbf{Hb} + (\mathbf{I} - \mathbf{HA})\mathbf{z}$$

for some vector $\mathbf{z}$. Accordingly, we have that

$$\begin{aligned}\|\mathbf{x}\|^2 &= [\mathbf{Hb} + (\mathbf{I} - \mathbf{HA})\mathbf{z}]'[\mathbf{Hb} + (\mathbf{I} - \mathbf{HA})\mathbf{z}] \\ &= \|\mathbf{Hb}\|^2 + \|(\mathbf{I} - \mathbf{HA})\mathbf{z}\|^2 + 2(\mathbf{Hb})'(\mathbf{I} - \mathbf{HA})\mathbf{z}.\end{aligned}$$

Moreover, upon observing that $\mathbf{b} = \mathbf{Ar}$ for some vector $\mathbf{r}$, we find that

$$(\mathbf{Hb})'(\mathbf{I} - \mathbf{HA})\mathbf{z} = \mathbf{r}'(\mathbf{HA})'(\mathbf{I} - \mathbf{HA})\mathbf{z} = \mathbf{r}'\mathbf{HA}(\mathbf{I} - \mathbf{HA})\mathbf{z} = \mathbf{r}'\mathbf{H}(\mathbf{A} - \mathbf{AHA})\mathbf{z} = 0.$$

Thus,

$$\|\mathbf{x}\|^2 = \|\mathbf{Hb}\|^2 + \|(\mathbf{I} - \mathbf{HA})\mathbf{z}\|^2,$$

so that $\|\mathbf{x}\|^2 \geq \|\mathbf{Hb}\|^2$, or equivalently $\|\mathbf{x}\| \geq \|\mathbf{Hb}\|$, with equality holding if and only if $\|(\mathbf{I} - \mathbf{HA})\mathbf{z}\| = 0$. Upon observing that $\|(\mathbf{I} - \mathbf{HA})\mathbf{z}\| = 0$ if and only if $(\mathbf{I} - \mathbf{HA})\mathbf{z} = \mathbf{0}$ and hence if and only if $\mathbf{x} = \mathbf{Hb}$ (and observing also that $\mathbf{Hb}$ is a solution to $\mathbf{Ax} = \mathbf{b}$), we conclude that $\|\mathbf{x}\|$ attains its minimum value over the set $\{\mathbf{x} : \mathbf{Ax} = \mathbf{b}\}$ uniquely at $\mathbf{x} = \mathbf{Hb}$.

EXERCISE 9. Let $\mathbf{X}$ represent an $N \times P$ matrix, and let $\mathbf{G}$ represent a $P \times N$ matrix that is subject to the following four conditions: (1) $\mathbf{XGX} = \mathbf{X}$; (2) $\mathbf{GXG} = \mathbf{G}$; (3) $(\mathbf{XG})' = \mathbf{XG}$; and (4) $(\mathbf{GX})' = \mathbf{GX}$.

(a) Show that if a $P \times P$ matrix $\mathbf{H}$ is a minimum norm generalized inverse of $\mathbf{X}'\mathbf{X}$, then conditions (1)–(4) can be satisfied by taking $\mathbf{G} = \mathbf{HX}'$.

(b) Use Part (a) and the result of Part (b) of Exercise 8 (or other means) to establish the existence of a $P \times N$ matrix $\mathbf{G}$ that satisfies conditions (1)–(4) and show that there is only one such matrix.

(c) Let $\mathbf{X}^+$ represent the unique $P \times N$ matrix $\mathbf{G}$ that satisfies conditions (1)–(4)—this matrix is customarily referred to as the *Moore-Penrose inverse*, and conditions (1)–(4) are customarily referred to as the *Moore-Penrose conditions*. Using Parts (a) and (b) and the results of Part (c) of Exercise 7 and Part (c) of Exercise 8 (or otherwise), show that $\mathbf{X}^+\mathbf{y}$ is a solution to the linear system $\mathbf{X}'\mathbf{Xb} = \mathbf{X}'\mathbf{y}$ (in $\mathbf{b}$) and that $\|\mathbf{b}\|$ attains its minimum value over the set $\{\mathbf{b} : \mathbf{X}'\mathbf{Xb} = \mathbf{X}'\mathbf{y}\}$ (comprising all solutions to the linear system) uniquely at $\mathbf{b} = \mathbf{X}^+\mathbf{y}$ (where $\|\cdot\|$ denotes the usual norm).

Solution. (a) Suppose that $\mathbf{H}$ is a minimum norm generalized inverse of $\mathbf{X}'\mathbf{X}$. Then,

(1) $\mathbf{X}(\mathbf{HX}')\mathbf{X} = \mathbf{X}$, as is evident from Part (1) of Theorem 2.12.2,

(2) $(\mathbf{HX}')\mathbf{X}(\mathbf{HX}') = \mathbf{HX}'$, as is evident from Part (2) of Theorem 2.12.2,

(3) $[\mathbf{X}(\mathbf{HX}')]' = \mathbf{XH}'\mathbf{X}' = \mathbf{X}(\mathbf{HX}')$, as is evident, for example, from Part (3) of Theorem 2.12.2 and from Corollary 2.10.11, and

(4) $[(\mathbf{HX}')\mathbf{X}]' = (\mathbf{HX}'\mathbf{X})' = \mathbf{HX}'\mathbf{X} = (\mathbf{HX}')\mathbf{X}$.

(b) Upon observing [in light of Part (b) of Exercise 8] that there exists a minimum norm generalized inverse of $\mathbf{X}'\mathbf{X}$, it follows from Part (a) (of the present exercise) that there exists a $P \times N$ matrix $\mathbf{G}$ that satisfies conditions (1)–(4). Now, let $\mathbf{G}_*$ represent any particular $P \times N$ matrix that satisfies conditions (1)–(4) [i.e., for which conditions (1)–(4) can be satisfied by setting $\mathbf{G} = \mathbf{G}_*$]. Then, for any matrix $\mathbf{G}$ that satisfies conditions (1)–(4),

$$\begin{aligned}\mathbf{G} = \mathbf{GXG} = \mathbf{G}(\mathbf{XG})' = \mathbf{GG}'\mathbf{X}' &= \mathbf{GG}'(\mathbf{XG}_*\mathbf{X})' = \mathbf{GG}'\mathbf{X}'(\mathbf{XG}_*)' \\ &= \mathbf{GG}'\mathbf{X}'\mathbf{XG}_* = \mathbf{G}(\mathbf{XG})'\mathbf{XG}_* = \mathbf{GXGXG}_* = \mathbf{GXG}_* \\ &= \mathbf{GXG}_*\mathbf{XG}_* = (\mathbf{GX})'(\mathbf{G}_*\mathbf{X})'\mathbf{G}_* = \mathbf{X}'\mathbf{G}'\mathbf{X}'\mathbf{G}_*'\mathbf{G}_* \\ &= (\mathbf{XGX})'\mathbf{G}_*'\mathbf{G}_* = \mathbf{X}'\mathbf{G}_*'\mathbf{G}_* = (\mathbf{G}_*\mathbf{X})'\mathbf{G}_* = \mathbf{G}_*\mathbf{XG}_* = \mathbf{G}_*.\end{aligned}$$

Thus, there is only one $P \times N$ matrix $\mathbf{G}$ that satisfies conditions (1)–(4).

(c) The Moore-Penrose inverse $\mathbf{X}^+$ of $\mathbf{X}$ is a least squares generalized inverse of $\mathbf{X}$. Accordingly, it follows from the result of Part (c) of Exercise 7 that $\mathbf{X}^+\mathbf{y}$ is a solution to the linear system $\mathbf{X}'\mathbf{X}\mathbf{b} = \mathbf{X}'\mathbf{y}$. Further, it follows from the results of Parts (a) and (b) (of the present exercise) that $\mathbf{X}^+ = \mathbf{H}\mathbf{X}'$, where $\mathbf{H}$ is a minimum norm generalized inverse of $\mathbf{X}'\mathbf{X}$. Thus, $\mathbf{X}^+\mathbf{y} = \mathbf{H}\mathbf{X}'\mathbf{y}$. And based on the result of Part (c) of Exercise 8, we conclude that $\mathbf{b}$ attains its minimum value over the set $\{\mathbf{b} : \mathbf{X}'\mathbf{X}\mathbf{b} = \mathbf{X}'\mathbf{y}\}$ uniquely at $\mathbf{b} = \mathbf{X}^+\mathbf{y}$.

EXERCISE 10. Consider further the alternative approach to the least squares computations, taking the formulation and the notation to be those of the final part of Section 5.4e.

(a) Let $\tilde{\mathbf{b}} = \mathbf{L}_1\tilde{\mathbf{h}}_1 + \mathbf{L}_2\tilde{\mathbf{h}}_2$, where $\tilde{\mathbf{h}}_2$ is an arbitrary $(P-K)$-dimensional column vector and $\tilde{\mathbf{h}}_1$ is the solution to the linear system $\mathbf{R}_1\mathbf{h}_1 = \mathbf{z}_1 - \mathbf{R}_2\tilde{\mathbf{h}}_2$. Show that $\|\tilde{\mathbf{b}}\|$ is minimized by taking

$$\tilde{\mathbf{h}}_2 = [\mathbf{I} + (\mathbf{R}_1^{-1}\mathbf{R}_2)'\mathbf{R}_1^{-1}\mathbf{R}_2]^{-1}(\mathbf{R}_1^{-1}\mathbf{R}_2)'\mathbf{R}_1^{-1}\mathbf{z}_1.$$

Do so by formulating this minimization problem as a least squares problem in which the role of $\mathbf{y}$ is played by the vector $\begin{pmatrix}\mathbf{R}_1^{-1}\mathbf{z}_1\\ \mathbf{0}\end{pmatrix}$, the role of $\mathbf{X}$ is played by the matrix $\begin{pmatrix}\mathbf{R}_1^{-1}\mathbf{R}_2\\ \mathbf{I}\end{pmatrix}$, and the role of $\mathbf{b}$ is played by $\tilde{\mathbf{h}}_2$.

(b) Let $\mathbf{O}_1$ represent a $P \times K$ matrix with orthonormal columns and $\mathbf{T}_1$ a $K \times K$ upper triangular matrix such that $\begin{pmatrix}\mathbf{R}_1'\\ \mathbf{R}_2'\end{pmatrix} = \mathbf{O}_1\mathbf{T}_1'$—the existence of a decomposition of this form can be established in much the same way as the existence of the QR decomposition (in which $\mathbf{T}_1$ would be lower triangular rather than upper triangular). Further, take $\mathbf{O}_2$ to be any $P \times (P-K)$ matrix such that the $P \times P$ matrix $\mathbf{O}$ defined by $\mathbf{O} = (\mathbf{O}_1, \mathbf{O}_2)$ is orthogonal.

(1) Show that $\mathbf{X} = \mathbf{Q}\mathbf{T}(\mathbf{L}\mathbf{O})'$, where $\mathbf{T} = \begin{pmatrix}\mathbf{T}_1 & \mathbf{0}\\ \mathbf{0} & \mathbf{0}\end{pmatrix}$.

(2) Show that $\mathbf{y} - \mathbf{X}\mathbf{b} = \mathbf{Q}_1(\mathbf{z}_1 - \mathbf{T}_1\mathbf{d}_1) + \mathbf{Q}_2\mathbf{z}_2$, where $\mathbf{d} = (\mathbf{L}\mathbf{O})'\mathbf{b}$ and $\mathbf{d}$ is partitioned as $\mathbf{d} = \begin{pmatrix}\mathbf{d}_1\\ \mathbf{d}_2\end{pmatrix}$.

(3) Show that $(\mathbf{y} - \mathbf{X}\mathbf{b})'(\mathbf{y} - \mathbf{X}\mathbf{b}) = (\mathbf{z}_1 - \mathbf{T}_1\mathbf{d}_1)'(\mathbf{z}_1 - \mathbf{T}_1\mathbf{d}_1) + \mathbf{z}_2'\mathbf{z}_2$.

(4) Taking $\tilde{\mathbf{d}}_1$ to be the solution to the linear system $\mathbf{T}_1\mathbf{d}_1 = \mathbf{z}_1$ (in $\mathbf{d}_1$), show that $(\mathbf{y} - \mathbf{X}\mathbf{b})'(\mathbf{y} - \mathbf{X}\mathbf{b})$ attains a minimum value of $\mathbf{z}_2'\mathbf{z}_2$ and that it does so at a value $\tilde{\mathbf{b}}$ of $\mathbf{b}$ if and only if $\tilde{\mathbf{b}} = \mathbf{L}\mathbf{O}\begin{pmatrix}\tilde{\mathbf{d}}_1\\ \tilde{\mathbf{d}}_2\end{pmatrix}$ for some $(P-K) \times 1$ vector $\tilde{\mathbf{d}}_2$.

(5) Letting $\tilde{\mathbf{b}}$ represent an arbitrary one of the values of $\mathbf{b}$ at which $(\mathbf{y} - \mathbf{X}\mathbf{b})'(\mathbf{y} - \mathbf{X}\mathbf{b})$ attains a minimum value [and, as in Part (4), taking $\tilde{\mathbf{d}}_1$ to be the solution to $\mathbf{T}_1\mathbf{d}_1 = \mathbf{z}_1$], show that $\|\tilde{\mathbf{b}}\|^2$ (where $\|\cdot\|$ denotes the usual norm) attains a minimum value of $\tilde{\mathbf{d}}_1'\tilde{\mathbf{d}}_1$ and that it does so uniquely at $\tilde{\mathbf{b}} = \mathbf{L}\mathbf{O}\begin{pmatrix}\tilde{\mathbf{d}}_1\\ \mathbf{0}\end{pmatrix}$.

Solution. (a) Clearly,

$$\|\tilde{\mathbf{b}}\|^2 = \tilde{\mathbf{b}}'\tilde{\mathbf{b}} = \begin{pmatrix}\tilde{\mathbf{h}}_1\\ \tilde{\mathbf{h}}_2\end{pmatrix}'\mathbf{L}'\mathbf{L}\begin{pmatrix}\tilde{\mathbf{h}}_1\\ \tilde{\mathbf{h}}_2\end{pmatrix} = \begin{pmatrix}\tilde{\mathbf{h}}_1\\ \tilde{\mathbf{h}}_2\end{pmatrix}'\begin{pmatrix}\tilde{\mathbf{h}}_1\\ \tilde{\mathbf{h}}_2\end{pmatrix} = \tilde{\mathbf{h}}_1'\tilde{\mathbf{h}}_1 + \tilde{\mathbf{h}}_2'\tilde{\mathbf{h}}_2.$$

Moreover, $\mathbf{h}_1 = \mathbf{R}_1^{-1}\mathbf{z}_1 - \mathbf{R}_1^{-1}\mathbf{R}_2\tilde{\mathbf{h}}_2$, so that

$$\begin{aligned}\tilde{\mathbf{h}}_1'\tilde{\mathbf{h}}_1 + \tilde{\mathbf{h}}_2'\tilde{\mathbf{h}}_2 &= (\mathbf{R}_1^{-1}\mathbf{z}_1 - \mathbf{R}_1^{-1}\mathbf{R}_2\tilde{\mathbf{h}}_2)'(\mathbf{R}_1^{-1}\mathbf{z}_1 - \mathbf{R}_1^{-1}\mathbf{R}_2\tilde{\mathbf{h}}_2) + \tilde{\mathbf{h}}_2'\tilde{\mathbf{h}}_2\\ &= \left[\begin{pmatrix}\mathbf{R}_1^{-1}\mathbf{z}_1\\ \mathbf{0}\end{pmatrix} - \begin{pmatrix}\mathbf{R}_1^{-1}\mathbf{R}_2\\ \mathbf{I}\end{pmatrix}\tilde{\mathbf{h}}_2\right]'\left[\begin{pmatrix}\mathbf{R}_1^{-1}\mathbf{z}_1\\ \mathbf{0}\end{pmatrix} - \begin{pmatrix}\mathbf{R}_1^{-1}\mathbf{R}_2\\ \mathbf{I}\end{pmatrix}\tilde{\mathbf{h}}_2\right].\end{aligned}$$

Accordingly, the problem of minimizing $\|\tilde{\mathbf{b}}\|$ with respect to $\tilde{\mathbf{h}}_2$ can be formulated as a least squares problem. More specifically, when the the roles of $\mathbf{y}$, $\mathbf{X}$, and $\mathbf{b}$ are played by $\begin{pmatrix} \mathbf{R}_1^{-1}\mathbf{z}_1 \\ \mathbf{0} \end{pmatrix}$, $\begin{pmatrix} \mathbf{R}_1^{-1}\mathbf{R}_2 \\ \mathbf{I} \end{pmatrix}$, and $\tilde{\mathbf{h}}_2$, respectively, it can be formulated as the (least squares) problem of minimizing $(\mathbf{y}-\mathbf{X}\mathbf{b})'(\mathbf{y}-\mathbf{X}\mathbf{b})$ with respect to $\mathbf{b}$. Thus, an expression for the value of $\tilde{\mathbf{h}}_2$ at which $\|\tilde{\mathbf{b}}\|$ attains its minimum value can be obtained from the solution to the normal equations for this least squares problem, that is, from the solution to the linear system

$$\begin{pmatrix} \mathbf{R}_1^{-1}\mathbf{R}_2 \\ \mathbf{I} \end{pmatrix}' \begin{pmatrix} \mathbf{R}_1^{-1}\mathbf{R}_2 \\ \mathbf{I} \end{pmatrix} \tilde{\mathbf{h}}_2 = \begin{pmatrix} \mathbf{R}_1^{-1}\mathbf{R}_2 \\ \mathbf{I} \end{pmatrix}' \begin{pmatrix} \mathbf{R}_1^{-1}\mathbf{z}_1 \\ \mathbf{0} \end{pmatrix}.$$

That solution is

$$\tilde{\mathbf{h}}_2 = [\mathbf{I} + (\mathbf{R}_1^{-1}\mathbf{R}_2)'\mathbf{R}_1^{-1}\mathbf{R}_2]^{-1}(\mathbf{R}_1^{-1}\mathbf{R}_2)'\mathbf{R}_1^{-1}\mathbf{z}_1.$$

(b) (1) Clearly,

$$(\mathbf{R}_1, \mathbf{R}_2) = \begin{pmatrix} \mathbf{R}_1' \\ \mathbf{R}_2' \end{pmatrix}' = \mathbf{T}_1\mathbf{O}_1' = (\mathbf{T}_1, \mathbf{0})\mathbf{O}',$$

so that

$$\mathbf{R} = \begin{pmatrix} \mathbf{R}_1 & \mathbf{R}_2 \\ \mathbf{0} & \mathbf{0} \end{pmatrix} = \begin{pmatrix} \mathbf{T}_1 & \mathbf{0} \\ \mathbf{0} & \mathbf{0} \end{pmatrix} \mathbf{O}' = \mathbf{T}\mathbf{O}'. \tag{S.2}$$

And upon replacing $\mathbf{R}$ in the expression $\mathbf{X} = \mathbf{Q}\mathbf{R}\mathbf{L}'$ with expression (S.2), we obtain the expression

$$\mathbf{X} = \mathbf{Q}\mathbf{T}(\mathbf{L}\mathbf{O})'.$$

(2) That $\mathbf{y} - \mathbf{X}\mathbf{b} = \mathbf{Q}_1(\mathbf{z}_1 - \mathbf{T}_1\mathbf{d}_1) + \mathbf{Q}_2\mathbf{z}_2$ is clear upon observing that

$$\mathbf{y} - \mathbf{X}\mathbf{b} = \mathbf{Q}(\mathbf{z} - \mathbf{T}\mathbf{d}) \tag{S.3}$$

and that

$$\mathbf{z} - \mathbf{T}\mathbf{d} = \begin{pmatrix} \mathbf{z}_1 - \mathbf{T}_1\mathbf{d}_1 \\ \mathbf{z}_2 \end{pmatrix}. \tag{S.4}$$

(3) Making use of results (S.3) and (S.4), we find that

$$\begin{aligned}(\mathbf{y} - \mathbf{X}\mathbf{b})'(\mathbf{y} - \mathbf{X}\mathbf{b}) &= (\mathbf{z} - \mathbf{T}\mathbf{d})'\mathbf{Q}'\mathbf{Q}(\mathbf{z} - \mathbf{T}\mathbf{d}) \\ &= (\mathbf{z} - \mathbf{T}\mathbf{d})'(\mathbf{z} - \mathbf{T}\mathbf{d}) \\ &= (\mathbf{z}_1 - \mathbf{T}_1\mathbf{d}_1)'(\mathbf{z}_1 - \mathbf{T}_1\mathbf{d}_1) + \mathbf{z}_2'\mathbf{z}_2. \end{aligned} \tag{S.5}$$

(4) Expression (S.5) attains a minimum value of $\mathbf{z}_2'\mathbf{z}_2$ and does so at those values of $\mathbf{d}$ for which the first term of expression (S.5) equals 0 or, equivalently, at those values for which $\mathbf{d}_1 = \tilde{\mathbf{d}}_1$. Thus, it follows from the results of Part (3) that $\mathbf{y} - \mathbf{X}\mathbf{b}'(\mathbf{y} - \mathbf{X}\mathbf{b})$ attains a minimum value of $\mathbf{z}_2'\mathbf{z}_2$ and that it does so at $\mathbf{b} = \tilde{\mathbf{b}}$ if and only if $\tilde{\mathbf{b}} = \mathbf{L}\mathbf{O}\begin{pmatrix} \tilde{\mathbf{d}}_1 \\ \tilde{\mathbf{d}}_2 \end{pmatrix}$ for some $(P-K) \times 1$ vector $\tilde{\mathbf{d}}_2$.

(5) In light of Part (4), it suffices to take $\tilde{\mathbf{b}} = \mathbf{L}\mathbf{O}\begin{pmatrix} \tilde{\mathbf{d}}_1 \\ \tilde{\mathbf{d}}_2 \end{pmatrix}$, where $\tilde{\mathbf{d}}_2$ is an arbitrary $(P-K) \times 1$ vector. Thus, recalling that the product of orthogonal matrices is orthogonal, we find that

$$\|\tilde{\mathbf{b}}\|^2 = \tilde{\mathbf{b}}'\tilde{\mathbf{b}} = \begin{pmatrix} \tilde{\mathbf{d}}_1 \\ \tilde{\mathbf{d}}_2 \end{pmatrix}' (\mathbf{L}\mathbf{O})'(\mathbf{L}\mathbf{O}) \begin{pmatrix} \tilde{\mathbf{d}}_1 \\ \tilde{\mathbf{d}}_2 \end{pmatrix} = \begin{pmatrix} \tilde{\mathbf{d}}_1 \\ \tilde{\mathbf{d}}_2 \end{pmatrix}' \begin{pmatrix} \tilde{\mathbf{d}}_1 \\ \tilde{\mathbf{d}}_2 \end{pmatrix} = \tilde{\mathbf{d}}_1'\tilde{\mathbf{d}}_1 + \tilde{\mathbf{d}}_2'\tilde{\mathbf{d}}_2.$$

Upon observing that $\tilde{\mathbf{d}}_1$ is uniquely determined—it is the unique solution to $\mathbf{T}_1\mathbf{d}_1 = \mathbf{z}_1$—and that $\tilde{\mathbf{d}}_2$ is arbitrary, it is clear that $\tilde{\mathbf{d}}_1'\tilde{\mathbf{d}}_1 + \tilde{\mathbf{d}}_2'\tilde{\mathbf{d}}_2$ attains a minimum value of $\tilde{\mathbf{d}}_1'\tilde{\mathbf{d}}_1$ and that it does so if and only if $\tilde{\mathbf{d}}_2 = \mathbf{0}$. We conclude that $\|\tilde{\mathbf{b}}\|^2$ attains a minimum value of $\tilde{\mathbf{d}}_1'\tilde{\mathbf{d}}_1$ and that it does so uniquely at $\tilde{\mathbf{b}} = \mathbf{L}\mathbf{O}\begin{pmatrix} \tilde{\mathbf{d}}_1 \\ \mathbf{0} \end{pmatrix}$.

EXERCISE 11. Verify that the difference (6.14) is a nonnegative definite matrix and that it equals $\mathbf{0}$ if and only if $\mathbf{c} + \mathbf{A}'\mathbf{y} = \boldsymbol{\ell}(\mathbf{y})$.

Solution. Clearly, $\mathrm{E}(\mathbf{A}'\mathbf{y}) = \boldsymbol{\Lambda}'\boldsymbol{\beta}$; that is, $\mathbf{A}'\mathbf{y}$ is an unbiased estimator of $\boldsymbol{\Lambda}'\boldsymbol{\beta}$. And as a consequence, the MSE matrix of $\mathbf{c} + \mathbf{A}'\mathbf{y}$ is

$$\begin{aligned}&\mathrm{E}[(\mathbf{c}+\mathbf{A}'\mathbf{y}-\boldsymbol{\Lambda}'\boldsymbol{\beta})(\mathbf{c}+\mathbf{A}'\mathbf{y}-\boldsymbol{\Lambda}'\boldsymbol{\beta})'] \\ &\quad = \mathbf{c}\mathbf{c}' + \mathrm{E}[(\mathbf{A}'\mathbf{y}-\boldsymbol{\Lambda}'\boldsymbol{\beta})(\mathbf{A}'\mathbf{y}-\boldsymbol{\Lambda}'\boldsymbol{\beta})'] + \mathrm{E}[\mathbf{c}(\mathbf{A}'\mathbf{y}-\boldsymbol{\Lambda}'\boldsymbol{\beta})'] + \mathrm{E}\{[\mathbf{c}(\mathbf{A}'\mathbf{y}-\boldsymbol{\Lambda}'\boldsymbol{\beta})']'\} \\ &\quad = \mathbf{c}\mathbf{c}' + \mathrm{var}(\mathbf{A}'\mathbf{y}).\end{aligned}$$

Thus,

$$\begin{aligned}&\mathrm{E}[(\mathbf{c}+\mathbf{A}'\mathbf{y}-\boldsymbol{\Lambda}'\boldsymbol{\beta})(\mathbf{c}+\mathbf{A}'\mathbf{y}-\boldsymbol{\Lambda}'\boldsymbol{\beta})'] - \mathrm{E}\{[\boldsymbol{\ell}(\mathbf{y})-\boldsymbol{\Lambda}'\boldsymbol{\beta}][\boldsymbol{\ell}(\mathbf{y})-\boldsymbol{\Lambda}'\boldsymbol{\beta}]'\} \\ &\qquad\qquad = \mathbf{c}\mathbf{c}' + \mathrm{var}(\mathbf{A}'\mathbf{y}) - \mathrm{var}[\boldsymbol{\ell}(\mathbf{y})]. \qquad \text{(S.6)}\end{aligned}$$

Moreover, it follows from Theorem 5.6.1 that $\mathrm{var}(\mathbf{A}'\mathbf{y}) - \mathrm{var}[\boldsymbol{\ell}(\mathbf{y})]$ is a nonnegative definite matrix. Accordingly, expression (S.6) is the sum of two nonnegative definite matrices and, consequently, is itself a nonnegative definite matrix.

It remains to show that if the difference (6.14) equals $\mathbf{0}$, then $\mathbf{c} + \mathbf{A}'\mathbf{y} = \boldsymbol{\ell}(\mathbf{y})$—that $\mathbf{c} + \mathbf{A}'\mathbf{y} = \boldsymbol{\ell}(\mathbf{y})$ implies that the difference (6.14) equals $\mathbf{0}$ is obvious. Accordingly, suppose that the difference (6.14) equals $\mathbf{0}$. Further, let $c_1, c_2, \ldots, c_M$ represent the elements of $\mathbf{c}$, and observe [in light of equality (S.6)] that

$$\mathrm{var}(\mathbf{A}'\mathbf{y}) - \mathrm{var}[\boldsymbol{\ell}(\mathbf{y})] = -\mathbf{c}\mathbf{c}'. \qquad \text{(S.7)}$$

Clearly, the diagonal elements of $-\mathbf{c}\mathbf{c}'$ are $-c_1^2, -c_2^2, \ldots, -c_M^2$, and because $\mathrm{var}(\mathbf{A}'\mathbf{y}) - \mathrm{var}[\boldsymbol{\ell}(\mathbf{y})]$ is nonnegative definite and because the diagonal elements of a nonnegative definite matrix are nonnegative, it follows from equality (S.7) that $-c_j^2 \geq 0$ $(j = 1, 2, \ldots, M)$ and hence that $c_1 = c_2 = \cdots = c_M = 0$ or, equivalently, $\mathbf{c} = \mathbf{0}$. Thus, $\mathbf{c} + \mathbf{A}'\mathbf{y}$ is a linear unbiased estimator of $\boldsymbol{\Lambda}'\boldsymbol{\beta}$, and

$$\mathrm{var}(\mathbf{c} + \mathbf{A}'\mathbf{y}) - \mathrm{var}[\boldsymbol{\ell}(\mathbf{y})] = \mathrm{var}(\mathbf{A}'\mathbf{y}) - \mathrm{var}[\boldsymbol{\ell}(\mathbf{y})] = \mathbf{0}.$$

And we conclude (on the basis of Theorem 5.6.1) that $\mathbf{c} + \mathbf{A}'\mathbf{y} = \boldsymbol{\ell}(\mathbf{y})$.

EXERCISE 12. Suppose that $\mathbf{y}$ is an $N \times 1$ observable random vector that follows a G–M, Aitken, or general linear model. And let $s(\mathbf{y})$ represent any particular translation-equivariant estimator of an estimable linear combination $\boldsymbol{\lambda}'\boldsymbol{\beta}$ of the elements of the parametric vector $\boldsymbol{\beta}$—e.g., $s(\mathbf{y})$ could be the least squares estimator of $\boldsymbol{\lambda}'\boldsymbol{\beta}$. Show that an estimator $t(\mathbf{y})$ of $\boldsymbol{\lambda}'\boldsymbol{\beta}$ is translation equivariant if and only if

$$t(\mathbf{y}) = s(\mathbf{y}) + d(\mathbf{y})$$

for some translation-invariant statistic $d(\mathbf{y})$.

Solution. Suppose that

$$t(\mathbf{y}) = s(\mathbf{y}) + d(\mathbf{y})$$

for some translation-invariant statistic $d(\mathbf{y})$. Then, for every $P \times 1$ vector $\mathbf{k}$ (and for every value of $\mathbf{y}$),

$$\begin{aligned}t(\mathbf{y} + \mathbf{X}\mathbf{k}) &= s(\mathbf{y} + \mathbf{X}\mathbf{k}) + d(\mathbf{y} + \mathbf{X}\mathbf{k}) \\ &= s(\mathbf{y}) + \boldsymbol{\lambda}'\mathbf{k} + d(\mathbf{y}) \\ &= t(\mathbf{y}) + \boldsymbol{\lambda}'\mathbf{k}.\end{aligned}$$

Thus, $t(\mathbf{y})$ is translation equivariant.

Conversely, suppose that $t(\mathbf{y})$ is translation equivariant. And define $d(\mathbf{y}) = t(\mathbf{y}) - s(\mathbf{y})$. Then, clearly,

$$t(\mathbf{y}) = s(\mathbf{y}) + d(\mathbf{y}).$$

Moreover, $d(\mathbf{y})$ is translation invariant, as is evident upon observing that, for every $P \times 1$ vector $\mathbf{k}$ (and for every value of $\mathbf{y}$),

$$\begin{aligned} d(\mathbf{y}+\mathbf{X}\mathbf{k}) &= t(\mathbf{y}+\mathbf{X}\mathbf{k}) - s(\mathbf{y}+\mathbf{X}\mathbf{k}) \\ &= t(\mathbf{y}) + \boldsymbol{\lambda}'\mathbf{k} - [s(\mathbf{y}) + \boldsymbol{\lambda}'\mathbf{k}] \\ &= t(\mathbf{y}) - s(\mathbf{y}) \\ &= d(\mathbf{y}). \end{aligned}$$

EXERCISE 13. Suppose that $\mathbf{y}$ is an $N \times 1$ observable random vector that follows a G–M model. And let $\mathbf{y}'\mathbf{A}\mathbf{y}$ represent a quadratic unbiased nonnegative-definite estimator of σ^2, that is, a quadratic form in $\mathbf{y}$ whose matrix $\mathbf{A}$ is a symmetric nonnegative definite matrix of constants and whose expected value is σ^2.

(a) Show that $\mathbf{y}'\mathbf{A}\mathbf{y}$ is translation invariant.

(b) Suppose that the fourth-order moments of the distribution of the vector $\mathbf{e} = (e_1, e_2, \ldots, e_N)'$ are such that (for $i, j, k, m = 1, 2, \ldots, N$) $\mathrm{E}(e_i e_j e_k e_m)$ satisfies condition (7.38). For what choice of $\mathbf{A}$ is the variance of the quadratic unbiased nonnegative-definite estimator $\mathbf{y}'\mathbf{A}\mathbf{y}$ a minimum? Describe your reasoning.

Solution. (a) As an application of formula (7.11), we have [as in result (7.57)] that

$$\mathrm{E}(\mathbf{y}'\mathbf{A}\mathbf{y}) = \sigma^2 \operatorname{tr}(\mathbf{A}) + \boldsymbol{\beta}'\mathbf{X}'\mathbf{A}\mathbf{X}\boldsymbol{\beta}.$$

And because $\mathbf{y}'\mathbf{A}\mathbf{y}$ estimates σ^2 unbiasedly, it follows that $\boldsymbol{\beta}'\mathbf{X}'\mathbf{A}\mathbf{X}\boldsymbol{\beta} = 0$ for all $\boldsymbol{\beta}$ [and that $\operatorname{tr}(\mathbf{A}) = 1$] and hence (in light of Corollary 2.13.4) that $\mathbf{X}'\mathbf{A}\mathbf{X} = \mathbf{0}$. Thus, based on Corollary 2.13.27, we conclude that $\mathbf{A}\mathbf{X} = \mathbf{0}$ and that, as a consequence, $\mathbf{y}'\mathbf{A}\mathbf{y}$ is translation invariant.

(b) It follows from Part (a) that every quadratic unbiased nonnegative-definite estimator of σ^2 is a quadratic unbiased translation-invariant estimator. And it follows from the results of Section 5.7d that the estimator of σ^2 that has minimum variance among quadratic unbiased translation-invariant estimators is the estimator $\hat{\sigma}^2 = \tilde{\mathbf{e}}'\tilde{\mathbf{e}}/(N - \operatorname{rank}\mathbf{X})$ [where $\tilde{\mathbf{e}} = (\mathbf{I} - \mathbf{P}_{\mathbf{X}})\mathbf{y}$]. Moreover, $\hat{\sigma}^2$ is expressible as

$$\hat{\sigma}^2 = \mathbf{y}'\left[\frac{1}{\sqrt{N - \operatorname{rank}\mathbf{X}}}(\mathbf{I} - \mathbf{P}_{\mathbf{X}})\right]' \frac{1}{\sqrt{N - \operatorname{rank}\mathbf{X}}}(\mathbf{I} - \mathbf{P}_{\mathbf{X}})\mathbf{y},$$

which is a quadratic form (in $\mathbf{y}$) whose matrix is (in light of Corollary 2.13.15) symmetric nonnegative definite. Thus, $\hat{\sigma}^2$ is a quadratic unbiased nonnegative-definite estimator. Since $\hat{\sigma}^2$ has minimum variance among quadratic unbiased translation-invariant estimators and since every quadratic unbiased nonnegative-definite estimator is a quadratic unbiased translation-invariant estimator, we conclude that $\hat{\sigma}^2$ has minimum variance among quadratic unbiased nonnegative-definite estimators.

EXERCISE 14. Suppose that $\mathbf{y}$ is an $N \times 1$ observable random vector that follows a G–M model. Suppose further that the distribution of the vector $\mathbf{e} = (e_1, e_2, \ldots, e_N)'$ has third-order moments $\lambda_{jkm} = \mathrm{E}(e_j e_k e_m)$ $(j, k, m = 1, 2, \ldots, N)$ and fourth-order moments $\gamma_{ijkm} = \mathrm{E}(e_i e_j e_k e_m)$ $(i, j, k, m = 1, 2, \ldots, N)$. And let $\mathbf{A} = \{a_{ij}\}$ represent an $N \times N$ symmetric matrix of constants.

(a) Show that in the special case where the elements $e_1, e_2, \ldots, e_N$ of $\mathbf{e}$ are statistically independent,

$$\operatorname{var}(\mathbf{y}'\mathbf{A}\mathbf{y}) = \mathbf{a}'\boldsymbol{\Omega}^*\mathbf{a} + 4\boldsymbol{\beta}'\mathbf{X}'\mathbf{A}\boldsymbol{\Lambda}^*\mathbf{a} + 2\sigma^4 \operatorname{tr}(\mathbf{A}^2) + 4\sigma^2\boldsymbol{\beta}'\mathbf{X}'\mathbf{A}^2\mathbf{X}\boldsymbol{\beta}, \tag{E.1}$$

where $\boldsymbol{\Omega}^*$ is the $N \times N$ diagonal matrix whose ith diagonal element is $\gamma_{iiii} - 3\sigma^4$, where $\boldsymbol{\Lambda}^*$ is the $N \times N$ diagonal matrix whose ith diagonal element is λ_{iii}, and where $\mathbf{a}$ is the $N \times 1$ vector whose elements are the diagonal elements $a_{11}, a_{22}, \ldots, a_{NN}$ of $\mathbf{A}$.

(b) Suppose that the elements $e_1, e_2, \ldots, e_N$ of $\mathbf{e}$ are statistically independent, that (for $i = 1, 2, \ldots, N$) $\gamma_{iiii} = \gamma$ (for some scalar γ), and that all N of the diagonal elements of the

$\mathbf{P_X}$ matrix are equal to each other. Show that the estimator $\hat{\sigma}^2 = \tilde{\mathbf{e}}'\tilde{\mathbf{e}}/(N - \text{rank}\,\mathbf{X})$ [where $\tilde{\mathbf{e}} = (\mathbf{I} - \mathbf{P_X})\mathbf{y}$] has minimum variance among all quadratic unbiased translation-invariant estimators of σ^2.

Solution. (a) Suppose that $e_1, e_2, \ldots, e_N$ are statistically independent. Then, unless $m = k = j$, $\lambda_{jkm} = 0$. And

$$\gamma_{ijkm} = \begin{cases} \sigma^4 & \text{if } j{=}i \text{ and } m{=}k{\neq}i, \text{ if } k{=}i \text{ and } m{=}j{\neq}i, \text{ or if } m{=}i \text{ and } k{=}j{\neq}i, \\ 0 & \text{unless } j{=}i \text{ and } m{=}k,\ k{=}i \text{ and } m{=}j, \text{ and/or } m{=}i \text{ and } k{=}j. \end{cases}$$

Thus, as an application of formula (7.17), we have that

$$\text{var}(\mathbf{y}'\mathbf{A}\mathbf{y}) = (\text{vec}\,\mathbf{A})'\boldsymbol{\Omega}\,\text{vec}\,\mathbf{A} + 4\boldsymbol{\beta}'\mathbf{X}'\boldsymbol{\Lambda}\,\text{vec}\,\mathbf{A} + 2\sigma^4\,\text{tr}(\mathbf{A}^2) + 4\sigma^2\boldsymbol{\beta}'\mathbf{X}'\mathbf{A}^2\mathbf{X}\boldsymbol{\beta},$$

where $\boldsymbol{\Omega}$ is an $N^2 \times N^2$ matrix whose entry for the ijth row [row $(i-1)N + j$] and kmth column [column $(k-1)N + m$] is $\gamma_{iiii} - 3\sigma^4$ if $m = k = j = i$ and is 0 otherwise and where $\boldsymbol{\Lambda}$ is an $N \times N^2$ matrix whose entry for the jth row and kmth column [column $(k-1)N + m$] is λ_{jjj} if $m = k = j$ and is 0 otherwise. Moreover, $(\text{vec}\,\mathbf{A})'\boldsymbol{\Omega}\,\text{vec}\,\mathbf{A} = \mathbf{a}'\boldsymbol{\Omega}^*\mathbf{a}$ and $\boldsymbol{\Lambda}\,\text{vec}\,\mathbf{A} = \boldsymbol{\Lambda}^*\mathbf{a}$.

(b) Suppose that $\mathbf{y}'\mathbf{A}\mathbf{y}$ is a quadratic unbiased translation-invariant estimator of σ^2. Then, upon applying formula (E.1) (with $\boldsymbol{\Omega}^*$, $\boldsymbol{\Lambda}^*$, and $\mathbf{a}$ being defined accordingly) and upon observing (in light of the results of Section 5.7d) that $\mathbf{AX} = \mathbf{0}$, we find that

$$\begin{aligned} \text{var}(\mathbf{y}'\mathbf{A}\mathbf{y}) &= \mathbf{a}'\boldsymbol{\Omega}^*\mathbf{a} + 4\boldsymbol{\beta}'(\mathbf{AX})'\boldsymbol{\Lambda}^*\mathbf{a} + 2\sigma^4\,\text{tr}(\mathbf{A}^2) + 4\sigma^2\boldsymbol{\beta}'\mathbf{X}'\mathbf{AAX}\boldsymbol{\beta} \\ &= (\gamma - 3\sigma^4)\mathbf{a}'\mathbf{a} + 0 + 2\sigma^4\,\text{tr}(\mathbf{A}^2) + 0 \\ &= (\gamma - 3\sigma^4)\mathbf{a}'\mathbf{a} + 2\sigma^4\,\text{tr}(\mathbf{A}^2). \end{aligned} \tag{S.8}$$

Let $\mathbf{R} = \mathbf{A} - \dfrac{1}{N - \text{rank}\,\mathbf{X}}(\mathbf{I} - \mathbf{P_X})$. Then, proceeding in much the same way as in the derivation of result (7.63), we find that

$$\text{tr}(\mathbf{R}) = 0 \tag{S.9}$$

and that

$$\text{tr}(\mathbf{A}^2) = \frac{1}{N - \text{rank}\,\mathbf{X}} + \text{tr}(\mathbf{R}'\mathbf{R}). \tag{S.10}$$

Now, let (for $i, j = 1, 2, \ldots, N$) r_{ij} represent the ijth element of $\mathbf{R}$, and take $\mathbf{r}$ to be the $N \times 1$ vector whose elements are the diagonal elements $r_{11}, r_{22}, \ldots, r_{NN}$ of $\mathbf{R}$. Because (by supposition) each of the diagonal elements of the $\mathbf{P_X}$ matrix has the same value, say c, we have that

$$\mathbf{r} = \mathbf{a} - \frac{1 - c}{N - \text{rank}\,\mathbf{X}}\mathbf{1}.$$

Moreover, it follows from result (7.35) that

$$c = \frac{1}{N}\,\text{tr}(\mathbf{P_X}) = \frac{1}{N}\,\text{rank}\,\mathbf{X}.$$

Thus, $\mathbf{r} = \mathbf{a} - \dfrac{1}{N}\mathbf{1}$, implying that $\mathbf{a}$ is expressible as $\mathbf{a} = \dfrac{1}{N}\mathbf{1} + \mathbf{r}$ and hence [in light of result (S.9)] that

$$\begin{aligned} \mathbf{a}'\mathbf{a} &= \frac{N}{N^2} + 2\frac{1}{N}\sum_i r_{ii} + \sum_i r_{ii}^2 \\ &= \frac{1}{N} + 2\frac{1}{N}\,\text{tr}(\mathbf{R}) + \sum_i r_{ii}^2 \\ &= \frac{1}{N} + 0 + \sum_i r_{ii}^2 = \frac{1}{N} + \sum_i r_{ii}^2. \end{aligned} \tag{S.11}$$

And upon replacing $\mathbf{a}'\mathbf{a}$ and $\operatorname{tr}(\mathbf{A}^2)$ in expression (S.8) with expressions (S.11) and (S.10), we find that

$$\begin{aligned}\operatorname{var}(\mathbf{y}'\mathbf{A}\mathbf{y}) &= (\gamma - 3\sigma^4)\Bigl(\frac{1}{N} + \sum_i r_{ii}^2\Bigr) + 2\sigma^4\Bigl[\frac{1}{N - \operatorname{rank}\mathbf{X}} + \operatorname{tr}(\mathbf{R}'\mathbf{R})\Bigr] \\ &= (\gamma - \sigma^4)\Bigl(\frac{1}{N} + \sum_i r_{ii}^2\Bigr) + 2\sigma^4\Bigl(\frac{1}{N - \operatorname{rank}\mathbf{X}} - \frac{1}{N} + \sum_{i,j} r_{ij}^2 - \sum_i r_{ii}^2\Bigr) \\ &= (\gamma - \sigma^4)\Bigl(\frac{1}{N} + \sum_i r_{ii}^2\Bigr) + 2\sigma^4\Bigl[\frac{\operatorname{rank}\mathbf{X}}{N(N - \operatorname{rank}\mathbf{X})} + \sum_{i,\,j\neq i} r_{ij}^2\Bigr],\end{aligned}$$

which (upon observing that $\gamma \geq \sigma^4$) leads to the conclusion that $\operatorname{var}(\mathbf{y}'\mathbf{A}\mathbf{y})$ attains its minimum value when $\mathbf{R} = \mathbf{0}$ or, equivalently, when $\mathbf{A} = \dfrac{1}{N - \operatorname{rank}\mathbf{X}}(\mathbf{I} - \mathbf{P}_{\mathbf{X}})$ (i.e., when $\mathbf{y}'\mathbf{A}\mathbf{y} = \hat{\sigma}^2$).

EXERCISE 15. Suppose that $\mathbf{y}$ is an $N \times 1$ observable random vector that follows a G–M model, and assume that the distribution of the vector $\mathbf{e}$ of residual effects is MVN.

(a) Letting $\boldsymbol{\lambda}'\boldsymbol{\beta}$ represent an estimable linear combination of the elements of the parametric vector $\boldsymbol{\beta}$, find a minimum-variance unbiased estimator of $(\boldsymbol{\lambda}'\boldsymbol{\beta})^2$.

(b) Find a minimum-variance unbiased estimator of σ^4.

Solution. Recall (from Section 5.8) that $\mathbf{X}'\mathbf{y}$ and $\mathbf{y}'(\mathbf{I} - \mathbf{P}_{\mathbf{X}})\mathbf{y}$ form a complete sufficient statistic and that, consequently, "any" function, say $t[\mathbf{X}'\mathbf{y},\, \mathbf{y}'(\mathbf{I} - \mathbf{P}_{\mathbf{X}})\mathbf{y}]$, of $\mathbf{X}'\mathbf{y}$ and $\mathbf{y}'(\mathbf{I} - \mathbf{P}_{\mathbf{X}})\mathbf{y}$ is a minimum-variance unbiased estimator of $\mathrm{E}\{t[\mathbf{X}'\mathbf{y},\, \mathbf{y}'(\mathbf{I} - \mathbf{P}_{\mathbf{X}})\mathbf{y}]\}$.

(a) Recall (from Section 5.4c) that the least squares estimator of $\boldsymbol{\lambda}'\boldsymbol{\beta}$ is $\tilde{\mathbf{r}}'\mathbf{X}'\mathbf{y}$, where $\tilde{\mathbf{r}}$ is any solution to the conjugate normal equations $\mathbf{X}'\mathbf{X}\mathbf{r} = \boldsymbol{\lambda}$. Recall also that

$$\mathrm{E}(\tilde{\mathbf{r}}'\mathbf{X}'\mathbf{y}) = \boldsymbol{\lambda}'\boldsymbol{\beta} \quad \text{and} \quad \operatorname{var}(\tilde{\mathbf{r}}'\mathbf{X}'\mathbf{y}) = \sigma^2\tilde{\mathbf{r}}'\boldsymbol{\lambda}.$$

And observe that

$$\mathrm{E}[(\tilde{\mathbf{r}}'\mathbf{X}'\mathbf{y})^2] = \operatorname{var}(\tilde{\mathbf{r}}'\mathbf{X}'\mathbf{y}) + [\mathrm{E}(\tilde{\mathbf{r}}'\mathbf{X}'\mathbf{y})]^2 = \sigma^2\tilde{\mathbf{r}}'\boldsymbol{\lambda} + (\boldsymbol{\lambda}'\boldsymbol{\beta})^2.$$

Further, let $\hat{\sigma}^2 = \mathbf{y}'(\mathbf{I} - \mathbf{P}_{\mathbf{X}})\mathbf{y}/(N - \operatorname{rank}\mathbf{X})$, and recall (from Section 5.7c) that $\mathrm{E}(\hat{\sigma}^2) = \sigma^2$. Thus,

$$\mathrm{E}[(\tilde{\mathbf{r}}'\mathbf{X}'\mathbf{y})^2] = \mathrm{E}(\hat{\sigma}^2\tilde{\mathbf{r}}'\boldsymbol{\lambda}) + (\boldsymbol{\lambda}'\boldsymbol{\beta})^2$$

or, equivalently,

$$\mathrm{E}[(\tilde{\mathbf{r}}'\mathbf{X}'\mathbf{y})^2 - \hat{\sigma}^2\tilde{\mathbf{r}}'\boldsymbol{\lambda}] = (\boldsymbol{\lambda}'\boldsymbol{\beta})^2.$$

We conclude that $(\tilde{\mathbf{r}}'\mathbf{X}'\mathbf{y})^2 - \hat{\sigma}^2\tilde{\mathbf{r}}'\boldsymbol{\lambda}$ is an unbiased estimator of $(\boldsymbol{\lambda}'\boldsymbol{\beta})^2$ and, upon observing that $(\tilde{\mathbf{r}}'\mathbf{X}'\mathbf{y})^2 - \hat{\sigma}^2\tilde{\mathbf{r}}'\boldsymbol{\lambda}$ depends on the value of $\mathbf{y}$ only through the values of $\mathbf{X}'\mathbf{y}$ and $\mathbf{y}'(\mathbf{I} - \mathbf{P}_{\mathbf{X}})\mathbf{y}$ (which form a complete sufficient statistic), that $(\tilde{\mathbf{r}}'\mathbf{X}'\mathbf{y})^2 - \hat{\sigma}^2\tilde{\mathbf{r}}'\boldsymbol{\lambda}$ has minimum variance among all unbiased estimators of $(\boldsymbol{\lambda}'\boldsymbol{\beta})^2$.

(b) Making use of the results of Section 5.7c, we find that

$$\begin{aligned}\mathrm{E}\{[\mathbf{y}'(\mathbf{I} - \mathbf{P}_{\mathbf{X}})\mathbf{y}]^2\} &= \operatorname{var}[\mathbf{y}'(\mathbf{I} - \mathbf{P}_{\mathbf{X}})\mathbf{y}] + \{\mathrm{E}[\mathbf{y}'(\mathbf{I} - \mathbf{P}_{\mathbf{X}})\mathbf{y}]\}^2 \\ &= 2\sigma^4(N - \operatorname{rank}\mathbf{X}) + \sigma^4(N - \operatorname{rank}\mathbf{X})^2 \\ &= \sigma^4(N - \operatorname{rank}\mathbf{X})[N - \operatorname{rank}(\mathbf{X}) + 2].\end{aligned}$$

Thus, σ^4 is estimated unbiasedly by $\dfrac{[\mathbf{y}'(\mathbf{I} - \mathbf{P}_{\mathbf{X}})\mathbf{y}]^2}{(N - \operatorname{rank}\mathbf{X})[N - \operatorname{rank}(\mathbf{X}) + 2]}$. And since this estimator of σ^4 depends on the value of $\mathbf{y}$ only through the value of $\mathbf{y}'(\mathbf{I} - \mathbf{P}_{\mathbf{X}})\mathbf{y}$ [which, in combination with $\mathbf{X}'\mathbf{y}$, forms a complete sufficient statistic], it has minimum variance among all unbiased estimators of σ^4.

EXERCISE 16. Suppose that $\mathbf{y}$ is an $N \times 1$ observable random vector that follows a G–M model, and assume that the distribution of the vector $\mathbf{e}$ of residual effects is MVN. Show that if σ^2 were known, $\mathbf{X}'\mathbf{y}$ would be a complete sufficient statistic.

Solution. Suppose that σ^2 were known.

Proceeding in much the same way as in Section 5.8, let us begin by letting $K = \text{rank}\,\mathbf{X}$ and by observing that there exists an $N \times K$ matrix, say $\mathbf{W}$, whose columns form a basis for $\mathcal{C}(\mathbf{X})$. Observe also that $\mathbf{W} = \mathbf{XR}$ for some matrix $\mathbf{R}$ and that $\mathbf{X} = \mathbf{WS}$ for some $(K \times P)$ matrix $\mathbf{S}$ (of rank K). Moreover, $\mathbf{X}'\mathbf{y}$ is expressible in terms of the $(K \times 1)$ vector $\mathbf{W}'\mathbf{y}$, and, conversely, $\mathbf{W}'\mathbf{y}$ is expressible in terms of $\mathbf{X}'\mathbf{y}$; we have that $\mathbf{X}'\mathbf{y} = \mathbf{S}'\mathbf{W}'\mathbf{y}$ and that $\mathbf{W}'\mathbf{y} = \mathbf{R}'\mathbf{X}'\mathbf{y}$. Thus, corresponding to any function $g(\mathbf{X}'\mathbf{y})$ of $\mathbf{X}'\mathbf{y}$, there is a function , say $g_*(\mathbf{W}'\mathbf{y})$, of $\mathbf{W}'\mathbf{y}$ such that $g_*(\mathbf{W}'\mathbf{y}) = g(\mathbf{X}'\mathbf{y})$ for every value of $\mathbf{y}$; namely, the function $g_*(\mathbf{W}'\mathbf{y})$ defined by $g_*(\mathbf{W}'\mathbf{y}) = g(\mathbf{S}'\mathbf{W}'\mathbf{y})$. Similary, corresponding to any function $h(\mathbf{W}'\mathbf{y})$ of $\mathbf{W}'\mathbf{y}$, there is a function, say $h_*(\mathbf{X}'\mathbf{y})$, of $\mathbf{X}'\mathbf{y}$ such that $h_*(\mathbf{X}'\mathbf{y}) = h(\mathbf{W}'\mathbf{y})$ for every value of $\mathbf{y}$; namely, the function $h_*(\mathbf{X}'\mathbf{y})$ defined by $h_*(\mathbf{X}'\mathbf{y}) = h(\mathbf{R}'\mathbf{X}'\mathbf{y})$.

Now, suppose that $\mathbf{W}'\mathbf{y}$ were a complete sufficient statistic. Then, since $\mathbf{W}'\mathbf{y} = \mathbf{R}'\mathbf{X}'\mathbf{y}$, $\mathbf{X}'\mathbf{y}$ would be a sufficient statistic. Moreover, if $\text{E}[g(\mathbf{X}'\mathbf{y})] = 0$, then $\text{E}[g_*(\mathbf{W}'\mathbf{y})] = 0$, implying that $\Pr[g_*(\mathbf{W}'\mathbf{y}) = 0] = 1$ and hence that $\Pr[g(\mathbf{X}'\mathbf{y}) = 0] = 1$. Thus, $\mathbf{X}'\mathbf{y}$ would be a complete statistic.

Conversely, suppose that $\mathbf{X}'\mathbf{y}$ were a complete sufficient statistic. Then, since $\mathbf{X}'\mathbf{y} = \mathbf{S}'\mathbf{W}'\mathbf{y}$, $\mathbf{W}'\mathbf{y}$ would be a sufficient statistic. Moreover, if $\text{E}[h(\mathbf{W}'\mathbf{y})] = 0$, then $\text{E}[h_*(\mathbf{X}'\mathbf{y})] = 0$, implying that $\Pr[h_*(\mathbf{X}'\mathbf{y}) = 0] = 1$ and hence that $\Pr[h(\mathbf{W}'\mathbf{y}) = 0] = 1$. Thus, $\mathbf{W}'\mathbf{y}$ would be a complete statistic.

At this point, we have established that $\mathbf{X}'\mathbf{y}$ would be a complete sufficient statistic if and only if $\mathbf{W}'\mathbf{y}$ would be a complete sufficient statistic. Thus, for purposes of verifying that $\mathbf{X}'\mathbf{y}$ would be a complete sufficient statistic, it suffices to consider the would-be sufficiency and completeness of the statistic $\mathbf{W}'\mathbf{y}$. In that regard, the probability density function of $\mathbf{y}$, say $f(\cdot)$, is expressible as follows:

$$f(\mathbf{y}) = \frac{1}{(2\pi\sigma^2)^{N/2}} \exp\left[-\frac{1}{2\sigma^2}\mathbf{y}'\mathbf{y}\right] \exp\left[-\frac{1}{2\sigma^2}\boldsymbol{\beta}'\mathbf{X}'\mathbf{X}\boldsymbol{\beta}\right] \exp\left[\left(\frac{1}{\sigma^2}\mathbf{S}\boldsymbol{\beta}\right)'\mathbf{W}'\mathbf{y}\right]. \tag{S.12}$$

Based on the same well-known result on complete sufficient statistics for exponential families that was employed in Section 5.8, it follows from result (S.12) that (under the supposition of normality) $\mathbf{W}'\mathbf{y}$ would be a complete sufficient statistic—to establish that the result on complete sufficient statistics for exponential families is applicable, it suffices to observe [in connection with expression (S.12)] that the $(K \times 1)$ vector $(1/\sigma^2)\mathbf{S}\boldsymbol{\beta}$ of parametric functions is such that, for any $K \times 1$ vector $\mathbf{d}$, $(1/\sigma^2)\mathbf{S}\boldsymbol{\beta} = \mathbf{d}$ for some value of $\boldsymbol{\beta}$ (as is evident upon noting that $\mathbf{S}$ contains K linearly independent columns). It remains only to observe that since $\mathbf{W}'\mathbf{y}$ would be a complete sufficient statistic, $\mathbf{X}'\mathbf{y}$ would be a complete sufficient statistic.

EXERCISE 17. Suppose that $\mathbf{y}$ is an $N \times 1$ observable random vector that follows a general linear model. Suppose further that the distribution of the vector $\mathbf{e}$ of residual effects is MVN or, more generally, that the distribution of $\mathbf{e}$ is known up to the value of the vector $\boldsymbol{\theta}$. And take $\mathbf{h}(\mathbf{y})$ to be any (possibly vector-valued) translation-invariant statistic.

(a) Show that if $\boldsymbol{\theta}$ were known, $\mathbf{h}(\mathbf{y})$ would be an ancillary statistic—for a definition of ancillarity, refer, e.g., to Casella and Berger (2002, def. 6.2.16) or to Lehmann and Casella (1998, p. 41).

(b) Suppose that $\mathbf{X}'\mathbf{y}$ would be a complete sufficient statistic if $\boldsymbol{\theta}$ were known. Show (1) that the least squares estimator of any estimable linear combination $\boldsymbol{\lambda}'\boldsymbol{\beta}$ of the elements of the parametric vector $\boldsymbol{\beta}$ has minimum variance among all unbiased estimators, (2) that any vector of least squares estimators of estimable linear combinations (of the elements of $\boldsymbol{\beta}$) is distributed independently of $\mathbf{h}(\mathbf{y})$, and (3) (using the result of Exercise 12 or otherwise) that the least squares estimator of any estimable linear combination $\boldsymbol{\lambda}'\boldsymbol{\beta}$ has minimum mean squared error among all translation-equivariant estimators. {*Hint* [for Part (2)]. Make use of Basu's theorem—refer, e.g., to Lehmann and Casella (1998, p. 42) for a statement of Basu's theorem.}

Solution. (a) It was established in Section 5.7d that any translation-invariant statistic depends on the value of $\mathbf{y}$ only through the value of $\mathbf{e}$. Thus, if $\boldsymbol{\theta}$ were known, the distribution of $\mathbf{h}(\mathbf{y})$ would not depend on the (unknown) parameters—the only (unknown) parameters would be the elements of $\boldsymbol{\beta}$. And it follows that $\mathbf{h}(\mathbf{y})$ would be an ancillary statistic.

(b) (1) Let $\ell(\mathbf{y})$ represent the least squares estimator of $\boldsymbol{\lambda}'\boldsymbol{\beta}$, and recall (from the results of Section 5.4c) that $\ell(\mathbf{y})$ is an unbiased estimator. And note that $\ell(\mathbf{y})$ and every other unbiased estimator of $\boldsymbol{\lambda}'\boldsymbol{\beta}$ would be among the estimators that would be unbiased if $\boldsymbol{\theta}$ were known. Note also that $\ell(\mathbf{y}) = \mathbf{r}'\mathbf{X}'\mathbf{y}$ for some $(P \times 1)$ vector $\mathbf{r}$. Thus, $\ell(\mathbf{y})$ depends on the value of $\mathbf{y}$ only through the value of what would be a complete sufficient statistic if $\boldsymbol{\theta}$ were known, and, consequently, $\ell(\mathbf{y})$ has minimum variance among unbiased estimators at the point $\boldsymbol{\theta}$. Moreover, this same line of reasoning is applicable and the same conclusion reached for every $\boldsymbol{\theta} \in \Theta$.

(2) As noted in Section 5.7c, any (column) vector of least squares estimators of estimable linear combinations of the elements of $\boldsymbol{\beta}$ is expressible as $\mathbf{R}'\mathbf{X}'\mathbf{y}$ for some matrix $\mathbf{R}$. Since $\mathbf{X}'\mathbf{y}$ would be a complete sufficient statistic and [in light of Part (a)] $\mathbf{h}(\mathbf{y})$ would be an ancillary statistic if $\boldsymbol{\theta}$ were known, it follows from Basu's theorem that $\mathbf{X}'\mathbf{y}$ is distributed independently of $\mathbf{h}(\mathbf{y})$ when the distribution of $\mathbf{e}$ is that corresponding to the point $\boldsymbol{\theta}$. Moreover, this same line of reasoning is applicable and the same conclusion reached for every $\boldsymbol{\theta} \in \Theta$. And, upon observing that $\mathbf{R}'\mathbf{X}'\mathbf{y}$ depends on the value of $\mathbf{y}$ only through the value of $\mathbf{X}'\mathbf{y}$, we conclude that $\mathbf{R}'\mathbf{X}'\mathbf{y}$ is distributed independently of $\mathbf{h}(\mathbf{y})$.

(3) Let $\ell(\mathbf{y})$ represent the least squares estimator of $\boldsymbol{\lambda}'\boldsymbol{\beta}$, and recall (from the results of Section 5.5) that $\ell(\mathbf{y})$ is translation-equivariant (as well as unbiased). Further, let $t(\mathbf{y})$ represent an arbitrary translation-equivariant estimator of $\boldsymbol{\lambda}'\boldsymbol{\beta}$, and observe (based on the result of Exercise 12) that

$$t(\mathbf{y}) = \ell(\mathbf{y}) + d(\mathbf{y})$$

for some translation-invariant statistic $d(\mathbf{y})$. Observe also [in light of Part (2)] that $\ell(\mathbf{y})$ and $d(\mathbf{y})$ are statistically independent. Thus, the mean squared error of $t(\mathbf{y})$ is expressible as

$$\begin{aligned} \mathrm{E}\{[t(\mathbf{y})-\boldsymbol{\lambda}'\boldsymbol{\beta}]^2\} &= \mathrm{E}\{[\ell(\mathbf{y})-\boldsymbol{\lambda}'\boldsymbol{\beta}+d(\mathbf{y})]^2\} \\ &= \mathrm{E}\{[\ell(\mathbf{y})-\boldsymbol{\lambda}'\boldsymbol{\beta}]^2\} + 2\,\mathrm{E}[\ell(\mathbf{y})d(\mathbf{y})] - 2\,\mathrm{E}[\boldsymbol{\lambda}'\boldsymbol{\beta} d(\mathbf{y})] + \mathrm{E}\{[d(\mathbf{y})]^2\} \\ &= \mathrm{var}[\ell(\mathbf{y})] + 2\,\mathrm{E}[\ell(\mathbf{y})]\,\mathrm{E}[d(\mathbf{y})] - 2\boldsymbol{\lambda}'\boldsymbol{\beta}\,\mathrm{E}[d(\mathbf{y})] + \mathrm{E}\{[d(\mathbf{y})]^2\} \\ &= \mathrm{var}[\ell(\mathbf{y})] + 2\boldsymbol{\lambda}'\boldsymbol{\beta}\,\mathrm{E}[d(\mathbf{y})] - 2\boldsymbol{\lambda}'\boldsymbol{\beta}\,\mathrm{E}[d(\mathbf{y})] + \mathrm{E}\{[d(\mathbf{y})]^2\} \\ &= \mathrm{var}[\ell(\mathbf{y})] + \mathrm{E}\{[d(\mathbf{y})]^2\} \\ &\geq \mathrm{var}[\ell(\mathbf{y})], \end{aligned} \tag{S.13}$$

with equality holding [in inequality (S.13)] if and only if $d(\mathbf{y}) = 0$ (with probability 1) or, equivalently, if and only if $t(\mathbf{y}) = \ell(\mathbf{y})$ (with probability 1). And it follows that $\ell(\mathbf{y})$ has minimum mean squared error among all translation-equivariant estimators of $\boldsymbol{\lambda}'\boldsymbol{\beta}$.

EXERCISE 18. Suppose that $\mathbf{y}$ is an $N \times 1$ observable random vector that follows a G–M model. Suppose further that the distribution of the vector $\mathbf{e}$ of residual effects is MVN. And, letting $\tilde{\mathbf{e}} = \mathbf{y} - \mathbf{P}_{\mathbf{X}}\mathbf{y}$, take $\tilde{\sigma}^2 = \tilde{\mathbf{e}}'\tilde{\mathbf{e}}/N$ to be the ML estimator of σ^2 and $\hat{\sigma}^2 = \tilde{\mathbf{e}}'\tilde{\mathbf{e}}/(N - \operatorname{rank}\mathbf{X})$ to be the unbiased estimator.

(a) Find the bias and the MSE of the ML estimator $\tilde{\sigma}^2$.

(b) Compare the MSE of the ML estimator $\tilde{\sigma}^2$ with that of the unbiased estimator $\hat{\sigma}^2$: for which values of N and of $\operatorname{rank}\mathbf{X}$ is the MSE of the ML estimator smaller than that of the unbiased estimator and for which values is it larger?

Solution. Let $K = \operatorname{rank} \mathbf{X}$.

(a) In light of result (7.36), the bias of $\tilde{\sigma}^2$ is

$$\mathrm{E}(\tilde{\sigma}^2) - \sigma^2 = \mathrm{E}\left(\frac{\tilde{\mathbf{e}}'\tilde{\mathbf{e}}}{N}\right) - \sigma^2 = \sigma^2 \frac{N-K}{N} - \sigma^2 = -\frac{K}{N}\sigma^2,$$

and, in light of result (7.42), the MSE of $\tilde{\sigma}^2$ is

$$\mathrm{E}[(\tilde{\sigma}^2 - \sigma^2)^2] = \frac{\sigma^4}{N^2}[2(N-K) + K^2]. \tag{S.14}$$

(b) Let D represent the difference between the MSE of $\hat{\sigma}^2$ and the MSE of $\tilde{\sigma}^2$. In light of results (7.40) and (S.14) [along with result (7.37)], we find that

$$\begin{aligned} D &= \frac{2\sigma^4}{N-K} - \frac{\sigma^4}{N^2}[2(N-K) + K^2] \\ &= \frac{\sigma^4}{N^2(N-K)}[2N^2 - 2(N-K)^2 - K^2(N-K)] \\ &= \frac{\sigma^4}{N^2(N-K)}[4NK - 2K^2 - K^2(N-K)] \\ &= \frac{K\sigma^4}{N^2(N-K)}[K(K-2) - N(K-4)]. \end{aligned}$$

Clearly, if $K = 1$, $K = 2$, $K = 3$, or $K = 4$, then $D > 0$ (regardless of the value of N). If $K \geq 5$, then $D > 0$, $D = 0$, or $D < 0$, depending on whether $N < \dfrac{K(K-2)}{K-4}$, $N = \dfrac{K(K-2)}{K-4}$, or $N > \dfrac{K(K-2)}{K-4}$ or, equivalently, on whether $N-K < \dfrac{2K}{K-4}$, $N-K = \dfrac{2K}{K-4}$, or $N-K > \dfrac{2K}{K-4}$. Thus, for K between 1 and 4, inclusive, the ML estimator has the smaller MSE. And, for $K \geq 5$, whether the MSE of the ML estimator is smaller than, equal to, or larger than that of the unbiased estimator depends on whether N is smaller than, equal to, or larger than $K(K-2)/(K-4)$ or, equivalently, on whether $N-K$ is smaller than, equal to, or larger than $2K/(K-4)$.

Note that, for $K > 4$, $2K/(K-4)$ is a decreasing function of K and that, for $K > 12$, $2K/(K-4) < 3$. The implication is that, for $K \geq 13$, the unbiased estimator has the smaller MSE whenever $N-K \geq 3$ or, equivalently, whenever $N \geq K+3$.

EXERCISE 19. Suppose that $\mathbf{y}$ is an $N \times 1$ observable random vector that follows a general linear model, that the distribution of the vector $\mathbf{e}$ of residual effects is MVN, and that the variance-covariance matrix $\mathbf{V}(\boldsymbol{\theta})$ of $\mathbf{e}$ is nonsingular (for all $\boldsymbol{\theta} \in \Theta$). And, letting $K = N - \operatorname{rank} \mathbf{X}$, take $\mathbf{R}$ to be any $N \times K$ matrix (of constants) of full column rank K such that $\mathbf{X}'\mathbf{R} = \mathbf{0}$, and (as in Section 5.9b) define $\mathbf{z} = \mathbf{R}'\mathbf{y}$. Further, let $\mathbf{w} = \mathbf{s}(\mathbf{z})$, where $\mathbf{s}(\cdot)$ is a $K \times 1$ vector of real-valued functions that defines a one-to-one mapping of $\mathcal{R}^K$ onto some set $\mathcal{W}$.

(a) Show that $\mathbf{w}$ is a maximal invariant.

(b) Let $f_1(\cdot\,;\boldsymbol{\theta})$ represent the pdf of the distribution of $\mathbf{z}$, and assume that $\mathbf{s}(\cdot)$ is such that the distribution of $\mathbf{w}$ has a pdf, say $f_2(\cdot\,;\boldsymbol{\theta})$, that is obtainable from $f_1(\cdot\,;\boldsymbol{\theta})$ via an application of the basic formula (e.g., Bickel and Doksum 2001, sec. B.2) for a change of variables. And, taking $L_1(\boldsymbol{\theta};\, \mathbf{R}'\underline{\mathbf{y}})$ and $L_2[\boldsymbol{\theta};\, \mathbf{s}(\mathbf{R}'\underline{\mathbf{y}})]$ (where $\underline{\mathbf{y}}$ denotes the observed value of $\mathbf{y}$) to be the likelihood functions defined by $L_1(\boldsymbol{\theta};\, \mathbf{R}'\underline{\mathbf{y}}) = f_1(\mathbf{R}'\underline{\mathbf{y}};\, \boldsymbol{\theta})$ and $L_2[\boldsymbol{\theta};\, \mathbf{s}(\mathbf{R}'\underline{\mathbf{y}})] = f_2[\mathbf{s}(\mathbf{R}'\underline{\mathbf{y}});\, \boldsymbol{\theta}]$, show that $L_1(\boldsymbol{\theta};\, \mathbf{R}'\underline{\mathbf{y}})$ and $L_2[\boldsymbol{\theta};\, \mathbf{s}(\mathbf{R}'\underline{\mathbf{y}})]$ differ from each other by no more than a multiplicative constant.

Solution. (a) As discussed in Section 5.9b, $\mathbf{z}$ is a maximal invariant, and any (possibly vector-valued) statistic that depends on $\mathbf{y}$ only through the value of a maximal invariant is translation invariant. Thus, $\mathbf{w}$ is translation invariant. Now, take $\mathbf{y}_1$ and $\mathbf{y}_2$ to be any pair of values of $\mathbf{y}$ such that $\mathbf{s}(\mathbf{R}'\mathbf{y}_2) = \mathbf{s}(\mathbf{R}'\mathbf{y}_1)$. Then, because the mapping defined by $\mathbf{s}(\cdot)$ is one-to-one, $\mathbf{R}'\mathbf{y}_2 = \mathbf{R}'\mathbf{y}_1$. And because $\mathbf{z}$ is a maximal invariant, it follows that $\mathbf{y}_2 = \mathbf{y}_1 + \mathbf{X}\mathbf{k}$ for some vector $\mathbf{k}$. Accordingly, we conclude that $\mathbf{w}$ is a maximal invariant.

(b) Let $J(\mathbf{z})$ represent the determinant of the $K \times K$ matrix whose itth element is the partial derivative of the ith element of $\mathbf{s}(\mathbf{z})$ with respect to the tth element of $\mathbf{z}$. Then,

$$L_2[\boldsymbol{\theta};\, \mathbf{s}(\mathbf{R}'\underline{\mathbf{y}})] = f_2[\mathbf{s}(\mathbf{R}'\underline{\mathbf{y}});\, \boldsymbol{\theta}] = f_1(\mathbf{R}'\underline{\mathbf{y}};\, \boldsymbol{\theta})|J(\mathbf{R}'\underline{\mathbf{y}})|^{-1} = |J(\mathbf{R}'\underline{\mathbf{y}})|^{-1} L_1(\boldsymbol{\theta};\, \mathbf{R}'\underline{\mathbf{y}}).$$

Thus, the two likelihood functions $L_1(\boldsymbol{\theta};\, \mathbf{R}'\underline{\mathbf{y}})$ and $L_2[\boldsymbol{\theta};\, \mathbf{s}(\mathbf{R}'\underline{\mathbf{y}})]$ differ from each other by no more than a multiplicative constant (which may depend on the observed value $\underline{\mathbf{y}}$ of $\mathbf{y}$).

EXERCISE 20. Suppose that $\mathbf{y}$ is an $N \times 1$ observable random vector that follows a general linear model, that the distribution of the vector $\mathbf{e}$ of residual effects is MVN, and that the variance-covariance matrix $\mathbf{V}(\boldsymbol{\theta})$ of $\mathbf{e}$ is nonsingular (for all $\boldsymbol{\theta} \in \Theta$). Further, let $\mathbf{z} = \mathbf{R}'\mathbf{y}$, where $\mathbf{R}$ is any $N \times (N - \operatorname{rank} \mathbf{X})$ matrix (of constants) of full column rank $N - \operatorname{rank} \mathbf{X}$ such that $\mathbf{X}'\mathbf{R} = \mathbf{0}$; and let $\mathbf{u} = \mathbf{X}_*'\mathbf{y}$, where $\mathbf{X}_*$ is any $N \times (\operatorname{rank} \mathbf{X})$ matrix (of constants) whose columns form a basis for $\mathcal{C}(\mathbf{X})$. And denote by $\underline{\mathbf{y}}$ the observed value of $\mathbf{y}$.

(a) Verify that the likelihood function that would result from regarding the observed value $(\mathbf{X}_*,\, \mathbf{R})'\underline{\mathbf{y}}$ of $\begin{pmatrix}\mathbf{u}\\ \mathbf{z}\end{pmatrix}$ as the data vector differs by no more than a multiplicative constant from that obtained by regarding the observed value $\underline{\mathbf{y}}$ of $\mathbf{y}$ as the data vector.

(b) Let $f_0(\cdot \,|\, \cdot\,;\boldsymbol{\beta},\boldsymbol{\theta})$ represent the pdf of the conditional distribution of $\mathbf{u}$ given $\mathbf{z}$. And take $L_0[\boldsymbol{\beta},\boldsymbol{\theta};\, (\mathbf{X}_*,\, \mathbf{R})'\underline{\mathbf{y}}]$ to be the function of $\boldsymbol{\beta}$ and $\boldsymbol{\theta}$ defined by $L_0[\boldsymbol{\beta},\boldsymbol{\theta};\, (\mathbf{X}_*,\, \mathbf{R})'\underline{\mathbf{y}}] = f_0(\mathbf{X}_*'\underline{\mathbf{y}} \,|\, \mathbf{R}'\underline{\mathbf{y}};\boldsymbol{\beta},\boldsymbol{\theta})$. Show that

$$L_0[\boldsymbol{\beta},\boldsymbol{\theta};\, (\mathbf{X}_*,\, \mathbf{R})'\underline{\mathbf{y}}] = (2\pi)^{-(\operatorname{rank}\mathbf{X})/2}\, |\mathbf{X}_*'\mathbf{X}_*|^{-1}\, |\mathbf{X}_*'[\mathbf{V}(\boldsymbol{\theta})]^{-1}\mathbf{X}_*|^{1/2} \times \exp\{-\tfrac{1}{2}[\tilde{\boldsymbol{\beta}}(\boldsymbol{\theta})-\boldsymbol{\beta}]'\mathbf{X}'[\mathbf{V}(\boldsymbol{\theta})]^{-1}\mathbf{X}[\tilde{\boldsymbol{\beta}}(\boldsymbol{\theta})-\boldsymbol{\beta}]\},$$

where $\tilde{\boldsymbol{\beta}}(\boldsymbol{\theta})$ is any solution to the linear system $\mathbf{X}'[\mathbf{V}(\boldsymbol{\theta})]^{-1}\mathbf{X}\mathbf{b} = \mathbf{X}'[\mathbf{V}(\boldsymbol{\theta})]^{-1}\underline{\mathbf{y}}$ (in the $P \times 1$ vector $\mathbf{b}$).

(c) In connection with Part (b), show (1) that

$$[\tilde{\boldsymbol{\beta}}(\boldsymbol{\theta})-\boldsymbol{\beta}]'\mathbf{X}'[\mathbf{V}(\boldsymbol{\theta})]^{-1}\mathbf{X}[\tilde{\boldsymbol{\beta}}(\boldsymbol{\theta})-\boldsymbol{\beta}] = (\underline{\mathbf{y}} - \mathbf{X}\boldsymbol{\beta})'[\mathbf{V}(\boldsymbol{\theta})]^{-1}\mathbf{X}\{\mathbf{X}'[\mathbf{V}(\boldsymbol{\theta})]^{-1}\mathbf{X}\}^{-}\mathbf{X}'[\mathbf{V}(\boldsymbol{\theta})]^{-1}(\underline{\mathbf{y}} - \mathbf{X}\boldsymbol{\beta})$$

and (2) that the distribution of the random variable s defined by

$$s = (\mathbf{y} - \mathbf{X}\boldsymbol{\beta})'[\mathbf{V}(\boldsymbol{\theta})]^{-1}\mathbf{X}\{\mathbf{X}'[\mathbf{V}(\boldsymbol{\theta})]^{-1}\mathbf{X}\}^{-}\mathbf{X}'[\mathbf{V}(\boldsymbol{\theta})]^{-1}(\mathbf{y} - \mathbf{X}\boldsymbol{\beta})$$

does not depend on $\boldsymbol{\beta}$.

Solution. (a) Let $f_1(\cdot\,;\boldsymbol{\beta},\boldsymbol{\theta})$ represent the pdf of the distribution of $\mathbf{y}$ and $f_2(\cdot\,;\boldsymbol{\beta},\boldsymbol{\theta})$ the pdf of the distribution of $\begin{pmatrix}\mathbf{u}\\ \mathbf{z}\end{pmatrix}$—the distribution of $\mathbf{y}$ is $N[\mathbf{X}\boldsymbol{\beta},\, \mathbf{V}(\boldsymbol{\theta})]$, and the distribution of $\begin{pmatrix}\mathbf{u}\\ \mathbf{z}\end{pmatrix}$ is $N[(\mathbf{X}_*,\, \mathbf{R})'\mathbf{X}\boldsymbol{\beta},\, (\mathbf{X}_*,\, \mathbf{R})'\mathbf{V}(\boldsymbol{\theta})(\mathbf{X}_*,\, \mathbf{R})]$. Then, the function $L_1(\boldsymbol{\beta},\boldsymbol{\theta};\underline{\mathbf{y}})$ of $\boldsymbol{\beta}$ and $\boldsymbol{\theta}$ defined by $L_1(\boldsymbol{\beta},\boldsymbol{\theta};\underline{\mathbf{y}}) = f_1(\underline{\mathbf{y}};\boldsymbol{\beta},\boldsymbol{\theta})$ is the likelihood function obtained by regarding the observed value $\underline{\mathbf{y}}$ of $\mathbf{y}$ as the data vector. And, similarly, the function $L_2[\boldsymbol{\beta},\boldsymbol{\theta};\, (\mathbf{X}_*,\, \mathbf{R})'\underline{\mathbf{y}}]$ of $\boldsymbol{\beta}$ and $\boldsymbol{\theta}$ defined by $L_2[\boldsymbol{\beta},\boldsymbol{\theta};\, (\mathbf{X}_*,\, \mathbf{R})'\underline{\mathbf{y}}] = f_2[(\mathbf{X}_*,\, \mathbf{R})'\underline{\mathbf{y}};\, \boldsymbol{\beta},\boldsymbol{\theta}]$ is the likelihood function obtained by regarding the observed value $(\mathbf{X}_*,\, \mathbf{R})'\underline{\mathbf{y}}$ of $\begin{pmatrix}\mathbf{u}\\ \mathbf{z}\end{pmatrix}$ as the data vector. Further,

$$f_1(\underline{\mathbf{y}};\boldsymbol{\beta},\boldsymbol{\theta}) = |\det(\mathbf{X}_*,\, \mathbf{R})|\, f_2[(\mathbf{X}_*,\, \mathbf{R})'\underline{\mathbf{y}};\, \boldsymbol{\beta},\boldsymbol{\theta}],$$

as can be verified directly from formula (3.5.32) for the pdf of a MVN distribution—clearly,

$$(2\pi)^{-N/2}\,|\mathbf{V}(\boldsymbol{\theta})|^{-1/2}\exp\{-\tfrac{1}{2}(\underline{\mathbf{y}}-\mathbf{X}\boldsymbol{\beta})'[\mathbf{V}(\boldsymbol{\theta})]^{-1}(\underline{\mathbf{y}}-\mathbf{X}\boldsymbol{\beta})\}$$
$$= |\det(\mathbf{X}_*,\,\mathbf{R})|\,(2\pi)^{-N/2}\,|(\mathbf{X}_*,\,\mathbf{R})'\mathbf{V}(\boldsymbol{\theta})(\mathbf{X}_*,\,\mathbf{R})|^{-1/2}$$
$$\times\exp\{-\tfrac{1}{2}[(\mathbf{X}_*,\,\mathbf{R})'(\underline{\mathbf{y}}-\mathbf{X}\boldsymbol{\beta})]'[(\mathbf{X}_*,\,\mathbf{R})'\mathbf{V}(\boldsymbol{\theta})(\mathbf{X}_*,\,\mathbf{R})]^{-1}(\mathbf{X}_*,\,\mathbf{R})'(\underline{\mathbf{y}}-\mathbf{X}\boldsymbol{\beta})\}$$

—or simply by observing that $\begin{pmatrix}\mathbf{u}\\ \mathbf{z}\end{pmatrix} = (\mathbf{X}_*,\,\mathbf{R})'\mathbf{y}$ and making use of standard results (e.g., Bickel and Doksum 2001, sec. B.2) on a change of variables. Thus,

$$\begin{aligned} L_2[\boldsymbol{\beta},\boldsymbol{\theta};\,(\mathbf{X}_*,\,\mathbf{R})'\underline{\mathbf{y}}] &= f_2[(\mathbf{X}_*,\,\mathbf{R})'\underline{\mathbf{y}};\,\boldsymbol{\beta},\boldsymbol{\theta}] \\ &= |\det(\mathbf{X}_*,\,\mathbf{R})|^{-1}f_1(\underline{\mathbf{y}};\boldsymbol{\beta},\boldsymbol{\theta}) \\ &= |\det(\mathbf{X}_*,\,\mathbf{R})|^{-1}L_1(\boldsymbol{\beta},\boldsymbol{\theta};\underline{\mathbf{y}}). \qquad \text{(S.15)}\end{aligned}$$

(b) Let $f_3(\cdot\,;\boldsymbol{\theta})$ represent the pdf of the (marginal) distribution of $\mathbf{z}$, and take $L_3(\boldsymbol{\theta};\,\mathbf{R}'\underline{\mathbf{y}})$ to be the function of $\boldsymbol{\theta}$ defined by $L_3(\boldsymbol{\theta};\,\mathbf{R}'\underline{\mathbf{y}}) = f_3(\mathbf{R}'\underline{\mathbf{y}};\boldsymbol{\theta})$. Further, define $L_1(\boldsymbol{\beta},\boldsymbol{\theta};\underline{\mathbf{y}})$ and $L_2[\boldsymbol{\beta},\boldsymbol{\theta};\,(\mathbf{X}_*,\,\mathbf{R})'\underline{\mathbf{y}}]$ as in the solution to Part (a). Then, making use of result (S.15), we find that

$$L_0[\boldsymbol{\beta},\boldsymbol{\theta};\,(\mathbf{X}_*,\,\mathbf{R})'\underline{\mathbf{y}}] = \frac{L_2[\boldsymbol{\beta},\boldsymbol{\theta};\,(\mathbf{X}_*,\,\mathbf{R})'\underline{\mathbf{y}}]}{L_3(\boldsymbol{\theta};\,\mathbf{R}'\underline{\mathbf{y}})} = \frac{|\det(\mathbf{X}_*,\,\mathbf{R})|^{-1}L_1(\boldsymbol{\beta},\boldsymbol{\theta};\underline{\mathbf{y}})}{L_3(\boldsymbol{\theta};\,\mathbf{R}'\underline{\mathbf{y}})}$$

and hence [in light of result (9.26)] that

$$L_0[\boldsymbol{\beta},\boldsymbol{\theta};\,(\mathbf{X}_*,\,\mathbf{R})'\underline{\mathbf{y}}]$$
$$= \frac{(2\pi)^{-N/2}\,|(\mathbf{X}_*,\,\mathbf{R})'\mathbf{V}(\boldsymbol{\theta})(\mathbf{X}_*,\,\mathbf{R})|^{-1/2}\exp\{-\frac{1}{2}(\underline{\mathbf{y}}-\mathbf{X}\boldsymbol{\beta})'[\mathbf{V}(\boldsymbol{\theta})]^{-1}(\underline{\mathbf{y}}-\mathbf{X}\boldsymbol{\beta})\}}{(2\pi)^{-(N-\operatorname{rank}\mathbf{X})/2}\,|\mathbf{R}'\mathbf{V}(\boldsymbol{\theta})\mathbf{R}|^{-1/2}\exp\{-\frac{1}{2}\underline{\mathbf{y}}'\mathbf{R}[\mathbf{R}'\mathbf{V}(\boldsymbol{\theta})\mathbf{R}]^{-1}\mathbf{R}'\underline{\mathbf{y}}\}}.$$

And in light of results (9.36), (9.32), and (9.40), it follows that

$$\begin{aligned} L_0[\boldsymbol{\beta},\boldsymbol{\theta};\,(\mathbf{X}_*,\,\mathbf{R})'\underline{\mathbf{y}}] &= (2\pi)^{-(\operatorname{rank}\mathbf{X})/2}\,|\mathbf{X}_*'\mathbf{X}_*|^{-1}\,|\mathbf{X}_*'[\mathbf{V}(\boldsymbol{\theta})]^{-1}\mathbf{X}_*|^{1/2} \\ &\quad\times\exp\{-\tfrac{1}{2}(\underline{\mathbf{y}}-\mathbf{X}\boldsymbol{\beta})'[\mathbf{V}(\boldsymbol{\theta})]^{-1}(\underline{\mathbf{y}}-\mathbf{X}\boldsymbol{\beta}) \\ &\qquad + \tfrac{1}{2}\underline{\mathbf{y}}'[\mathbf{V}(\boldsymbol{\theta})]^{-1}\underline{\mathbf{y}} - \tfrac{1}{2}[\tilde{\boldsymbol{\beta}}(\boldsymbol{\theta})]'\mathbf{X}'[\mathbf{V}(\boldsymbol{\theta})]^{-1}\underline{\mathbf{y}}\} \\ &= (2\pi)^{-(\operatorname{rank}\mathbf{X})/2}\,|\mathbf{X}_*'\mathbf{X}_*|^{-1}\,|\mathbf{X}_*'[\mathbf{V}(\boldsymbol{\theta})]^{-1}\mathbf{X}_*|^{1/2} \\ &\quad\times\exp\{-\tfrac{1}{2}[\tilde{\boldsymbol{\beta}}(\boldsymbol{\theta})-\boldsymbol{\beta}]'\mathbf{X}'[\mathbf{V}(\boldsymbol{\theta})]^{-1}\mathbf{X}[\tilde{\boldsymbol{\beta}}(\boldsymbol{\theta})-\boldsymbol{\beta}]\}.\end{aligned}$$

(c) (1) Clearly,

$$[\tilde{\boldsymbol{\beta}}(\boldsymbol{\theta})-\boldsymbol{\beta}]'\mathbf{X}'[\mathbf{V}(\boldsymbol{\theta})]^{-1}\mathbf{X}[\tilde{\boldsymbol{\beta}}(\boldsymbol{\theta})-\boldsymbol{\beta}]$$
$$= \{\mathbf{X}'[\mathbf{V}(\boldsymbol{\theta})]^{-1}\mathbf{X}[\tilde{\boldsymbol{\beta}}(\boldsymbol{\theta})-\boldsymbol{\beta}]\}'\{\mathbf{X}'[\mathbf{V}(\theta)]^{-1}\mathbf{X}\}^{-}\mathbf{X}'[\mathbf{V}(\theta)]^{-1}\mathbf{X}[\tilde{\boldsymbol{\beta}}(\boldsymbol{\theta})-\boldsymbol{\beta}].$$

Moreover,

$$\begin{aligned}\mathbf{X}'[\mathbf{V}(\theta)]^{-1}\mathbf{X}[\tilde{\boldsymbol{\beta}}(\boldsymbol{\theta})-\boldsymbol{\beta}] &= \mathbf{X}'[\mathbf{V}(\theta)]^{-1}\mathbf{X}\tilde{\boldsymbol{\beta}}(\boldsymbol{\theta})-\mathbf{X}'[\mathbf{V}(\theta)]^{-1}\mathbf{X}\boldsymbol{\beta} \\ &= \mathbf{X}'[\mathbf{V}(\theta)]^{-1}\underline{\mathbf{y}}-\mathbf{X}'[\mathbf{V}(\theta)]^{-1}\mathbf{X}\boldsymbol{\beta} = \mathbf{X}'[\mathbf{V}(\theta)]^{-1}(\underline{\mathbf{y}}-\mathbf{X}\boldsymbol{\beta}).\end{aligned}$$

Thus,

$$[\tilde{\boldsymbol{\beta}}(\boldsymbol{\theta})-\boldsymbol{\beta}]'\mathbf{X}'[\mathbf{V}(\boldsymbol{\theta})]^{-1}\mathbf{X}[\tilde{\boldsymbol{\beta}}(\boldsymbol{\theta})-\boldsymbol{\beta}]$$
$$= (\underline{\mathbf{y}}-\mathbf{X}\boldsymbol{\beta})'[\mathbf{V}(\boldsymbol{\theta})]^{-1}\mathbf{X}\{\mathbf{X}'[\mathbf{V}(\boldsymbol{\theta})]^{-1}\mathbf{X}\}^{-}\mathbf{X}'[\mathbf{V}(\boldsymbol{\theta})]^{-1}(\underline{\mathbf{y}}-\mathbf{X}\boldsymbol{\beta}).$$

(2) Clearly, the random variable s is a function of the random vector $\mathbf{y} - \mathbf{X}\boldsymbol{\beta}$. Moreover, the distribution of $\mathbf{y} - \mathbf{X}\boldsymbol{\beta}$ is $N[\mathbf{0}, \mathbf{V}(\boldsymbol{\theta})]$, which does not depend on $\boldsymbol{\beta}$. Thus, the distribution of s does not depend on $\boldsymbol{\beta}$.

EXERCISE 21. Suppose that $\mathbf{z}$ is an $S \times 1$ observable random vector and that $\mathbf{z} \sim N(\mathbf{0}, \sigma^2\mathbf{I})$, where σ is a (strictly) positive unknown parameter.

(a) Show that $\mathbf{z}'\mathbf{z}$ is a complete sufficient statistic.

(b) Take $\mathbf{w}(\mathbf{z})$ to be the S-dimensional vector-valued statistic defined by $\mathbf{w}(\mathbf{z}) = (\mathbf{z}'\mathbf{z})^{-1/2}\mathbf{z}$—$\mathbf{w}(\mathbf{z})$ is defined for $\mathbf{z} \neq \mathbf{0}$ and hence with probability 1. Show that $\mathbf{z}'\mathbf{z}$ and $\mathbf{w}(\mathbf{z})$ are statistically independent. (*Hint.* Make use of Basu's theorem.)

(c) Show that any estimator of σ^2 of the form $\mathbf{z}'\mathbf{z}/k$ (where k is a nonzero constant) is scale equivariant—an estimator, say $t(\mathbf{z})$, of σ^2 is to be regarded as *scale equivariant* if for every (strictly) positive scalar c (and for every nonnull value of $\mathbf{z}$) $t(c\mathbf{z}) = c^2 t(\mathbf{z})$.

(d) Let $t_0(\mathbf{z})$ represent any particular scale-equivariant estimator of σ^2 such that $t_0(\mathbf{z}) \neq 0$ for $\mathbf{z} \neq \mathbf{0}$. Show that an estimator $t(\mathbf{z})$ of σ^2 is scale equivariant if and only if, for some function $u(\cdot)$ such that $u(c\mathbf{z}) = u(\mathbf{z})$ (for every strictly positive constant c and every nonnull value of $\mathbf{z}$),

$$t(\mathbf{z}) = u(\mathbf{z})t_0(\mathbf{z}) \quad \text{for} \quad \mathbf{z} \neq \mathbf{0}. \tag{E.2}$$

(e) Show that a function $u(\mathbf{z})$ of $\mathbf{z}$ is such that $u(c\mathbf{z}) = u(\mathbf{z})$ (for every strictly positive constant c and every nonnull value of $\mathbf{z}$) if and only if $u(\mathbf{z})$ depends on the value of $\mathbf{z}$ only through $\mathbf{w}(\mathbf{z})$ [where $\mathbf{w}(\mathbf{z})$ is as defined in Part (b)].

(f) Show that the estimator $\mathbf{z}'\mathbf{z}/(S+2)$ has minimum MSE among all scale-equivariant estimators of σ^2.

Solution. (a) The distribution of $\mathbf{z}$ has a pdf, say $f(\cdot)$, that is expressible as follows:

$$f(\mathbf{z}) = (2\pi)^{-S/2}\sigma^{-S}\exp\left(-\frac{1}{2\sigma^2}\mathbf{z}'\mathbf{z}\right). \tag{S.16}$$

Based on the same well-known result on complete sufficient statistics for exponential families that was employed in Section 5.8, it follows from result (S.16) that $\mathbf{z}'\mathbf{z}$ is a complete sufficient statistic.

(b) Clearly,

$$\mathbf{w}(\mathbf{z}) = \{[(1/\sigma)\mathbf{z}]'[(1/\sigma)\mathbf{z}]\}^{-1/2}(1/\sigma)\mathbf{z} = \mathbf{w}[(1/\sigma)\mathbf{z}].$$

And upon observing that the distribution of $(1/\sigma)\mathbf{z}$ [which is $N(\mathbf{0}, \mathbf{I})$] does not depend on σ, it follows that $\mathbf{w}(\mathbf{z})$ is an ancillary statistic. Since [according to Part (a)] $\mathbf{z}'\mathbf{z}$ is a complete sufficient statistic, we conclude (on the basis of Basu's theorem) that $\mathbf{z}'\mathbf{z}$ and $\mathbf{w}(\mathbf{z})$ are statistically independent.

(c) Clearly, for every (strictly) positive scalar c (and for every value of $\mathbf{z}$), $(c\mathbf{z})'(c\mathbf{z})/k = c^2\mathbf{z}'\mathbf{z}/k$. Thus, $\mathbf{z}'\mathbf{z}/k$ is a scale-equivariant estimator of σ^2.

(d) Suppose that, for some function $u(\cdot)$ such that $u(c\mathbf{z}) = u(\mathbf{z})$ (for $c > 0$ and for $\mathbf{z} \neq \mathbf{0}$), $t(\mathbf{z}) = u(\mathbf{z})t_0(\mathbf{z})$ for $\mathbf{z} \neq \mathbf{0}$. Then, for $c > 0$ and $\mathbf{z} \neq \mathbf{0}$,

$$t(c\mathbf{z}) = u(c\mathbf{z})t_0(c\mathbf{z}) = u(\mathbf{z})[c^2 t_0(\mathbf{z})] = c^2 u(\mathbf{z})t_0(\mathbf{z}) = c^2 t(\mathbf{z}).$$

Thus, $t(\mathbf{z})$ is scale equivariant.

Conversely, suppose that $t(\mathbf{z})$ is scale equivariant. Then, taking $u(\cdot)$ to be the function defined (for $\mathbf{z} \neq \mathbf{0}$) by $u(\mathbf{z}) = t(\mathbf{z})/t_0(\mathbf{z})$, we find that (for $c > 0$ and $\mathbf{z} \neq \mathbf{0}$)

$$u(c\mathbf{z}) = t(c\mathbf{z})/t_0(c\mathbf{z}) = c^2 t(\mathbf{z})/[c^2 t_0(\mathbf{z})] = t(\mathbf{z})/t_0(\mathbf{z}) = u(\mathbf{z}).$$

Moreover, $u(\cdot)$ is such that (for $\mathbf{z} \neq \mathbf{0}$) $t(\mathbf{z}) = u(\mathbf{z})t_0(\mathbf{z})$.

(e) Suppose that there exists a function $h(\cdot)$ such that $u(\mathbf{z}) = h[\mathbf{w}(\mathbf{z})]$ (for every $\mathbf{z} \neq \mathbf{0}$). Then, for every $c > 0$ and every $\mathbf{z} \neq \mathbf{0}$, it is clear [upon observing that $\mathbf{w}(c\mathbf{z}) = \mathbf{w}(\mathbf{z})$] that

$$u(c\mathbf{z}) = h[\mathbf{w}(c\mathbf{z})] = h[\mathbf{w}(\mathbf{z})] = u(\mathbf{z}).$$

Conversely, suppose that $u(c\mathbf{z}) = u(\mathbf{z})$ for every $c > 0$ and every $\mathbf{z} \neq \mathbf{0}$. Then, for every nonnull value of $\mathbf{z}$,

$$u(\mathbf{z}) = u[(\mathbf{z}'\mathbf{z})^{-1/2}\mathbf{z}] = u[\mathbf{w}(\mathbf{z})].$$

(f) Let $t(\mathbf{z})$ represent an arbitrary scale-equivariant estimator of σ^2, or equivalently [in light of Parts (c), (d), and (e)] take $t(\mathbf{z})$ to be any estimator of σ^2 that (for $\mathbf{z} \neq \mathbf{0}$) is expressible in the form

$$t(\mathbf{z}) = h[\mathbf{w}(\mathbf{z})]\,\mathbf{z}'\mathbf{z} \tag{S.17}$$

for some function $h(\cdot)$. The MSE of an estimator of the form (S.17) is

$$\mathrm{E}\big(\{h[\mathbf{w}(\mathbf{z})]\,\mathbf{z}'\mathbf{z} - \sigma^2\}^2\big).$$

Recall [from Part (b)] that $\mathbf{z}'\mathbf{z}$ and $\mathbf{w}(\mathbf{z})$ are statistically independent, and consider the minimization of the MSE conditional on the value of $\mathbf{w}(\mathbf{z})$. Upon employing essentially the same line of reasoning as in the derivation (in Section 5.7c) of the Hodges-Lehmann estimator, we find that the conditional MSE attains its minimum value when the value of $h[\mathbf{w}(\mathbf{z})]$ is taken to be $1/(S+2)$. And we conclude that the estimator $\mathbf{z}'\mathbf{z}/(S+2)$ has minimum MSE (both conditionally and unconditionally) among all scale-equivariant estimators of σ^2.

EXERCISE 22. Suppose that $\mathbf{y}$ is an $N \times 1$ observable random vector that follows a G–M model and that the distribution of the vector $\mathbf{e}$ of residual effects is MVN. Using the result of Part (f) of Exercise 21 (or otherwise), show that the Hodges-Lehmann estimator $\mathbf{y}'(\mathbf{I}-\mathbf{P_X})\mathbf{y}/[N-\mathrm{rank}(\mathbf{X})+2]$ has minimum MSE among all translation-invariant estimators of σ^2 that are scale equivariant—a translation-invariant estimator, say $t(\mathbf{y})$, of σ^2 is to be regarded as *scale equivariant* if $t(c\mathbf{y}) = c^2 t(\mathbf{y})$ for every (strictly) positive scalar c and for every nonnull value of $\mathbf{y}$ in $\mathcal{N}(\mathbf{X}')$.

Solution. As discussed in Section 5.7d, the Hodges-Lehmann estimator is translation invariant. Further, upon observing that (for every scalar c and for every value of $\mathbf{y}$) $(c\mathbf{y})'(\mathbf{I}-\mathbf{P_X})(c\mathbf{y}) = c^2\mathbf{y}'(\mathbf{I}-\mathbf{P_X})\mathbf{y}$, it is clear that the Hodges-Lehmann estimator is scale equivariant.

Now, let $\mathbf{z} = \mathbf{R}'\mathbf{y}$, where $\mathbf{R}$ is any $N \times (N - \mathrm{rank}\,\mathbf{X})$ matrix (of constants) such that $\mathbf{X}'\mathbf{R} = \mathbf{0}$, $\mathbf{R}'\mathbf{R} = \mathbf{I}$, and $\mathbf{R}\mathbf{R}' = \mathbf{I}-\mathbf{P_X}$—the existence of such a matrix was established in Part 2 of Section 5.9b. Then, in light of the results of the introductory part of Section 5.9b, $\mathbf{z}$ is a maximal invariant, and, as a consequence, any translation-invariant estimator of σ^2 is expressible as a function, say $t(\mathbf{z})$, of $\mathbf{z}$. Moreover, if this estimator is scale equivariant, then, for every strictly positive scalar c and for every nonnull value of $\mathbf{y}$ in $\mathcal{N}(\mathbf{X}')$, $t[\mathbf{R}'(c\mathbf{y})] = c^2 t(\mathbf{R}'\mathbf{y})$, implying [since the columns of $\mathbf{R}$ form a basis for $\mathcal{N}(\mathbf{X}')$] that, for every strictly positive scalar c and for every nonnull value of $\mathbf{z}$,

$$t(c\mathbf{z}) = t(c\mathbf{R}'\mathbf{R}\mathbf{z}) = t[\mathbf{R}'(c\mathbf{R}\mathbf{z})] = c^2 t[\mathbf{R}'(\mathbf{R}\mathbf{z})] = c^2 t(\mathbf{R}'\mathbf{R}\mathbf{z}) = c^2 t(\mathbf{z}).$$

Also, the Hodges-Lehmann estimator is expressible as $\mathbf{z}'\mathbf{z}/[N - \mathrm{rank}(\mathbf{X}) + 2]$, and the distribution of $\mathbf{z}$ is $N(\mathbf{0}, \sigma^2\mathbf{I})$. Thus, upon regarding $\mathbf{z}$ as the data vector and applying the result of Part (f) of Exercise 21 (with $S = N - \mathrm{rank}\,\mathbf{X}$), it follows that the Hodges-Lehmann estimator has minimum MSE among all translation-invariant estimators of σ^2 that are scale equivariant.

EXERCISE 23. Let $\mathbf{z} = (z_1, z_2, \ldots, z_M)'$ represent an M-dimensional random (column) vector that has an absolutely continuous distribution with a pdf $f(\cdot)$. And suppose that for some (nonnegative) function $g(\cdot)$ (of a single nonnegative variable), $f(\mathbf{z}) \propto g(\mathbf{z}'\mathbf{z})$ (in which case the distribution of $\mathbf{z}$ is spherical). Show (for $i = 1, 2, \ldots, M$) that $\mathrm{E}(z_i^2)$ exists if and only if $\int_0^\infty s^{M+1} g(s^2)\, ds < \infty$, in which case

$$\mathrm{var}(z_i) = \mathrm{E}(z_i^2) = \frac{1}{M}\,\frac{\int_0^\infty s^{M+1} g(s^2)\, ds}{\int_0^\infty s^{M-1} g(s^2)\, ds}.$$

Solution. Clearly, $\int_{\mathbb{R}^M} g(\mathbf{z}'\mathbf{z})\, d\mathbf{z} < \infty$, and

$$f(\mathbf{z}) = c^{-1} g(\mathbf{z}'\mathbf{z}),$$

where $c = \int_{\mathbb{R}^M} g(\mathbf{z}'\mathbf{z})\, d\mathbf{z}$. Accordingly, it follows from the results of Section 5.9c that

$\int_0^\infty s^{M-1} g(s^2)\,ds < \infty$ and that c is expressible in the form

$$c = \frac{2\pi^{M/2}}{\Gamma(M/2)} \int_0^\infty s^{M-1} g(s^2)\,ds.$$

Moreover, since [in light of result (9.48)] $z_j \sim z_i$ for $j \neq i$,

$$\begin{aligned}\mathrm{E}(z_i^2) = \frac{1}{M}\sum_{j=1}^{M} \mathrm{E}(z_j^2) = \frac{1}{M}\mathrm{E}\Bigl(\sum_{j=1}^{M} z_j^2\Bigr) = \frac{1}{M}\mathrm{E}(\mathbf{z}'\mathbf{z}) &= \frac{1}{M}\int_{\mathbb{R}^M} (\mathbf{z}'\mathbf{z}) f(\mathbf{z})\,d\mathbf{z} \\ &= \frac{1}{Mc}\int_{\mathbb{R}^M} g_*(\mathbf{z}'\mathbf{z})\,d\mathbf{z},\end{aligned}$$

where $g_*(\cdot)$ is the (nonnegative) function (of a single nonnegative variable, say u) defined by $g_*(u) = ug(u)$. And in light of the results of Section 5.9c, $\int_{\mathbb{R}^M} g_*(\mathbf{z}'\mathbf{z})\,d\mathbf{z} < \infty$ if and only if $\int_0^\infty s^{M-1} g_*(s^2)\,ds < \infty$ or, equivalently, if and only if $\int_0^\infty s^{M+1} g(s^2)\,ds < \infty$, in which case

$$\int_{\mathbb{R}^M} g_*(\mathbf{z}'\mathbf{z})\,d\mathbf{z} = \frac{2\pi^{M/2}}{\Gamma(M/2)} \int_0^\infty s^{M-1} g_*(s^2)\,ds = \frac{2\pi^{M/2}}{\Gamma(M/2)} \int_0^\infty s^{M+1} g(s^2)\,ds.$$

Thus, $\mathrm{E}(z_i^2)$ exists if and only if $\int_0^\infty s^{M+1} g(s^2)\,ds < \infty$, in which case

$$\mathrm{E}(z_i^2) = \frac{1}{M}\frac{\int_{\mathbb{R}^M} g_*(\mathbf{z}'\mathbf{z})\,d\mathbf{z}}{c} = \frac{1}{M}\frac{\int_0^\infty s^{M+1} g(s^2)\,ds}{\int_0^\infty s^{M-1} g(s^2)\,ds}.$$

It remains only to observe that if $\mathrm{E}(z_i^2)$ exists [in which case $\mathrm{E}(z_i)$ also exists], then [in light of result (9.46)] $\mathrm{E}(z_i) = 0$ and hence $\mathrm{var}(z_i) = \mathrm{E}(z_i^2)$.

EXERCISE 24. Let $\mathbf{z}$ represent an N-dimensional random column vector, and let $\mathbf{z}_*$ represent an M-dimensional subvector of $\mathbf{z}$ (where $M < N$). And suppose that the distributions of $\mathbf{z}$ and $\mathbf{z}_*$ are absolutely continuous with pdfs $f(\cdot)$ and $f_*(\cdot)$, respectively. Suppose also that there exist (nonnegative) functions $g(\cdot)$ and $g_*(\cdot)$ (of a single nonnegative variable) such that (for every value of $\mathbf{z}$) $f(\mathbf{z}) = g(\mathbf{z}'\mathbf{z})$ and (for every value of $\mathbf{z}_*$) $f_*(\mathbf{z}_*) = g_*(\mathbf{z}_*'\mathbf{z}_*)$ (in which case the distributions of $\mathbf{z}$ and $\mathbf{z}_*$ are spherical).

(a) Show that (for $v \geq 0$)

$$g_*(v) = \frac{\pi^{(N-M)/2}}{\Gamma[(N-M)/2]} \int_v^\infty (u-v)^{[(N-M)/2]-1} g(u)\,du.$$

(b) Show that if $N - M = 2$, then (for $v > 0$)

$$g(v) = -\frac{1}{\pi} g_*'(v),$$

where $g_*'(\cdot)$ is the derivative of $g_*(\cdot)$.

Solution. (a) The result of Part (a) is an immediate consequence of result (9.67).

(b) Suppose that $N - M = 2$. Then, since $\Gamma(1) = 1$, the result of Part (a) simplifies to

$$g_*(v) = \pi \int_v^\infty g(u)\,du.$$

And it follows that (for $v > 0$)

$$g_*'(v) = \pi[-g(v)] = -\pi g(v)$$

and hence that

$$g(v) = -\frac{1}{\pi} g_*'(v).$$

EXERCISE 25. Let $\mathbf{y}$ represent an $N \times 1$ random vector and $\mathbf{w}$ an $M \times 1$ random vector. Suppose that the second-order moments of the joint distribution of $\mathbf{y}$ and $\mathbf{w}$ exist, and adopt the following notation: $\boldsymbol{\mu}_y = \mathrm{E}(\mathbf{y})$, $\boldsymbol{\mu}_w = \mathrm{E}(\mathbf{w})$, $\mathbf{V}_y = \mathrm{var}(\mathbf{y})$, $\mathbf{V}_{yw} = \mathrm{cov}(\mathbf{y}, \mathbf{w})$, and $\mathbf{V}_w = \mathrm{var}(\mathbf{w})$. Further, assume that $\mathbf{V}_y$ is nonsingular.

(a) Show that the matrix $\mathbf{V}_w - \mathbf{V}_{yw}'\mathbf{V}_y^{-1}\mathbf{V}_{yw} - \mathrm{E}[\mathrm{var}(\mathbf{w} \mid \mathbf{y})]$ is nonnegative definite and that it equals $\mathbf{0}$ if and only if (for some nonrandom vector $\mathbf{c}$ and some nonrandom matrix $\mathbf{A}$) $\mathrm{E}(\mathbf{w} \mid \mathbf{y}) = \mathbf{c} + \mathbf{A}'\mathbf{y}$ (with probability 1).

(b) Show that the matrix $\mathrm{var}[\mathrm{E}(\mathbf{w} \mid \mathbf{y})] - \mathbf{V}_{yw}'\mathbf{V}_y^{-1}\mathbf{V}_{yw}$ is nonnegative definite and that it equals $\mathbf{0}$ if and only if (for some nonrandom vector $\mathbf{c}$ and some nonrandom matrix $\mathbf{A}$) $\mathrm{E}(\mathbf{w} \mid \mathbf{y}) = \mathbf{c} + \mathbf{A}'\mathbf{y}$ (with probability 1).

Solution. Let $\boldsymbol{\eta}(\mathbf{y}) = \boldsymbol{\tau} + \mathbf{V}_{yw}'\mathbf{V}_y^{-1}\mathbf{y}$, where $\boldsymbol{\tau} = \boldsymbol{\mu}_w - \mathbf{V}_{yw}'\mathbf{V}_y^{-1}\boldsymbol{\mu}_y$.

(a) Making use of results (10.16) and (10.15), we find that

$$\begin{aligned}\mathbf{V}_w - \mathbf{V}_{yw}'\mathbf{V}_y^{-1}\mathbf{V}_{yw} - \mathrm{E}[\mathrm{var}(\mathbf{w} \mid \mathbf{y})] &= \mathrm{var}[\boldsymbol{\eta}(\mathbf{y}) - \mathbf{w}] - \mathrm{E}[\mathrm{var}(\mathbf{w} \mid \mathbf{y})] \\ &= \mathrm{var}[\boldsymbol{\eta}(\mathbf{y}) - \mathrm{E}(\mathbf{w} \mid \mathbf{y})].\end{aligned}$$

Clearly, $\mathrm{var}[\boldsymbol{\eta}(\mathbf{y}) - \mathrm{E}(\mathbf{w} \mid \mathbf{y})]$ is nonnegative definite, and $\mathrm{var}[\boldsymbol{\eta}(\mathbf{y}) - \mathrm{E}(\mathbf{w} \mid \mathbf{y})] = \mathbf{0}$ if and only if $\mathrm{E}(\mathbf{w} \mid \mathbf{y}) = \boldsymbol{\eta}(\mathbf{y})$ (with probability 1).

If $\mathrm{E}(\mathbf{w} \mid \mathbf{y}) = \boldsymbol{\eta}(\mathbf{y})$ (with probability 1), then it is obviously the case that (for some $\mathbf{c}$ and some $\mathbf{A}$, specifically, for $\mathbf{c} = \boldsymbol{\tau}$ and $\mathbf{A} = \mathbf{V}_y^{-1}\mathbf{V}_{yw}$) $\mathrm{E}(\mathbf{w} \mid \mathbf{y}) = \mathbf{c} + \mathbf{A}'\mathbf{y}$ (with probability 1). Conversely, if (for some $\mathbf{c}$ and some $\mathbf{A}$) $\mathrm{E}(\mathbf{w} \mid \mathbf{y}) = \mathbf{c} + \mathbf{A}'\mathbf{y}$ (with probability 1), then because $\boldsymbol{\eta}(\mathbf{y})$ is the (unique) best linear approximation to $\mathrm{E}(\mathbf{w} \mid \mathbf{y})$, $\boldsymbol{\eta}(\mathbf{y}) = \mathbf{c} + \mathbf{A}'\mathbf{y}$ and hence $\mathrm{E}(\mathbf{w} \mid \mathbf{y}) = \boldsymbol{\eta}(\mathbf{y})$ (with probability 1).

(b) Clearly,

$$\mathbf{V}_w = \mathrm{E}[\mathrm{var}(\mathbf{w} \mid \mathbf{y})] + \mathrm{var}[\mathrm{E}(\mathbf{w} \mid \mathbf{y})]$$

or, equivalently,

$$\mathrm{var}[\mathrm{E}(\mathbf{w} \mid \mathbf{y})] = \mathbf{V}_w - \mathrm{E}[\mathrm{var}(\mathbf{w} \mid \mathbf{y})].$$

Thus,

$$\mathrm{var}[\mathrm{E}(\mathbf{w} \mid \mathbf{y})] - \mathbf{V}_{yw}'\mathbf{V}_y^{-1}\mathbf{V}_{yw} = \mathbf{V}_w - \mathbf{V}_{yw}'\mathbf{V}_y^{-1}\mathbf{V}_{yw} - \mathrm{E}[\mathrm{var}(\mathbf{w} \mid \mathbf{y})],$$

so that Part (b) follows from the result of Part (a).

EXERCISE 26. Let $\mathbf{y}$ represent an $N \times 1$ observable random vector and $\mathbf{w}$ an $M \times 1$ unobservable random vector. Suppose that the second-order moments of the joint distribution of $\mathbf{y}$ and $\mathbf{w}$ exist, and adopt the following notation: $\boldsymbol{\mu}_y = \mathrm{E}(\mathbf{y})$, $\boldsymbol{\mu}_w = \mathrm{E}(\mathbf{w})$, $\mathbf{V}_y = \mathrm{var}(\mathbf{y})$, $\mathbf{V}_{yw} = \mathrm{cov}(\mathbf{y}, \mathbf{w})$, and $\mathbf{V}_w = \mathrm{var}(\mathbf{w})$. Assume that $\boldsymbol{\mu}_y$, $\boldsymbol{\mu}_w$, $\mathbf{V}_y$, $\mathbf{V}_{yw}$, and $\mathbf{V}_w$ are known. Further, define $\boldsymbol{\eta}(\mathbf{y}) = \boldsymbol{\mu}_w + \mathbf{V}_{yw}'\mathbf{V}_y^{-}(\mathbf{y} - \boldsymbol{\mu}_y)$, and take $\mathbf{t}(\mathbf{y})$ to be an $(M \times 1)$-dimensional vector-valued function of the form $\mathbf{t}(\mathbf{y}) = \mathbf{c} + \mathbf{A}'\mathbf{y}$, where $\mathbf{c}$ is a vector of constants and $\mathbf{A}$ is an $N \times M$ matrix of constants. Extend various of the results of Section 5.10a (to the case where $\mathbf{V}_y$ may be singular) by using Theorem 3.5.11 to show (1) that $\boldsymbol{\eta}(\mathbf{y})$ is the best linear predictor of $\mathbf{w}$ in the sense that the difference between the matrix $\mathrm{E}\{[\mathbf{t}(\mathbf{y}) - \mathbf{w}][\mathbf{t}(\mathbf{y}) - \mathbf{w}]'\}$ and the matrix $\mathrm{var}[\boldsymbol{\eta}(\mathbf{y}) - \mathbf{w}]$ [which is the MSE matrix of $\boldsymbol{\eta}(\mathbf{y})$] equals the matrix $\mathrm{E}\{[\mathbf{t}(\mathbf{y}) - \boldsymbol{\eta}(\mathbf{y})][\mathbf{t}(\mathbf{y}) - \boldsymbol{\eta}(\mathbf{y})]'\}$, which is nonnegative definite and which equals $\mathbf{0}$ if and only if $\mathbf{t}(\mathbf{y}) = \boldsymbol{\eta}(\mathbf{y})$ for every value of $\mathbf{y}$ such that $\mathbf{y} - \boldsymbol{\mu}_y \in \mathcal{C}(\mathbf{V}_y)$, (2) that $\Pr[\mathbf{y} - \boldsymbol{\mu}_y \in \mathcal{C}(\mathbf{V}_y)] = 1$, and (3) that $\mathrm{var}[\boldsymbol{\eta}(\mathbf{y}) - \mathbf{w}] = \mathbf{V}_w - \mathbf{V}_{yw}'\mathbf{V}_y^{-}\mathbf{V}_{yw}$.

Solution. (1) Clearly, $\boldsymbol{\eta}(\mathbf{y})$ is unbiased. It is also linear, as is evident upon observing that it is reexpressible as $\boldsymbol{\eta}(\mathbf{y}) = \boldsymbol{\tau} + \mathbf{V}_{yw}'\mathbf{V}_y^{-}\mathbf{y}$, where $\boldsymbol{\tau} = \boldsymbol{\mu}_w - \mathbf{V}_{yw}'\mathbf{V}_y^{-}\boldsymbol{\mu}_y$.

Now, decompose the difference between the vector-valued function $\mathbf{t}(\mathbf{y})$ and the vector $\mathbf{w}$ into two components as follows:

$$\mathbf{t}(\mathbf{y}) - \mathbf{w} = [\mathbf{t}(\mathbf{y}) - \boldsymbol{\eta}(\mathbf{y})] + [\boldsymbol{\eta}(\mathbf{y}) - \mathbf{w}].$$

And upon observing [in light of Part (2) of Theorem 3.5.11] that $\text{cov}[\mathbf{w} - \boldsymbol{\eta}(\mathbf{y}),\, \mathbf{y}] = \mathbf{0}$ or, equivalently, that $\text{cov}[\mathbf{y},\, \boldsymbol{\eta}(\mathbf{y}) - \mathbf{w}] = \mathbf{0}$ and hence that

$$\begin{aligned} \text{E}\{[\mathbf{t}(\mathbf{y}) - \boldsymbol{\eta}(\mathbf{y})][\boldsymbol{\eta}(\mathbf{y}) - \mathbf{w}]'\} &= \text{cov}[\mathbf{t}(\mathbf{y}) - \boldsymbol{\eta}(\mathbf{y}),\, \boldsymbol{\eta}(\mathbf{y}) - \mathbf{w}] \\ &= (\mathbf{A}' - \mathbf{V}_{yw}'\mathbf{V}_y^-)\,\text{cov}[\mathbf{y},\, \boldsymbol{\eta}(\mathbf{y}) - \mathbf{w}] = \mathbf{0}, \end{aligned}$$

we find that

$$\begin{aligned} &\text{E}\{[\mathbf{t}(\mathbf{y}) - \mathbf{w}][\mathbf{t}(\mathbf{y}) - \mathbf{w}]'\} \\ &\quad = \text{E}\big(\{[\mathbf{t}(\mathbf{y}) - \boldsymbol{\eta}(\mathbf{y})] + [\boldsymbol{\eta}(\mathbf{y}) - \mathbf{w}]\}\{[\mathbf{t}(\mathbf{y}) - \boldsymbol{\eta}(\mathbf{y})]' + [\boldsymbol{\eta}(\mathbf{y}) - \mathbf{w}]'\}\big) \\ &\quad = \text{E}\{[\mathbf{t}(\mathbf{y}) - \boldsymbol{\eta}(\mathbf{y})][\mathbf{t}(\mathbf{y}) - \boldsymbol{\eta}(\mathbf{y})]'\} + \text{var}[\boldsymbol{\eta}(\mathbf{y}) - \mathbf{w}]. \end{aligned} \tag{S.18}$$

It follows from result (S.18) that $\boldsymbol{\eta}(\mathbf{y})$ is a best linear predictor in the sense that the difference between the matrix $\text{E}\{[\mathbf{t}(\mathbf{y}) - \mathbf{w}][\mathbf{t}(\mathbf{y}) - \mathbf{w}]'\}$ and the matrix $\text{var}[\boldsymbol{\eta}(\mathbf{y}) - \mathbf{w}]$ [which is the MSE matrix of $\boldsymbol{\eta}(\mathbf{y})$] equals the matrix $\text{E}\{[\mathbf{t}(\mathbf{y}) - \boldsymbol{\eta}(\mathbf{y})][\mathbf{t}(\mathbf{y}) - \boldsymbol{\eta}(\mathbf{y})]'\}$, which is nonnegative definite. It remains to show that $\text{E}\{[\mathbf{t}(\mathbf{y}) - \boldsymbol{\eta}(\mathbf{y})][\mathbf{t}(\mathbf{y}) - \boldsymbol{\eta}(\mathbf{y})]'\} = \mathbf{0}$ if and only if $\mathbf{t}(\mathbf{y}) = \boldsymbol{\eta}(\mathbf{y})$ for every value of $\mathbf{y}$ such that $\mathbf{y} - \boldsymbol{\mu}_y \in \mathcal{C}(\mathbf{V}_y)$.

Clearly,

$$\begin{aligned} \mathbf{t}(\mathbf{y}) - \boldsymbol{\eta}(\mathbf{y}) &= \mathbf{c} + \mathbf{A}'\boldsymbol{\mu}_y - \boldsymbol{\tau} + (\mathbf{A}' - \mathbf{V}_{yw}'\mathbf{V}_y^-)(\mathbf{y} - \boldsymbol{\mu}_y), \\ \text{E}[\mathbf{t}(\mathbf{y}) - \boldsymbol{\eta}(\mathbf{y})] &= \mathbf{c} + \mathbf{A}'\boldsymbol{\mu}_y - \boldsymbol{\tau}, \quad \text{and} \\ \text{var}[\mathbf{t}(\mathbf{y}) - \boldsymbol{\eta}(\mathbf{y})] &= (\mathbf{A}' - \mathbf{V}_{yw}'\mathbf{V}_y^-)\mathbf{V}_y(\mathbf{A}' - \mathbf{V}_{yw}'\mathbf{V}_y^-)'. \end{aligned}$$

And, making use of Corollary 2.13.27, we find that

$$\begin{aligned} &\text{E}\{[\mathbf{t}(\mathbf{y}) - \boldsymbol{\eta}(\mathbf{y})][\mathbf{t}(\mathbf{y}) - \boldsymbol{\eta}(\mathbf{y})]'\} = \mathbf{0} \\ &\qquad \Leftrightarrow \text{E}[\mathbf{t}(\mathbf{y}) - \boldsymbol{\eta}(\mathbf{y})]\{\text{E}[\mathbf{t}(\mathbf{y}) - \boldsymbol{\eta}(\mathbf{y})]\}' + \text{var}[\mathbf{t}(\mathbf{y}) - \boldsymbol{\eta}(\mathbf{y})] = \mathbf{0} \\ &\qquad \Leftrightarrow \text{E}[\mathbf{t}(\mathbf{y}) - \boldsymbol{\eta}(\mathbf{y})] = \mathbf{0} \text{ and } \text{var}[\mathbf{t}(\mathbf{y}) - \boldsymbol{\eta}(\mathbf{y})] = \mathbf{0} \\ &\qquad \Leftrightarrow \mathbf{c} + \mathbf{A}'\boldsymbol{\mu}_y - \boldsymbol{\tau} = \mathbf{0} \text{ and } (\mathbf{A}' - \mathbf{V}_{yw}'\mathbf{V}_y^-)\mathbf{V}_y = \mathbf{0}. \end{aligned}$$

Moreover, $(\mathbf{A}' - \mathbf{V}_{yw}'\mathbf{V}_y^-)\mathbf{V}_y = \mathbf{0}$ if and only if $(\mathbf{A}' - \mathbf{V}_{yw}'\mathbf{V}_y^-)(\mathbf{y} - \boldsymbol{\mu}_y) = \mathbf{0}$ for every value of $\mathbf{y}$ such that $\mathbf{y} - \boldsymbol{\mu}_y \in \mathcal{C}(\mathbf{V}_y)$. Thus, $\text{E}\{[\mathbf{t}(\mathbf{y}) - \boldsymbol{\eta}(\mathbf{y})][\mathbf{t}(\mathbf{y}) - \boldsymbol{\eta}(\mathbf{y})]'\} = \mathbf{0}$ if and only if $\mathbf{t}(\mathbf{y}) = \boldsymbol{\eta}(\mathbf{y})$ for every value of $\mathbf{y}$ such that $\mathbf{y} - \boldsymbol{\mu}_y \in \mathcal{C}(\mathbf{V}_y)$.

(2) That $\Pr[\mathbf{y} - \boldsymbol{\mu}_y \in \mathcal{C}(\mathbf{V}_y)] = 1$ follows from Part (0) of Theorem 3.5.11.

(3) That $\text{var}[\boldsymbol{\eta}(\mathbf{y}) - \mathbf{w}] = \mathbf{V}_w - \mathbf{V}_{yw}'\mathbf{V}_y^-\mathbf{V}_{yw}$ follows from Part (2) of Theorem 3.5.11.

EXERCISE 27. Suppose that $\mathbf{y}$ is an $N \times 1$ observable random vector that follows a G–M model, and take $\mathbf{w}$ to be an $M \times 1$ unobservable random vector whose value is to be predicted. Suppose further that $\text{E}(\mathbf{w})$ is of the form $\text{E}(\mathbf{w}) = \boldsymbol{\Lambda}'\boldsymbol{\beta}$ (where $\boldsymbol{\Lambda}$ is a matrix of known constants) and that $\text{cov}(\mathbf{y},\, \mathbf{w})$ is of the form $\text{cov}(\mathbf{y},\, \mathbf{w}) = \sigma^2\mathbf{H}_{yw}$ (where $\mathbf{H}_{yw}$ is a known matrix). Let $\boldsymbol{\tau} = (\boldsymbol{\Lambda}' - \mathbf{H}_{yw}'\mathbf{X})\boldsymbol{\beta}$, denote by $\tilde{\mathbf{w}}(\mathbf{y})$ an arbitrary predictor (of $\mathbf{w}$), and define $\tilde{\boldsymbol{\tau}}(\mathbf{y}) = \tilde{\mathbf{w}}(\mathbf{y}) - \mathbf{H}_{yw}'\mathbf{y}$. Verify that $\tilde{\mathbf{w}}(\mathbf{y})$ is a translation-equivariant predictor (of $\mathbf{w}$) if and only if $\tilde{\boldsymbol{\tau}}(\mathbf{y})$ is a translation-equivariant estimator of $\boldsymbol{\tau}$.

Solution. For an arbitrary $P \times 1$ vector $\mathbf{k}$ (and for an arbitrary value of $\mathbf{y}$),

$$\begin{aligned} \tilde{\mathbf{w}}(\mathbf{y} + \mathbf{X}\mathbf{k}) = \tilde{\mathbf{w}}(\mathbf{y}) + \boldsymbol{\Lambda}'\mathbf{k} \;&\Leftrightarrow\; \tilde{\mathbf{w}}(\mathbf{y} + \mathbf{X}\mathbf{k}) - \mathbf{H}_{yw}'(\mathbf{y} + \mathbf{X}\mathbf{k}) = \tilde{\boldsymbol{\tau}}(\mathbf{y}) + (\boldsymbol{\Lambda}' - \mathbf{H}_{yw}'\mathbf{X})\mathbf{k} \\ &\Leftrightarrow\; \tilde{\boldsymbol{\tau}}(\mathbf{y} + \mathbf{X}\mathbf{k}) = \tilde{\boldsymbol{\tau}}(\mathbf{y}) + (\boldsymbol{\Lambda}' - \mathbf{H}_{yw}'\mathbf{X})\mathbf{k}. \end{aligned}$$

Thus, $\tilde{\mathbf{w}}(\mathbf{y})$ is a translation-equivariant predictor (of $\mathbf{w}$) if and only if $\tilde{\boldsymbol{\tau}}(\mathbf{y})$ is a translation-equivariant estimator of $\boldsymbol{\tau}$.

6

Some Relevant Distributions and Their Properties

EXERCISE 1. Let x represent a random variable whose distribution is $Ga(\alpha, \beta)$, and let c represent a (strictly) positive constant. Show that $cx \sim Ga(\alpha, c\beta)$ [thereby verifying result (1.2)].

Solution. Let $y = cx$. And denote by $f(\cdot)$ the pdf of the $Ga(\alpha, \beta)$ distribution and by $h(\cdot)$ the pdf of the distribution of the random variable y. Upon observing that $x = y/c$ and that $dx/dy = 1/c$ and upon making use of standard results on a change of variable (e.g., Casella and Berger 2002, sec. 2.1), we find that, for $0 < y < \infty$,

$$h(y) = f(y/c)\,(1/c) = \frac{1}{\Gamma(\alpha)\beta^\alpha}\,(y/c)^{\alpha-1}e^{-y/(c\beta)}(1/c) = \frac{1}{\Gamma(\alpha)(c\beta)^\alpha}\,y^{\alpha-1}e^{-y/(c\beta)}$$

—for $-\infty < y \le 0$, $h(y) = 0$. Thus, the pdf $h(\cdot)$ is identical to the pdf of the $Ga(\alpha, c\beta)$ distribution, so that $y \sim Ga(\alpha, c\beta)$.

EXERCISE 2. Let w represent a random variable whose distribution is $Ga(\alpha, \beta)$, where α is a (strictly positive) integer. Show that (for any strictly positive scalar t)

$$\Pr(w \le t) = \Pr(u \ge \alpha),$$

where u is a random variable whose distribution is Poisson with parameter t/β [so that $\Pr(u = s) = e^{-t/\beta}(t/\beta)^s/s!$ for $s = 0, 1, 2, \ldots$].

Solution. The equality

$$\Pr(w \le t) = \Pr(u \ge \alpha) \tag{S.1}$$

can be verified by mathematical induction.

For $\alpha = 1$,

$$\begin{aligned}
\Pr(w \le t) &= \int_0^t \frac{1}{\beta}e^{-x/\beta}\,dx \\
&= \int_0^{t/\beta} e^{-y}\,dy \\
&= -e^{-y}\Big|_{y=0}^{y=t/\beta} \\
&= 1 - e^{-t/\beta} \\
&= 1 - \Pr(u = 0) \\
&= \Pr(u \ge 1) \\
&= \Pr(u \ge \alpha).
\end{aligned}$$

Thus, equality (S.1) holds for $\alpha = 1$.

Moreover, upon observing that $d(-\beta e^{-x/\beta})/dx = e^{-x/\beta}$ and employing integration by parts, we find that (for $\alpha \ge 2$)

$$\begin{aligned}
\Pr(w \le t) &= \frac{1}{(\alpha-1)!\beta^\alpha}\int_0^t x^{\alpha-1}e^{-x/\beta}\,dx \\
&= \frac{1}{(\alpha-1)!\beta^\alpha}\left[t^{\alpha-1}(-\beta e^{-t/\beta}) - 0^{\alpha-1}(-\beta e^{-0/\beta}) - \int_0^t (\alpha-1)x^{\alpha-2}(-\beta e^{-x/\beta})\,dx\right] \\
&= -\frac{1}{(\alpha-1)!\beta^{\alpha-1}}t^{\alpha-1}e^{-t/\beta} + \frac{1}{(\alpha-2)!\beta^{\alpha-1}}\int_0^t x^{\alpha-2}e^{-x/\beta}\,dx \\
&= -\Pr(u=\alpha-1) + \Pr(w^* \le t), \qquad \text{(S.2)}
\end{aligned}$$

where w^* is a random variable whose distribution is $Ga(\alpha\text{–}1,\ \beta)$. And upon observing that expression (S.2) equals

$$\Pr(u \ge \alpha) + \Pr(w^* \le t) - \Pr(u \ge \alpha - 1),$$

it is clear that if equality (S.1) holds for $\alpha = \alpha' - 1$ (where $\alpha' \ge 2$), then it also holds for $\alpha = \alpha'$. By mathematical induction, it follows that equality (S.1) holds for every value of α.

EXERCISE 3. Let u and w represent random variables that are distributed independently as $Be(\alpha, \delta)$ and $Be(\alpha+\delta,\ \lambda)$, respectively. Show that $uw \sim Be(\alpha,\ \lambda+\delta)$.

Solution. Define $x = uw$ and $y = u$. Let us find the pdf of the joint distribution of x and y.

Denote by $f(\cdot\,,\cdot)$ the pdf of the joint distribution of u and w, and by $h(\cdot\,,\cdot)$ the pdf of the joint distribution of x and y. Further, denote by A the set $\{(u, w) : 0 < u < 1,\ 0 < w < 1\}$, and by B the set $\{(x, y) : 0 < x < y < 1\}$. For $(u, w) \in A$,

$$f(u,w) = \frac{\Gamma(\alpha+\delta)}{\Gamma(\alpha)\Gamma(\delta)}u^{\alpha-1}(1-u)^{\delta-1}\frac{\Gamma(\alpha+\delta+\lambda)}{\Gamma(\alpha+\delta)\Gamma(\lambda)}w^{\alpha+\delta-1}(1-w)^{\lambda-1};$$

for $(u, w) \notin A,\ f(u, w) = 0$.

The mapping $(u, w) \to (uw,\ u)$ defines a one-to-one transformation from the set A onto the set B. The inverse transformation is defined by the two equalities $u = y$ and $w = x/y$. Clearly,

$$\det\begin{pmatrix}\partial u/\partial x & \partial u/\partial y \\ \partial w/\partial x & \partial w/\partial y\end{pmatrix} = \det\begin{pmatrix}0 & 1 \\ 1/y & -x/y^2\end{pmatrix} = -1/y.$$

Thus, for $(x, y) \in B$,

$$h(x,y) = \frac{\Gamma(\alpha+\delta+\lambda)}{\Gamma(\alpha)\Gamma(\delta)\Gamma(\lambda)}y^{\alpha-1}(1-y)^{\delta-1}\left(\frac{x}{y}\right)^{\alpha+\delta-1}\left(1-\frac{x}{y}\right)^{\lambda-1}\frac{1}{y};$$

for $(x, y) \notin B,\ h(x, y) = 0$.

It is now clear that the distribution of x is the distribution with pdf $g(\cdot)$, where (for $0 < x < 1$)

$$g(x) = \int_x^1 h(x,y)\,dy = \frac{\Gamma(\alpha+\delta+\lambda)}{\Gamma(\alpha)\Gamma(\delta)\Gamma(\lambda)}x^{\alpha-1}\int_x^1\left(\frac{x}{y}-x\right)^{\delta-1}\left(1-\frac{x}{y}\right)^{\lambda-1}\frac{x}{y^2}\,dy$$

—for $x \le 0$ and $x \ge 1,\ g(x) = 0$. And upon introducing the change of variable $z = (1-x)^{-1}[1-(x/y)]$ and upon observing that $1 - z = (1-x)^{-1}[(x/y) - x]$ and that $dz/dy = (1-x)^{-1}(x/y^2)$, we find that (for $0 < x < 1$)

$$\begin{aligned}
g(x) &= \frac{\Gamma(\alpha+\delta+\lambda)}{\Gamma(\alpha)\Gamma(\delta)\Gamma(\lambda)}x^{\alpha-1}(1-x)^{\lambda+\delta-1}\int_0^1 z^{\lambda-1}(1-z)^{\delta-1}\,dz \\
&= \frac{\Gamma(\alpha+\delta+\lambda)}{\Gamma(\alpha)\Gamma(\delta)\Gamma(\lambda)}x^{\alpha-1}(1-x)^{\lambda+\delta-1}\frac{\Gamma(\lambda)\Gamma(\delta)}{\Gamma(\lambda+\delta)} \\
&= \frac{1}{B(\alpha,\ \lambda+\delta)}x^{\alpha-1}(1-x)^{\lambda+\delta-1}.
\end{aligned}$$

Thus, $x \sim Be(\alpha,\ \lambda+\delta)$.

EXERCISE 4. Let x represent a random variable whose distribution is $Be(\alpha, \lambda)$.

(a) Show that, for $r > -\alpha$,

$$\mathrm{E}(x^r) = \frac{\Gamma(\alpha + r)\Gamma(\alpha + \lambda)}{\Gamma(\alpha)\Gamma(\alpha + \lambda + r)}.$$

(b) Show that

$$\mathrm{E}(x) = \frac{\alpha}{\alpha + \lambda} \quad \text{and} \quad \mathrm{var}(x) = \frac{\alpha\lambda}{(\alpha + \lambda)^2(\alpha + \lambda + 1)}.$$

Solution. (a) For $r > -\alpha$,

$$\begin{aligned}
\mathrm{E}(x^r) &= \int_0^1 s^r \frac{1}{B(\alpha, \lambda)} s^{\alpha-1}(1-s)^{\lambda-1}\, ds \\
&= \frac{1}{B(\alpha, \lambda)} \int_0^1 s^{\alpha+r-1}(1-s)^{\lambda-1}\, ds \\
&= \frac{B(\alpha + r, \lambda)}{B(\alpha, \lambda)} = \frac{\Gamma(\alpha + r)\Gamma(\lambda)\Gamma(\alpha + \lambda)}{\Gamma(\alpha + \lambda + r)\Gamma(\alpha)\Gamma(\lambda)} = \frac{\Gamma(\alpha + r)\Gamma(\alpha + \lambda)}{\Gamma(\alpha)\Gamma(\alpha + \lambda + r)}.
\end{aligned} \tag{S.3}$$

(b) Upon setting $r = 1$ in expression (S.3), we find [in light of result (3.5.6)] that

$$\mathrm{E}(x) = \frac{\Gamma(\alpha + 1)}{\Gamma(\alpha)} \frac{\Gamma(\alpha + \lambda)}{\Gamma(\alpha + \lambda + 1)} = \alpha \frac{1}{\alpha + \lambda} = \frac{\alpha}{\alpha + \lambda}.$$

And upon setting $r = 2$ in expression (S.3), we find [in light of result (1.23)] that

$$\mathrm{E}(x^2) = \frac{\Gamma(\alpha + 2)}{\Gamma(\alpha)} \frac{\Gamma(\alpha + \lambda)}{\Gamma(\alpha + \lambda + 2)} = (\alpha + 1)\alpha \frac{1}{(\alpha + \lambda + 1)(\alpha + \lambda)} = \frac{\alpha(\alpha + 1)}{(\alpha + \lambda)(\alpha + \lambda + 1)},$$

so that

$$\begin{aligned}
\mathrm{var}(x) = \mathrm{E}(x^2) - [\mathrm{E}(x)]^2 &= \frac{\alpha(\alpha + 1)}{(\alpha + \lambda)(\alpha + \lambda + 1)} - \frac{\alpha^2}{(\alpha + \lambda)^2} \\
&= \frac{\alpha(\alpha + 1)(\alpha + \lambda) - \alpha^2(\alpha + \lambda + 1)}{(\alpha + \lambda)^2(\alpha + \lambda + 1)} \\
&= \frac{\alpha(\alpha + \lambda) - \alpha^2}{(\alpha + \lambda)^2(\alpha + \lambda + 1)} = \frac{\alpha\lambda}{(\alpha + \lambda)^2(\alpha + \lambda + 1)}.
\end{aligned}$$

EXERCISE 5. Take $x_1, x_2, \ldots, x_K$ to be K random variables whose joint distribution is $Di(\alpha_1, \alpha_2, \ldots, \alpha_K, \alpha_{K+1}; K)$, define $x_{K+1} = 1 - \sum_{k=1}^{K} x_k$, and let $\alpha = \alpha_1 + \cdots + \alpha_K + \alpha_{K+1}$.

(a) Generalize the results of Part (a) of Exercise 4 by showing that, for $r_1 > -\alpha_1, \ldots, r_K > -\alpha_K$, and $r_{K+1} > -\alpha_{K+1}$,

$$\mathrm{E}\big(x_1^{r_1} \cdots x_K^{r_K} x_{K+1}^{r_{K+1}}\big) = \frac{\Gamma(\alpha)}{\Gamma\big(\alpha + \sum_{k=1}^{K+1} r_k\big)} \prod_{k=1}^{K+1} \frac{\Gamma(\alpha_k + r_k)}{\Gamma(\alpha_k)}.$$

(b) Generalize the results of Part (b) of Exercise 4 by showing that (for an arbitrary integer k between 1 and $K+1$, inclusive)

$$\mathrm{E}(x_k) = \frac{\alpha_k}{\alpha} \quad \text{and} \quad \mathrm{var}(x_k) = \frac{\alpha_k(\alpha - \alpha_k)}{\alpha^2(\alpha + 1)}$$

and that (for any 2 distinct integers j and k between 1 and $K+1$, inclusive)

$$\mathrm{cov}(x_j, x_k) = \frac{-\alpha_j \alpha_k}{\alpha^2(\alpha + 1)}.$$

Solution. (a) For $r_1 > -\alpha_1, \ldots, r_K > -\alpha_K$, and $r_{K+1} > -\alpha_{K+1}$,

$$\begin{aligned} &\mathrm{E}\big(x_1^{r_1} \cdots x_K^{r_K} x_{K+1}^{r_{K+1}}\big) \\ &\quad = \int_S \Big(\prod_{k=1}^{K} s_k^{r_k}\Big)\Big(1 - \sum_{k=1}^{K} s_k\Big)^{r_{K+1}} \frac{\Gamma(\alpha)}{\prod_{k=1}^{K+1} \Gamma(\alpha_k)} \Big(\prod_{k=1}^{K} s_k^{\alpha_k - 1}\Big)\Big(1 - \sum_{k=1}^{K} s_k\Big)^{\alpha_{K+1}-1} d\mathbf{s}, \end{aligned}$$

where $\mathbf{s} = (s_1, s_2, \ldots, s_K)'$ and $S = \{\mathbf{s} : s_k > 0 \ (k = 1, 2, \ldots, K), \ \sum_{k=1}^{K} s_k < 1\}$. Thus,

$$\mathrm{E}\big(x_1^{r_1} \cdots x_K^{r_K} x_{K+1}^{r_{K+1}}\big) = \frac{\Gamma(\alpha)}{\prod_{k=1}^{K+1} \Gamma(\alpha_k)} \frac{\prod_{k=1}^{K+1} \Gamma(\alpha_k + r_k)}{\Gamma\big(\alpha + \sum_{k=1}^{K+1} r_k\big)} \int_{\mathbb{R}^K} h(\mathbf{s})\, d\mathbf{s},$$

where $h(\cdot)$ is a function defined as follows: for $\mathbf{s} \in S$,

$$h(\mathbf{s}) = \frac{\Gamma\big[\sum_{k=1}^{K+1} (\alpha_k + r_k)\big]}{\prod_{k=1}^{K+1} \Gamma(\alpha_k + r_k)} \Big(\prod_{k=1}^{K} s_k^{\alpha_k + r_k - 1}\Big)\Big(1 - \sum_{k=1}^{K} s_k\Big)^{\alpha_{K+1} + r_{K+1} - 1}$$

and, for $\mathbf{s} \notin S$, $h(\mathbf{s}) = 0$. And upon observing that $h(\cdot)$ is the pdf of the $Di(\alpha_1 + r_1, \ldots, \alpha_K + r_K, \alpha_{K+1} + r_{K+1}; K)$ distribution and hence that $\int_{\mathbb{R}^K} h(\mathbf{s})\, d\mathbf{s} = 1$, we conclude that

$$\mathrm{E}\big(x_1^{r_1} \cdots x_K^{r_K} x_{K+1}^{r_{K+1}}\big) = \frac{\Gamma(\alpha)}{\Gamma\big(\alpha + \sum_{k=1}^{K+1} r_k\big)} \prod_{k=1}^{K+1} \frac{\Gamma(\alpha_k + r_k)}{\Gamma(\alpha_k)}. \tag{S.4}$$

(b) Upon applying formula (S.4) and recalling result (3.5.6), we find that

$$\mathrm{E}(x_k) = \mathrm{E}\big(x_1^0 \cdots x_{k-1}^0 x_k^1 x_{k+1}^0 \cdots x_{K+1}^0\big) = \frac{\Gamma(\alpha)}{\Gamma(\alpha + 1)} \frac{\Gamma(\alpha_k + 1)}{\Gamma(\alpha_k)} = \frac{1}{\alpha} \alpha_k = \frac{\alpha_k}{\alpha}.$$

And upon applying formula (S.4) and recalling result (1.23), we find that

$$\begin{aligned} \mathrm{E}\big(x_k^2\big) &= \mathrm{E}\big(x_1^0 \cdots x_{k-1}^0 x_k^2 x_{k+1}^0 \cdots x_{K+1}^0\big) \\ &= \frac{\Gamma(\alpha)}{\Gamma(\alpha + 2)} \frac{\Gamma(\alpha_k + 2)}{\Gamma(\alpha_k)} = \frac{1}{(\alpha + 1)\,\alpha} (\alpha_k + 1)\, \alpha_k = \frac{\alpha_k}{\alpha} \frac{\alpha_k + 1}{\alpha + 1}, \end{aligned}$$

so that

$$\begin{aligned} \mathrm{var}(x_k) &= \mathrm{E}\big(x_k^2\big) - \big[\mathrm{E}(x_k)\big]^2 \\ &= \frac{\alpha_k}{\alpha} \frac{\alpha_k + 1}{\alpha + 1} - \Big(\frac{\alpha_k}{\alpha}\Big)^2 \\ &= \frac{\alpha_k}{\alpha} \Big(\frac{\alpha_k + 1}{\alpha + 1} - \frac{\alpha_k}{\alpha}\Big) = \frac{\alpha_k}{\alpha} \frac{\alpha(\alpha_k + 1) - \alpha_k(\alpha + 1)}{\alpha(\alpha + 1)} = \frac{\alpha_k(\alpha - \alpha_k)}{\alpha^2(\alpha + 1)}. \end{aligned}$$

Similarly,

$$\mathrm{E}(x_j x_k) == \frac{\Gamma(\alpha)}{\Gamma(\alpha + 2)} \frac{\Gamma(\alpha_j + 1)}{\Gamma(\alpha_j)} \frac{\Gamma(\alpha_k + 1)}{\Gamma(\alpha_k)} = \frac{1}{(\alpha + 1)\,\alpha} \alpha_j \alpha_k = \frac{\alpha_j \alpha_k}{\alpha\,(\alpha + 1)},$$

so that

$$\begin{aligned} \mathrm{cov}(x_j, x_k) &= \mathrm{E}(x_j x_k) - \mathrm{E}(x_j)\,\mathrm{E}(x_k) \\ &= \frac{\alpha_j \alpha_k}{\alpha\,(\alpha + 1)} - \frac{\alpha_j}{\alpha} \frac{\alpha_k}{\alpha} = \alpha_j \alpha_k \frac{\alpha - (\alpha + 1)}{\alpha^2(\alpha + 1)} = \frac{-\alpha_j \alpha_k}{\alpha^2(\alpha + 1)}. \end{aligned}$$

EXERCISE 6. Verify that the function $b(\cdot)$ defined by expression (1.48) is a pdf of the chi distribution with N degrees of freedom.

Solution. Let $x = \sqrt{v}$, where v is a random variable whose distribution is $\chi^2(N)$. And recall that a pdf, say $f(\cdot)$, of the $\chi^2(N)$ distribution is obtainable from expression (1.16), and observe that $v = x^2$ and that $dv/dx = 2x$. Then, using standard results on a change of variable, we find that a pdf, say $b(\cdot)$, of the distribution of x is obtained by taking (for $0 < x < \infty$)

$$b(x) = f(x^2)\,2x = \frac{1}{\Gamma(N/2)\,2^{N/2-1}}\,x^{N-1}e^{-x^2/2}$$

and, ultimately, by taking

$$b(x) = \begin{cases} \dfrac{1}{\Gamma(N/2)\,2^{(N/2)-1}}\,x^{N-1}e^{-x^2/2}, & \text{for } 0 < x < \infty, \\ 0, & \text{elsewhere.} \end{cases}$$

EXERCISE 7. For strictly positive integers J and K, let $s_1, \ldots, s_J, s_{J+1}, \ldots, s_{J+K}$ represent $J+K$ random variables whose joint distribution is $Di(\alpha_1, \ldots, \alpha_J, \alpha_{J+1}, \ldots, \alpha_{J+K}, \alpha_{J+K+1}; J+K)$. Further, for $k = 1, 2, \ldots, K$, let $x_k = s_{J+k}/\bigl(1 - \sum_{j=1}^J s_j\bigr)$. Show that the conditional distribution of $x_1, x_2, \ldots, x_K$ given $s_1, s_2, \ldots, s_J$ is $Di(\alpha_{J+1}, \ldots, \alpha_{J+K}, \alpha_{J+K+1}; K)$.

Solution. Let $u_j = s_j$ $(j = 1, 2, \ldots, J)$, and take S to be the set

$$S = \bigl\{(s_1, \ldots, s_J, s_{J+1}, \ldots, s_{J+K}) : s_j > 0\ (j = 1, \ldots, J, J+1, \ldots, J+K),\ \textstyle\sum_{j=1}^{J+K} s_j < 1\bigr\}$$

and S^* to be the set

$$S^* = \bigl\{(u_1, \ldots, u_J, x_1, \ldots, x_K) : u_j > 0\ (j = 1, \ldots, J),\ x_k > 0\ (k = 1, \ldots, K), \\ \textstyle\sum_{j=1}^J u_j < 1,\ \sum_{k=1}^K x_k < 1\bigr\}.$$

Clearly, the $J+K$ equalities

$$u_j = s_j \quad (j = 1, 2, \ldots, J) \qquad \text{and} \qquad x_k = \frac{s_{J+k}}{1 - \sum_{j=1}^J s_j} \quad (k = 1, 2, \ldots, K)$$

define a one-to-one transformation from the set S onto the set S^*. The inverse transformation is the transformation defined by the $J+K$ equalities

$$s_j = u_j \quad (j = 1, 2, \ldots, J) \quad \text{and} \quad s_{J+k} = x_k\bigl(1 - \textstyle\sum_{j=1}^J u_j\bigr) \quad (k = 1, 2, \ldots, K).$$

Further, the $(J+K) \times (J+K)$ matrix whose jth row is (for $j = 1, \ldots, J, J+1, \ldots, J+K$) $(\partial s_j/\partial u_1, \ldots, \partial s_j/\partial u_J, \partial s_j/\partial x_1, \ldots, \partial s_j/\partial x_K)$ equals

$$\begin{bmatrix} \mathbf{I} & \mathbf{0} \\ \mathbf{A} & (1 - \sum_{j=1}^J u_j)\mathbf{I} \end{bmatrix}, \tag{S.5}$$

where $\mathbf{A}$ is the $K \times J$ matrix whose kth row is $-x_k\mathbf{1}'$. The determinant of the matrix (S.5) equals $\bigl(1 - \sum_{j=1}^J u_j\bigr)^K$, as is evident upon applying Theorem 2.14.14.

Now, let $f(\cdot, \ldots, \cdot, \cdot, \ldots, \cdot)$ represent the pdf of the joint distribution of $u_1, \ldots, u_J, x_1, \ldots, x_K$, and let $h(\cdot, \ldots, \cdot, \cdot, \ldots, \cdot)$ represent the pdf of the joint distribution of $s_1, \ldots, s_J, s_{J+1}, \ldots, s_{J+K}$. Then, making use of standard results on a change of variables and also making use of expression

(1.38) for the pdf of a Dirichlet distribution, we find that, for $(u_1, \ldots, u_J, x_1, \ldots, x_K) \in S^*$,

$$
\begin{aligned}
f(u_1, &\ldots, u_J, x_1, \ldots, x_K) \\
&= \frac{\Gamma(\alpha_1 + \cdots + \alpha_{J+K} + \alpha_{J+K+1})}{\Gamma(\alpha_1) \cdots \Gamma(\alpha_{J+K}) \Gamma(\alpha_{J+K+1})} \\
&\quad \times \prod_{j=1}^{J} u_j^{\alpha_j - 1} \prod_{k=1}^{K} x_k^{\alpha_{J+k} - 1} \Bigl(1 - \sum_{j=1}^{J} u_j\Bigr)^{\sum_{k=1}^{K} (\alpha_{J+k} - 1)} \\
&\quad \times \Bigl[1 - \sum_{j=1}^{J} u_j - \sum_{k=1}^{K} x_k \Bigl(1 - \sum_{j=1}^{J} u_j\Bigr)\Bigr]^{\alpha_{J+K+1} - 1} \Bigl(1 - \sum_{j=1}^{J} u_j\Bigr)^{K} \\
&= \frac{\Gamma(\alpha_1 + \cdots + \alpha_{J+K} + \alpha_{J+K+1})}{\Gamma(\alpha_1) \cdots \Gamma(\alpha_{J+K}) \Gamma(\alpha_{J+K+1})} \\
&\quad \times \prod_{j=1}^{J} u_j^{\alpha_j - 1} \Bigl(1 - \sum_{j'=1}^{J} u_{j'}\Bigr)^{\left(\sum_{k=1}^{K+1} \alpha_{J+k}\right) - 1} \\
&\quad \times \prod_{k=1}^{K} x_k^{\alpha_{J+k} - 1} \Bigl(1 - \sum_{k'=1}^{K} x_{k'}\Bigr)^{\alpha_{J+K+1} - 1}.
\end{aligned}
$$

Moreover, marginally, $u_1, \ldots, u_J$ have a $Di\bigl(\alpha_1, \ldots, \alpha_J, \sum_{k=1}^{K+1} \alpha_{J+k}; J\bigr)$ distribution, the pdf of which, say $g(\cdot, \ldots, \cdot)$, is [for $u_1, \ldots, u_J$ such that $u_j > 0$ $(j = 1, \ldots, J)$ and $\sum_{j=1}^{J} u_j < 1$] expressible as

$$
g(u_1, \ldots, u_J) = \frac{\Gamma(\alpha_1 + \cdots + \alpha_{J+K} + \alpha_{J+K+1})}{\Gamma(\alpha_1) \cdots \Gamma(\alpha_J) \Gamma\bigl(\sum_{k=1}^{K+1} \alpha_{J+k}\bigr)} \prod_{j=1}^{J} u_j^{\alpha_j - 1} \Bigl(1 - \sum_{j'=1}^{J} u_{j'}\Bigr)^{\left(\sum_{k=1}^{K+1} \alpha_{J+k}\right) - 1}.
$$

Thus, for $(u_1, \ldots, u_J, x_1, \ldots, x_K) \in S^*$,

$$
\begin{aligned}
\frac{f(u_1, \ldots, u_J, x_1, \ldots, x_K)}{g(u_1, \ldots, u_J)} &= \frac{\Gamma(\alpha_{J+1} + \cdots + \alpha_{J+K} + \alpha_{J+K+1})}{\Gamma(\alpha_{J+1}) \cdots \Gamma(\alpha_{J+K}) \Gamma(\alpha_{J+K+1})} \\
&\quad \times \prod_{k=1}^{K} x_k^{\alpha_{J+k} - 1} \Bigl(1 - \sum_{k'=1}^{K} x_{k'}\Bigr)^{\alpha_{J+K+1} - 1}.
\end{aligned}
$$

And it follows that the conditional distribution of $x_1, x_2, \ldots, x_K$ given $s_1, s_2, \ldots, s_J$ has the same pdf as the $Di(\alpha_{J+1}, \ldots, \alpha_{J+K}, \alpha_{J+K+1}; K)$ distribution (and hence that these two distributions are the same).

EXERCISE 8. Let $z_1, z_2, \ldots, z_M$ represent random variables whose joint distribution is absolutely continuous with a pdf $f(\cdot, \cdot, \ldots, \cdot)$ of the form $f(z_1, z_2, \ldots, z_M) = g\bigl(\sum_{i=1}^{M} z_i^2\bigr)$ [where $g(\cdot)$ is a nonnegative function of a single nonnegative variable]. Verify that the function $b(\cdot)$ defined by expression (1.53) is a pdf of the distribution of the random variable $\bigl(\sum_{i=1}^{M} z_i^2\bigr)^{1/2}$. (*Note.* This exercise can be regarded as a more general version of Exercise 6.)

Solution. Let $s = \sqrt{v}$, where $v = \sum_{i=1}^{M} z_i^2$. And observe that a pdf, say $r(\cdot)$, of the distribution of v is obtainable from expression (1.51), and observe also that $v = s^2$ and that $dv/ds = 2s$. Then, using standard results on a change of variable, we find that a pdf, say $b(\cdot)$, of the distribution of s is obtained by taking (for $0 < s < \infty$)

$$
b(s) = r(s^2) 2s = \frac{2\pi^{M/2}}{\Gamma(M/2)} s^{M-1} g(s^2)
$$

and, ultimately, by taking

$$b(s) = \begin{cases} \dfrac{2\pi^{M/2}}{\Gamma(M/2)} s^{M-1} g(s^2), & \text{for } 0 < s < \infty, \\ 0, & \text{elsewhere.} \end{cases}$$

EXERCISE 9. Use the procedure described in Section 6.2a to construct a 6×6 orthogonal matrix whose first row is proportional to the vector $(0, -3, 4, 2, 0, -1)$.

Solution. Formulas (2.2) and (2.3) can be used to construct a 4×4 orthogonal matrix $\mathbf{Q}$ whose first row is proportional to the vector $(-3, 4, 2, -1)$. Upon observing that $(-3)^2 = 9$, $(-3)^2 + 4^2 = 25$, $(-3)^2 + 4^2 + 2^2 = 29$, and $(-3)^2 + 4^2 + 2^2 + (-1)^2 = 30$, the four rows of $\mathbf{Q}$ are found to be as follows:

$$30^{-1/2}(-3, 4, 2, -1) = \left(\frac{-3}{\sqrt{30}}, \frac{4}{\sqrt{30}}, \frac{2}{\sqrt{30}}, \frac{-1}{\sqrt{30}}\right);$$

$$\left[\frac{16}{9(25)}\right]^{1/2}(-3, -9/4, 0, 0) = \left(\frac{-4}{5}, \frac{-3}{5}, 0, 0\right);$$

$$\left[\frac{4}{25(29)}\right]^{1/2}(-3, 4, -25/2, 0) = \left(\frac{-6}{5\sqrt{29}}, \frac{8}{5\sqrt{29}}, \frac{-5}{\sqrt{29}}, 0\right); \text{ and}$$

$$\left[\frac{1}{29(30)}\right]^{1/2}[-3, 4, 2, -29/(-1)] = \left(\frac{-3}{\sqrt{870}}, \frac{4}{\sqrt{870}}, \frac{2}{\sqrt{870}}, \frac{\sqrt{29}}{\sqrt{30}}\right).$$

Thus, the matrix

$$\begin{pmatrix} 0 & \frac{-3}{\sqrt{30}} & \frac{4}{\sqrt{30}} & \frac{2}{\sqrt{30}} & 0 & \frac{-1}{\sqrt{30}} \\ 0 & \frac{-4}{5} & \frac{-3}{5} & 0 & 0 & 0 \\ 0 & \frac{-6}{5\sqrt{29}} & \frac{8}{5\sqrt{29}} & \frac{-5}{\sqrt{29}} & 0 & 0 \\ 0 & \frac{-3}{\sqrt{870}} & \frac{4}{\sqrt{870}} & \frac{2}{\sqrt{870}} & 0 & \frac{\sqrt{29}}{\sqrt{30}} \\ 1 & 0 & 0 & 0 & 0 & 0 \\ 0 & 0 & 0 & 0 & 1 & 0 \end{pmatrix}$$

is a 6×6 orthogonal matrix whose first row is proportional to the vector

$$(0, -3, 4, 2, 0, -1).$$

EXERCISE 10. Let $\mathbf{x}_1$ and $\mathbf{x}_2$ represent M-dimensional column vectors.

(a) Use the results of Section 6.2a (pertaining to Helmert matrices) to show that if $\mathbf{x}_2'\mathbf{x}_2 = \mathbf{x}_1'\mathbf{x}_1$, then there exist orthogonal matrices $\mathbf{O}_1$ and $\mathbf{O}_2$ such that $\mathbf{O}_2\mathbf{x}_2 = \mathbf{O}_1\mathbf{x}_1$.

(b) Use the result of Part (a) to devise an alternative proof of the "only if" part of Lemma 5.9.9.

Solution. (a) Suppose that $\mathbf{x}_2'\mathbf{x}_2 = \mathbf{x}_1'\mathbf{x}_1$. And assume that both $\mathbf{x}_1$ and $\mathbf{x}_2$ are nonnull—if either $\mathbf{x}_1$ or $\mathbf{x}_2$ is null, then they are both null, in which case $\mathbf{O}_2\mathbf{x}_2 = \mathbf{O}_1\mathbf{x}_1$ for any $M \times M$ matrices $\mathbf{O}_1$ and $\mathbf{O}_2$ (orthogonal or not). Then, it follows from the results of Section 6.2a that there exist an $M \times M$ orthogonal matrix $\mathbf{O}_1$ whose first row is proportional to $\mathbf{x}_1'$ and an $M \times M$ orthogonal matrix $\mathbf{O}_2$ whose first row is proportional to $\mathbf{x}_2'$. Moreover,

$$\mathbf{O}_2\mathbf{x}_2 = [(\mathbf{x}_2'\mathbf{x}_2)^{1/2}, 0, 0, \ldots, 0]' = [(\mathbf{x}_1'\mathbf{x}_1)^{1/2}, 0, 0, \ldots, 0]' = \mathbf{O}_1\mathbf{x}_1.$$

(b) Suppose [as in Part (a)] that $\mathbf{x}_2'\mathbf{x}_2 = \mathbf{x}_1'\mathbf{x}_1$. And take $\mathbf{O}_1$ and $\mathbf{O}_2$ to be $M \times M$ orthogonal matrices such that $\mathbf{O}_2\mathbf{x}_2 = \mathbf{O}_1\mathbf{x}_1$—the existence of such matrices follows from Part (a). Then, to prove the "only if" part of Lemma 5.9.9, it suffices (upon recalling that the transpose of an orthogonal matrix equals its inverse and that the transposes and products of orthogonal matrices are orthogonal) to observe that

$$\mathbf{x}_2 = \mathbf{O}_2'\mathbf{O}_1\mathbf{x}_1$$

and that $\mathbf{O}_2'\mathbf{O}_1$ is orthogonal.

EXERCISE 11. Let w represent a random variable whose distribution is $\chi^2(N, \lambda)$. Verify that the expressions for $\mathrm{E}(w)$ and $\mathrm{E}(w^2)$ provided by formula (2.36) are in agreement with those provided by results (2.33) and (2.34) [or, equivalently, by results (2.28) and (2.30)].

Solution. Upon applying formula (2.36) in the special case where $r = 1$, we find that

$$\mathrm{E}(w) = \lambda^1 + N\binom{1}{0}\lambda^0 = N + \lambda,$$

in agreement with result (2.33) [and result (2.28)]. And upon applying formula (2.36) in the special case where $r = 2$, we find that

$$\begin{aligned}\mathrm{E}(w^2) &= \lambda^2 + N(N+2)\binom{2}{0}\lambda^0 + (N+2)\binom{2}{1}\lambda^1 \\ &= \lambda^2 + N(N+2) + (N+2)2\lambda \\ &= (N+2)(N+2\lambda) + \lambda^2,\end{aligned}$$

in agreement with result (2.34) [and result (2.30)].

EXERCISE 12. Let w_1 and w_2 represent random variables that are distributed independently as $Ga(\alpha_1, \beta, \delta_1)$ and $Ga(\alpha_2, \beta, \delta_2)$, respectively, and define $w = w_1 + w_2$. Derive the pdf of the distribution of w by starting with the pdf of the joint distribution of w_1 and w_2 and introducing a suitable change of variables. [*Note*. This derivation serves the purpose of verifying that $w \sim Ga(\alpha_1+\alpha_2, \beta, \delta_1+\delta_2)$ and (when coupled with a mathematical-induction argument) represents an alternative way of establishing Theorem 6.2.2 (and Theorem 6.2.1).]

Solution. Define $s = w_1/(w_1+w_2)$, and denote by $h(\cdot)$ the pdf of the distribution of w_1 and by $b(\cdot)$ the pdf of the distribution of w_2. Then,

$$h(w_1) = \sum_{j=0}^{\infty} \frac{\delta_1^j e^{-\delta_1}}{j!} h_j(w_1),$$

where (for $j = 0, 1, 2, \ldots$) $h_j(\cdot)$ is the pdf of a $Ga(\alpha_1+j, \beta)$ distribution, and, similarly,

$$b(w_2) = \sum_{k=0}^{\infty} \frac{\delta_2^k e^{-\delta_2}}{k!} b_k(w_2),$$

where (for $k = 0, 1, 2, \ldots$) $b_k(\cdot)$ is the pdf of a $Ga(\alpha_2+k, \beta)$ distribution. And, proceeding in essentially the same way as in Section 6.2c [in the derivation of result (2.10)], we find that a pdf, say

$f(\cdot\,,\cdot)$, of the joint distribution of w and s is obtained by taking, for $w > 0$ and $0 < s < 1$,

$$\begin{aligned}
f(w,s) &= h(sw)\,b[(1-s)w]\,|-w| \\
&= w\sum_{j=0}^{\infty}\frac{\delta_1^j e^{-\delta_1}}{j!}h_j(sw)\sum_{k=0}^{\infty}\frac{\delta_2^k e^{-\delta_2}}{k!}b_k[(1-s)w] \\
&= \sum_{r=0}^{\infty} w\sum_{j=0}^{r}\frac{\delta_1^j e^{-\delta_1}}{j!}h_j(sw)\frac{\delta_2^{r-j}e^{-\delta_2}}{(r-j)!}b_{r-j}[(1-s)w] \\
&= \sum_{r=0}^{\infty} e^{-(\delta_1+\delta_2)}\beta^{-(\alpha_1+\alpha_2+r)}w^{\alpha_1+\alpha_2+r-1}e^{-w/\beta} \\
&\qquad \times\sum_{j=0}^{r}\frac{1}{j!\,(r-j)!\,\Gamma(\alpha_1+j)\Gamma(\alpha_2+r-j)}\delta_1^j\delta_2^{r-j}s^{\alpha_1+j-1}(1-s)^{\alpha_2+r-j-1}
\end{aligned}$$

—for $w \le 0$ and for s such that $s \le 0$ or $s \ge 1$, $f(w,s) = 0$.

Now, let $q(\cdot)$ represent the pdf of the (marginal) distribution of w. Then, making use of result (1.10), we find that, for $w > 0$,

$$\begin{aligned}
q(w) &= \int_0^1 f(w,s)\,ds \\
&= \sum_{r=0}^{\infty} e^{-(\delta_1+\delta_2)}\beta^{-(\alpha_1+\alpha_2+r)}w^{\alpha_1+\alpha_2+r-1}e^{-w/\beta}\sum_{j=0}^{r}\frac{1}{j!\,(r-j)!\,\Gamma(\alpha_1+\alpha_2+r)}\delta_1^j\delta_2^{r-j} \\
&= \sum_{r=0}^{\infty} e^{-(\delta_1+\delta_2)}\beta^{-(\alpha_1+\alpha_2+r)}w^{\alpha_1+\alpha_2+r-1}e^{-w/\beta}\frac{(\delta_1+\delta_2)^r}{r!\,\Gamma(\alpha_1+\alpha_2+r)} \\
&= \sum_{r=0}^{\infty}\frac{(\delta_1+\delta_2)^r e^{-(\delta_1+\delta_2)}}{r!}\frac{1}{\Gamma(\alpha_1+\alpha_2+r)\,\beta^{\alpha_1+\alpha_2+r}}w^{\alpha_1+\alpha_2+r-1}e^{-w/\beta}
\end{aligned}$$

—for $w \le 0$, $q(w) = 0$. Thus, $q(\cdot)$ is the pdf of the $Ga(\alpha_1+\alpha_2,\ \beta,\ \delta_1+\delta_2)$ distribution [implying that $w \sim Ga(\alpha_1+\alpha_2,\ \beta,\ \delta_1+\delta_2)$].

EXERCISE 13. Let $\mathbf{x} = \boldsymbol{\mu} + \mathbf{z}$, where $\mathbf{z}$ is an N-dimensional random column vector that has an absolutely continuous spherical distribution and where $\boldsymbol{\mu}$ is an N-dimensional nonrandom column vector. Verify that in the special case where $\mathbf{z} \sim N(\mathbf{0}, \mathbf{I})$, the pdf $q(\cdot)$ derived in Section 6.2h for the distribution of $\mathbf{x}'\mathbf{x}$ "simplifies to" (i.e., is reexpressible in the form of) the expression (2.15) given in Section 6.2c for the pdf of the noncentral chi-square distribution [with N degrees of freedom and with noncentrality parameter λ $(= \boldsymbol{\mu}'\boldsymbol{\mu})$].

Solution. Adopt the notation employed in Section 6.2h. And suppose that $\mathbf{z} \sim N(\mathbf{0}, \mathbf{I})$, in which case

$$g(c) = (2\pi)^{-N/2}\exp(-c/2) \qquad \text{(for every nonnegative scalar } c\text{)}.$$

Then, the pdf $q(\cdot)$ is such that, for $0 < y < \infty$,

$$q(y) = \frac{\pi^{(N-1)/2}}{\Gamma[(N-1)/2]}y^{(N/2)-1}(2\pi)^{-N/2}e^{-y/2}e^{-\lambda/2}\int_{-1}^{1}(1-s^2)^{(N-3)/2}e^{\tau y^{1/2}s}\,ds. \tag{S.6}$$

Further, upon expanding $e^{\tau y^{1/2}s}$ in the power series

$$e^{\tau y^{1/2}s} = \sum_{j=0}^{\infty}(\tau y^{1/2}s)^j/j!,$$

we find that

$$\int_{-1}^{1} (1-s^2)^{(N-3)/2} e^{\tau y^{1/2} s}\, ds = \sum_{j=0}^{\infty} [(\tau y^{1/2})^j / j!] \int_{-1}^{1} s^j (1-s^2)^{(N-3)/2}\, ds.$$

Since $\int_{-1}^{1} s^j (1-s^2)^{(N-3)/2}\, ds = 0$ for $j = 1, 3, 5, \ldots$, it follows that

$$\begin{aligned}\int_{-1}^{1} (1-s^2)^{(N-3)/2} e^{\tau y^{1/2} s}\, ds &= \sum_{r=0}^{\infty} [(\tau y^{1/2})^{2r}/(2r)!] \int_{-1}^{1} s^{2r} (1-s^2)^{(N-3)/2}\, ds \\ &= \sum_{r=0}^{\infty} [(\lambda y)^r/(2r)!] \int_{0}^{1} t^{r+(1/2)-1} (1-t)^{[(N-1)/2]-1}\, dt \\ &= \sum_{r=0}^{\infty} [(\lambda y)^r/(2r)!] \frac{\Gamma[r+(1/2)]\Gamma[(N-1)/2]}{\Gamma[(N/2)+r]}\end{aligned}$$

and hence [in light of result (3.5.11)] that

$$\int_{-1}^{1} (1-s^2)^{(N-3)/2} e^{\tau y^{1/2} s}\, ds = \sum_{r=0}^{\infty} \frac{\pi^{1/2} (\lambda y)^r}{4^r r!\, \Gamma[(N/2)+r]}. \tag{S.7}$$

Upon substituting expression (S.7) for $\int_{-1}^{1} (1-s^2)^{(N-3)/2} e^{\tau y^{1/2} s}\, ds$ in expression (S.6), we find that, for $0 < y < \infty$,

$$\begin{aligned} q(y) &= y^{(N/2)-1} 2^{-N/2} e^{-y/2} e^{-\lambda/2} \sum_{r=0}^{\infty} \frac{(\lambda/2)^r y^r}{2^r r!\, \Gamma[(N/2)+r]} \\ &= \sum_{r=0}^{\infty} \frac{(\lambda/2)^r e^{-\lambda/2}}{r!} \frac{1}{\Gamma[(2r+N)/2]\, 2^{(2r+N)/2}} y^{[(2r+N)/2]-1} e^{-y/2}. \end{aligned}$$

Moreover, $q(y) = 0$ for $-\infty < y \le 0$. Thus, $q(\cdot)$ is reexpressible in the form of expression (2.15).

EXERCISE 14. Let u and v represent random variables that are distributed independently as $\chi^2(M)$ and $\chi^2(N)$, respectively. And define $w = (u/M)/(v/N)$. Devise an alternative derivation of the pdf of the $SF(M, N)$ distribution by (1) deriving the pdf of the joint distribution of w and v and by (2) determining the pdf of the marginal distribution of w from the pdf of the joint distribution of w and v.

Solution. (1) Let $r = v$. And observe that the two equalities $w = (u/M)/(v/N)$ and $r = v$ define a one-to-one transformation from the rectangular region $\{u,\ v\ :\ u > 0,\ v > 0\}$ onto the rectangular region

$$\{w,\ r\ :\ w > 0,\ r > 0\}.$$

Observe also that the inverse transformation is defined by the two equalities

$$u = (M/N)rw \quad \text{and} \quad v = r.$$

We find that

$$\det\begin{pmatrix} \partial u/\partial w & \partial u/\partial r \\ \partial v/\partial w & \partial v/\partial r \end{pmatrix} = \det\begin{bmatrix} (M/N)r & (M/N)w \\ 0 & 1 \end{bmatrix} = (M/N)r.$$

Thus, letting $g(\cdot)$ represent a pdf of the $\chi^2(M)$ distribution and $h(\cdot)$ a pdf of the $\chi^2(N)$ distribution, we find that a pdf, say $f(\cdot,\cdot)$, of the joint distribution of w and r (or, equivalently, of the joint distribution of w and v) is obtained by taking, for $w > 0$ and $r > 0$,

$$f(w, r) = g[(M/N)rw]h(r)(M/N)r$$

—for w and r such that $w \le 0$ or $r \le 0$, $f(w,r) = 0$. And upon taking $g(\cdot)$ and $h(\cdot)$ to be the pdfs obtained by applying formula (1.16), we find that, for $w > 0$ and $r > 0$,

$$f(w,r) = \frac{1}{\Gamma(M/2)\Gamma(N/2)2^{(M+N)/2}} (M/N)^{M/2} r^{[(M+N)/2]-1} w^{(M/2)-1} e^{-[1+(M/N)w]r/2}.$$

(2) A pdf, say $f^*(\cdot)$, of the marginal distribution of w is obtained from the joint distribution of w and v by taking, for $w > 0$,

$$f^*(w) = \int_0^\infty f(w,r)\,dr \tag{S.8}$$

—for $-\infty < w \le 0$, $f^*(w) = 0$. And upon introducing the change of variable $s = [1+(M/N)w]r$, we find that (for $w > 0$)

$$\begin{aligned} f^*(w) &= \int_0^\infty f\{w,\,[1+(M/N)w]^{-1}s\}[1+(M/N)w]^{-1}\,ds \\ &= \int_0^\infty \frac{1}{\Gamma(M/2)\Gamma(N/2)2^{(M+N)/2}}(M/N)^{M/2} \\ &\qquad\qquad \times w^{(M/2)-1}[1+(M/N)w]^{-(M+N)/2} s^{[(M+N)/2]-1} e^{-s/2}\,ds \\ &= \frac{\Gamma[(M+N)/2]}{\Gamma(M/2)\Gamma(N/2)}(M/N)^{M/2} w^{(M/2)-1}[1+(M/N)w]^{-(M+N)/2} \\ &\qquad\qquad \times \int_0^\infty \frac{1}{\Gamma[(M+N)/2]2^{(M+N)/2}} s^{[(M+N)/2]-1} e^{-s/2}\,ds. \end{aligned} \tag{S.9}$$

Moreover, the integrand of the integral in expression (S.9) is a pdf—it is the pdf of a $\chi^2(M+N)$ distribution—and hence the integral equals 1. Thus, for $w > 0$,

$$f^*(w) = \frac{\Gamma[(M+N)/2]}{\Gamma(M/2)\Gamma(N/2)}(M/N)^{M/2} w^{(M/2)-1}[1+(M/N)w]^{-(M+N)/2},$$

in agreement with result (3.12).

EXERCISE 15. Let $t = z/\sqrt{v/N}$, where z and v are random variables that are statistically independent with $z \sim N(0,1)$ and $v \sim \chi^2(N)$ [in which case $t \sim St(N)$].

(a) Starting with the pdf of the joint distribution of z and v, derive the pdf of the joint distribution of t and v.

(b) Derive the pdf of the $St(N)$ distribution from the pdf of the joint distribution of t and v, thereby providing an alternative to the derivation given in Part 2 of Section 6.4a.

Solution. (a) Let $w = v$. And observe that the equalities $t = z/\sqrt{v/N}$ and $w = v$ define a one-to-one transformation from the region defined by the inequalities $-\infty < z < \infty$ and $0 < v < \infty$ onto the region defined by the inequalities $-\infty < t < \infty$ and $0 < w < \infty$. Observe also that the inverse of this transformation is the transformation defined by the equalities

$$z = (w/N)^{1/2} t \quad \text{and} \quad v = w.$$

Further,

$$\begin{vmatrix} \partial z/\partial t & \partial z/\partial w \\ \partial v/\partial t & \partial v/\partial w \end{vmatrix} = \begin{vmatrix} (w/N)^{1/2} & (1/2)(Nw)^{-1/2}t \\ 0 & 1 \end{vmatrix} = (w/N)^{1/2}.$$

Thus, denoting by $b(\cdot)$ the pdf of the $N(0,1)$ distribution and by $d(\cdot)$ the pdf of the $\chi^2(N)$ distribution and making use of standard results on a change of variables, the pdf of the joint distribution of t

and w (and hence the pdf of the joint distribution of t and v) is the function $f(\cdot,\cdot)$ (of 2 variables) obtained by taking (for $-\infty < t < \infty$ and $0 < w < \infty$)

$$f(t,w) = b[(w/N)^{1/2}t]\,d(w)\,(w/N)^{1/2}$$
$$= \frac{1}{\Gamma(N/2)\,2^{(N+1)/2}\pi^{1/2}} N^{-1/2} w^{[(N+1)/2]-1} e^{-w[1+(t^2/N)]/2}$$

—for $-\infty < t < \infty$ and $-\infty < w \le 0$, $f(t,w) = 0$.

(b) Clearly, a pdf of the (marginal) distribution of t is given by the function $h(\cdot)$ obtained by taking (for all t)

$$h(t) = \int_0^\infty f(t,w)\,dw.$$

Moreover, upon introducing the change of variable $s = w[1+(t^2/N)]/2$ and recalling (from Section 3.5b) the definition of the gamma function, we find that

$$h(t) = \frac{1}{\Gamma(N/2)\,\pi^{1/2}} N^{-1/2}\left(1+\frac{t^2}{N}\right)^{-(N+1)/2} \int_0^\infty s^{[(N+1)/2]-1} e^{-s}\,ds$$
$$= \frac{\Gamma[(N+1)/2]}{\Gamma(N/2)\,\pi^{1/2}} N^{-1/2}\left(1+\frac{t^2}{N}\right)^{-(N+1)/2},$$

in agreement with expression (4.8).

EXERCISE 16. Let $t = (x_1 + x_2)/|x_1 - x_2|$, where x_1 and x_2 are random variables that are distributed independently and identically as $N(\mu, \sigma^2)$ (with $\sigma > 0$). Show that t has a noncentral t distribution, and determine the values of the parameters (the degrees of freedom and the noncentrality parameter) of this distribution.

Solution. Let $y_1 = (x_1 + x_2)/(\sqrt{2}\sigma)$ and $y_2 = (x_1 - x_2)/(\sqrt{2}\sigma)$, and define $\mathbf{x} = (x_1, x_2)'$ and $\mathbf{y} = (y_1, y_2)'$. Clearly,

$$\mathbf{x} \sim N(\mu\mathbf{1}, \sigma^2\mathbf{I}),$$

and

$$\mathbf{y} = \mathbf{A}\mathbf{x}, \qquad \text{where } \mathbf{A} = (\sqrt{2}\sigma)^{-1}\begin{pmatrix} 1 & 1 \\ 1 & -1 \end{pmatrix}.$$

Thus, $\mathbf{y}$ has a multivariate normal distribution with

$$\mathrm{E}(\mathbf{y}) = \mathbf{A}(\mu\mathbf{1}) = \begin{pmatrix} \sqrt{2}\mu/\sigma \\ 0 \end{pmatrix}$$

and

$$\mathrm{var}(\mathbf{y}) = \mathbf{A}(\sigma^2\mathbf{I})\mathbf{A}' = \mathbf{I}.$$

And it follows that $y_1 \sim N(\sqrt{2}\mu/\sigma,\, 1)$ and also that y_2 is distributed independently of y_1 as $N(0,1)$ and hence that y_2^2 is distributed independently of y_1 as $\chi^2(1)$. Further,

$$t = \frac{y_1}{|y_2|} = \frac{y_1}{\sqrt{y_2^2/1}},$$

leading to the conclusion that $t \sim St(1,\, \sqrt{2}\mu/\sigma)$.

EXERCISE 17. Let t represent a random variable that has an $St(N,\mu)$ distribution. And take r to be an arbitrary one of the integers $1, 2, \ldots < N$. Generalize expressions (4.38) and (4.39) [for $\mathrm{E}(t^1)$ and $\mathrm{E}(t^2)$, respectively] by obtaining an expression for $\mathrm{E}(t^r)$ (in terms of μ). (*Note.* This exercise is closely related to Exercise 3.12.)

Solution. Let x represent a random variable that is distributed as $N(\mu, 1)$, and take z to be a random variable that is distributed as $N(0, 1)$. Then, clearly,

$$\mathrm{E}(x^r) = \mathrm{E}[(\mu + z)^r].$$

And based on the binomial theorem (e.g., Casella and Berger 2002, sec. 3.2), we have that

$$(\mu + z)^r = \sum_{k=0}^{r} \binom{r}{k} \mu^{r-k} z^k$$

—interpret 0^0 as 1. Thus, letting r_* represent the largest even integer that does not exceed r (so that $r_* = r$, if r is even, and $r_* = r - 1$, if r is odd) and making use of results (3.5.17) and (3.5.18), we find that

$$\begin{aligned}
\mathrm{E}(x^r) &= \sum_{k=0}^{r} \binom{r}{k} \mu^{r-k}\, \mathrm{E}(z^k) \\
&= \sum_{s=0}^{r_*/2} \binom{r}{2s} \mu^{r-2s}\, \mathrm{E}(z^{2s}) \\
&= \mu^r + \sum_{s=1}^{r_*/2} \frac{r(r-1)(r-2)\cdots(r-2s+1)}{2s(2s-2)(2s-4)\cdots 6\cdot 4\cdot 2}\, \mu^{r-2s}.
\end{aligned} \tag{S.10}$$

Further, upon substituting expression (S.10) for $\mathrm{E}(x^r)$ in expression (4.36) or (4.37), we obtain the expression

$$\mathrm{E}(t^r) = (N/2)^{r/2} \frac{\Gamma[(N-r)/2]}{\Gamma(N/2)} \left[\mu^r + \sum_{s=1}^{r_*/2} \frac{r(r-1)(r-2)\cdots(r-2s+1)}{2s(2s-2)(2s-4)\cdots 6\cdot 4\cdot 2}\, \mu^{r-2s}\right]$$

or, in the special case where r is even, the expression

$$\mathrm{E}(t^r) = \frac{N^{r/2}}{(N-2)(N-4)\cdots(N-r)} \left[\mu^r + \sum_{s=1}^{r/2} \frac{r(r-1)(r-2)\cdots(r-2s+1)}{2s(2s-2)(2s-4)\cdots 6\cdot 4\cdot 2}\, \mu^{r-2s}\right].$$

EXERCISE 18. Let $\mathbf{t}$ represent an M-dimensional random column vector that has an $MVt(N, \mathbf{I}_M)$ distribution. And let $w = \mathbf{t}'\mathbf{t}$. Derive the pdf of the distribution of w in each of the following two ways: (1) as a special case of the pdf (1.51) and (2) by making use of the relationship (4.50).

Solution. (1) In light of result (4.65), a pdf, say $r(\cdot)$, of the distribution of w is obtainable as a special case of the pdf (1.51); it is obtainable upon taking the function $g(\cdot)$ in expression (1.51) to be as follows: for $0 < w < \infty$,

$$g(w) = \frac{\Gamma[(N+M)/2]}{\Gamma(N/2)\,\pi^{M/2}}\, N^{-M/2} \left(1 + \frac{w}{N}\right)^{-(N+M)/2}.$$

Thus, for $0 < w < \infty$,

$$\begin{aligned}
r(w) &= \frac{\pi^{M/2}}{\Gamma(M/2)}\, w^{(M/2)-1}\, \frac{\Gamma[(N+M)/2]}{\Gamma(N/2)\,\pi^{M/2}}\, N^{-M/2} \left(1 + \frac{w}{N}\right)^{-(N+M)/2} \\
&= \frac{\Gamma[(N+M)/2]}{\Gamma(N/2)\Gamma(M/2)}\, N^{-1} \left(\frac{w}{N}\right)^{(M/2)-1} \left(1 + \frac{w}{N}\right)^{-(N+M)/2}
\end{aligned}$$

—for $-\infty < w \le 0$, $r(w) = 0$.

(2) Let $u = w/M$. Then, according to the relationship (4.50), $u \sim SF(M, N)$. And upon observing that $du/dw = 1/M$, upon denoting the pdf of the $SF(M, N)$ distribution by $f(\cdot)$ and recalling expression (3.12), and upon making use of standard results on a change of variable, we find that the distribution of w has as a pdf the function $r(\cdot)$ obtained by taking, for $0 < w < \infty$,

$$\begin{aligned} r(w) &= f(w/M)(1/M) \\ &= \frac{\Gamma[(M+N)/2]}{\Gamma(M/2)\Gamma(N/2)}(M/N)^{M/2}\left(\frac{w}{M}\right)^{(M/2)-1}\left(1+\frac{M}{N}\frac{w}{M}\right)^{-(M+N)/2}\frac{1}{M} \\ &= \frac{\Gamma[(N+M)/2]}{\Gamma(N/2)\Gamma(M/2)}N^{-1}\left(\frac{w}{N}\right)^{(M/2)-1}\left(1+\frac{w}{N}\right)^{-(N+M)/2} \end{aligned}$$

—for $-\infty < w \le 0$, $r(w) = 0$.

EXERCISE 19. Let $\mathbf{x}$ represent an M-dimensional random column vector whose distribution has as a pdf a function $f(\cdot)$ that is expressible in the following form: for all $\mathbf{x}$,

$$f(\mathbf{x}) = \int_0^\infty h(\mathbf{x} \mid u) g(u)\, du,$$

where $g(\cdot)$ is the pdf of the distribution of a strictly positive random variable u and where (for every u) $h(\cdot \mid u)$ is the pdf of the $N(\mathbf{0}, u^{-1}\mathbf{I}_M)$ distribution.

(a) Show that the distribution of $\mathbf{x}$ is spherical.

(b) Show that the distribution of u can be chosen in such a way that $f(\cdot)$ is the pdf of the $MVt(N, \mathbf{I}_M)$ distribution.

Solution. (a) For all $\mathbf{x}$ (and for $0 < u < \infty$),

$$h(\mathbf{x} \mid u) = (2\pi)^{-M/2} u^{M/2} e^{-(u/2)\mathbf{x}'\mathbf{x}}. \tag{S.11}$$

Thus, $h(\mathbf{x} \mid u)$ depends on the value of $\mathbf{x}$ only through $\mathbf{x}'\mathbf{x}$ and, consequently, $f(\mathbf{x})$ depends on the value of $\mathbf{x}$ only through $\mathbf{x}'\mathbf{x}$. And it follows that the distribution of $\mathbf{x}$ is spherical.

(b) Take $u \sim Ga(N/2,\, 2/N)$. Then, for $0 < u < \infty$,

$$g(u) = \frac{N^{N/2}}{\Gamma(N/2)\, 2^{N/2}} u^{(N/2)-1} e^{-Nu/2}.$$

And in light of result (S.11), we find that (for all $\mathbf{x}$)

$$\begin{aligned} f(\mathbf{x}) &= \int_0^\infty \frac{N^{N/2}}{\Gamma(N/2)\, 2^{(N+M)/2} \pi^{M/2}} u^{[(N+M)/2]-1} e^{-(u/2)(N+\mathbf{x}'\mathbf{x})}\, du \\ &= \frac{\Gamma[(N+M)/2]}{\Gamma(N/2)\,\pi^{M/2}} N^{-M/2}\left(1+\frac{\mathbf{x}'\mathbf{x}}{N}\right)^{-(N+M)/2} \\ &\quad \times \int_0^\infty \frac{1}{\Gamma[(N+M)/2][2/(N+\mathbf{x}'\mathbf{x})]^{(N+M)/2}} \\ &\qquad\qquad u^{[(N+M)/2]-1} e^{-u/[2/(N+\mathbf{x}'\mathbf{x})]}\, du. \end{aligned} \tag{S.12}$$

Moreover, the integrand of the integral in expression (S.12) is the pdf of a $Ga[(N+M)/2,\, 2/(N+\mathbf{x}'\mathbf{x})]$ distribution. Thus, for all $\mathbf{x}$,

$$f(\mathbf{x}) = \frac{\Gamma[(N+M)/2]}{\Gamma(N/2)\,\pi^{M/2}} N^{-M/2}\left(1+\frac{\mathbf{x}'\mathbf{x}}{N}\right)^{-(N+M)/2},$$

and it follows that $f(\cdot)$ is the pdf of the $MVt(N, \mathbf{I}_M)$ distribution.

EXERCISE 20. Show that if condition (6.7) of Theorem 6.6.2 is replaced by the condition

$$\boldsymbol{\Sigma}(\mathbf{b}+2\mathbf{A}\boldsymbol{\mu}) \in \mathcal{C}(\boldsymbol{\Sigma}\mathbf{A}\boldsymbol{\Sigma}),$$

the theorem is still valid.

Solution. Condition (6.7) clearly implies that $\boldsymbol{\Sigma}(\mathbf{b}+2\mathbf{A}\boldsymbol{\mu}) \in \mathcal{C}(\boldsymbol{\Sigma}\mathbf{A}\boldsymbol{\Sigma})$. Thus, to show that Theorem 6.6.2 remains valid when condition (6.7) is replaced by the condition $\boldsymbol{\Sigma}(\mathbf{b}+2\mathbf{A}\boldsymbol{\mu}) \in \mathcal{C}(\boldsymbol{\Sigma}\mathbf{A}\boldsymbol{\Sigma})$, it suffices to show that the condition $\boldsymbol{\Sigma}(\mathbf{b}+2\mathbf{A}\boldsymbol{\mu}) \in \mathcal{C}(\boldsymbol{\Sigma}\mathbf{A}\boldsymbol{\Sigma})$, in combination with condition (6.6), implies condition (6.7). If $\boldsymbol{\Sigma}(\mathbf{b}+2\mathbf{A}\boldsymbol{\mu}) \in \mathcal{C}(\boldsymbol{\Sigma}\mathbf{A}\boldsymbol{\Sigma})$, then $\boldsymbol{\Sigma}(\mathbf{b}+2\mathbf{A}\boldsymbol{\mu}) = \boldsymbol{\Sigma}\mathbf{A}\boldsymbol{\Sigma}\mathbf{r}$ for some column vector $\mathbf{r}$ and if, in addition, condition (6.6) is satisfied, then

$$\boldsymbol{\Sigma}\mathbf{A}\boldsymbol{\Sigma}(\mathbf{b}+2\mathbf{A}\boldsymbol{\mu}) = \boldsymbol{\Sigma}\mathbf{A}\boldsymbol{\Sigma}\mathbf{A}\boldsymbol{\Sigma}\mathbf{r} = \boldsymbol{\Sigma}\mathbf{A}\boldsymbol{\Sigma}\mathbf{r} = \boldsymbol{\Sigma}(\mathbf{b}+2\mathbf{A}\boldsymbol{\mu}),$$

so that condition (6.7) is satisfied.

EXERCISE 21. Let $\mathbf{x}$ represent an M-dimensional random column vector that has an $N(\boldsymbol{\mu}, \boldsymbol{\Sigma})$ distribution (where $\boldsymbol{\Sigma} \neq \mathbf{0}$), and take $\mathbf{G}$ to be a symmetric generalized inverse of $\boldsymbol{\Sigma}$. Show that

$$\mathbf{x}'\mathbf{G}\mathbf{x} \sim \chi^2(\text{rank}\,\boldsymbol{\Sigma},\ \boldsymbol{\mu}'\mathbf{G}\boldsymbol{\mu})$$

if $\boldsymbol{\mu} \in \mathcal{C}(\boldsymbol{\Sigma})$ or $\mathbf{G}\boldsymbol{\Sigma}\mathbf{G} = \mathbf{G}$. [*Note.* A symmetric generalized inverse $\mathbf{G}$ is obtainable from a possibly nonsymmetric generalized inverse, say $\mathbf{H}$, by taking $\mathbf{G} = \frac{1}{2}\mathbf{H} + \frac{1}{2}\mathbf{H}'$; the condition $\mathbf{G}\boldsymbol{\Sigma}\mathbf{G} = \mathbf{G}$ is the second of the so-called Moore-Penrose conditions—refer, e.g., to Harville (1997, chap. 20) for a discussion of the Moore-Penrose conditions.]

Solution. Clearly, $\mathbf{x}'\mathbf{G}\mathbf{x} = 0 + \mathbf{0}'\mathbf{x} + \mathbf{x}'\mathbf{G}\mathbf{x}$. And since $\boldsymbol{\Sigma}\mathbf{G}\boldsymbol{\Sigma} = \boldsymbol{\Sigma}$,

$$\boldsymbol{\Sigma}\mathbf{G}\boldsymbol{\Sigma}\mathbf{G}\boldsymbol{\Sigma} = \boldsymbol{\Sigma}\mathbf{G}\boldsymbol{\Sigma} \quad \text{and} \quad \boldsymbol{\Sigma}(2\mathbf{G}\boldsymbol{\mu}) = \boldsymbol{\Sigma}\mathbf{G}\boldsymbol{\Sigma}(2\mathbf{G}\boldsymbol{\mu}).$$

And if $\mathbf{G}\boldsymbol{\Sigma}\mathbf{G} = \mathbf{G}$, then

$$\boldsymbol{\mu}'\mathbf{G}\boldsymbol{\mu} = \boldsymbol{\mu}'\mathbf{G}\boldsymbol{\Sigma}\mathbf{G}\boldsymbol{\mu} = \tfrac{1}{4}(2\mathbf{G}\boldsymbol{\mu})'\boldsymbol{\Sigma}(2\mathbf{G}\boldsymbol{\mu}).$$

Similarly, if $\boldsymbol{\mu} \in \mathcal{C}(\boldsymbol{\Sigma})$, then $\boldsymbol{\mu} = \boldsymbol{\Sigma}\mathbf{k}$ for some column vector $\mathbf{k}$ and hence

$$\boldsymbol{\mu}'\mathbf{G}\boldsymbol{\mu} = \boldsymbol{\mu}'\mathbf{G}\boldsymbol{\Sigma}\mathbf{k} = \boldsymbol{\mu}'\mathbf{G}\boldsymbol{\Sigma}\mathbf{G}\boldsymbol{\Sigma}\mathbf{k} = \boldsymbol{\mu}'\mathbf{G}\boldsymbol{\Sigma}\mathbf{G}\boldsymbol{\mu} = \tfrac{1}{4}(2\mathbf{G}\boldsymbol{\mu})'\boldsymbol{\Sigma}(2\mathbf{G}\boldsymbol{\mu}).$$

Thus, upon observing that $\text{rank}(\boldsymbol{\Sigma}\mathbf{G}\boldsymbol{\Sigma}) = \text{rank}\,\boldsymbol{\Sigma}$, it follows from Theorem 6.6.2 that

$$\mathbf{x}'\mathbf{G}\mathbf{x} \sim \chi^2(\text{rank}\,\boldsymbol{\Sigma},\ \boldsymbol{\mu}'\mathbf{G}\boldsymbol{\mu}).$$

EXERCISE 22. Let $\mathbf{z}$ represent an N-dimensional random column vector. And suppose that the distribution of $\mathbf{z}$ is an absolutely continuous spherical distribution, so that the distribution of $\mathbf{z}$ has as a pdf a function $f(\cdot)$ such that (for all $\mathbf{z}$) $f(\mathbf{z}) = g(\mathbf{z}'\mathbf{z})$, where $g(\cdot)$ is a (nonnegative) function of a single nonnegative variable. Further, take $\mathbf{z}_*$ to be an M-dimensional subvector of $\mathbf{z}$ (where $M < N$), and let $v = \mathbf{z}_*'\mathbf{z}_*$.

(a) Show that the distribution of v has as a pdf the function $h(\cdot)$ defined as follows: for $v > 0$,

$$h(v) = \frac{\pi^{N/2}}{\Gamma(M/2)\,\Gamma[(N-M)/2]}\, v^{(M/2)-1} \int_0^\infty w^{[(N-M)/2]-1} g(v+w)\, dw;$$

for $v \le 0$, $h(v) = 0$.

(b) Verify that in the special case where $\mathbf{z} \sim N(\mathbf{0}, \mathbf{I}_N)$, $h(\cdot)$ simplifies to the pdf of the $\chi^2(M)$ distribution.

Solution. (a) Let $g_*(\cdot)$ represent the (nonnegative) function of a single nonnegative variable obtained by taking (for $v \ge 0$)

$$g_*(v) = \frac{\pi^{(N-M)/2}}{\Gamma[(N-M)/2]} \int_0^\infty w^{[(N-M)/2]-1} g(v+w)\, dw.$$

Then, according to result (5.9.66), the distribution of $\mathbf{z}_*$ has as a pdf the function $f_*(\cdot)$ defined by taking (for all $\mathbf{z}_*$) $f_*(\mathbf{z}_*) = g_*(\mathbf{z}_*'\mathbf{z}_*)$. And based on the results of Section 6.1g, we conclude that the distribution of v has as a pdf the function $h(\cdot)$ defined by taking, for $v > 0$,

$$\begin{aligned} h(v) &= \frac{\pi^{M/2}}{\Gamma(M/2)}\, v^{(M/2)-1} g_*(v) \\ &= \frac{\pi^{N/2}}{\Gamma(M/2)\,\Gamma[(N-M)/2]}\, v^{(M/2)-1} \int_0^\infty w^{[(N-M)/2]-1} g(v+w)\, dw \end{aligned}$$

and by taking, for $v \le 0$, $h(v) = 0$.

(b) In the special case where $\mathbf{z} \sim N(\mathbf{0}, \mathbf{I}_N)$, we have that (for $v \ge 0$)

$$g(v) = (2\pi)^{-N/2} e^{-v/2}$$

and hence that (for $v > 0$)

$$h(v) = \frac{\pi^{N/2}}{\Gamma(M/2)\,\Gamma[(N-M)/2]}\, v^{(M/2)-1} e^{-v/2} (2\pi)^{-N/2} \int_0^\infty w^{[(N-M)/2]-1} e^{-w/2}\, dw.$$

Moreover, upon changing the variable of integration from w to $u = w/2$, we find that

$$\begin{aligned} \int_0^\infty w^{[(N-M)/2]-1} e^{-w/2}\, dw &= 2^{(N-M)/2} \int_0^\infty u^{[(N-M)/2]-1} e^{-u}\, du \\ &= 2^{(N-M)/2}\, \Gamma[(N-M)/2]. \end{aligned}$$

Thus, in the special case where $\mathbf{z} \sim N(\mathbf{0}, \mathbf{I}_N)$, we find that (for $v > 0$)

$$h(v) = \frac{1}{\Gamma(M/2)\, 2^{M/2}}\, v^{(M/2)-1} e^{-v/2},$$

so that (in that special case) $h(\cdot)$ simplifies to the pdf of the $\chi^2(M)$ distribution.

EXERCISE 23. Let $\mathbf{z} = (z_1, z_2, \ldots, z_M)'$ represent an M-dimensional random (column) vector that has a spherical distribution. And take $\mathbf{A}$ to be an $M \times M$ symmetric idempotent matrix of rank R (where $R \ge 1$).

(a) Starting from first principles (i.e., from the definition of a spherical distribution), use the results of Theorems 5.9.5 and 6.6.6 to show that (1) $\mathbf{z}'\mathbf{A}\mathbf{z} \sim \sum_{i=1}^R z_i^2$ and [assuming that $\Pr(\mathbf{z} \ne \mathbf{0}) = 1$] that (2) $\mathbf{z}'\mathbf{A}\mathbf{z}/\mathbf{z}'\mathbf{z} \sim \sum_{i=1}^R z_i^2 / \sum_{i=1}^M z_i^2$.

(b) Provide an alternative "derivation" of results (1) and (2) of Part (a); do so by showing that (when $\mathbf{z}$ has an absolutely continuous spherical distribution) these two results can be obtained by applying Theorem 6.6.7 (and by making use of the results of Sections 6.1f and 6.1g).

Solution. (a) According to Theorem 5.9.5, there exists an $M \times R$ matrix $\mathbf{Q}_1$ such that $\mathbf{A} = \mathbf{Q}_1\mathbf{Q}_1'$; and, necessarily, this matrix is such that $\mathbf{Q}_1'\mathbf{Q}_1 = \mathbf{I}_R$ (and hence is such that its columns are orthonormal). Now, take $\mathbf{Q}$ to be the $M \times M$ matrix defined as follows: if $R = M$, take $\mathbf{Q} = \mathbf{Q}_1$; if $R < M$, take $\mathbf{Q} = (\mathbf{Q}_1, \mathbf{Q}_2)$, where $\mathbf{Q}_2$ is an $M \times (M-R)$ matrix whose columns consist of any $M-R$ vectors that, in combination with the R columns of $\mathbf{Q}_1$, form an orthonormal basis for $\mathcal{R}^M$—the existence of such vectors follows from Theorem 6.6.6. Further, define $\mathbf{y}_1 = \mathbf{Q}_1'\mathbf{z}$ and $\mathbf{y} = \mathbf{Q}'\mathbf{z}$, and observe that the elements of $\mathbf{y}_1$ are the first R elements of $\mathbf{y}$. Observe also that $\mathbf{Q}$ is orthogonal and that, as a consequence, $\mathbf{y} \sim \mathbf{z}$. Accordingly, we find that

$$\mathbf{z}'\mathbf{A}\mathbf{z} = \mathbf{z}'\mathbf{Q}_1\mathbf{Q}_1'\mathbf{z} = (\mathbf{Q}_1'\mathbf{z})'\mathbf{Q}_1'\mathbf{z} = \mathbf{y}_1'\mathbf{y}_1 \sim \textstyle\sum_{i=1}^R z_i^2.$$

And upon assuming that $\Pr(\mathbf{z} \neq \mathbf{0}) = 1$ and upon observing that, for $\mathbf{z} \neq \mathbf{0}$,

$$\frac{\mathbf{z}'\mathbf{A}\mathbf{z}}{\mathbf{z}'\mathbf{z}} = \frac{\mathbf{z}'\mathbf{Q}_1\mathbf{Q}_1'\mathbf{z}}{\mathbf{z}'\mathbf{Q}\mathbf{Q}'\mathbf{z}} = \frac{(\mathbf{Q}_1'\mathbf{z})'\mathbf{Q}_1'\mathbf{z}}{(\mathbf{Q}'\mathbf{z})'\mathbf{Q}'\mathbf{z}} = \frac{\mathbf{y}_1'\mathbf{y}_1}{\mathbf{y}'\mathbf{y}},$$

we find that

$$\frac{\mathbf{z}'\mathbf{A}\mathbf{z}}{\mathbf{z}'\mathbf{z}} \sim \frac{\sum_{i=1}^{R} z_i^2}{\sum_{i=1}^{M} z_i^2}.$$

(b) Assume that $\mathbf{z}$ has an absolutely continuous spherical distribution. Further, let $\mathbf{y} = (y_1, y_2, \ldots, y_M)'$ represent an M-dimensional random (column) vector that has an $N(\mathbf{0}, \mathbf{I}_M)$ distribution. Then, it follows from Theorem 6.6.7 that

$$\frac{\mathbf{y}'\mathbf{A}\mathbf{y}}{\mathbf{y}'\mathbf{y}} \sim \frac{\sum_{i=1}^{R} y_i^2}{\sum_{i=1}^{M} y_i^2}.$$

And upon letting $\mathbf{v} = (v_1, v_2, \ldots, v_M)' = (\mathbf{z}'\mathbf{z})^{-1/2}\mathbf{z}$ and $\mathbf{u} = (u_1, u_2, \ldots, u_M)' = (\mathbf{y}'\mathbf{y})^{-1/2}\mathbf{y}$ and observing (in light of the results of Sections 6.1f and 6.1g) that the vector $\mathbf{z}$ has the same distribution as the vector $\mathbf{u}$—each of these 2 vectors is distributed uniformly on the surface of an M-dimensional unit sphere—we find that

$$\frac{\mathbf{z}'\mathbf{A}\mathbf{z}}{\mathbf{z}'\mathbf{z}} = \mathbf{v}'\mathbf{A}\mathbf{v} \sim \mathbf{u}'\mathbf{A}\mathbf{u} = \frac{\mathbf{y}'\mathbf{A}\mathbf{y}}{\mathbf{y}'\mathbf{y}} \sim \frac{\sum_{i=1}^{R} y_i^2}{\sum_{i=1}^{M} y_i^2} = \sum_{i=1}^{R} u_i^2 \sim \sum_{i=1}^{R} v_i^2 = \frac{\sum_{i=1}^{R} z_i^2}{\sum_{i=1}^{M} z_i^2}. \tag{S.13}$$

It remains only to observe (in light of the results of Section 6.1g) that $\mathbf{z}'\mathbf{z}$ is distributed independently of $\mathbf{v}$, implying that $\mathbf{z}'\mathbf{z}$ is distributed independently of $\mathbf{v}'\mathbf{A}\mathbf{v}$ and of $\sum_{i=1}^{R} v_i^2$, or equivalently of $\mathbf{z}'\mathbf{A}\mathbf{z}/\mathbf{z}'\mathbf{z}$ and of $\sum_{i=1}^{R} z_i^2 / \sum_{i=1}^{M} z_i^2$, and hence [in light of result (S.13)] that

$$\mathbf{z}'\mathbf{A}\mathbf{z} = (\mathbf{z}'\mathbf{z})\frac{\mathbf{z}'\mathbf{A}\mathbf{z}}{\mathbf{z}'\mathbf{z}} \sim (\mathbf{z}'\mathbf{z})\frac{\sum_{i=1}^{R} z_i^2}{\sum_{i=1}^{M} z_i^2} = \sum_{i=1}^{R} z_i^2.$$

EXERCISE 24. Let $\mathbf{A}$ represent an $N \times N$ symmetric matrix. And take $\mathbf{Q}$ to be an $N \times N$ orthogonal matrix and $\mathbf{D}$ an $N \times N$ diagonal matrix such that $\mathbf{A} = \mathbf{Q}\mathbf{D}\mathbf{Q}'$—the decomposition $\mathbf{A} = \mathbf{Q}\mathbf{D}\mathbf{Q}'$ is the spectral decomposition, the existence and properties of which are established in Section 6.7a. Further, denote by $d_1, d_2, \ldots, d_N$ the diagonal elements of $\mathbf{D}$ (which are the not-necessarily-distinct eigenvalues of $\mathbf{A}$), and taking $\mathbf{D}^+$ to be the $N \times N$ diagonal matrix whose ith diagonal element is d_i^+, where $d_i^+ = 0$ if $d_i = 0$ and where $d_i^+ = 1/d_i$ if $d_i \neq 0$, define $\mathbf{A}^+ = \mathbf{Q}\mathbf{D}^+\mathbf{Q}'$. Show that (1) $\mathbf{A}\mathbf{A}^+\mathbf{A} = \mathbf{A}$ (i.e., $\mathbf{A}^+$ is a generalized inverse of $\mathbf{A}$) and also that (2) $\mathbf{A}^+\mathbf{A}\mathbf{A}^+ = \mathbf{A}^+$, (3) $\mathbf{A}\mathbf{A}^+$ is symmetric, and (4) $\mathbf{A}^+\mathbf{A}$ is symmetric—as discussed, e.g., by Harville (1997, chap. 20), these four conditions are known as the Moore-Penrose conditions, and they serve to determine a unique matrix $\mathbf{A}^+$ that is known as the Moore-Penrose inverse.

Solution. Observe that $\mathbf{D}\mathbf{D}^+\mathbf{D}$, $\mathbf{D}^+\mathbf{D}\mathbf{D}^+$, $\mathbf{D}\mathbf{D}^+$, and $\mathbf{D}^+\mathbf{D}$ are diagonal matrices. Observe also that (for $i = 1, 2, \ldots, N$) the ith diagonal element of $\mathbf{D}\mathbf{D}^+\mathbf{D}$ equals d_i and the ith diagonal element of $\mathbf{D}^+\mathbf{D}\mathbf{D}^+$ equals d_i^+. Thus,

(1) $\mathbf{A}\mathbf{A}^+\mathbf{A} = \mathbf{Q}\mathbf{D}\mathbf{Q}'\mathbf{Q}\mathbf{D}^+\mathbf{Q}'\mathbf{Q}\mathbf{D}\mathbf{Q}' = \mathbf{Q}\mathbf{D}\mathbf{D}^+\mathbf{D}\mathbf{Q}' = \mathbf{Q}\mathbf{D}\mathbf{Q}' = \mathbf{A}$;

(2) $\mathbf{A}^+\mathbf{A}\mathbf{A}^+ = \mathbf{Q}\mathbf{D}^+\mathbf{Q}'\mathbf{Q}\mathbf{D}\mathbf{Q}'\mathbf{Q}\mathbf{D}^+\mathbf{Q}' = \mathbf{Q}\mathbf{D}^+\mathbf{D}\mathbf{D}^+\mathbf{Q}' = \mathbf{Q}\mathbf{D}^+\mathbf{Q}' = \mathbf{A}^+$;

(3) $\mathbf{A}\mathbf{A}^+ = \mathbf{Q}\mathbf{D}\mathbf{Q}'\mathbf{Q}\mathbf{D}^+\mathbf{Q}' = \mathbf{Q}\mathbf{D}\mathbf{D}^+\mathbf{Q}'$ (a symmetric matrix); and

(4) $\mathbf{A}^+\mathbf{A} = \mathbf{Q}\mathbf{D}^+\mathbf{Q}'\mathbf{Q}\mathbf{D}\mathbf{Q}' = \mathbf{Q}\mathbf{D}^+\mathbf{D}\mathbf{Q}'$ (a symmetric matrix).

EXERCISE 25. Let $\boldsymbol{\Sigma}$ represent an $N \times N$ symmetric nonnegative definite matrix, and take $\boldsymbol{\Gamma}_1$ to be a $P_1 \times N$ matrix and $\boldsymbol{\Gamma}_2$ a $P_2 \times N$ matrix such that $\boldsymbol{\Sigma} = \boldsymbol{\Gamma}_1'\boldsymbol{\Gamma}_1 = \boldsymbol{\Gamma}_2'\boldsymbol{\Gamma}_2$. Further, take $\mathbf{A}$ to be an $N \times N$ symmetric matrix. And assuming that $P_2 \geq P_1$ (as can be done without any essential loss of generality), show that the P_2 not-necessarily-distinct eigenvalues of the $P_2 \times P_2$ matrix $\mathbf{\Gamma}_2\mathbf{A}\mathbf{\Gamma}_2'$ consist of the P_1 not-necessarily-distinct eigenvalues of the $P_1 \times P_1$ matrix $\mathbf{\Gamma}_1\mathbf{A}\mathbf{\Gamma}_1'$ and of $P_2 - P_1$ zeroes. (*Hint*. Make use of Corollary 6.4.2.)

Solution. Let $p(\cdot)$ represent the characteristic polynomial of the $P_2 \times P_2$ matrix $\mathbf{\Gamma}_2\mathbf{A}\mathbf{\Gamma}_2'$. Then, making use of Corollary 6.4.2 (and of Corollary 2.14.6), we find that, for any nonzero scalar λ,

$$\begin{aligned}
p(\lambda) &= |\mathbf{\Gamma}_2\mathbf{A}\mathbf{\Gamma}_2' - \lambda\mathbf{I}_{P_2}| \\
&= (-\lambda)^{P_2}|\mathbf{I}_{P_2} - (1/\lambda)\mathbf{\Gamma}_2\mathbf{A}\mathbf{\Gamma}_2'| \\
&= (-\lambda)^{P_2}|\mathbf{I}_N - (1/\lambda)\mathbf{A}\mathbf{\Gamma}_2'\mathbf{\Gamma}_2| \\
&= (-\lambda)^{P_2}|\mathbf{I}_N - (1/\lambda)\mathbf{A}\mathbf{\Sigma}| \\
&= (-\lambda)^{P_2}|\mathbf{I}_N - (1/\lambda)\mathbf{A}\mathbf{\Gamma}_1'\mathbf{\Gamma}_1| \\
&= (-\lambda)^{P_2}|\mathbf{I}_{P_1} - (1/\lambda)\mathbf{\Gamma}_1\mathbf{A}\mathbf{\Gamma}_1'| \\
&= (-\lambda)^{P_2}[-(1/\lambda)]^{P_1}|\mathbf{\Gamma}_1\mathbf{A}\mathbf{\Gamma}_1' - \lambda\mathbf{I}_{P_1}| \\
&= (-1)^{P_2-P_1}\lambda^{P_2-P_1}q(\lambda),
\end{aligned}$$

where $q(\cdot)$ is the characteristic polynomial of the $P_1 \times P_1$ matrix $\mathbf{\Gamma}_1\mathbf{A}\mathbf{\Gamma}_1'$. Based, for example, on the observation that two polynomials that are equal over some nondegenerate interval are equal everywhere, we conclude that, for every scalar λ,

$$p(\lambda) = (-1)^{P_2-P_1}\lambda^{P_2-P_1}q(\lambda).$$

Thus, the P_2 not-necessarily-distinct eigenvalues of $\mathbf{\Gamma}_2\mathbf{A}\mathbf{\Gamma}_2'$ [which coincide with the P_2 not-necessarily-distinct roots of the polynomial $p(\cdot)$] consist of the P_1 not-necessarily-distinct eigenvalues of $\mathbf{\Gamma}_1\mathbf{A}\mathbf{\Gamma}_1'$ and of $P_2 - P_1$ zeroes.

EXERCISE 26. Let $\mathbf{A}$ represent an $M \times M$ symmetric matrix and $\mathbf{\Sigma}$ an $M \times M$ symmetric nonnegative definite matrix. Show that the condition $\mathbf{\Sigma}\mathbf{A}\mathbf{\Sigma}\mathbf{A}\mathbf{\Sigma} = \mathbf{\Sigma}\mathbf{A}\mathbf{\Sigma}$ (which appears in Theorem 6.6.2) is equivalent to *each* of the following three conditions:

(1) $(\mathbf{A}\mathbf{\Sigma})^3 = (\mathbf{A}\mathbf{\Sigma})^2$;

(2) $\text{tr}[(\mathbf{A}\mathbf{\Sigma})^2] = \text{tr}[(\mathbf{A}\mathbf{\Sigma})^3] = \text{tr}[(\mathbf{A}\mathbf{\Sigma})^4]$; and

(3) $\text{tr}[(\mathbf{A}\mathbf{\Sigma})^2] = \text{tr}(\mathbf{A}\mathbf{\Sigma}) = \text{rank}(\mathbf{\Sigma}\mathbf{A}\mathbf{\Sigma})$.

Solution. Take $\mathbf{\Gamma}$ to be any matrix (with M columns) such that $\mathbf{\Sigma} = \mathbf{\Gamma}'\mathbf{\Gamma}$.

(1) That $\mathbf{\Sigma}\mathbf{A}\mathbf{\Sigma}\mathbf{A}\mathbf{\Sigma} = \mathbf{\Sigma}\mathbf{A}\mathbf{\Sigma} \Rightarrow (\mathbf{A}\mathbf{\Sigma})^3 = (\mathbf{A}\mathbf{\Sigma})^2$ is obvious. Now, for purposes of establishing the converse, suppose that $(\mathbf{A}\mathbf{\Sigma})^3 = (\mathbf{A}\mathbf{\Sigma})^2$ or, equivalently, that

$$\mathbf{A}\mathbf{\Gamma}'\mathbf{\Gamma}\mathbf{A}\mathbf{\Gamma}'\mathbf{\Gamma}\mathbf{A}\mathbf{\Gamma}'\mathbf{\Gamma} = \mathbf{A}\mathbf{\Gamma}'\mathbf{\Gamma}\mathbf{A}\mathbf{\Gamma}'\mathbf{\Gamma}.$$

Then, upon applying Corollary 2.3.4, we find that

$$\mathbf{A}\mathbf{\Gamma}'\mathbf{\Gamma}\mathbf{A}\mathbf{\Gamma}'\mathbf{\Gamma}\mathbf{A}\mathbf{\Gamma}' = \mathbf{A}\mathbf{\Gamma}'\mathbf{\Gamma}\mathbf{A}\mathbf{\Gamma}',$$

implying that

$$\mathbf{\Gamma}\mathbf{A}\mathbf{\Gamma}'\mathbf{\Gamma}\mathbf{A}\mathbf{\Gamma}'\mathbf{\Gamma}\mathbf{A}\mathbf{\Gamma}' = \mathbf{\Gamma}\mathbf{A}\mathbf{\Gamma}'\mathbf{\Gamma}\mathbf{A}\mathbf{\Gamma}'$$

or, equivalently, that

$$(\mathbf{\Gamma}\mathbf{A}\mathbf{\Gamma}')'(\mathbf{\Gamma}\mathbf{A}\mathbf{\Gamma}')\mathbf{\Gamma}\mathbf{A}\mathbf{\Gamma}' = (\mathbf{\Gamma}\mathbf{A}\mathbf{\Gamma}')'(\mathbf{\Gamma}\mathbf{A}\mathbf{\Gamma}').$$

And upon making further use of Corollary 2.3.4, it follows that

$$\mathbf{\Gamma}\mathbf{A}\mathbf{\Gamma}'\mathbf{\Gamma}\mathbf{A}\mathbf{\Gamma}' = \mathbf{\Gamma}\mathbf{A}\mathbf{\Gamma}'$$

and hence that

$$\mathbf{\Sigma}\mathbf{A}\mathbf{\Sigma}\mathbf{A}\mathbf{\Sigma} = \mathbf{\Gamma}'\mathbf{\Gamma}\mathbf{A}\mathbf{\Gamma}'\mathbf{\Gamma}\mathbf{A}\mathbf{\Gamma}'\mathbf{\Gamma} = \mathbf{\Gamma}'\mathbf{\Gamma}\mathbf{A}\mathbf{\Gamma}'\mathbf{\Gamma} = \mathbf{\Sigma}\mathbf{A}\mathbf{\Sigma}.$$

(2) Suppose that $\boldsymbol{\Sigma A \Sigma A \Sigma} = \boldsymbol{\Sigma A \Sigma}$. Then, clearly (in light of what has already been established),

$$(\mathbf{A\Sigma})^2 = (\mathbf{A\Sigma})^3 = \mathbf{A\Sigma}(\mathbf{A\Sigma})^2 = \mathbf{A\Sigma}(\mathbf{A\Sigma})^3 = (\mathbf{A\Sigma})^4,$$

implying that $\mathrm{tr}[(\mathbf{A\Sigma})^2] = \mathrm{tr}[(\mathbf{A\Sigma})^3] = \mathrm{tr}[(\mathbf{A\Sigma})^4]$.

Conversely, suppose that

$$\mathrm{tr}[(\mathbf{A\Sigma})^2] = \mathrm{tr}[(\mathbf{A\Sigma})^3] = \mathrm{tr}[(\mathbf{A\Sigma})^4]. \tag{S.14}$$

And denote by P the number of rows in the matrix $\boldsymbol{\Gamma}$, and take $\mathbf{O}$ to be a $P \times P$ orthogonal matrix such that

$$\mathbf{O}'\boldsymbol{\Gamma}\mathbf{A}\boldsymbol{\Gamma}'\mathbf{O} = \mathrm{diag}(d_1, d_2, \ldots, d_P)$$

for some scalars $d_1, d_2, \ldots, d_P$—the existence of such a $P \times P$ orthogonal matrix follows from Theorem 6.7.4. Further, observe (in light of Lemma 2.3.1) that

$$\begin{aligned}\mathrm{tr}[(\mathbf{A\Sigma})^2] = \mathrm{tr}(\boldsymbol{\Gamma}\mathbf{A}\boldsymbol{\Gamma}'\boldsymbol{\Gamma}\mathbf{A}\boldsymbol{\Gamma}') &= \mathrm{tr}(\boldsymbol{\Gamma}\mathbf{A}\boldsymbol{\Gamma}'\mathbf{O}\mathbf{O}'\boldsymbol{\Gamma}\mathbf{A}\boldsymbol{\Gamma}'\mathbf{O}\mathbf{O}') \\ &= \mathrm{tr}[(\mathbf{O}'\boldsymbol{\Gamma}\mathbf{A}\boldsymbol{\Gamma}'\mathbf{O})^2] = \textstyle\sum_{i=1}^P d_i^2\end{aligned} \tag{S.15}$$

and, similarly,

$$\mathrm{tr}[(\mathbf{A\Sigma})^3] = \textstyle\sum_{i=1}^P d_i^3 \quad \text{and} \quad \mathrm{tr}[(\mathbf{A\Sigma})^4] = \textstyle\sum_{i=1}^P d_i^4,$$

implying [in light of the supposition (S.14)] that

$$\textstyle\sum_{i=1}^P d_i^2 = \sum_{i=1}^P d_i^3 = \sum_{i=1}^P d_i^4$$

and hence that

$$\textstyle\sum_{i=1}^P d_i^2(1-d_i)^2 = 0.$$

Thus, either $d_i = 0$ or $d_i = 1$ $(i = 1, 2, \ldots, P)$; and upon observing that $d_1, d_2, \ldots, d_P$ are the not-necessarily-distinct eigenvalues of $\boldsymbol{\Gamma}\mathbf{A}\boldsymbol{\Gamma}'$, it follows from Theorem 6.7.7 that $\boldsymbol{\Gamma}\mathbf{A}\boldsymbol{\Gamma}'$ is idempotent, that is,

$$\boldsymbol{\Gamma}\mathbf{A}\boldsymbol{\Gamma}'\boldsymbol{\Gamma}\mathbf{A}\boldsymbol{\Gamma}' = \boldsymbol{\Gamma}\mathbf{A}\boldsymbol{\Gamma}',$$

in which case

$$\boldsymbol{\Sigma A \Sigma A \Sigma} = \boldsymbol{\Gamma}'\boldsymbol{\Gamma}\mathbf{A}\boldsymbol{\Gamma}'\boldsymbol{\Gamma}\mathbf{A}\boldsymbol{\Gamma}'\boldsymbol{\Gamma} = \boldsymbol{\Gamma}'\boldsymbol{\Gamma}\mathbf{A}\boldsymbol{\Gamma}'\boldsymbol{\Gamma} = \boldsymbol{\Sigma A \Sigma}.$$

(3) The condition $\boldsymbol{\Sigma A \Sigma A \Sigma} = \boldsymbol{\Sigma A \Sigma}$ is reexpressible as $\boldsymbol{\Gamma}'\boldsymbol{\Gamma}\mathbf{A}\boldsymbol{\Gamma}'\boldsymbol{\Gamma}\mathbf{A}\boldsymbol{\Gamma}'\boldsymbol{\Gamma} = \boldsymbol{\Gamma}'\boldsymbol{\Gamma}\mathbf{A}\boldsymbol{\Gamma}'\boldsymbol{\Gamma}$. Thus, in light of Corollary 2.3.4,

$$\boldsymbol{\Sigma A \Sigma A \Sigma} = \boldsymbol{\Sigma A \Sigma} \quad \Leftrightarrow \quad \boldsymbol{\Gamma}\mathbf{A}\boldsymbol{\Gamma}'\boldsymbol{\Gamma}\mathbf{A}\boldsymbol{\Gamma}' = \boldsymbol{\Gamma}\mathbf{A}\boldsymbol{\Gamma}',$$

that is, $\boldsymbol{\Sigma A \Sigma A \Sigma} = \boldsymbol{\Sigma A \Sigma}$ if and only if $\boldsymbol{\Gamma}\mathbf{A}\boldsymbol{\Gamma}'$ is idempotent. Moreover, in light of Lemma 2.3.1,

$$\mathrm{tr}[(\mathbf{A\Sigma})^2] = \mathrm{tr}(\boldsymbol{\Gamma}\mathbf{A}\boldsymbol{\Gamma}'\boldsymbol{\Gamma}\mathbf{A}\boldsymbol{\Gamma}') \quad \text{and} \quad \mathrm{tr}(\mathbf{A\Sigma}) = \mathrm{tr}(\boldsymbol{\Gamma}\mathbf{A}\boldsymbol{\Gamma}');$$

and in light of Lemma 2.12.3,

$$\mathrm{rank}(\boldsymbol{\Sigma A \Sigma}) = \mathrm{rank}(\boldsymbol{\Sigma}\mathbf{A}\boldsymbol{\Gamma}') = \mathrm{rank}(\boldsymbol{\Gamma}\mathbf{A}\boldsymbol{\Gamma}').$$

Accordingly, to show that the condition $\boldsymbol{\Sigma A \Sigma A \Sigma} = \boldsymbol{\Sigma A \Sigma}$ is equivalent to the condition $\mathrm{tr}[(\mathbf{A\Sigma})^2] = \mathrm{tr}(\mathbf{A\Sigma}) = \mathrm{rank}(\boldsymbol{\Sigma A \Sigma})$, it suffices to show that $\boldsymbol{\Gamma}\mathbf{A}\boldsymbol{\Gamma}'$ is idempotent if and only if

$$\mathrm{tr}[(\boldsymbol{\Gamma}\mathbf{A}\boldsymbol{\Gamma}')^2] = \mathrm{tr}(\boldsymbol{\Gamma}\mathbf{A}\boldsymbol{\Gamma}') = \mathrm{rank}(\boldsymbol{\Gamma}\mathbf{A}\boldsymbol{\Gamma}'). \tag{S.16}$$

That the idempotency of $\boldsymbol{\Gamma}\mathbf{A}\boldsymbol{\Gamma}'$ implies that condition (S.16) is satisfied is an obvious consequence of Corollary 2.8.3. Conversely, suppose that condition (S.16) is satisfied. And [as in Part (2)] denote by P the number of rows in the matrix $\boldsymbol{\Gamma}$, and take $\mathbf{O}$ to be a $P \times P$ orthogonal matrix such that $\mathbf{O}'\boldsymbol{\Gamma}\mathbf{A}\boldsymbol{\Gamma}'\mathbf{O} = \mathrm{diag}(d_1, d_2, \ldots, d_P)$ for some scalars $d_1, d_2, \ldots, d_P$ (in which case $d_1, d_2, \ldots, d_P$ are the not-necessarily-distinct eigenvalues of $\boldsymbol{\Gamma}\mathbf{A}\boldsymbol{\Gamma}'$). Then, as in the case of result (S.15),

$$\mathrm{tr}[(\boldsymbol{\Gamma}\mathbf{A}\boldsymbol{\Gamma}')^2] = \textstyle\sum_{i=1}^P d_i^2.$$

Moreover,

$$\operatorname{tr}(\boldsymbol{\Gamma}\mathbf{A}\boldsymbol{\Gamma}') = \textstyle\sum_{i=1}^{P} d_i \quad \text{and} \quad \operatorname{rank}(\boldsymbol{\Gamma}\mathbf{A}\boldsymbol{\Gamma}') = \textstyle\sum_{i=1\,(i\,:\,d_i \neq 0)}^{P} 1,$$

as is evident from Theorem 6.7.5. Thus,

$$\begin{aligned}\textstyle\sum_{i=1\,(i\,:\,d_i\neq 0)}^{P} (d_i - 1)^2 &= \textstyle\sum_{i=1}^{P} d_i^2 - 2\sum_{i=1}^{P} d_i + \sum_{i=1\,(i\,:\,d_i\neq 0)}^{P} 1 \\ &= \operatorname{tr}[(\boldsymbol{\Gamma}\mathbf{A}\boldsymbol{\Gamma}')^2] - 2\operatorname{tr}(\boldsymbol{\Gamma}\mathbf{A}\boldsymbol{\Gamma}') + \operatorname{rank}(\boldsymbol{\Gamma}\mathbf{A}\boldsymbol{\Gamma}') = 0,\end{aligned}$$

leading to the conclusion that all of the nonzero eigenvalues of $\boldsymbol{\Gamma}\mathbf{A}\boldsymbol{\Gamma}'$ equal 1 and hence (in light of Theorem 6.7.7) that $\boldsymbol{\Gamma}\mathbf{A}\boldsymbol{\Gamma}'$ is idempotent.

EXERCISE 27. Let $\mathbf{z}$ represent an M-dimensional random column vector that has an $N(\mathbf{0}, \mathbf{I}_M)$ distribution, and take $q = c + \mathbf{b}'\mathbf{z} + \mathbf{z}'\mathbf{A}\mathbf{z}$, where c is a constant, $\mathbf{b}$ an M-dimensional column vector of constants, and $\mathbf{A}$ an $M \times M$ (nonnull) symmetric matrix of constants. Further, denote by $m(\cdot)$ the moment generating function of q. Provide an alternative derivation of the "sufficiency part" of Theorem 6.6.1 by showing that if $\mathbf{A}^2 = \mathbf{A}$, $\mathbf{b} = \mathbf{A}\mathbf{b}$, and $c = \frac{1}{4}\mathbf{b}'\mathbf{b}$, then, for every scalar t in some neighborhood of 0, $m(t) = m_*(t)$, where $m_*(\cdot)$ is the moment generating function of a $\chi^2(R, c)$ distribution and where $R = \operatorname{rank}\mathbf{A} = \operatorname{tr}(\mathbf{A})$.

Solution. Proceeding as in the proof of the "necessity part" of Theorem 6.6.1, define $K = \operatorname{rank}\mathbf{A}$. And take $\mathbf{O}$ to be an $M \times M$ orthogonal matrix such that

$$\mathbf{O}'\mathbf{A}\mathbf{O} = \operatorname{diag}(d_1, d_2, \ldots, d_M)$$

for some scalars $d_1, d_2, \ldots, d_M$—the existence of such an $M \times M$ orthogonal matrix follows from Theorem 6.7.4—and define $\mathbf{u} = \mathbf{O}'\mathbf{b}$. In light of Theorem 6.7.5, K of the scalars $d_1, d_2, \ldots, d_M$ (which are the not-necessarily-distinct eigenvalues of $\mathbf{A}$) are nonzero, and the rest of them equal 0. Assume (without loss of generality) that it is the first K of the scalars $d_1, d_2, \ldots, d_M$ that are nonzero, so that $d_{K+1} = d_{K+2} = \cdots = d_M = 0$—this assumption can always be satisfied by reordering d_1, d_2, $\ldots$, d_M and the corresponding columns of $\mathbf{O}$ (as necessary). Further, letting $\mathbf{D}_1 = \operatorname{diag}(d_1, d_2, \ldots, d_K)$ and partitioning $\mathbf{O}$ as $\mathbf{O} = (\mathbf{O}_1, \mathbf{O}_2)$ (where the dimensions of $\mathbf{O}_1$ are $M \times K$), observe that

$$\mathbf{A} = \mathbf{O}_1\mathbf{D}_1\mathbf{O}_1'.$$

Upon applying result (7.20), we find that, for every scalar t in some neighborhood S of 0,

$$m(t) = \prod_{i=1}^{K} (1-2td_i)^{-1/2} \exp\!\left[tc + \frac{t^2}{2}\left(\sum_{i=1}^{K} \frac{u_i^2}{1-2td_i} + \sum_{i=K+1}^{M} u_i^2\right)\right],$$

where $u_1, u_2, \ldots, u_M$ represent the elements of $\mathbf{u}$.

Now, suppose that $\mathbf{A}^2 = \mathbf{A}$, $\mathbf{b} = \mathbf{A}\mathbf{b}$, and $c = \frac{1}{4}\mathbf{b}'\mathbf{b}$. Then, in light of Theorem 6.7.7, $d_1 = d_2 = \cdots = d_K = 1$, so that $\mathbf{D}_1 = \mathbf{I}_K$; and in light of Theorem 6.7.5, $K = \operatorname{tr}(\mathbf{A})$. Further,

$$\mathbf{O}_2\mathbf{O}_2'\mathbf{b} = (\mathbf{I} - \mathbf{O}_1\mathbf{O}_1')\mathbf{b} = (\mathbf{I} - \mathbf{O}_1\mathbf{I}\mathbf{O}_1')\mathbf{b} = (\mathbf{I} - \mathbf{O}_1\mathbf{D}_1\mathbf{O}_1')\mathbf{b} = (\mathbf{I} - \mathbf{A})\mathbf{b} = \mathbf{b} - \mathbf{b} = \mathbf{0},$$

implying that $\mathbf{O}_2'\mathbf{b} = \mathbf{0}$ and hence that

$$\textstyle\sum_{i=K+1}^{M} u_i^2 = (\mathbf{O}_2'\mathbf{b})'\mathbf{O}_2'\mathbf{b} = 0.$$

And

$$\textstyle\sum_{i=1}^{K} u_i^2 = \sum_{i=1}^{M} u_i^2 = \mathbf{u}'\mathbf{u} = (\mathbf{O}'\mathbf{b})'\mathbf{O}'\mathbf{b} = \mathbf{b}'\mathbf{b} = 4c.$$

Thus, for $t \in S$,

$$\begin{aligned} m(t) &= (1-2t)^{-K/2} \exp[tc + (t^2/2)(4c)/(1-2t)] \\ &= (1-2t)^{-K/2} \exp[tc/(1-2t)]. \end{aligned}$$

We conclude [in light of result (2.18)] that, for every scalar t in some neighborhood of 0,

$$m(t) = m_*(t).$$

EXERCISE 28. Let $\mathbf{x}$ represent an M-dimensional random column vector that has an $N(\mathbf{0}, \boldsymbol{\Sigma})$ distribution, and denote by $\mathbf{A}$ an $M \times M$ symmetric matrix of constants. Construct an example where $M = 3$ and where $\boldsymbol{\Sigma}$ and $\mathbf{A}$ are such that $\mathbf{A}\boldsymbol{\Sigma}$ is *not* idempotent but are nevertheless such that $\mathbf{x}'\mathbf{A}\mathbf{x}$ has a chi-square distribution.

Solution. Take $\boldsymbol{\Sigma} = \begin{pmatrix} 1 & 1 & 0 \\ 1 & 1 & 0 \\ 0 & 0 & 1 \end{pmatrix}$ and $\mathbf{A} = \begin{pmatrix} 1 & 0 & 0 \\ 0 & -1 & 0 \\ 0 & 0 & 1 \end{pmatrix}$. Then,

$$(\mathbf{A}\boldsymbol{\Sigma})^2 = \begin{pmatrix} 0 & 0 & 0 \\ 0 & 0 & 0 \\ 0 & 0 & 1 \end{pmatrix} \neq \begin{pmatrix} 1 & 1 & 0 \\ -1 & -1 & 0 \\ 0 & 0 & 1 \end{pmatrix} = \mathbf{A}\boldsymbol{\Sigma},$$

but

$$\boldsymbol{\Sigma}\mathbf{A}\boldsymbol{\Sigma}\mathbf{A}\boldsymbol{\Sigma} = \begin{pmatrix} 0 & 0 & 0 \\ 0 & 0 & 0 \\ 0 & 0 & 1 \end{pmatrix} = \boldsymbol{\Sigma}\mathbf{A}\boldsymbol{\Sigma}.$$

Moreover, $\text{tr}(\mathbf{A}\boldsymbol{\Sigma})$ [$= \text{rank}(\boldsymbol{\Sigma}\mathbf{A}\boldsymbol{\Sigma})$] $= 1$. Thus, it follows from Theorem 6.6.2 that $\mathbf{x}'\mathbf{A}\mathbf{x} \sim \chi^2(1)$.

EXERCISE 29. Let $\mathbf{x}$ represent an M-dimensional random column vector that has an $N(\mathbf{0}, \boldsymbol{\Sigma})$ distribution. Further, partition $\mathbf{x}$ and $\boldsymbol{\Sigma}$ as

$$\mathbf{x} = \begin{pmatrix} \mathbf{x}_1 \\ \mathbf{x}_2 \end{pmatrix} \quad \text{and} \quad \boldsymbol{\Sigma} = \begin{pmatrix} \boldsymbol{\Sigma}_{11} & \boldsymbol{\Sigma}_{12} \\ \boldsymbol{\Sigma}_{21} & \boldsymbol{\Sigma}_{22} \end{pmatrix}$$

(where the dimensions of $\boldsymbol{\Sigma}_{11}$ are the same as the dimension of $\mathbf{x}_1$). And take $\mathbf{G}_1$ to be a generalized inverse of $\boldsymbol{\Sigma}_{11}$ and $\mathbf{G}_2$ a generalized inverse of $\boldsymbol{\Sigma}_{22}$. Show that $\mathbf{x}_1'\mathbf{G}_1\mathbf{x}_1$ and $\mathbf{x}_2'\mathbf{G}_2\mathbf{x}_2$ are distributed independently if and only if $\boldsymbol{\Sigma}_{12} = \mathbf{0}$.

Solution. Clearly, $\mathbf{x}_1'\mathbf{G}_1\mathbf{x}_1 = \mathbf{x}_1'\mathbf{G}_1^*\mathbf{x}_1$ and $\mathbf{x}_2'\mathbf{G}_2\mathbf{x}_2 = \mathbf{x}_2'\mathbf{G}_2^*\mathbf{x}_2$, where $\mathbf{G}_1^* = (1/2)(\mathbf{G}_1 + \mathbf{G}_1')$ and $\mathbf{G}_2^* = (1/2)(\mathbf{G}_2 + \mathbf{G}_2')$. Moreover, $\mathbf{G}_1^*$ is a symmetric generalized inverse of $\boldsymbol{\Sigma}_{11}$ and $\mathbf{G}_2^*$ a symmetric generalized inverse of $\boldsymbol{\Sigma}_{22}$. Accordingly, it follows from Theorem 6.8.1 that $\mathbf{x}_1'\mathbf{G}_1\mathbf{x}_1$ and $\mathbf{x}_2'\mathbf{G}_2\mathbf{x}_2$ are distributed independently if and only if

$$\boldsymbol{\Sigma}\mathbf{A}_1\boldsymbol{\Sigma}\mathbf{A}_2\boldsymbol{\Sigma} = \mathbf{0},$$

where $\mathbf{A}_1 = \begin{pmatrix} \mathbf{G}_1^* & \mathbf{0} \\ \mathbf{0} & \mathbf{0} \end{pmatrix}$ and $\mathbf{A}_2 = \begin{pmatrix} \mathbf{0} & \mathbf{0} \\ \mathbf{0} & \mathbf{G}_2^* \end{pmatrix}$. Thus, it suffices to show that $\boldsymbol{\Sigma}\mathbf{A}_1\boldsymbol{\Sigma}\mathbf{A}_2\boldsymbol{\Sigma} = \mathbf{0}$ if and only if $\boldsymbol{\Sigma}_{12} = \mathbf{0}$.

We find that

$$\boldsymbol{\Sigma}\mathbf{A}_1\boldsymbol{\Sigma}\mathbf{A}_2\boldsymbol{\Sigma} = \begin{pmatrix} \boldsymbol{\Sigma}_{11}\mathbf{G}_1^*\boldsymbol{\Sigma}_{12}\mathbf{G}_2^*\boldsymbol{\Sigma}_{12}' & \boldsymbol{\Sigma}_{11}\mathbf{G}_1^*\boldsymbol{\Sigma}_{12}\mathbf{G}_2^*\boldsymbol{\Sigma}_{22} \\ \boldsymbol{\Sigma}_{12}'\mathbf{G}_1^*\boldsymbol{\Sigma}_{12}\mathbf{G}_2^*\boldsymbol{\Sigma}_{12}' & \boldsymbol{\Sigma}_{12}'\mathbf{G}_1^*\boldsymbol{\Sigma}_{12}\mathbf{G}_2^*\boldsymbol{\Sigma}_{22} \end{pmatrix}. \tag{S.17}$$

That $\boldsymbol{\Sigma}\mathbf{A}_1\boldsymbol{\Sigma}\mathbf{A}_2\boldsymbol{\Sigma} = \mathbf{0}$ if $\boldsymbol{\Sigma}_{12} = \mathbf{0}$ follows immediately from expression (S.17). Now, for purposes of establishing the converse, suppose that $\boldsymbol{\Sigma}\mathbf{A}_1\boldsymbol{\Sigma}\mathbf{A}_2\boldsymbol{\Sigma} = \mathbf{0}$. Then,

$$\boldsymbol{\Sigma}_{11}\mathbf{G}_1^*\boldsymbol{\Sigma}_{12}\mathbf{G}_2^*\boldsymbol{\Sigma}_{22} = \mathbf{0} \tag{S.18}$$

[as is evident from expression (S.17)]. Moreover, since $\boldsymbol{\Sigma}$ is symmetric and nonnegative definite, $\boldsymbol{\Sigma} = \boldsymbol{\Gamma}'\boldsymbol{\Gamma}$ for some matrix $\boldsymbol{\Gamma}$, and upon partitioning $\boldsymbol{\Gamma}$ as $\boldsymbol{\Gamma} = (\boldsymbol{\Gamma}_1, \boldsymbol{\Gamma}_2)$ (where $\boldsymbol{\Gamma}_1$ has the same number of columns as $\boldsymbol{\Sigma}_{11}$), we find that $\boldsymbol{\Sigma}_{11} = \boldsymbol{\Gamma}_1'\boldsymbol{\Gamma}_1$, $\boldsymbol{\Sigma}_{12} = \boldsymbol{\Gamma}_1'\boldsymbol{\Gamma}_2$, and $\boldsymbol{\Sigma}_{22} = \boldsymbol{\Gamma}_2'\boldsymbol{\Gamma}_2$. And in light of results (2.12.1) and (2.12.2), it follows that

$$\boldsymbol{\Sigma}_{11}\mathbf{G}_1^*\boldsymbol{\Sigma}_{12}\mathbf{G}_2^*\boldsymbol{\Sigma}_{22} = \boldsymbol{\Gamma}_1'\boldsymbol{\Gamma}_1\mathbf{G}_1^*\boldsymbol{\Gamma}_1'\boldsymbol{\Gamma}_2\mathbf{G}_2^*\boldsymbol{\Gamma}_2'\boldsymbol{\Gamma}_2 = \boldsymbol{\Gamma}_1'\boldsymbol{\Gamma}_2 = \boldsymbol{\Sigma}_{12},$$

leading [in light of equality (S.18)] to the conclusion that $\boldsymbol{\Sigma}_{12} = \mathbf{0}$.

EXERCISE 30. Let $\mathbf{x}$ represent an M-dimensional random column vector that has an $N(\boldsymbol{\mu}, \mathbf{I}_M)$ distribution. And, for $i = 1, 2, \ldots, K$, take $q_i = c_i + \mathbf{b}_i'\mathbf{x} + \mathbf{x}'\mathbf{A}_i\mathbf{x}$, where c_i is a constant, $\mathbf{b}_i$ an M-dimensional column vector of constants, and $\mathbf{A}_i$ an $M \times M$ symmetric matrix of constants. Further, denote by $m(\cdot, \cdot, \ldots, \cdot)$ the moment generating function of the joint distribution of $q_1, q_2, \ldots, q_K$. Provide an alternative derivation of the "sufficiency part" of Theorem 6.8.3 by showing that if, for $j \neq i = 1, 2, \ldots, K$, $\mathbf{A}_i\mathbf{A}_j = \mathbf{0}$, $\mathbf{A}_i\mathbf{b}_j = \mathbf{0}$, and $\mathbf{b}_i'\mathbf{b}_j = 0$, then there exist (strictly) positive scalars $h_1, h_2, \ldots, h_K$ such that, for any scalars $t_1, t_2, \ldots, t_K$ for which $|t_1| < h_1$, $|t_2| < h_2$, ..., $|t_K| < h_K$,

$$m(t_1, t_2, \ldots, t_K) = m(t_1, 0, 0, \ldots, 0)\, m(0, t_2, 0, 0, \ldots, 0) \cdots m(0, \ldots, 0, 0, t_K).$$

Solution. For $i = 1, 2, \ldots, K$, let $\mathbf{d}_i = \mathbf{b}_i + 2\mathbf{A}_i\boldsymbol{\mu}$. And observe (in light of Lemma 6.5.2) that there exist (strictly) positive scalars $h_1, h_2, \ldots, h_K$ such that the matrix $\mathbf{I}_M - \sum_{i=1}^K t_i\mathbf{A}_i$ is positive definite for any scalars $t_1, t_2, \ldots, t_K$ for which $|t_1| < h_1$, $|t_2| < h_2$, ..., $|t_K| < h_K$. Accordingly, it follows from result (5.12) or (5.16) that, for any scalars $t_1, t_2, \ldots, t_K$ for which $|t_1| < h_1$, $|t_2| < h_2$, ..., $|t_K| < h_K$,

$$\begin{aligned} m(t_1, t_2, \ldots, t_K) = \big|\mathbf{I} - 2\textstyle\sum_i t_i\mathbf{A}_i\big|^{-1/2} \exp\big[\textstyle\sum_i t_i(c_i + \mathbf{b}_i'\boldsymbol{\mu} + \boldsymbol{\mu}'\mathbf{A}_i\boldsymbol{\mu})\big] \\ \times \exp\big[\tfrac{1}{2}\big(\textstyle\sum_i t_i\mathbf{d}_i\big)'\big(\mathbf{I} - 2\textstyle\sum_i t_i\mathbf{A}_i\big)^{-1}\textstyle\sum_i t_i\mathbf{d}_i\big]. \end{aligned}$$

Now, suppose that, for $j \neq i = 1, 2, \ldots, K$, $\mathbf{A}_i\mathbf{A}_j = \mathbf{0}$, $\mathbf{A}_i\mathbf{b}_j = \mathbf{0}$, and $\mathbf{b}_i'\mathbf{b}_j = 0$. Then,

$$\mathbf{I} - 2\textstyle\sum_i t_i\mathbf{A}_i = \prod_i (\mathbf{I} - 2t_i\mathbf{A}_i),$$

as can be readily verified by mathematical induction. Moreover, for $j \neq i = 1, 2, \ldots, K$,

$$\mathbf{d}_i = (\mathbf{I} - 2t_j\mathbf{A}_j)\mathbf{d}_i \quad \text{or, equivalently (provided } |t_j| < h_j\text{)}, \quad (\mathbf{I} - 2t_j\mathbf{A}_j)^{-1}\mathbf{d}_i = \mathbf{d}_i.$$

Thus, for $t_1, t_2, \ldots, t_K$ for which $|t_1| < h_1$, $|t_2| < h_2$, ..., $|t_K| < h_K$,

$$(\mathbf{I} - 2\textstyle\sum_j t_j\mathbf{A}_j)^{-1}\mathbf{d}_i = \prod_j (\mathbf{I} - 2t_j\mathbf{A}_j)^{-1}\mathbf{d}_i = (\mathbf{I} - 2t_i\mathbf{A}_i)^{-1}\mathbf{d}_i,$$

so that

$$\mathbf{d}_i'(\mathbf{I} - 2\textstyle\sum_j t_j\mathbf{A}_j)^{-1}\mathbf{d}_i = \mathbf{d}_i'(\mathbf{I} - 2t_i\mathbf{A}_i)^{-1}\mathbf{d}_i$$

and (if $s \neq i$)

$$\mathbf{d}_s'(\mathbf{I} - 2\textstyle\sum_j t_j\mathbf{A}_j)^{-1}\mathbf{d}_i = \mathbf{d}_s'(\mathbf{I} - 2t_i\mathbf{A}_i)^{-1}\mathbf{d}_i = [(\mathbf{I} - 2t_i\mathbf{A}_i)^{-1}\mathbf{d}_s]'\mathbf{d}_i = \mathbf{d}_s'\mathbf{d}_i = 0.$$

And it follows that

$$\begin{aligned} m(t_1, t_2, \ldots, t_K) &= \textstyle\prod_i |\mathbf{I} - 2t_i\mathbf{A}_i|^{-1/2} \prod_i \exp[t_i(c_i + \mathbf{b}_i'\boldsymbol{\mu} + \boldsymbol{\mu}'\mathbf{A}_i\boldsymbol{\mu}] \\ &\qquad \times \textstyle\prod_i \exp\big[\tfrac{1}{2}t_i^2\mathbf{d}_i'(\mathbf{I} - 2t_i\mathbf{A}_i)^{-1}\mathbf{d}_i\big] \\ &= m(t_1, 0, 0, \ldots, 0)\, m(0, t_2, 0, 0, \ldots, 0) \cdots m(0, \ldots, 0, 0, t_K). \end{aligned}$$

EXERCISE 31. Let $\mathbf{x}$ represent an M-dimensional random column vector that has an $N(\mathbf{0}, \boldsymbol{\Sigma})$ distribution. And take $\mathbf{A}_1$ and $\mathbf{A}_2$ to be $M \times M$ symmetric *nonnegative definite* matrices of constants. Show that the two quadratic forms $\mathbf{x}'\mathbf{A}_1\mathbf{x}$ and $\mathbf{x}'\mathbf{A}_2\mathbf{x}$ are statistically independent if and only if they are uncorrelated.

Solution. That $\mathbf{x}'\mathbf{A}_1\mathbf{x}$ and $\mathbf{x}'\mathbf{A}_2\mathbf{x}$ are uncorrelated if they are statistically independent is obvious—refer to Lemma 3.2.1. Now, for purposes of establishing the converse, suppose that $\mathbf{x}'\mathbf{A}_1\mathbf{x}$ and $\mathbf{x}'\mathbf{A}_2\mathbf{x}$ are uncorrelated. Then, in light of result (5.7.19) or as an application of condition (8.24), we have that

$$\operatorname{tr}(\mathbf{A}_1\boldsymbol{\Sigma}\mathbf{A}_2\boldsymbol{\Sigma}) = 0. \tag{S.19}$$

Moreover, $\mathbf{A}_1 = \mathbf{R}_1'\mathbf{R}_1$ and $\mathbf{A}_2 = \mathbf{R}_2'\mathbf{R}_2$ for some matrices $\mathbf{R}_1$ and $\mathbf{R}_2$. And making use of Lemma 2.3.1, we find that

$$\mathrm{tr}(\mathbf{A}_1\boldsymbol{\Sigma}\mathbf{A}_2\boldsymbol{\Sigma}) = \mathrm{tr}(\mathbf{R}_1'\mathbf{R}_1\boldsymbol{\Sigma}\mathbf{R}_2'\mathbf{R}_2\boldsymbol{\Sigma}) = \mathrm{tr}(\mathbf{R}_2\boldsymbol{\Sigma}\mathbf{R}_1'\mathbf{R}_1\boldsymbol{\Sigma}\mathbf{R}_2') = \mathrm{tr}[(\mathbf{R}_1\boldsymbol{\Sigma}\mathbf{R}_2')'\mathbf{R}_1\boldsymbol{\Sigma}\mathbf{R}_2']. \quad \text{(S.20)}$$

Together, results (S.19) and (S.20) imply (in light of Lemma 2.3.2) that

$$\mathbf{R}_1\boldsymbol{\Sigma}\mathbf{R}_2' = \mathbf{0},$$

and it follows that

$$\mathbf{A}_1\boldsymbol{\Sigma}\mathbf{A}_2 = \mathbf{R}_1'\mathbf{R}_1\boldsymbol{\Sigma}\mathbf{R}_2'\mathbf{R}_2 = \mathbf{0}$$

[and also that $\mathbf{A}_2\boldsymbol{\Sigma}\mathbf{A}_1 = [\mathbf{A}_1\boldsymbol{\Sigma}\mathbf{A}_2]' = \mathbf{0}$], leading (in light of Theorem 6.8.1) to the conclusion that $\mathbf{x}'\mathbf{A}_1\mathbf{x}$ and $\mathbf{x}'\mathbf{A}_2\mathbf{x}$ are statistically independent.

EXERCISE 32. Let $\mathbf{x}$ represent an M-dimensional random column vector that has an $N(\boldsymbol{\mu}, \mathbf{I}_M)$ distribution. Show (by producing an example) that there exist quadratic forms $\mathbf{x}'\mathbf{A}_1\mathbf{x}$ and $\mathbf{x}'\mathbf{A}_2\mathbf{x}$ (where $\mathbf{A}_1$ and $\mathbf{A}_2$ are $M \times M$ symmetric matrices of constants) that are uncorrelated for *every* $\boldsymbol{\mu} \in \mathcal{R}^M$ but that are not statistically independent for *any* $\boldsymbol{\mu} \in \mathcal{R}^M$.

Solution. Take $\mathbf{A}_1 = \mathrm{diag}(\mathbf{B}_1, \mathbf{0})$ and $\mathbf{A}_2 = \mathrm{diag}(\mathbf{B}_2, \mathbf{0})$, where

$$\mathbf{B}_1 = \begin{pmatrix} 1 & 1 \\ 1 & -1 \end{pmatrix} \quad \text{and} \quad \mathbf{B}_2 = \begin{pmatrix} 1 & -1 \\ -1 & -1 \end{pmatrix}.$$

Then,

$$\mathbf{A}_1\mathbf{A}_2 = \mathrm{diag}(\mathbf{C}_1, \mathbf{0}),$$

where

$$\mathbf{C}_1 = \mathbf{B}_1\mathbf{B}_2 = \begin{pmatrix} 0 & -2 \\ 2 & 0 \end{pmatrix}.$$

Thus, $\mathbf{A}_1\mathbf{A}_2 \neq \mathbf{0}$, implying (in light of Corollary 6.8.2) that $\mathbf{x}'\mathbf{A}_1\mathbf{x}$ and $\mathbf{x}'\mathbf{A}_2\mathbf{x}$ are not statistically independent for any $\boldsymbol{\mu} \in \mathcal{R}^M$. Now, consider the covariance of $\mathbf{x}'\mathbf{A}_1\mathbf{x}$ and $\mathbf{x}'\mathbf{A}_2\mathbf{x}$. Denoting by $\mu_1, \mu_2, \ldots, \mu_M$ the elements of $\boldsymbol{\mu}$ and applying formula (5.7.19), we find that

$$\begin{aligned} \mathrm{cov}(\mathbf{x}'\mathbf{A}_1\mathbf{x}, \ \mathbf{x}'\mathbf{A}_2\mathbf{x}) &= 2\,\mathrm{tr}(\mathbf{A}_1\mathbf{A}_2) + 4\boldsymbol{\mu}'\mathbf{A}_1\mathbf{A}_2\boldsymbol{\mu} \\ &= 0 + \begin{pmatrix} \mu_1 \\ \mu_2 \end{pmatrix}' \mathbf{C}_1 \begin{pmatrix} \mu_1 \\ \mu_2 \end{pmatrix} \\ &= 2\mu_2\mu_1 - 2\mu_1\mu_2 = 0, \end{aligned}$$

so that $\mathbf{x}'\mathbf{A}_1\mathbf{x}$ and $\mathbf{x}'\mathbf{A}_2\mathbf{x}$ are uncorrelated for every $\boldsymbol{\mu} \in \mathcal{R}^M$.

7

Confidence Intervals (or Sets) and Tests of Hypotheses

EXERCISE 1. Take the context to be that of Section 7.1 (where a second-order G–M model is applied to the results of an experimental study of the yield of lettuce plants for purposes of making inferences about the response surface and various of its characteristics). Assume that the second-order regression coefficients β_{11}, β_{12}, β_{13}, β_{22}, β_{23}, and β_{33} are such that the matrix $\mathbf{A}$ is nonsingular. Assume also that the distribution of the vector $\mathbf{e}$ of residual effects in the G–M model is MVN (in which case the matrix $\hat{\mathbf{A}}$ is nonsingular with probability 1). Show that the large-sample distribution of the estimator $\hat{\mathbf{u}}_0$ of the stationary point $\mathbf{u}_0$ of the response surface is MVN with mean vector $\mathbf{u}_0$ and variance-covariance matrix

$$(1/4)\,\mathbf{A}^{-1}\mathrm{var}(\hat{\mathbf{a}} + 2\hat{\mathbf{A}}\mathbf{u}_0)\,\mathbf{A}^{-1}. \tag{E.1}$$

Do so by applying standard results on multi-parameter maximum likelihood estimation—refer, e.g., to McCulloch, Searle, and Neuhaus (2008, sec. S.4) and Zacks (1971, chap. 5).

Solution. As discussed in Section 7.1, the matrix $\hat{\mathbf{A}}$ is nonsingular (with probability 1), and $\hat{\mathbf{u}}_0$ $[= -(1/2)\hat{\mathbf{A}}^{-1}\hat{\mathbf{a}}]$ is the ML estimator of $\mathbf{u}_0$. Define (column) vectors $\boldsymbol{\theta}$ and $\boldsymbol{\phi}$ of parameters as follows:

$$\boldsymbol{\theta} = \begin{pmatrix} \mathbf{a} \\ \beta_1 \\ \boldsymbol{\beta}_2 \end{pmatrix} \quad \text{and} \quad \boldsymbol{\phi} = \begin{pmatrix} \mathbf{u}_0 \\ \beta_1 \\ \boldsymbol{\beta}_2 \end{pmatrix}$$

[where $\boldsymbol{\beta}_2 = (\beta_{11}, \beta_{12}, \beta_{13}, \beta_{22}, \beta_{23}, \beta_{33})'$]. Further, letting $\mathbf{a}_1$, $\mathbf{a}_2$, and $\mathbf{a}_3$ represent the columns of $\mathbf{A}$, we find that

$$\mathbf{a} = -2\mathbf{A}\mathbf{u}_0 = -2\begin{pmatrix} \mathbf{u}_0'\mathbf{a}_1 \\ \mathbf{u}_0'\mathbf{a}_2 \\ \mathbf{u}_0'\mathbf{a}_3 \end{pmatrix} = -2\,\mathrm{diag}(\mathbf{u}_0', \mathbf{u}_0', \mathbf{u}_0')\begin{pmatrix} \mathbf{a}_1 \\ \mathbf{a}_2 \\ \mathbf{a}_3 \end{pmatrix} = \mathbf{H}\boldsymbol{\beta}_2, \tag{S.1}$$

where

$$\mathbf{H} = -\,\mathrm{diag}(\mathbf{u}_0', \mathbf{u}_0', \mathbf{u}_0')\begin{pmatrix} 2 & 0 & 0 & 0 & 0 & 0 \\ 0 & 1 & 0 & 0 & 0 & 0 \\ 0 & 0 & 1 & 0 & 0 & 0 \\ 0 & 1 & 0 & 0 & 0 & 0 \\ 0 & 0 & 0 & 2 & 0 & 0 \\ 0 & 0 & 0 & 0 & 1 & 0 \\ 0 & 0 & 1 & 0 & 0 & 0 \\ 0 & 0 & 0 & 0 & 1 & 0 \\ 0 & 0 & 0 & 0 & 0 & 2 \end{pmatrix}.$$

And letting $\hat{\boldsymbol{\beta}}_2 = (\hat{\beta}_{11}, \hat{\beta}_{12}, \hat{\beta}_{13}, \hat{\beta}_{22}, \hat{\beta}_{23}, \hat{\beta}_{33})'$, the ML estimators of $\boldsymbol{\theta}$ and $\boldsymbol{\phi}$ are respectively the vectors $\hat{\boldsymbol{\theta}}$ and $\hat{\boldsymbol{\phi}}$ defined as follows:

$$\hat{\boldsymbol{\theta}} = \begin{pmatrix} \hat{\mathbf{a}} \\ \hat{\beta}_1 \\ \hat{\boldsymbol{\beta}}_2 \end{pmatrix} \quad \text{and} \quad \hat{\boldsymbol{\phi}} = \begin{pmatrix} \hat{\mathbf{u}}_0 \\ \hat{\beta}_1 \\ \hat{\boldsymbol{\beta}}_2 \end{pmatrix}.$$

In light of standard results on multi-parameter ML estimation, it suffices to show that the variance-covariance matrix of the large-sample distribution of $\hat{\mathbf{u}}_0$ is given by expression (E.1). The vector $\boldsymbol{\theta}$

can be regarded as a (vector-valued) function of the vector $\boldsymbol{\phi}$; the Jacobian matrix of this function is the matrix

$$\frac{\partial \boldsymbol{\theta}}{\partial \boldsymbol{\phi}'} = \begin{pmatrix} -2\mathbf{A} & \mathbf{0} & \mathbf{H} \\ \mathbf{0}' & 1 & \mathbf{0}' \\ \mathbf{0} & \mathbf{0} & \mathbf{I} \end{pmatrix}$$

[as can be readily verified by using result (S.1) together with result (5.4.12)]. Accordingly, it follows from standard results on multi-parameter ML estimation that the variance-covariance matrix of the large-sample distribution of $\hat{\boldsymbol{\phi}}$ is

$$\left(\frac{\partial \boldsymbol{\theta}}{\partial \boldsymbol{\phi}'}\right)^{-1} \mathrm{var}(\hat{\boldsymbol{\theta}}) \left[\left(\frac{\partial \boldsymbol{\theta}}{\partial \boldsymbol{\phi}'}\right)^{-1}\right]'.$$

And upon applying Lemma 2.6.4, we find that

$$\left(\frac{\partial \boldsymbol{\theta}}{\partial \boldsymbol{\phi}'}\right)^{-1} = \begin{pmatrix} -\frac{1}{2}\mathbf{A}^{-1} & \mathbf{0} & \frac{1}{2}\mathbf{A}^{-1}\mathbf{H} \\ \mathbf{0}' & 1 & \mathbf{0}' \\ \mathbf{0} & \mathbf{0} & \mathbf{I} \end{pmatrix}.$$

Thus, the variance-covariance matrix of the large-sample distribution of $\hat{\mathbf{u}}_0$ is

$$\begin{aligned} (-\tfrac{1}{2}\mathbf{A}^{-1},\ \mathbf{0},\ \tfrac{1}{2}\mathbf{A}^{-1}\mathbf{H})\, \mathrm{var}(\hat{\boldsymbol{\theta}})\, (-\tfrac{1}{2}\mathbf{A}^{-1},\ \mathbf{0},\ \tfrac{1}{2}\mathbf{A}^{-1}\mathbf{H})' &= \mathrm{var}\big[(-\tfrac{1}{2}\mathbf{A}^{-1},\ \mathbf{0},\ \tfrac{1}{2}\mathbf{A}^{-1}\mathbf{H})\,\hat{\boldsymbol{\theta}}\big] \\ &= \mathrm{var}\big[-\tfrac{1}{2}\mathbf{A}^{-1}(\mathbf{I},\ \mathbf{0},\ -\mathbf{H})\,\hat{\boldsymbol{\theta}}\big] \\ &= (1/4)\,\mathbf{A}^{-1}\mathrm{var}(\hat{\mathbf{a}} - \mathbf{H}\hat{\boldsymbol{\beta}}_2)\,\mathbf{A}^{-1}. \end{aligned}$$

Moreover, analogous to the equality $\mathbf{H}\boldsymbol{\beta}_2 = -2\mathbf{A}\mathbf{u}_0$ [from result (S.1)], we have that

$$\mathbf{H}\hat{\boldsymbol{\beta}}_2 = -2\hat{\mathbf{A}}\mathbf{u}_0$$

and hence that

$$(1/4)\,\mathbf{A}^{-1}\mathrm{var}(\hat{\mathbf{a}} - \mathbf{H}\hat{\boldsymbol{\beta}}_2)\,\mathbf{A}^{-1} = (1/4)\,\mathbf{A}^{-1}\mathrm{var}(\hat{\mathbf{a}} + 2\hat{\mathbf{A}}\mathbf{u}_0)\,\mathbf{A}^{-1}.$$

EXERCISE 2. Taking the context to be that of Section 7.2a (and adopting the same notation and terminology as in Section 7.2a), show that for $\tilde{\mathbf{R}}'\mathbf{X}'\mathbf{y}$ to have the same expected value under the augmented G–M model as under the original G–M model, it is necessay (as well as sufficient) that $\mathbf{X}'\mathbf{Z} = \mathbf{0}$.

Solution. The vector $\tilde{\mathbf{R}}'\mathbf{X}'\mathbf{y}$ has the same expected value under the augmented G–M model as under the original G–M model if and only if

$$\tilde{\mathbf{R}}'\mathbf{X}'\mathbf{Z} = \mathbf{0}$$

[as is evident from expression (2.5)]. Thus, it suffices to show that

$$\tilde{\mathbf{R}}'\mathbf{X}'\mathbf{Z} = \mathbf{0} \ \Rightarrow\ \mathbf{X}'\mathbf{Z} = \mathbf{0}$$

—clearly, $\mathbf{X}'\mathbf{Z} = \mathbf{0} \Rightarrow \tilde{\mathbf{R}}'\mathbf{X}'\mathbf{Z} = \mathbf{0}$.

By definition, $\mathbf{X}'\mathbf{X}\tilde{\mathbf{R}} = \boldsymbol{\Lambda}$ and $\mathcal{R}(\boldsymbol{\Lambda}') = \mathcal{R}(\mathbf{X})$, implying (in light of Theorem 2.12.2) that

$$\tilde{\mathbf{R}}'\mathbf{X}'\mathbf{Z} = \tilde{\mathbf{R}}'\mathbf{X}'\mathbf{X}(\mathbf{X}'\mathbf{X})^{-}\mathbf{X}'\mathbf{Z} = \boldsymbol{\Lambda}'(\mathbf{X}'\mathbf{X})^{-}\mathbf{X}'\mathbf{Z}$$

and (in light of Lemma 2.4.3) that

$$\mathbf{X} = \mathbf{L}\boldsymbol{\Lambda}'$$

for some matrix $\mathbf{L}$. Thus,

$$\mathbf{L}\tilde{\mathbf{R}}'\mathbf{X}'\mathbf{Z} = \mathbf{L}\boldsymbol{\Lambda}'(\mathbf{X}'\mathbf{X})^{-}\mathbf{X}'\mathbf{Z} = \mathbf{X}(\mathbf{X}'\mathbf{X})^{-}\mathbf{X}'\mathbf{Z} = \mathbf{P}_{\mathbf{X}}\mathbf{Z},$$

so that

$$\tilde{\mathbf{R}}'\mathbf{X}'\mathbf{Z} = \mathbf{0} \ \Rightarrow\ \mathbf{L}\tilde{\mathbf{R}}'\mathbf{X}'\mathbf{Z} = \mathbf{0} \ \Leftrightarrow\ \mathbf{P}_{\mathbf{X}}\mathbf{Z} = \mathbf{0}$$

and hence [since (according to Theorem 2.12.2) $\mathbf{X}'\mathbf{P}_{\mathbf{X}} = \mathbf{X}'$]

$$\tilde{\mathbf{R}}'\mathbf{X}'\mathbf{Z} = \mathbf{0} \;\Rightarrow\; \mathbf{X}'\mathbf{P}_{\mathbf{X}}\mathbf{Z} = \mathbf{0} \;\Leftrightarrow\; \mathbf{X}'\mathbf{Z} = \mathbf{0}.$$

EXERCISE 3. Adopting the same notation and terminology as in Section 7.2, consider the expected value of the usual estimator $\mathbf{y}'(\mathbf{I}-\mathbf{P}_{\mathbf{X}})\mathbf{y}/[N-\text{rank}(\mathbf{X})]$ of the variance of the residual effects of the (original) G–M model $\mathbf{y} = \mathbf{X}\boldsymbol{\beta} + \mathbf{e}$. How is the expected value of this estimator affected when the model equation is augmented via the inclusion of the additional "term" $\mathbf{Z}\boldsymbol{\tau}$? That is, what is the expected value of this estimator when its expected value is determined under the augmented G–M model (rather than under the original G–M model)?

Solution. Suppose that the $N \times 1$ observable random vector $\mathbf{y}$ follows the augmented G–M model $\mathbf{y} = \mathbf{X}\boldsymbol{\beta} + \mathbf{Z}\boldsymbol{\tau} + \mathbf{e}$. Then,

$$\text{E}(\mathbf{y}) = \mathbf{X}\boldsymbol{\beta} + \mathbf{Z}\boldsymbol{\tau} \qquad \text{and} \qquad \text{var}(\mathbf{y}) = \sigma^2\mathbf{I}.$$

And upon applying formula (5.7.11) for the expected value of a quadratic form, we find that

$$\text{E}\left[\frac{\mathbf{y}'(\mathbf{I}-\mathbf{P}_{\mathbf{X}})\mathbf{y}}{N-\text{rank}(\mathbf{X})}\right] = \frac{\text{tr}[(\mathbf{I}-\mathbf{P}_{\mathbf{X}})(\sigma^2\mathbf{I})]+(\mathbf{X}\boldsymbol{\beta} + \mathbf{Z}\boldsymbol{\tau})'(\mathbf{I}-\mathbf{P}_{\mathbf{X}})(\mathbf{X}\boldsymbol{\beta}+\mathbf{Z}\boldsymbol{\tau})}{N-\text{rank}(\mathbf{X})}.$$

Moreover, in light of Lemma 2.8.4 and Theorem 2.12.2,

$$\text{tr}[(\mathbf{I}-\mathbf{P}_{\mathbf{X}})(\sigma^2\mathbf{I})] = \sigma^2\,\text{tr}(\mathbf{I}-\mathbf{P}_{\mathbf{X}}) = \sigma^2[N-\text{rank}(\mathbf{P}_{\mathbf{X}})] = \sigma^2[N-\text{rank}(\mathbf{X})]$$

and

$$(\mathbf{X}\boldsymbol{\beta} + \mathbf{Z}\boldsymbol{\tau})'(\mathbf{I}-\mathbf{P}_{\mathbf{X}})(\mathbf{X}\boldsymbol{\beta}+\mathbf{Z}\boldsymbol{\tau}) = \boldsymbol{\tau}'\mathbf{Z}'(\mathbf{I}-\mathbf{P}_{\mathbf{X}})\mathbf{Z}\boldsymbol{\tau}.$$

Thus,

$$\text{E}\left[\frac{\mathbf{y}'(\mathbf{I}-\mathbf{P}_{\mathbf{X}})\mathbf{y}}{N-\text{rank}(\mathbf{X})}\right] = \sigma^2 + \frac{\boldsymbol{\tau}'\mathbf{Z}'(\mathbf{I}-\mathbf{P}_{\mathbf{X}})\mathbf{Z}\boldsymbol{\tau}}{N-\text{rank}(\mathbf{X})}.$$

EXERCISE 4. Adopting the same notation and terminology as in Sections 7.1 and 7.2, regard the lettuce yields as the observed values of the N $(= 20)$ elements of the random column vector $\mathbf{y}$, and take the model to be the "reduced" model derived from the second-order G–M model (in the three variables Cu, Mo, and Fe) by deleting the four terms involving the variable Mo—such a model would be consistent with an assumption that Mo is "more-or-less inert," i.e., has no discernible effect on the yield of lettuce.

(a) Compute the values of the least squares estimators of the regression coefficients (β_1, β_2, β_4, β_{11}, β_{13}, and β_{33}) of the reduced model, and determine the standard errors, estimated standard errors, and correlation matrix of these estimators.

(b) Determine the expected values of the least squares estimators (of β_1, β_2, β_4, β_{11}, β_{13}, and β_{33}) from Part (a) under the complete second-order G–M model (i.e., the model that includes the 4 terms involving the variable Mo), and determine (on the basis of the complete model) the estimated standard errors of these estimators.

(c) Find four linearly independent linear combinations of the four deleted regression coefficients (β_3, β_{12}, β_{22}, and β_{23}) that, under the complete second-order G–M model, would be estimable and whose least squares estimators would be uncorrelated, each with a standard error of σ; and compute the values of the least squares estimators of these linearly independent linear combinations, and determine the estimated standard errors of the least squares estimators.

Solution. (a) The residual sum of squares equals 139.4432, and $N-P = 14$. Thus, the value of the usual unbiased estimator of σ^2 is $139.4432/14 = 9.96$, the square root of which equals 3.16. And the least squares estimates, the standard errors, and the estimated standard errors are as follows:

Coefficient of	Regression coefficient	Least squares estimate	Std. error of the estimator	Estimated std. error of the estimator
1	β_1	23.68	0.333σ	1.05
u_1	β_2	−4.57	0.279σ	0.88
u_3	β_4	−0.11	0.317σ	1.00
u_1^2	β_{11}	−5.18	0.265σ	0.84
u_1u_3	β_{13}	−1.29	0.462σ	1.46
u_3^2	β_{33}	−5.20	0.271σ	0.86

The correlation matrix of the vector of least squares estimators [which is expressible as $\mathbf{S}^{-1}(\mathbf{X}'\mathbf{X})^{-1}\mathbf{S}^{-1}$, where $\mathbf{S}$ is the diagonal matrix of order $P = 6$ whose first through Pth diagonal elements are respectively the square roots of the first through Pth diagonal elements of $(\mathbf{X}'\mathbf{X})^{-1}$] is

$$\begin{pmatrix} 1.00 & 0.00 & 0.04 & -0.61 & 0.00 & -0.51 \\ 0.00 & 1.00 & 0.00 & 0.00 & -0.24 & 0.00 \\ 0.04 & 0.00 & 1.00 & -0.09 & 0.00 & -0.21 \\ -0.61 & 0.00 & -0.09 & 1.00 & 0.00 & 0.17 \\ 0.00 & -0.24 & 0.00 & 0.00 & 1.00 & 0.00 \\ -0.51 & 0.00 & -0.21 & 0.17 & 0.00 & 1.00 \end{pmatrix}.$$

(b) Let $\tilde{\beta}_1$, $\tilde{\beta}_2$, $\tilde{\beta}_4$, $\tilde{\beta}_{11}$, $\tilde{\beta}_{13}$, and $\tilde{\beta}_{33}$ represent the least squares estimators (of β_1, β_2, β_4, β_{11}, β_{13}, and β_{33}) from Part (a). Then, taking $\mathbf{X}$ to be the model matrix of the reduced second-order G–M model and applying result (2.5) with $\tilde{\mathbf{R}} = (\mathbf{X}'\mathbf{X})^{-1}$ (corresponding to $\mathbf{\Lambda} = \mathbf{I}$), we find that, under the complete second-order G–M model,

$$\begin{aligned} \mathrm{E}(\tilde{\beta}_1) &= \beta_1 + 0.861\beta_{22}, \\ \mathrm{E}(\tilde{\beta}_2) &= \beta_2, \\ \mathrm{E}(\tilde{\beta}_4) &= \beta_4 + 0.114\beta_{22}, \\ \mathrm{E}(\tilde{\beta}_{11}) &= \beta_{11} - 0.126\beta_{22}, \\ \mathrm{E}(\tilde{\beta}_{13}) &= \beta_{13}, \qquad \text{and} \\ \mathrm{E}(\tilde{\beta}_{33}) &= \beta_{33} - 0.195\beta_{22}. \end{aligned}$$

Moreover, when the model is taken to be the complete second-order model, the value obtained for the usual estimator of σ^2 is (as determined in Section 7.2c) 10.89, the square root of which equals 3.30. Accordingly, the estimated value of σ is 4.6% higher than the estimated value (3.16) obtained when the model is taken to be the reduced second-order model. And (when determined under the complete second-order model) the estimated standard errors of $\tilde{\beta}_1$, $\tilde{\beta}_2$, $\tilde{\beta}_4$, $\tilde{\beta}_{11}$, $\tilde{\beta}_{13}$, and $\tilde{\beta}_{33}$ are 1.10, 0.92, 1.05, 0.87, 1.53, and 0.89, respectively, which are 4.6% higher than those determined [in Part (a)] under the reduced model.

(c) Take $\mathbf{X}$ to be the 20×6 matrix whose first through sixth columns are the columns of the model matrix of the complete second-order G–M model corresponding to the parameters β_1, β_2, β_4, β_{11}, β_{13}, and β_{33}, and take $\mathbf{Z}$ to be the 20×4 matrix whose first through fourth columns are the columns of the model matrix of the complete second-order model corresponding to the parameters β_3, β_{12}, β_{22}, and β_{23}. Further, take $(\mathbf{X}, \mathbf{Z}) = \mathbf{OU}$ to be the QR decomposition of the partitioned matrix $(\mathbf{X}, \mathbf{Z})$ (so that $\mathbf{O}$ is a 20×10 matrix with orthonormal columns and $\mathbf{U}$ a 10×10 upper triangular matrix with positive diagonal elements), define $\boldsymbol{\tau} = (\beta_3, \beta_{12}, \beta_{22}, \beta_{23})'$, and partition $\mathbf{O}$ as $\mathbf{O} = (\mathbf{O}_1, \mathbf{O}_2)$ and $\mathbf{U}$ as $\mathbf{U} = \begin{pmatrix} \mathbf{U}_{11} & \mathbf{U}_{12} \\ \mathbf{0} & \mathbf{U}_{22} \end{pmatrix}$ (where $\mathbf{O}_2$ is of dimensions 20×4 and $\mathbf{U}_{22}$ of dimensions 4×4). Then, the elements of the vector $\mathbf{U}_{22}\boldsymbol{\tau}$ are linearly independent linear combinations of β_3, β_{12}, β_{22}, and β_{23}. And (when the model is taken to be the complete second-order model) these linear combinations are estimable, their least squares estimators equal the (corresponding) elements of the vector $\mathbf{O}_2'\mathbf{y}$ (and are uncorrelated, each with a standard error of σ), and the estimated standard errors of the least squares estimators equal the estimate of the standard deviation σ. The elements of $\mathbf{U}_{22}\boldsymbol{\tau}$ are found to be

$$3.696\beta_3 + 0.545\beta_{23}, \quad 2.828\beta_{12}, \quad 3.740\beta_{22}, \quad \text{and} \quad 2.165\beta_{23},$$

and their least squares estimates are found to be

$$-2.68, \quad -3.72, \quad -2.74, \quad \text{and} \quad 1.43,$$

respectively; the estimated standard errors of the least squares estimators equal 3.30.

EXERCISE 5. Suppose that $\mathbf{y}$ is an $N \times 1$ observable random vector that follows the G–M model. Show that

$$\mathrm{E}(\mathbf{y}) = \mathbf{X}\tilde{\mathbf{R}}\mathbf{S}\boldsymbol{\alpha} + \mathbf{U}\boldsymbol{\eta},$$

where $\tilde{\mathbf{R}}$, $\mathbf{S}$, $\mathbf{U}$, $\boldsymbol{\alpha}$, and $\boldsymbol{\eta}$ are as defined in Section 7.3a.

Solution. Employing the same notation as in Section 7.3a and making use of result (3.17), we find that

$$\begin{aligned}
\mathrm{E}(\mathbf{y}) &= \mathrm{E}(\mathbf{O}\mathbf{z}) \\
&= \mathbf{O}\,\mathrm{E}(\mathbf{z}) \\
&= (\mathbf{X}\tilde{\mathbf{R}}\mathbf{S},\ \mathbf{U},\ \mathbf{L}) \begin{pmatrix} \boldsymbol{\alpha} \\ \boldsymbol{\eta} \\ \mathbf{0} \end{pmatrix} \\
&= \mathbf{X}\tilde{\mathbf{R}}\mathbf{S}\boldsymbol{\alpha} + \mathbf{U}\boldsymbol{\eta}.
\end{aligned}$$

EXERCISE 6. Taking the context to be that of Section 7.3, adopting the notation employed therein, supposing that the distribution of the vector $\mathbf{e}$ of residual effects (in the G–M model) is MVN, and assuming that $N > P_*+2$, show that

$$\begin{aligned}
\mathrm{E}[\tilde{F}(\boldsymbol{\alpha}^{(0)})] &= \frac{N-P_*}{N-P_*-2}\left[1 + \frac{(\boldsymbol{\alpha}-\boldsymbol{\alpha}^{(0)})'(\boldsymbol{\alpha}-\boldsymbol{\alpha}^{(0)})}{M_*\sigma^2}\right] \\
&= \frac{N-P_*}{N-P_*-2}\left[1 + \frac{(\boldsymbol{\tau}-\boldsymbol{\tau}^{(0)})'\mathbf{C}^-(\boldsymbol{\tau}-\boldsymbol{\tau}^{(0)})}{M_*\sigma^2}\right].
\end{aligned}$$

Solution. According to result (3.46), $\tilde{F}(\boldsymbol{\alpha}^{(0)}) \sim SF[M_*,\ N-P_*,\ (1/\sigma^2)(\boldsymbol{\alpha}-\boldsymbol{\alpha}^{(0)})'(\boldsymbol{\alpha}-\boldsymbol{\alpha}^{(0)})]$. Thus, as an application of formula (6.3.31), we have that

$$\mathrm{E}[\tilde{F}(\boldsymbol{\alpha}^{(0)})] = \frac{N-P_*}{N-P_*-2}\left[1 + \frac{(\boldsymbol{\alpha}-\boldsymbol{\alpha}^{(0)})'(\boldsymbol{\alpha}-\boldsymbol{\alpha}^{(0)})}{M_*\sigma^2}\right].$$

And upon applying result (3.34) (with $\dot{\boldsymbol{\tau}} = \boldsymbol{\tau}^{(0)}$) or result (3.48), we find that

$$1 + \frac{(\boldsymbol{\alpha}-\boldsymbol{\alpha}^{(0)})'(\boldsymbol{\alpha}-\boldsymbol{\alpha}^{(0)})}{M_*\sigma^2} = 1 + \frac{(\boldsymbol{\tau}-\boldsymbol{\tau}^{(0)})'\mathbf{C}^-(\boldsymbol{\tau}-\boldsymbol{\tau}^{(0)})}{M_*\sigma^2}.$$

EXERCISE 7. Take the context to be that of Section 7.3, adopt the notation employed therein, and suppose that the distribution of the vector $\mathbf{e}$ of residual effects (in the G–M model) is MVN. For $\dot{\boldsymbol{\tau}} \in \mathcal{C}(\boldsymbol{\Lambda}')$, the distribution of $F(\dot{\boldsymbol{\tau}})$ is obtainable (upon setting $\dot{\boldsymbol{\alpha}} = \mathbf{S}'\dot{\boldsymbol{\tau}}$) from that of $\tilde{F}(\dot{\boldsymbol{\alpha}})$: in light of the relationship (3.32) and results (3.24) and (3.34),

$$F(\dot{\boldsymbol{\tau}}) \sim SF[M_*,\ N-P_*,\ (1/\sigma^2)(\dot{\boldsymbol{\tau}}-\boldsymbol{\tau})'\mathbf{C}^-(\dot{\boldsymbol{\tau}}-\boldsymbol{\tau})].$$

Provide an alternative derivation of the distribution of $F(\dot{\boldsymbol{\tau}})$ by (1) taking $\mathbf{b}$ to be a $P \times 1$ vector such that $\dot{\boldsymbol{\tau}} = \boldsymbol{\Lambda}'\mathbf{b}$ and establishing that $F(\dot{\boldsymbol{\tau}})$ is expressible in the form

$$F(\dot{\tau}) = \frac{(1/\sigma^2)(\mathbf{y}-\mathbf{Xb})'\mathbf{P}_{\mathbf{X}\tilde{\mathbf{R}}}(\mathbf{y}-\mathbf{Xb})/M_*}{(1/\sigma^2)(\mathbf{y}-\mathbf{Xb})'(\mathbf{I}-\mathbf{P}_{\mathbf{X}})(\mathbf{y}-\mathbf{Xb})/(N-P_*)}$$

and by (2) regarding $(1/\sigma^2)(\mathbf{y}-\mathbf{Xb})'\mathbf{P}_{\mathbf{X}\tilde{\mathbf{R}}}(\mathbf{y}-\mathbf{Xb})$ and $(1/\sigma^2)(\mathbf{y}-\mathbf{Xb})'(\mathbf{I}-\mathbf{P}_{\mathbf{X}})(\mathbf{y}-\mathbf{Xb})$ as quadratic forms (in $\mathbf{y}-\mathbf{Xb}$) and making use of Corollaries 6.6.4 and 6.8.2.

Solution. (1) By definition,

$$F(\dot{\tau}) = \frac{(\hat{\tau}-\dot{\tau})'\mathbf{C}^-(\hat{\tau}-\dot{\tau})/M_*}{\mathbf{y}'(\mathbf{I}-\mathbf{P}_{\mathbf{X}})\mathbf{y}/(N-P_*)}.$$

And upon taking $\mathbf{b}$ to be a $P \times 1$ vector such that $\dot{\tau} = \mathbf{\Lambda}'\mathbf{b}$ and recalling (from Section 7.3a) that $\mathbf{X}'\mathbf{X}\tilde{\mathbf{R}} = \mathbf{\Lambda}$, that $\hat{\tau} = \tilde{\mathbf{R}}'\mathbf{X}'\mathbf{y}$, and that $\mathbf{C} = \tilde{\mathbf{R}}'\mathbf{X}'\mathbf{X}\tilde{\mathbf{R}}$ [$= (\mathbf{X}\tilde{\mathbf{R}})'\mathbf{X}\tilde{\mathbf{R}}$], we find that

$$\hat{\tau}-\dot{\tau} = \tilde{\mathbf{R}}'\mathbf{X}'\mathbf{y} - (\mathbf{X}'\mathbf{X}\tilde{\mathbf{R}})'\mathbf{b} = \tilde{\mathbf{R}}'\mathbf{X}'(\mathbf{y}-\mathbf{Xb})$$

and hence that

$$(\hat{\tau}-\dot{\tau})'\mathbf{C}^-(\hat{\tau}-\dot{\tau}) = (\mathbf{y}-\mathbf{Xb})'\mathbf{X}\tilde{\mathbf{R}}\mathbf{C}^-(\mathbf{X}\tilde{\mathbf{R}})'(\mathbf{y}-\mathbf{Xb}) = (\mathbf{y}-\mathbf{Xb})'\mathbf{P}_{\mathbf{X}\tilde{\mathbf{R}}}(\mathbf{y}-\mathbf{Xb}).$$

Moreover, since $(\mathbf{I}-\mathbf{P}_{\mathbf{X}})\mathbf{X} = \mathbf{0}$ and $\mathbf{X}'(\mathbf{I}-\mathbf{P}_{\mathbf{X}}) = \mathbf{0}$ (as is evident from Theorem 2.12.2),

$$\mathbf{y}'(\mathbf{I}-\mathbf{P}_{\mathbf{X}})\mathbf{y} = (\mathbf{y}-\mathbf{Xb})'(\mathbf{I}-\mathbf{P}_{\mathbf{X}})(\mathbf{y}-\mathbf{Xb}).$$

Thus,

$$\begin{aligned} F(\dot{\tau}) &= \frac{(\mathbf{y}-\mathbf{Xb})'\mathbf{P}_{\mathbf{X}\tilde{\mathbf{R}}}(\mathbf{y}-\mathbf{Xb})/M_*}{(\mathbf{y}-\mathbf{Xb})'(\mathbf{I}-\mathbf{P}_{\mathbf{X}})(\mathbf{y}-\mathbf{Xb})/(N-P_*)} \\ &= \frac{(1/\sigma^2)(\mathbf{y}-\mathbf{Xb})'\mathbf{P}_{\mathbf{X}\tilde{\mathbf{R}}}(\mathbf{y}-\mathbf{Xb})/M_*}{(1/\sigma^2)(\mathbf{y}-\mathbf{Xb})'(\mathbf{I}-\mathbf{P}_{\mathbf{X}})(\mathbf{y}-\mathbf{Xb})/(N-P_*)}. \end{aligned} \tag{S.2}$$

(2) We have that $\mathbf{e} \sim N(\mathbf{0}, \sigma^2\mathbf{I})$, so that $\mathbf{y} \sim N(\mathbf{X}\boldsymbol{\beta}, \sigma^2\mathbf{I})$ and hence $\mathbf{y}-\mathbf{Xb} \sim N[\mathbf{X}(\boldsymbol{\beta}-\mathbf{b}), \sigma^2\mathbf{I}]$. Thus, upon regarding $(1/\sigma^2)(\mathbf{y}-\mathbf{Xb})'\mathbf{P}_{\mathbf{X}\tilde{\mathbf{R}}}(\mathbf{y}-\mathbf{Xb})$ and $(1/\sigma^2)(\mathbf{y}-\mathbf{Xb})'(\mathbf{I}-\mathbf{P}_{\mathbf{X}})(\mathbf{y}-\mathbf{Xb})$ as quadratic forms (in $\mathbf{y}-\mathbf{Xb}$) and upon observing (in light of Theorem 2.12.2) that

$$\begin{aligned} [(1/\sigma^2)\mathbf{P}_{\mathbf{X}\tilde{\mathbf{R}}}](\sigma^2\mathbf{I})(1/\sigma^2)(\mathbf{I}-\mathbf{P}_{\mathbf{X}}) &= (1/\sigma^2)\mathbf{X}\tilde{\mathbf{R}}\mathbf{C}^-\tilde{\mathbf{R}}'\mathbf{X}'(\mathbf{I}-\mathbf{P}_{\mathbf{X}}) = \mathbf{0}, \\ [(1/\sigma^2)\mathbf{P}_{\mathbf{X}\tilde{\mathbf{R}}}](\sigma^2\mathbf{I})(1/\sigma^2)\mathbf{P}_{\mathbf{X}\tilde{\mathbf{R}}} &= (1/\sigma^2)\mathbf{P}_{\mathbf{X}\tilde{\mathbf{R}}}, \quad \text{and} \\ [(1/\sigma^2)(\mathbf{I}-\mathbf{P}_{\mathbf{X}})](\sigma^2\mathbf{I})(1/\sigma^2)(\mathbf{I}-\mathbf{P}_{\mathbf{X}}) &= (1/\sigma^2)(\mathbf{I}-\mathbf{P}_{\mathbf{X}}), \end{aligned}$$

it follows from Corollary 6.8.2 that $(1/\sigma^2)(\mathbf{y}-\mathbf{Xb})'\mathbf{P}_{\mathbf{X}\tilde{\mathbf{R}}}(\mathbf{y}-\mathbf{Xb})$ and $(1/\sigma^2)(\mathbf{y}-\mathbf{Xb})'(\mathbf{I}-\mathbf{P}_{\mathbf{X}})(\mathbf{y}-\mathbf{Xb})$ are statistically independent and from Corollary 6.6.4 that

$$(1/\sigma^2)(\mathbf{y}-\mathbf{Xb})'\mathbf{P}_{\mathbf{X}\tilde{\mathbf{R}}}(\mathbf{y}-\mathbf{Xb}) \sim \chi^2[\text{rank}(\mathbf{P}_{\mathbf{X}\tilde{\mathbf{R}}}),\ (1/\sigma^2)(\boldsymbol{\beta}-\mathbf{b})'\mathbf{X}'\mathbf{P}_{\mathbf{X}\tilde{\mathbf{R}}}\mathbf{X}(\boldsymbol{\beta}-\mathbf{b})]$$

and

$$(1/\sigma^2)(\mathbf{y}-\mathbf{Xb})'(\mathbf{I}-\mathbf{P}_{\mathbf{X}})(\mathbf{y}-\mathbf{Xb}) \sim \chi^2[\text{rank}(\mathbf{I}-\mathbf{P}_{\mathbf{X}}),\ (1/\sigma^2)(\boldsymbol{\beta}-\mathbf{b})'\mathbf{X}'(\mathbf{I}-\mathbf{P}_{\mathbf{X}})\mathbf{X}(\boldsymbol{\beta}-\mathbf{b})].$$

Moreover, in light of Theorem 2.12.2, result (3.1), and Lemma 2.8.4,

$$\text{rank}(\mathbf{P}_{\mathbf{X}\tilde{\mathbf{R}}}) = \text{rank}(\mathbf{X}\tilde{\mathbf{R}}) = M_*,$$

$$\text{rank}(\mathbf{I}-\mathbf{P}_{\mathbf{X}}) = N - \text{rank}(\mathbf{P}_{\mathbf{X}}) = N - P_*,$$

$$\begin{aligned} (1/\sigma^2)(\boldsymbol{\beta}-\mathbf{b})'\mathbf{X}'\mathbf{P}_{\mathbf{X}\tilde{\mathbf{R}}}\mathbf{X}(\boldsymbol{\beta}-\mathbf{b}) &= (1/\sigma^2)(\boldsymbol{\beta}-\mathbf{b})'\mathbf{X}'\mathbf{X}\tilde{\mathbf{R}}\mathbf{C}^-(\mathbf{X}\tilde{\mathbf{R}})'\mathbf{X}(\boldsymbol{\beta}-\mathbf{b}) \\ &= (1/\sigma^2)(\boldsymbol{\beta}-\mathbf{b})'\mathbf{\Lambda}\mathbf{C}^-\mathbf{\Lambda}'(\boldsymbol{\beta}-\mathbf{b}) \\ &= (1/\sigma^2)(\dot{\tau}-\tau)'\mathbf{C}^-(\dot{\tau}-\tau), \end{aligned}$$

and

$$(1/\sigma^2)(\boldsymbol{\beta}-\mathbf{b})'\mathbf{X}'(\mathbf{I}-\mathbf{P}_{\mathbf{X}})\mathbf{X}(\boldsymbol{\beta}-\mathbf{b}) = 0.$$

And in combination with result (S.2), these results imply that

$$F(\dot{\tau}) \sim SF[M_*,\, N-P_*,\, (1/\sigma^2)(\dot{\tau}-\tau)'\mathbf{C}^-(\dot{\tau}-\tau)].$$

EXERCISE 8. Take the context to be that of Section 7.3, and adopt the notation employed therein. Taking the model to be the canonical form of the G–M model and taking the distribution of the vector of residual effects to be $N(\mathbf{0}, \sigma^2\mathbf{I})$, derive (in terms of the transformed vector $\mathbf{z}$) the size-$\dot{\gamma}$ likelihood ratio test of the null hypothesis $\tilde{H}_0 : \boldsymbol{\alpha} = \boldsymbol{\alpha}^{(0)}$ (versus the alternative hypothesis $\tilde{H}_1 : \boldsymbol{\alpha} \neq \boldsymbol{\alpha}^{(0)}$)—refer, e.g., to Casella and Berger (2002, sec. 8.2) for a discussion of likelihood ratio tests. Show that the size-$\dot{\gamma}$ likelihood ratio test is identical to the size-$\dot{\gamma}$ F test.

Solution. The distribution of $\mathbf{z}$ is $N\left[\begin{pmatrix}\boldsymbol{\alpha}\\ \boldsymbol{\eta}\\ \mathbf{0}\end{pmatrix},\, \sigma^2\mathbf{I}\right]$. Accordingly, the likelihood function, say $\ell(\boldsymbol{\alpha}, \boldsymbol{\eta}, \sigma;\, \mathbf{z})$, is expressible as follows:

$$\ell(\boldsymbol{\alpha}, \boldsymbol{\eta}, \sigma;\, \mathbf{z}) = (2\pi\sigma^2)^{-N/2}\exp\{-[1/(2\sigma^2)][(\hat{\boldsymbol{\alpha}}-\boldsymbol{\alpha})'(\hat{\boldsymbol{\alpha}}-\boldsymbol{\alpha}) + (\hat{\boldsymbol{\eta}}-\boldsymbol{\eta})'(\hat{\boldsymbol{\eta}}-\boldsymbol{\eta}) + \mathbf{d}'\mathbf{d}]\}.$$

Upon recalling the results of Section 5.9a, it becomes clear that $\ell(\boldsymbol{\alpha}, \boldsymbol{\eta}, \sigma;\, \mathbf{z})$ attains its maximum value (with respect to $\boldsymbol{\alpha}$, $\boldsymbol{\eta}$, and σ) at $\boldsymbol{\alpha} = \hat{\boldsymbol{\alpha}}$, $\boldsymbol{\eta} = \hat{\boldsymbol{\eta}}$, and $\sigma = (\mathbf{d}'\mathbf{d}/N)^{1/2}$ and that $\ell(\boldsymbol{\alpha}^{(0)}, \boldsymbol{\eta}, \sigma;\, \mathbf{z})$ attains its maximum value (with respect to $\boldsymbol{\eta}$ and σ) at $\boldsymbol{\eta} = \hat{\boldsymbol{\eta}}$ and $\sigma = \{[(\hat{\boldsymbol{\alpha}}-\boldsymbol{\alpha}^{(0)})'(\hat{\boldsymbol{\alpha}}-\boldsymbol{\alpha}^{(0)}) + \mathbf{d}'\mathbf{d}]/N\}^{1/2}$. Thus, the likelihood ratio test statistic is

$$\frac{\max_{\boldsymbol{\eta},\sigma} \ell(\boldsymbol{\alpha}^{(0)}, \boldsymbol{\eta}, \sigma;\, \mathbf{z})}{\max_{\boldsymbol{\alpha},\boldsymbol{\eta},\sigma} \ell(\boldsymbol{\alpha}, \boldsymbol{\eta}, \sigma;\, \mathbf{z})} = \left[\frac{\mathbf{d}'\mathbf{d}}{(\hat{\boldsymbol{\alpha}}-\boldsymbol{\alpha}^{(0)})'(\hat{\boldsymbol{\alpha}}-\boldsymbol{\alpha}^{(0)}) + \mathbf{d}'\mathbf{d}}\right]^{N/2}$$

$$= \left[1 + \frac{(\hat{\boldsymbol{\alpha}}-\boldsymbol{\alpha}^{(0)})'(\hat{\boldsymbol{\alpha}}-\boldsymbol{\alpha}^{(0)})}{\mathbf{d}'\mathbf{d}}\right]^{-N/2}.$$

This statistic is a decreasing function of $(\hat{\boldsymbol{\alpha}}-\boldsymbol{\alpha}^{(0)})'(\hat{\boldsymbol{\alpha}}-\boldsymbol{\alpha}^{(0)})/\mathbf{d}'\mathbf{d}$ and, since

$$\tilde{F}(\boldsymbol{\alpha}^{(0)}) = [(N-P_*)/M_*]\,(\hat{\boldsymbol{\alpha}}-\boldsymbol{\alpha}^{(0)})'(\hat{\boldsymbol{\alpha}}-\boldsymbol{\alpha}^{(0)})/\mathbf{d}'\mathbf{d},$$

a decreasing function of $\tilde{F}(\boldsymbol{\alpha}^{(0)})$. Thus, the size-$\dot{\gamma}$ likelihood ratio test is identical to the size-$\dot{\gamma}$ F test.

EXERCISE 9. Verify result (3.67).

Solution. Let us employ the same notation as in Sections 7.3a and 7.3b. Making use of equality (3.64) and recalling [from result (3.8)] that $\boldsymbol{\tau} = \mathbf{W}'\boldsymbol{\alpha}$, we find that

$$\tilde{F}_1(\boldsymbol{\alpha}) = \mathbf{S}'F_1(\mathbf{W}'\boldsymbol{\alpha}) = \mathbf{S}'F_1(\boldsymbol{\tau}).$$

Thus,

$$\mathbf{O}\begin{bmatrix}\tilde{F}_1(\boldsymbol{\alpha})\\ \tilde{F}_2(\boldsymbol{\eta})\\ \mathbf{0}\end{bmatrix} = \mathbf{X}\tilde{\mathbf{R}}\mathbf{S}\tilde{F}_1(\boldsymbol{\alpha}) + \mathbf{U}\tilde{F}_2(\boldsymbol{\eta}) = \mathbf{X}\tilde{\mathbf{R}}\mathbf{S}\mathbf{S}'F_1(\boldsymbol{\tau}) + \mathbf{U}\tilde{F}_2(\boldsymbol{\eta}).$$

Moreover, as noted earlier [and as is apparent from result (3.29)], $\mathbf{S}\mathbf{S}'$ is a generalized inverse of the matrix $\mathbf{C}$.

It remains only to establish that the value of the product $\mathbf{X}\tilde{\mathbf{R}}\mathbf{C}^-F_1(\boldsymbol{\tau})$ is invariant to the choice of the generalized inverse $\mathbf{C}^-$. For any $M \times 1$ vector $\dot{\boldsymbol{\tau}}$ and, in particular, for $\dot{\boldsymbol{\tau}} = \boldsymbol{\tau}$, $F_1(\dot{\boldsymbol{\tau}}) \in \mathcal{C}(\boldsymbol{\Lambda}')$ [which in light of result (3.6), is consistent with expression (3.63)], implying that $F_1(\boldsymbol{\tau}) = \boldsymbol{\Lambda}'\mathbf{b}$ for some (column) vector $\mathbf{b}$. And upon recalling that $\mathbf{C} = \tilde{\mathbf{R}}'\mathbf{X}'\mathbf{X}\tilde{\mathbf{R}}$ and that $\boldsymbol{\Lambda} = \mathbf{X}'\mathbf{X}\tilde{\mathbf{R}}$, we find that

$$\mathbf{X}\tilde{\mathbf{R}}\mathbf{C}^-F_1(\boldsymbol{\tau}) = \mathbf{X}\tilde{\mathbf{R}}\mathbf{C}^-\boldsymbol{\Lambda}'\mathbf{b} = \mathbf{X}\tilde{\mathbf{R}}\mathbf{C}^-(\mathbf{X}\tilde{\mathbf{R}})'\mathbf{X}\mathbf{b}.$$

Moreover, $\mathbf{X\tilde{R}C^-(X\tilde{R})'}$ is invariant to the choice of the generalized inverse $\mathbf{C}^-$, as is evident upon applying Part (3) of Theorem 2.12.2 (with $\mathbf{X\tilde{R}}$ in place of $\mathbf{X}$).

EXERCISE 10. Verify the equivalence of conditions (3.59) and (3.68) and the equivalence of conditions (3.61) and (3.69).

Solution. Let $\dot{\boldsymbol{\tau}}$ represent an arbitrary vector in $\mathcal{C}(\boldsymbol{\Lambda}')$. And recalling [from result (3.6)] that $\mathcal{C}(\boldsymbol{\Lambda}') = \mathcal{C}(\mathbf{W}')$, observe that there exists an $M_*\times 1$ vector $\dot{\boldsymbol{\alpha}}$ such that $\dot{\boldsymbol{\tau}} = \mathbf{W}'\dot{\boldsymbol{\alpha}}$. Moreover, since [according to result (3.6)] $\mathbf{WS} = \mathbf{I}$,

$$\dot{\boldsymbol{\alpha}} = (\mathbf{WS})'\dot{\boldsymbol{\alpha}} = \mathbf{S}'\dot{\boldsymbol{\tau}} \quad \text{and} \quad \dot{\boldsymbol{\tau}} = \mathbf{W}'\mathbf{S}'\dot{\boldsymbol{\tau}}.$$

Now, suppose that $\tilde{F}_1(\dot{\boldsymbol{\alpha}}) = \dot{\boldsymbol{\alpha}}$, as would be the case if condition (3.59) is satisfied. Then, $\tilde{F}_1(\mathbf{S}'\dot{\boldsymbol{\tau}}) = \mathbf{S}'\dot{\boldsymbol{\tau}}$, so that [in light of relationship (3.63)]

$$F_1(\dot{\boldsymbol{\tau}}) = \mathbf{W}'\tilde{F}_1(\mathbf{S}'\dot{\boldsymbol{\tau}}) = \mathbf{W}'\mathbf{S}'\dot{\boldsymbol{\tau}} = \dot{\boldsymbol{\tau}}.$$

Conversely, let $\dot{\boldsymbol{\alpha}}$ represent an arbitrary $M_* \times 1$ vector. Further, let $\dot{\boldsymbol{\tau}} = \mathbf{W}'\dot{\boldsymbol{\alpha}}$, in which case $\dot{\boldsymbol{\tau}} \in \mathcal{C}(\mathbf{W}') = \mathcal{C}(\boldsymbol{\Lambda}')$. And supposing that $F_1(\dot{\boldsymbol{\tau}}) = \dot{\boldsymbol{\tau}}$ [as would be the case if condition (3.68) is satisfied], we find that $F_1(\mathbf{W}'\dot{\boldsymbol{\alpha}}) = \mathbf{W}'\dot{\boldsymbol{\alpha}}$, so that [in light of relationship (3.64)]

$$\tilde{F}_1(\dot{\boldsymbol{\alpha}}) = \mathbf{S}'F_1(\mathbf{W}'\dot{\boldsymbol{\alpha}}) = \mathbf{S}'\mathbf{W}'\dot{\boldsymbol{\alpha}} = (\mathbf{WS})'\dot{\boldsymbol{\alpha}} = \dot{\boldsymbol{\alpha}}.$$

Thus, condition (3.59) implies condition (3.68) and vice versa; and we conclude that conditions (3.59) and (3.68) are equivalent. Moreover, since $\boldsymbol{\tau}^{(0)} = \mathbf{W}'\boldsymbol{\alpha}^{(0)}$, we conclude also that condition (3.61) implies condition (3.69), and vice versa, and hence that conditions (3.61) and (3.69) are equivalent.

EXERCISE 11. Taking the context to be that of Section 7.3 and adopting the notation employed therein, show that, corresponding to any two choices $\mathbf{S}_1$ and $\mathbf{S}_2$ for the matrix $\mathbf{S}$ (i.e., any two $M \times M_*$ matrices $\mathbf{S}_1$ and $\mathbf{S}_2$ such that $\mathbf{TS}_1$ and $\mathbf{TS}_2$ are orthogonal), there exists a unique $M_* \times M_*$ matrix $\mathbf{Q}$ such that

$$\mathbf{X\tilde{R}S}_2 = \mathbf{X\tilde{R}S}_1\mathbf{Q},$$

and show that this matrix is orthogonal.

Solution. Let $\mathbf{Q}$ represent an arbitrary $M_*\times M_*$ matrix. Then, recalling Corollary 2.3.4 and observing that

$$(\mathbf{X\tilde{R}})'\mathbf{X\tilde{R}} = \mathbf{C} = \mathbf{T}'\mathbf{T},$$

we find that

$$\mathbf{X\tilde{R}S}_2 = \mathbf{X\tilde{R}S}_1\mathbf{Q} \tag{S.3}$$

if and only if

$$(\mathbf{X\tilde{R}})'\mathbf{X\tilde{R}S}_2 = (\mathbf{X\tilde{R}})'\mathbf{X\tilde{R}S}_1\mathbf{Q},$$

or equivalently if and only if

$$\mathbf{T}'\mathbf{TS}_2 = \mathbf{T}'\mathbf{TS}_1\mathbf{Q},$$

and hence if and only if

$$\mathbf{TS}_2 = \mathbf{TS}_1\mathbf{Q}.$$

And (since $\mathbf{TS}_1$ is orthogonal) it follows that equality (S.3) is satisfied uniquely by taking $\mathbf{Q} = (\mathbf{TS}_1)'\mathbf{TS}_2$ and (since $\mathbf{TS}_2$, like $\mathbf{TS}_1$, is orthogonal) that

$$[(\mathbf{TS}_1)'\mathbf{TS}_2]'(\mathbf{TS}_1)'\mathbf{TS}_2 = (\mathbf{TS}_2)'\mathbf{TS}_1(\mathbf{TS}_1)'\mathbf{TS}_2 = (\mathbf{TS}_2)'\mathbf{ITS}_2 = \mathbf{I}$$

[so that $(\mathbf{TS}_1)'\mathbf{TS}_2$ is orthogonal].

EXERCISE 12. Taking the context to be that of Section 7.3, adopting the notation employed therein, and making use of the results of Exercise 11 (or otherwise), verify that none of the groups $G_0, G_1, G_{01}, G_2(\boldsymbol{\tau}^{(0)}), G_3(\boldsymbol{\tau}^{(0)})$, and G_4 (of transformations of $\mathbf{y}$) introduced in the final three parts of Subsection b of Section 7.3 vary with the choice of the matrices $\mathbf{S}$, $\mathbf{U}$, and $\mathbf{L}$.

Solution. Letting $\mathbf{L}_1$ and $\mathbf{L}_2$ represent $N \times (N-P_*)$ matrices and supposing that $\mathbf{L}_1$ is among the choices for the matrix $\mathbf{L}$ [whose columns form an orthonormal basis for $\mathcal{N}(\mathbf{X}')$], observe that , for $\mathbf{L}_2$ to be among the choices for $\mathbf{L}$, it is necessary and sufficient that

$$\mathbf{L}_2 = \mathbf{L}_1\mathbf{M} \tag{S.4}$$

for some orthogonal matrix $\mathbf{M}$. And observe that

$$\mathcal{N}[(\mathbf{X}\tilde{\mathbf{R}}\mathbf{S}_2,\, \mathbf{L}_2)'] = \mathcal{N}[(\mathbf{X}\tilde{\mathbf{R}}\mathbf{S}_1,\, \mathbf{L}_1)'] \tag{S.5}$$

for any two choices $\mathbf{S}_1$ and $\mathbf{S}_2$ for the matrix $\mathbf{S}$ (i.e., any two $M \times M_*$ matrices $\mathbf{S}_1$ and $\mathbf{S}_2$ such that $\mathbf{T}\mathbf{S}_1$ and $\mathbf{T}\mathbf{S}_2$ are orthogonal) and for any two choices $\mathbf{L}_1$ and $\mathbf{L}_2$ for the matrix $\mathbf{L}$ [as is evident from the results of Exercise 11 and from result (S.4)]. Further, letting $\mathbf{U}_1$ and $\mathbf{U}_2$ represent $N \times (P_*-M_*)$ matrices and supposing that $\mathbf{U}_1$ is among the choices for the matrix $\mathbf{U}$ (whose columns form an orthonormal basis for $\mathcal{N}[(\mathbf{X}\tilde{\mathbf{R}}\mathbf{S},\, \mathbf{L})']$), observe that , for $\mathbf{U}_2$ to be among the choices for $\mathbf{U}$, it is necessary and sufficient that

$$\mathbf{U}_2 = \mathbf{U}_1\mathbf{A} \tag{S.6}$$

for some orthogonal matrix $\mathbf{A}$.

That the groups G_0, G_1, and G_{01} do not vary with the choice of the matrices $\mathbf{S}$, $\mathbf{U}$, and $\mathbf{L}$ is apparent from result (3.72) upon observing [in light of result (3.1) and Corollary 2.4.17] that

$$\mathcal{C}(\mathbf{X}\tilde{\mathbf{R}}\mathbf{S}) = \mathcal{C}(\mathbf{X}\tilde{\mathbf{R}})$$

and upon observing [in light of results (S.5) and (S.6)] that $\mathcal{C}(\mathbf{U})$ does not vary with the choice of $\mathbf{U}$.

Turning now to the group $G_2(\boldsymbol{\tau}^{(0)})$, we find that

$$\begin{aligned} T_2(\mathbf{y}\,;\boldsymbol{\tau}^{(0)}) &= \mathbf{O}\,\tilde{T}_2(\mathbf{O}'\mathbf{y}\,;\mathbf{S}'\boldsymbol{\tau}^{(0)}) \\ &= \mathbf{O}\begin{bmatrix} \mathbf{S}'\boldsymbol{\tau}^{(0)} + k(\mathbf{S}'\tilde{\mathbf{R}}'\mathbf{X}'\mathbf{y} - \mathbf{S}'\boldsymbol{\tau}^{(0)}) \\ k\mathbf{U}'\mathbf{y} \\ k\mathbf{L}'\mathbf{y} \end{bmatrix} \\ &= \mathbf{X}\tilde{\mathbf{R}}\mathbf{S}\mathbf{S}'\boldsymbol{\tau}^{(0)} + k\mathbf{X}\tilde{\mathbf{R}}\mathbf{S}\mathbf{S}'(\tilde{\mathbf{R}}'\mathbf{X}'\mathbf{y} - \boldsymbol{\tau}^{(0)}) + k\mathbf{U}\mathbf{U}'\mathbf{y} + k\mathbf{L}\mathbf{L}'\mathbf{y}. \end{aligned}$$

And since $\boldsymbol{\tau}^{(0)} \in \mathcal{C}(\boldsymbol{\Lambda}')$ (and since $\boldsymbol{\Lambda}' = \tilde{\mathbf{R}}'\mathbf{X}'\mathbf{X}$),

$$\boldsymbol{\tau}^{(0)} = \tilde{\mathbf{R}}'\mathbf{X}'\mathbf{X}\boldsymbol{\ell} \tag{S.7}$$

for some vector $\boldsymbol{\ell}$. Thus,

$$\mathbf{X}\tilde{\mathbf{R}}\mathbf{S}\mathbf{S}'\boldsymbol{\tau}^{(0)} + k\mathbf{X}\tilde{\mathbf{R}}\mathbf{S}\mathbf{S}'(\tilde{\mathbf{R}}'\mathbf{X}'\mathbf{y} - \boldsymbol{\tau}^{(0)}) = \mathbf{X}\tilde{\mathbf{R}}\mathbf{S}\mathbf{S}'\tilde{\mathbf{R}}'\mathbf{X}'[k\mathbf{y} + (1-k)\mathbf{X}\boldsymbol{\ell}]. \tag{S.8}$$

Moreover, it follows from the results of Exercise 11 that, corresponding to any two choices $\mathbf{S}_1$ and $\mathbf{S}_2$ for the matrix $\mathbf{S}$, there exists an orthogonal matrix $\mathbf{Q}$ such that

$$\mathbf{X}\tilde{\mathbf{R}}\mathbf{S}_2 = \mathbf{X}\tilde{\mathbf{R}}\mathbf{S}_1\mathbf{Q}, \tag{S.9}$$

in which case

$$\mathbf{X}\tilde{\mathbf{R}}\mathbf{S}_2\mathbf{S}_2'\tilde{\mathbf{R}}'\mathbf{X}' = \mathbf{X}\tilde{\mathbf{R}}\mathbf{S}_2(\mathbf{X}\tilde{\mathbf{R}}\mathbf{S}_2)' = \mathbf{X}\tilde{\mathbf{R}}\mathbf{S}_1\mathbf{Q}\mathbf{Q}'(\mathbf{X}\tilde{\mathbf{R}}\mathbf{S}_1)' = \mathbf{X}\tilde{\mathbf{R}}\mathbf{S}_1\mathbf{S}_1'\tilde{\mathbf{R}}'\mathbf{X}'. \tag{S.10}$$

Based on result (S.10) and on results (S.4), (S.5), and (S.6), it is clear that expression (S.8) and the quantities $k\mathbf{U}\mathbf{U}'\mathbf{y}$ and $k\mathbf{L}\mathbf{L}'\mathbf{y}$ do not vary with the choice of the matrices $\mathbf{S}$, $\mathbf{U}$, and $\mathbf{L}$, leading to the conclusion that $T_2(\mathbf{y}\,;\boldsymbol{\tau}^{(0)})$ and hence the group $G_2(\boldsymbol{\tau}^{(0)})$ do not vary with that choice.

Finally, consider the groups $G_3(\boldsymbol{\tau}^{(0)})$ and G_4. Clearly,

$$\begin{aligned} T_3(\mathbf{y}\,;\boldsymbol{\tau}^{(0)}) &= \mathbf{O}\,\tilde{T}_3(\mathbf{O}'\mathbf{y}\,;\mathbf{S}'\boldsymbol{\tau}^{(0)}) \\ &= \mathbf{O}\begin{bmatrix} \mathbf{S}'\boldsymbol{\tau}^{(0)} + \mathbf{P}'(\mathbf{S}'\tilde{\mathbf{R}}'\mathbf{X}'\mathbf{y} - \mathbf{S}'\boldsymbol{\tau}^{(0)}) \\ \mathbf{U}'\mathbf{y} \\ \mathbf{L}'\mathbf{y} \end{bmatrix} \\ &= \mathbf{X}\tilde{\mathbf{R}}\mathbf{S}\mathbf{S}'\boldsymbol{\tau}^{(0)} + \mathbf{X}\tilde{\mathbf{R}}\mathbf{S}\mathbf{P}'\mathbf{S}'(\tilde{\mathbf{R}}'\mathbf{X}'\mathbf{y} - \boldsymbol{\tau}^{(0)}) + \mathbf{U}\mathbf{U}'\mathbf{y} + \mathbf{L}\mathbf{L}'\mathbf{y} \end{aligned} \tag{S.11}$$

and

$$T_4(\mathbf{y}) = \mathbf{O}\tilde{T}_4(\mathbf{O}'\mathbf{y}) = \mathbf{X}\tilde{\mathbf{R}}\mathbf{S}\mathbf{S}'\tilde{\mathbf{R}}'\mathbf{X}'\mathbf{y} + \mathbf{U}\mathbf{U}'\mathbf{y} + \mathbf{L}\mathbf{B}'\mathbf{L}'\mathbf{y}. \tag{S.12}$$

Moreover, in light of result (S.7),

$$\mathbf{X}\tilde{\mathbf{R}}\mathbf{S}\mathbf{S}'\boldsymbol{\tau}^{(0)} + \mathbf{X}\tilde{\mathbf{R}}\mathbf{S}\mathbf{P}'\mathbf{S}'(\tilde{\mathbf{R}}'\mathbf{X}'\mathbf{y} - \boldsymbol{\tau}^{(0)}) = \mathbf{X}\tilde{\mathbf{R}}\mathbf{S}\mathbf{S}'\tilde{\mathbf{R}}'\mathbf{X}'\mathbf{X}\boldsymbol{\ell} + \mathbf{X}\tilde{\mathbf{R}}\mathbf{S}\mathbf{P}'(\mathbf{X}\tilde{\mathbf{R}}\mathbf{S})'(\mathbf{y} - \mathbf{X}\boldsymbol{\ell}). \tag{S.13}$$

And (for the two choices $\mathbf{S}_1$ and $\mathbf{S}_2$ for the matrix $\mathbf{S}$)

$$\mathbf{X}\tilde{\mathbf{R}}\mathbf{S}_2\mathbf{P}'(\mathbf{X}\tilde{\mathbf{R}}\mathbf{S}_2)' = \mathbf{X}\tilde{\mathbf{R}}\mathbf{S}_1(\mathbf{Q}\mathbf{P}\mathbf{Q}')'(\mathbf{X}\tilde{\mathbf{R}}\mathbf{S}_1)' \tag{S.14}$$

[where $\mathbf{Q}$ is the orthogonal matrix that satisfies equality (S.9)].

In light of result (S.10) and results (S.4), (S.5), and (S.6), the first term of expression (S.13) and the quantities $\mathbf{U}\mathbf{U}'\mathbf{y}$ and $\mathbf{L}\mathbf{L}'\mathbf{y}$ do not vary with the choice of the matrices $\mathbf{S}$, $\mathbf{U}$, and $\mathbf{L}$. And as is apparent from result (S.14), the second term of expression (S.13) is affected by the choice of $\mathbf{S}$, but the collection of values obtained for the second term as $\mathbf{P}$ ranges over all $M_* \times M_*$ orthogonal matrices is unaffected—the mapping $\mathbf{P} \to \mathbf{Q}\mathbf{P}\mathbf{Q}'$ is a 1-to-1 mapping of the set of all $M_* \times M_*$ orthogonal matrices onto itself. Thus, the group $G_3(\boldsymbol{\tau}^{(0)})$ does not vary with the choice of $\mathbf{S}$, $\mathbf{U}$, and $\mathbf{L}$.

Similarly, the first two terms of expression (S.12) do not vary with the choice of $\mathbf{S}$, $\mathbf{U}$, and $\mathbf{L}$ [as is evident from result (S.10) and results (S.5) and (S.6)]. The last term [of expression (S.12)] varies with the choice of $\mathbf{L}$; for the two choices $\mathbf{L}_1$ and $\mathbf{L}_2$,

$$\mathbf{L}_2\mathbf{B}'\mathbf{L}_2' = \mathbf{L}_1(\mathbf{M}\mathbf{B}\mathbf{M}')'\mathbf{L}_1'$$

[where $\mathbf{M}$ is the orthogonal matrix that satisfies equality (S.4)]. However, the collection of matrices generated by the expression $\mathbf{L}_1(\mathbf{M}\mathbf{B}\mathbf{M}')'\mathbf{L}_1'$ as $\mathbf{B}$ ranges over all $(N-P_*) \times (N-P_*)$ orthogonal matrices is the same as that generated by the expression $\mathbf{L}_1\mathbf{B}'\mathbf{L}_1'$. Thus, the group G_4 does not vary with the choice of $\mathbf{S}$, $\mathbf{U}$, and $\mathbf{L}$.

EXERCISE 13. Consider the set $\tilde{A}_{\tilde{\boldsymbol{\delta}}}$ (of $\tilde{\boldsymbol{\delta}}'\boldsymbol{\alpha}$-values) defined (for $\tilde{\boldsymbol{\delta}} \in \tilde{\Delta}$) by expression (3.89) or (3.90). Underlying this definition is an implicit assumption that (for any $M_* \times 1$ vector $\dot{\mathbf{t}}$) the function $f(\tilde{\boldsymbol{\delta}}) = |\tilde{\boldsymbol{\delta}}'\dot{\mathbf{t}}|/(\tilde{\boldsymbol{\delta}}'\tilde{\boldsymbol{\delta}})^{1/2}$, with domain $\{\tilde{\boldsymbol{\delta}} \in \tilde{\Delta} : \tilde{\boldsymbol{\delta}} \neq \mathbf{0}\}$, attains a maximum value. Show (1) that this function has a supremum and (2) that if the set

$$\ddot{\Delta} = \{\ddot{\boldsymbol{\delta}} \in \mathbb{R}^{M_*} : \exists \text{ a nonnull vector } \tilde{\boldsymbol{\delta}} \text{ in } \tilde{\Delta} \text{ such that } \ddot{\boldsymbol{\delta}} = (\tilde{\boldsymbol{\delta}}'\tilde{\boldsymbol{\delta}})^{-1/2}\tilde{\boldsymbol{\delta}}\}$$

is closed, then there exists a nonnull vector in $\tilde{\Delta}$ at which this function attains a maximum value.

Solution. (1) For any $M_* \times 1$ nonnull vector $\tilde{\boldsymbol{\delta}}$,

$$\frac{|\tilde{\boldsymbol{\delta}}'\dot{\mathbf{t}}|}{(\tilde{\boldsymbol{\delta}}'\tilde{\boldsymbol{\delta}})^{1/2}} \leq (\dot{\mathbf{t}}'\dot{\mathbf{t}})^{1/2},$$

as is evident from inequality (2.4.10) (which is a special case of the Cauchy-Schwarz inequality). Thus, the function $f(\tilde{\boldsymbol{\delta}})$ is bounded and hence has a supremum—refer, e.g., to Bartle (1976, sec. 17) or to Bartle and Sherbert (2011).

(2) Let $g(\ddot{\boldsymbol{\delta}})$ represent the function defined (for $\ddot{\boldsymbol{\delta}} \in \mathbb{R}^{M_*}$) as follows: $g(\ddot{\boldsymbol{\delta}}) = \ddot{\boldsymbol{\delta}}'\dot{\mathbf{t}}$. And let $r(\ddot{\boldsymbol{\delta}})$ represent the restriction of the function $g(\ddot{\boldsymbol{\delta}})$ to the set $\ddot{\Delta}$, and define (for $\ddot{\boldsymbol{\delta}} \in \ddot{\Delta}$) $h(\ddot{\boldsymbol{\delta}}) = |r(\ddot{\boldsymbol{\delta}})|$. Since $g(\cdot)$ is a linear function, it is continuous; and since a restriction of a continuous function is continuous and since the absolute value of a continuous function is continuous, $r(\cdot)$ is continuous and hence $h(\cdot)$ is continuous—refer, e.g., to Bartle (1976, secs. 20 and 21). Moreover, the set $\ddot{\Delta}$ is bounded (as is evident upon observing that every vector in $\ddot{\Delta}$ has a norm of 1).

Now, suppose that the set $\ddot{\Delta}$ is closed. Then, since $\ddot{\Delta}$ is bounded, it follows from the Heine-Borel theorem (e.g., Bartle 1976, sec. 11; Bartle and Sherbert 2011, sec 11.2) that $\ddot{\Delta}$ is compact. And upon applying the extreme (maximum and minimum) value theorem (e.g., Bartle 1976, sec. 22), we conclude that $\ddot{\Delta}$ contains a vector, say $\ddot{\boldsymbol{\delta}}_*$, at which the function $h(\cdot)$ attains a maximum value. It remains only to observe that the set $\{\tilde{\boldsymbol{\delta}} \in \tilde{\Delta} : \tilde{\boldsymbol{\delta}} \neq \mathbf{0}\}$ contains a vector, say $\tilde{\boldsymbol{\delta}}_*$, for which $\ddot{\boldsymbol{\delta}}_* = (\tilde{\boldsymbol{\delta}}_*'\tilde{\boldsymbol{\delta}}_*)^{-1/2}\tilde{\boldsymbol{\delta}}_*$ and that $f(\cdot)$ attains a maximum value at $\tilde{\boldsymbol{\delta}}_*$.

EXERCISE 14. Take the context to be that of Section 7.3, and adopt the notation employed therein. Further, let $r = \hat{\sigma} c_{\dot{\gamma}}$, and for $\tilde{\boldsymbol{\delta}} \in \mathbb{R}^{M_*}$, let

$$\dot{A}_{\tilde{\boldsymbol{\delta}}} = \{\dot{\tau} \in \mathbb{R}^1 \, : \, \dot{\tau} = \tilde{\boldsymbol{\delta}}' \dot{\boldsymbol{\alpha}}, \, \dot{\boldsymbol{\alpha}} \in \dot{A}\},$$

where $\dot{A}$ is the set (3.121) or (3.122) and is expressible in the form

$$\dot{A} = \left\{ \dot{\boldsymbol{\alpha}} \in \mathbb{R}^{M_*} \, : \, \frac{|\dot{\boldsymbol{\delta}}'(\dot{\boldsymbol{\alpha}} - \hat{\boldsymbol{\alpha}})|}{(\dot{\boldsymbol{\delta}}'\dot{\boldsymbol{\delta}})^{1/2}} \leq r \text{ for every nonnull } \dot{\boldsymbol{\delta}} \in \tilde{\Delta} \right\}.$$

For $\tilde{\boldsymbol{\delta}} \notin \tilde{\Delta}$, $\dot{A}_{\tilde{\boldsymbol{\delta}}}$ is identical to the set $\tilde{A}_{\tilde{\boldsymbol{\delta}}}$ defined by expression (3.123). Show that for $\tilde{\boldsymbol{\delta}} \in \tilde{\Delta}$, $\dot{A}_{\tilde{\boldsymbol{\delta}}}$ is identical to the set $\tilde{A}_{\tilde{\boldsymbol{\delta}}}$ defined by expression (3.89) or (3.90) or, equivalently, by the expression

$$\tilde{A}_{\tilde{\boldsymbol{\delta}}} = \{\dot{\tau} \in \mathbb{R}^1 \, : \, |\dot{\tau} - \tilde{\boldsymbol{\delta}}'\hat{\boldsymbol{\alpha}})| \leq (\tilde{\boldsymbol{\delta}}'\tilde{\boldsymbol{\delta}})^{1/2} r\}. \tag{E.2}$$

Solution. Let $\dot{\tau}$ represent an arbitrary scalar, and let $\tilde{\boldsymbol{\delta}}$ represent any nonnull $M_* \times 1$ vector in $\tilde{\Delta}$. It suffices to show that $\dot{\tau}$ is contained in the set (E.2) if and only if it is contained in the set $\dot{A}_{\tilde{\boldsymbol{\delta}}}$—when $\tilde{\boldsymbol{\delta}} = \mathbf{0}$, both of these sets equal $\{\mathbf{0}\}$.

Suppose that $\dot{\tau} \in \dot{A}_{\tilde{\boldsymbol{\delta}}}$. Then, $\dot{\tau} = \tilde{\boldsymbol{\delta}}'\dot{\boldsymbol{\alpha}}$ for some $M_* \times 1$ vector $\dot{\boldsymbol{\alpha}}$ such that $|\tilde{\boldsymbol{\delta}}'(\dot{\boldsymbol{\alpha}} - \hat{\boldsymbol{\alpha}})|/(\tilde{\boldsymbol{\delta}}'\tilde{\boldsymbol{\delta}})^{1/2} \leq r$, in which case $|\dot{\tau} - \tilde{\boldsymbol{\delta}}'\hat{\boldsymbol{\alpha}}| \leq (\tilde{\boldsymbol{\delta}}'\tilde{\boldsymbol{\delta}})^{1/2} r$, implying that $\dot{\tau}$ is contained in the set (E.2).

Conversely, suppose that $\dot{\tau}$ is contained in the set (E.2). Then,

$$\dot{\tau} - \tilde{\boldsymbol{\delta}}'\hat{\boldsymbol{\alpha}} = (\tilde{\boldsymbol{\delta}}'\tilde{\boldsymbol{\delta}})^{1/2} k \tag{S.15}$$

for some scalar k such that $0 \leq |k| \leq r$. Now, let $\dot{\boldsymbol{\alpha}} = \hat{\boldsymbol{\alpha}} + k(\tilde{\boldsymbol{\delta}}'\tilde{\boldsymbol{\delta}})^{-1/2}\tilde{\boldsymbol{\delta}}$. And observe [in light of result (2.4.10) (which is a special case of the Cauchy-Schwarz inequality)] that for any nonnull $M_* \times 1$ vector $\dot{\boldsymbol{\delta}}$,

$$\frac{|\dot{\boldsymbol{\delta}}'(\dot{\boldsymbol{\alpha}} - \hat{\boldsymbol{\alpha}})|}{(\dot{\boldsymbol{\delta}}'\dot{\boldsymbol{\delta}})^{1/2}} \leq [(\dot{\boldsymbol{\alpha}} - \hat{\boldsymbol{\alpha}})'(\dot{\boldsymbol{\alpha}} - \hat{\boldsymbol{\alpha}})]^{1/2} = |k| \leq r,$$

so that $\dot{\boldsymbol{\alpha}} \in \dot{A}$. Observe also that

$$\tilde{\boldsymbol{\delta}}'\dot{\boldsymbol{\alpha}} - \tilde{\boldsymbol{\delta}}'\hat{\boldsymbol{\alpha}} = (\tilde{\boldsymbol{\delta}}'\tilde{\boldsymbol{\delta}})^{1/2} k,$$

which in combination with result (S.15) implies that $\dot{\tau} = \tilde{\boldsymbol{\delta}}'\dot{\boldsymbol{\alpha}}$, and leads to the conclusion that $\dot{\tau} \in \dot{A}_{\tilde{\boldsymbol{\delta}}}$.

EXERCISE 15. Taking the sets Δ and $\tilde{\Delta}$, the matrix $\mathbf{C}$, and the random vector $\mathbf{t}$ to be as defined in Section 7.3, supposing that the set $\{\boldsymbol{\delta} \in \Delta \, : \, \boldsymbol{\delta}'\mathbf{C}\boldsymbol{\delta} \neq 0\}$ consists of a finite number of vectors $\boldsymbol{\delta}_1, \boldsymbol{\delta}_2, \ldots, \boldsymbol{\delta}_Q$, and letting $\mathbf{K}$ represent a $Q \times Q$ (correlation) matrix with ijth element $(\boldsymbol{\delta}_i'\mathbf{C}\boldsymbol{\delta}_i)^{-1/2}(\boldsymbol{\delta}_j'\mathbf{C}\boldsymbol{\delta}_j)^{-1/2}\boldsymbol{\delta}_i'\mathbf{C}\boldsymbol{\delta}_j$, show that

$$\max_{\{\tilde{\boldsymbol{\delta}} \in \tilde{\Delta} \, : \, \tilde{\boldsymbol{\delta}} \neq \mathbf{0}\}} \frac{|\tilde{\boldsymbol{\delta}}'\mathbf{t}|}{(\tilde{\boldsymbol{\delta}}'\tilde{\boldsymbol{\delta}})^{1/2}} = \max(|u_1|, |u_2|, \ldots, |u_Q|),$$

where $u_1, u_2, \ldots, u_Q$ are the elements of a random vector $\mathbf{u}$ that has an $MVt(N-P_*, \mathbf{K})$ distribution.

Solution. Adopting the notation employed in Section 7.3, recalling that $\tilde{\Delta} = \{\tilde{\boldsymbol{\delta}} \, : \, \tilde{\boldsymbol{\delta}} = \mathbf{W}\boldsymbol{\delta}, \, \boldsymbol{\delta} \in \boldsymbol{\Delta}\}$ (and that $\mathbf{W}$ is a matrix such that $\mathbf{C} = \mathbf{W}'\mathbf{W}$), and observing that [since $(\mathbf{W}\boldsymbol{\delta})'\mathbf{W}\boldsymbol{\delta} = \boldsymbol{\delta}'\mathbf{C}\boldsymbol{\delta}$] $\mathbf{W}\boldsymbol{\delta} = \mathbf{0}$ if and only if $\boldsymbol{\delta}'\mathbf{C}\boldsymbol{\delta} = 0$, it follows from the supposition about the nature of the set Δ that

$$\{\tilde{\boldsymbol{\delta}} \in \tilde{\Delta} \, : \, \tilde{\boldsymbol{\delta}} \neq \mathbf{0}\} = \{\tilde{\boldsymbol{\delta}}_1, \tilde{\boldsymbol{\delta}}_2, \ldots, \tilde{\boldsymbol{\delta}}_Q\},$$

where (for $i = 1, 2, \ldots, Q$) $\tilde{\boldsymbol{\delta}}_i = \mathbf{W}\boldsymbol{\delta}_i$. Thus,

$$\begin{aligned} \max_{\{\tilde{\boldsymbol{\delta}} \in \tilde{\Delta} \, : \, \tilde{\boldsymbol{\delta}} \neq \mathbf{0}\}} \frac{|\tilde{\boldsymbol{\delta}}'\mathbf{t}|}{(\tilde{\boldsymbol{\delta}}'\tilde{\boldsymbol{\delta}})^{1/2}} &= \max\left(\frac{|\tilde{\boldsymbol{\delta}}_1'\mathbf{t}|}{(\tilde{\boldsymbol{\delta}}_1'\tilde{\boldsymbol{\delta}}_1)^{1/2}}, \frac{|\tilde{\boldsymbol{\delta}}_2'\mathbf{t}|}{(\tilde{\boldsymbol{\delta}}_2'\tilde{\boldsymbol{\delta}}_2)^{1/2}}, \ldots, \frac{|\tilde{\boldsymbol{\delta}}_Q'\mathbf{t}|}{(\tilde{\boldsymbol{\delta}}_Q'\tilde{\boldsymbol{\delta}}_Q)^{1/2}} \right) \\ &= \max(|u_1|, |u_2|, \ldots, |u_Q|), \end{aligned}$$

where (for $i = 1, 2, \ldots, Q$) $u_i = (\tilde{\boldsymbol{\delta}}_i'\tilde{\boldsymbol{\delta}}_i)^{-1/2}\tilde{\boldsymbol{\delta}}_i'\mathbf{t}$.

Now, let $\mathbf{u} = (u_1, u_2, \ldots, u_Q)'$. And observe that

$$\mathbf{u} = \mathbf{D}\mathbf{B}'\mathbf{t},$$

where $\mathbf{D} = \operatorname{diag}\big[(\tilde{\boldsymbol{\delta}}_1'\tilde{\boldsymbol{\delta}}_1)^{-1/2}, (\tilde{\boldsymbol{\delta}}_2'\tilde{\boldsymbol{\delta}}_2)^{-1/2}, \ldots, (\tilde{\boldsymbol{\delta}}_Q'\tilde{\boldsymbol{\delta}}_Q)^{-1/2}\big]$ and $\mathbf{B} = \big(\tilde{\boldsymbol{\delta}}_1, \tilde{\boldsymbol{\delta}}_2, \ldots, \tilde{\boldsymbol{\delta}}_Q\big)$. Observe also that (by definition)

$$\mathbf{t} \sim \frac{1}{\sqrt{v/(N-P_*)}}\mathbf{z},$$

where v is a random variable that is distributed as $\chi^2(N-P_*)$ and $\mathbf{z}$ is a random vector that is distributed independently of v as $N(\mathbf{0}, \mathbf{I})$. Then,

$$\mathbf{u} \sim \frac{1}{\sqrt{v/(N-P_*)}}\mathbf{D}\mathbf{B}'\mathbf{z} \quad \text{and} \quad \mathbf{D}\mathbf{B}'\mathbf{z} \sim N(\mathbf{0}, \mathbf{D}\mathbf{B}'\mathbf{B}\mathbf{D}).$$

Further, for $i, j = 1, 2, \ldots, Q$, the ijth element of the $Q \times Q$ matrix $\mathbf{D}\mathbf{B}'\mathbf{B}\mathbf{D}$ equals $(\tilde{\boldsymbol{\delta}}_i'\tilde{\boldsymbol{\delta}}_i)^{-1/2}(\tilde{\boldsymbol{\delta}}_j'\tilde{\boldsymbol{\delta}}_j)^{-1/2}\tilde{\boldsymbol{\delta}}_i'\tilde{\boldsymbol{\delta}}_j$, so that $\mathbf{D}\mathbf{B}'\mathbf{B}\mathbf{D}$ is a correlation matrix, and

$$\begin{aligned}(\tilde{\boldsymbol{\delta}}_i'\tilde{\boldsymbol{\delta}}_i)^{-1/2}(\tilde{\boldsymbol{\delta}}_j'\tilde{\boldsymbol{\delta}}_j)^{-1/2}\tilde{\boldsymbol{\delta}}_i'\tilde{\boldsymbol{\delta}}_j &= (\boldsymbol{\delta}_i'\mathbf{W}'\mathbf{W}\boldsymbol{\delta}_i)^{-1/2}(\boldsymbol{\delta}_j'\mathbf{W}'\mathbf{W}\boldsymbol{\delta}_j)^{-1/2}\boldsymbol{\delta}_i'\mathbf{W}'\mathbf{W}\boldsymbol{\delta}_j \\ &= (\boldsymbol{\delta}_i'\mathbf{C}\boldsymbol{\delta}_i)^{-1/2}(\boldsymbol{\delta}_j'\mathbf{C}\boldsymbol{\delta}_j)^{-1/2}\boldsymbol{\delta}_i'\mathbf{C}\boldsymbol{\delta}_j.\end{aligned}$$

It remains only to observe that $\mathbf{u} \sim MVt(N-P_*, \mathbf{D}\mathbf{B}'\mathbf{B}\mathbf{D})$.

EXERCISE 16. Define $\tilde{\Delta}$, $c_{\dot{\gamma}}$, and $\mathbf{t}$ as in Section 7.3c [so that $\mathbf{t}$ is an $M_* \times 1$ random vector that has an $MVt(N-P_*, \mathbf{I}_{M_*})$ distribution]. Show that

$$\Pr\Bigg[\max_{\{\tilde{\boldsymbol{\delta}} \in \tilde{\Delta}\,:\,\tilde{\boldsymbol{\delta}} \neq \mathbf{0}\}} \frac{\tilde{\boldsymbol{\delta}}'\mathbf{t}}{(\tilde{\boldsymbol{\delta}}'\tilde{\boldsymbol{\delta}})^{1/2}} > c_{\dot{\gamma}}\Bigg] \geq \dot{\gamma}/2,$$

with equality holding if and only if there exists a nonnull $M_* \times 1$ vector $\ddot{\boldsymbol{\delta}}$ (of norm 1) such that $(\tilde{\boldsymbol{\delta}}'\tilde{\boldsymbol{\delta}})^{-1/2}\tilde{\boldsymbol{\delta}} = \ddot{\boldsymbol{\delta}}$ for every nonnull vector $\tilde{\boldsymbol{\delta}}$ in $\tilde{\Delta}$.

Solution. Clearly, $c_{\dot{\gamma}} > 0$, and [in light of result (3.96)]

$$\begin{aligned}\dot{\gamma} &= \Pr\Bigg[\max_{\{\tilde{\boldsymbol{\delta}} \in \tilde{\Delta}\,:\,\tilde{\boldsymbol{\delta}} \neq \mathbf{0}\}} \frac{|\tilde{\boldsymbol{\delta}}'\mathbf{t}|}{(\tilde{\boldsymbol{\delta}}'\tilde{\boldsymbol{\delta}})^{1/2}} > c_{\dot{\gamma}}\Bigg] \\ &= \Pr\Bigg[\max_{\{\tilde{\boldsymbol{\delta}} \in \tilde{\Delta}\,:\,\tilde{\boldsymbol{\delta}} \neq \mathbf{0}\}} \frac{\tilde{\boldsymbol{\delta}}'\mathbf{t}}{(\tilde{\boldsymbol{\delta}}'\tilde{\boldsymbol{\delta}})^{1/2}} > c_{\dot{\gamma}}\Bigg] + \Pr\Bigg[-\min_{\{\tilde{\boldsymbol{\delta}} \in \tilde{\Delta}\,:\,\tilde{\boldsymbol{\delta}} \neq \mathbf{0}\}} \frac{\tilde{\boldsymbol{\delta}}'\mathbf{t}}{(\tilde{\boldsymbol{\delta}}'\tilde{\boldsymbol{\delta}})^{1/2}} > c_{\dot{\gamma}}\Bigg] \\ &\qquad - \Pr\Bigg[\max_{\{\tilde{\boldsymbol{\delta}} \in \tilde{\Delta}\,:\,\tilde{\boldsymbol{\delta}} \neq \mathbf{0}\}} \frac{\tilde{\boldsymbol{\delta}}'\mathbf{t}}{(\tilde{\boldsymbol{\delta}}'\tilde{\boldsymbol{\delta}})^{1/2}} > c_{\dot{\gamma}} \text{ and } -\min_{\{\tilde{\boldsymbol{\delta}} \in \tilde{\Delta}\,:\,\tilde{\boldsymbol{\delta}} \neq \mathbf{0}\}} \frac{\tilde{\boldsymbol{\delta}}'\mathbf{t}}{(\tilde{\boldsymbol{\delta}}'\tilde{\boldsymbol{\delta}})^{1/2}} > c_{\dot{\gamma}}\Bigg].\end{aligned}$$

Moreover, upon recalling result (3.145) and observing that $-\mathbf{t} \sim \mathbf{t}$, we find that

$$-\min_{\{\tilde{\boldsymbol{\delta}} \in \tilde{\Delta}\,:\,\tilde{\boldsymbol{\delta}} \neq \mathbf{0}\}} \frac{\tilde{\boldsymbol{\delta}}'\mathbf{t}}{(\tilde{\boldsymbol{\delta}}'\tilde{\boldsymbol{\delta}})^{1/2}} > c_{\dot{\gamma}} \quad \Leftrightarrow \quad \max_{\{\tilde{\boldsymbol{\delta}} \in \tilde{\Delta}\,:\,\tilde{\boldsymbol{\delta}} \neq \mathbf{0}\}} \frac{\tilde{\boldsymbol{\delta}}'(-\mathbf{t})}{(\tilde{\boldsymbol{\delta}}'\tilde{\boldsymbol{\delta}})^{1/2}} > c_{\dot{\gamma}}$$

and that

$$\max_{\{\tilde{\boldsymbol{\delta}} \in \tilde{\Delta}\,:\,\tilde{\boldsymbol{\delta}} \neq \mathbf{0}\}} \frac{\tilde{\boldsymbol{\delta}}'(-\mathbf{t})}{(\tilde{\boldsymbol{\delta}}'\tilde{\boldsymbol{\delta}})^{1/2}} \sim \max_{\{\tilde{\boldsymbol{\delta}} \in \tilde{\Delta}\,:\,\tilde{\boldsymbol{\delta}} \neq \mathbf{0}\}} \frac{\tilde{\boldsymbol{\delta}}'\mathbf{t}}{(\tilde{\boldsymbol{\delta}}'\tilde{\boldsymbol{\delta}})^{1/2}}.$$

Thus,

$$\begin{aligned}\dot{\gamma} &= 2\Pr\Bigg[\max_{\{\tilde{\boldsymbol{\delta}} \in \tilde{\Delta}\,:\,\tilde{\boldsymbol{\delta}} \neq \mathbf{0}\}} \frac{\tilde{\boldsymbol{\delta}}'\mathbf{t}}{(\tilde{\boldsymbol{\delta}}'\tilde{\boldsymbol{\delta}})^{1/2}} > c_{\dot{\gamma}}\Bigg] \\ &\qquad - \Pr\Bigg[\max_{\{\tilde{\boldsymbol{\delta}} \in \tilde{\Delta}\,:\,\tilde{\boldsymbol{\delta}} \neq \mathbf{0}\}} \frac{\tilde{\boldsymbol{\delta}}'\mathbf{t}}{(\tilde{\boldsymbol{\delta}}'\tilde{\boldsymbol{\delta}})^{1/2}} > c_{\dot{\gamma}} \text{ and } -\min_{\{\tilde{\boldsymbol{\delta}} \in \tilde{\Delta}\,:\,\tilde{\boldsymbol{\delta}} \neq \mathbf{0}\}} \frac{\tilde{\boldsymbol{\delta}}'\mathbf{t}}{(\tilde{\boldsymbol{\delta}}'\tilde{\boldsymbol{\delta}})^{1/2}} > c_{\dot{\gamma}}\Bigg].\end{aligned}$$

or, equivalently,

$$\Pr\left[\max_{\{\tilde{\delta}\in\tilde{\Delta}:\,\tilde{\delta}\neq\mathbf{0}\}} \frac{\tilde{\delta}'\mathbf{t}}{(\tilde{\delta}'\tilde{\delta})^{1/2}} > c_{\dot{\gamma}}\right]$$
$$= \dot{\gamma}/2 + \Pr\left[\max_{\{\tilde{\delta}\in\tilde{\Delta}:\,\tilde{\delta}\neq\mathbf{0}\}} \frac{\tilde{\delta}'\mathbf{t}}{(\tilde{\delta}'\tilde{\delta})^{1/2}} > c_{\dot{\gamma}} \text{ and } -\min_{\{\tilde{\delta}\in\tilde{\Delta}:\,\tilde{\delta}\neq\mathbf{0}\}} \frac{\tilde{\delta}'\mathbf{t}}{(\tilde{\delta}'\tilde{\delta})^{1/2}} > c_{\dot{\gamma}}\right],$$

so that

$$\Pr\left[\max_{\{\tilde{\delta}\in\tilde{\Delta}:\,\tilde{\delta}\neq\mathbf{0}\}} \frac{\tilde{\delta}'\mathbf{t}}{(\tilde{\delta}'\tilde{\delta})^{1/2}} > c_{\dot{\gamma}}\right] \geq \dot{\gamma}/2,$$

with equality holding if and only if

$$\Pr\left[\max_{\{\tilde{\delta}\in\tilde{\Delta}:\,\tilde{\delta}\neq\mathbf{0}\}} \frac{\tilde{\delta}'\mathbf{t}}{(\tilde{\delta}'\tilde{\delta})^{1/2}} > c_{\dot{\gamma}} \text{ and } -\min_{\{\tilde{\delta}\in\tilde{\Delta}:\,\tilde{\delta}\neq\mathbf{0}\}} \frac{\tilde{\delta}'\mathbf{t}}{(\tilde{\delta}'\tilde{\delta})^{1/2}} > c_{\dot{\gamma}}\right] = 0$$

or, equivalently, if and only if

$$\Pr\left[\min_{\{\tilde{\delta}\in\tilde{\Delta}:\,\tilde{\delta}\neq\mathbf{0}\}} \frac{\tilde{\delta}'\mathbf{t}}{(\tilde{\delta}'\tilde{\delta})^{1/2}} < -c_{\dot{\gamma}} \text{ and } \max_{\{\tilde{\delta}\in\tilde{\Delta}:\,\tilde{\delta}\neq\mathbf{0}\}} \frac{\tilde{\delta}'\mathbf{t}}{(\tilde{\delta}'\tilde{\delta})^{1/2}} > c_{\dot{\gamma}}\right] = 0.$$

Now, suppose that there exists a nonnull $M_* \times 1$ vector $\ddot{\delta}$ such that $(\tilde{\delta}'\tilde{\delta})^{-1/2}\tilde{\delta} = \ddot{\delta}$ for every nonnull vector $\tilde{\delta}$ in $\tilde{\Delta}$. Then,

$$\max_{\{\tilde{\delta}\in\tilde{\Delta}:\,\tilde{\delta}\neq\mathbf{0}\}} \frac{\tilde{\delta}'\mathbf{t}}{(\tilde{\delta}'\tilde{\delta})^{1/2}} = \min_{\{\tilde{\delta}\in\tilde{\Delta}:\,\tilde{\delta}\neq\mathbf{0}\}} \frac{\tilde{\delta}'\mathbf{t}}{(\tilde{\delta}'\tilde{\delta})^{1/2}} = \ddot{\delta}'\mathbf{t}.$$

And it follows that

$$\Pr\left[\min_{\{\tilde{\delta}\in\tilde{\Delta}:\,\tilde{\delta}\neq\mathbf{0}\}} \frac{\tilde{\delta}'\mathbf{t}}{(\tilde{\delta}'\tilde{\delta})^{1/2}} < -c_{\dot{\gamma}} \text{ and } \max_{\{\tilde{\delta}\in\tilde{\Delta}:\,\tilde{\delta}\neq\mathbf{0}\}} \frac{\tilde{\delta}'\mathbf{t}}{(\tilde{\delta}'\tilde{\delta})^{1/2}} > c_{\dot{\gamma}}\right]$$
$$= \Pr[\ddot{\delta}'\mathbf{t} < -c_{\dot{\gamma}} \text{ and } \ddot{\delta}'\mathbf{t} > c_{\dot{\gamma}}] = 0.$$

Alternatively, suppose that there does not exist a nonnull $M_*\times 1$ vector $\ddot{\delta}$ such that $(\tilde{\delta}'\tilde{\delta})^{-1/2}\tilde{\delta} = \ddot{\delta}$ for every nonnull vector $\tilde{\delta}$ in $\tilde{\Delta}$. Then, $\tilde{\Delta}$ contains two nonnull vectors $\tilde{\delta}_1$ and $\tilde{\delta}_2$ such that $(\tilde{\delta}_1'\tilde{\delta}_1)^{-1/2}\tilde{\delta}_1 \neq (\tilde{\delta}_2'\tilde{\delta}_2)^{-1/2}\tilde{\delta}_2$. And the joint distribution of the random variables $(\tilde{\delta}_1'\tilde{\delta}_1)^{-1/2}\tilde{\delta}_1'\mathbf{t}$ and $(\tilde{\delta}_2'\tilde{\delta}_2)^{-1/2}\tilde{\delta}_2'\mathbf{t}$ is bivariate t (with $N - P_*$ degrees of freedom); the correlation of these two random variables equals $(\tilde{\delta}_1'\tilde{\delta}_1)^{-1/2}(\tilde{\delta}_2'\tilde{\delta}_2)^{-1/2}\tilde{\delta}_1'\tilde{\delta}_2$, the absolute value of which is [in light of the Cauchy-Schwarz inequality (2.4.10)] strictly less than 1. Thus,

$$\Pr\left[\min_{\{\tilde{\delta}\in\tilde{\Delta}:\,\tilde{\delta}\neq\mathbf{0}\}} \frac{\tilde{\delta}'\mathbf{t}}{(\tilde{\delta}'\tilde{\delta})^{1/2}} < -c_{\dot{\gamma}} \text{ and } \max_{\{\tilde{\delta}\in\tilde{\Delta}:\,\tilde{\delta}\neq\mathbf{0}\}} \frac{\tilde{\delta}'\mathbf{t}}{(\tilde{\delta}'\tilde{\delta})^{1/2}} > c_{\dot{\gamma}}\right]$$
$$\geq \Pr\left\{\min\left[\frac{\tilde{\delta}_1'\mathbf{t}}{(\tilde{\delta}_1'\tilde{\delta}_1)^{1/2}}, \frac{\tilde{\delta}_2'\mathbf{t}}{(\tilde{\delta}_2'\tilde{\delta}_2)^{1/2}}\right] < -c_{\dot{\gamma}} \text{ and } \max\left[\frac{\tilde{\delta}_1'\mathbf{t}}{(\tilde{\delta}_1'\tilde{\delta}_1)^{1/2}}, \frac{\tilde{\delta}_2'\mathbf{t}}{(\tilde{\delta}_2'\tilde{\delta}_2)^{1/2}}\right] > c_{\dot{\gamma}}\right\}$$
$$= \Pr\left[\frac{\tilde{\delta}_1'\mathbf{t}}{(\tilde{\delta}_1'\tilde{\delta}_1)^{1/2}} < -c_{\dot{\gamma}} \text{ and } \frac{\tilde{\delta}_2'\mathbf{t}}{(\tilde{\delta}_2'\tilde{\delta}_2)^{1/2}} > c_{\dot{\gamma}}\right] + \Pr\left[\frac{\tilde{\delta}_2'\mathbf{t}}{(\tilde{\delta}_2'\tilde{\delta}_2)^{1/2}} < -c_{\dot{\gamma}} \text{ and } \frac{\tilde{\delta}_1'\mathbf{t}}{(\tilde{\delta}_1'\tilde{\delta}_1)^{1/2}} > c_{\dot{\gamma}}\right]$$
$$> 0.$$

EXERCISE 17.

(a) Letting $E_1, E_2, \ldots, E_L$ represent any events in a probability space and (for any event E) denoting by $\overline{E}$ the complement of E, verify the following (Bonferroni) inequality:

$$\Pr(E_1 \cap E_2 \cap \cdots \cap E_L) \geq 1 - \textstyle\sum_{i=1}^{L} \Pr(\overline{E_i}).$$

(b) Take the context to be that of Section 7.3c, where $\mathbf{y}$ is an $N \times 1$ observable random vector that follows a G–M model with $N \times P$ model matrix $\mathbf{X}$ of rank P_* and where $\boldsymbol{\tau} = \boldsymbol{\Lambda}'\boldsymbol{\beta}$ is an $M \times 1$ vector of estimable linear combinations of the elements of $\boldsymbol{\beta}$ (such that $\boldsymbol{\Lambda} \neq \mathbf{0}$). Further, suppose that the distribution of the vector $\mathbf{e}$ of residual effects is $N(\mathbf{0}, \sigma^2\mathbf{I})$ (or is some other spherically symmetric distribution with mean vector $\mathbf{0}$ and variance-covariance matrix $\sigma^2\mathbf{I}$), let $\hat{\boldsymbol{\tau}}$ represent the least squares estimator of $\boldsymbol{\tau}$, let $\mathbf{C} = \boldsymbol{\Lambda}'(\mathbf{X}'\mathbf{X})^{-}\boldsymbol{\Lambda}$, let $\boldsymbol{\delta}_1, \boldsymbol{\delta}_2, \ldots, \boldsymbol{\delta}_L$ represent $M \times 1$ vectors of constants such that (for $i = 1, 2, \ldots, L$) $\boldsymbol{\delta}_i'\mathbf{C}\boldsymbol{\delta}_i > 0$, let $\hat{\sigma}$ represent the positive square root of the usual estimator of σ^2 (i.e., the estimator obtained upon dividing the residual sum of squares by $N - P_*$), and let $\dot{\gamma}_1, \dot{\gamma}_2, \ldots, \dot{\gamma}_L$ represent positive scalars such that $\sum_{i=1}^{L} \dot{\gamma}_i = \dot{\gamma}$. And (for $i = 1, 2, \ldots, L$) denote by $A_i(\mathbf{y})$ a confidence interval for $\boldsymbol{\delta}_i'\boldsymbol{\tau}$ with end points

$$\boldsymbol{\delta}_i'\hat{\boldsymbol{\tau}} \pm (\boldsymbol{\delta}_i'\mathbf{C}\boldsymbol{\delta}_i)^{1/2}\hat{\sigma}\,\bar{t}_{\dot{\gamma}_i/2}(N - P_*).$$

Use the result of Part (a) to show that

$$\Pr[\boldsymbol{\delta}_i'\boldsymbol{\tau} \in A_i(\mathbf{y}) \ \ (i = 1, 2, \ldots, L)] \geq 1 - \dot{\gamma}$$

and hence that the intervals $A_1(\mathbf{y}), A_2(\mathbf{y}), \ldots, A_L(\mathbf{y})$ are conservative in the sense that their probability of simultaneous coverage is greater than or equal to $1-\dot{\gamma}$—when $\dot{\gamma}_1 = \dot{\gamma}_2 = \cdots = \dot{\gamma}_L$, the end points of interval $A_i(\mathbf{y})$ become

$$\boldsymbol{\delta}_i'\hat{\boldsymbol{\tau}} \pm (\boldsymbol{\delta}_i'\mathbf{C}\boldsymbol{\delta}_i)^{1/2}\hat{\sigma}\,\bar{t}_{\dot{\gamma}/(2L)}(N - P_*),$$

and the intervals $A_1(\mathbf{y}), A_2(\mathbf{y}), \ldots, A_L(\mathbf{y})$ are referred to as Bonferroni t-intervals.

Solution. (a) Since $\Pr(E_1 \cap E_2 \cap \cdots \cap E_L) = 1 - \Pr(\overline{E_1 \cap E_2 \cap \cdots \cap E_L})$, it suffices to prove that

$$\Pr(\overline{E_1 \cap E_2 \cap \cdots \cap E_L}) \leq \Pr(\overline{E_1}) + \Pr(\overline{E_2}) + \cdots + \Pr(\overline{E_L}). \tag{S.16}$$

To establish the validity of inequality (S.16), let us employ mathematical induction.

Clearly, $\overline{E_1 \cap E_2} = \overline{E_1} \cup \overline{E_2}$, and, consequently,

$$\begin{aligned} \Pr(\overline{E_1 \cap E_2}) &= \Pr(\overline{E_1} \cup \overline{E_2}) \\ &= \Pr(\overline{E_1}) + \Pr(\overline{E_2}) - \Pr(\overline{E_1} \cap \overline{E_2}) \\ &\leq \Pr(\overline{E_1}) + \Pr(\overline{E_2}), \end{aligned} \tag{S.17}$$

which establishes the validity of inequality (S.16) when applied to any two events. To complete the induction argument, suppose that inequality (S.16) is valid when applied to any $L-1$ events. Then, upon applying inequality (S.17) (with $E_1 \cap E_2 \cap \cdots \cap E_{L-1}$ in place of E_1 and E_L in place of E_2), we find that

$$\begin{aligned} \Pr(\overline{E_1 \cap E_2 \cap \cdots \cap E_L}) &= \Pr[\overline{(E_1 \cap E_2 \cap \cdots \cap E_{L-1}) \cap E_L}] \\ &\leq \Pr(\overline{E_1 \cap E_2 \cap \cdots \cap E_{L-1}}) + \Pr(\overline{E_L}) \\ &\leq \Pr(\overline{E_1}) + \Pr(\overline{E_2}) + \cdots + \Pr(\overline{E_{L-1}}) + \Pr(\overline{E_L}). \end{aligned}$$

And we conclude that inequality (S.16) is valid when applied to any L events (for every positive integer $L \geq 2$).

(b) For $i = 1, 2, \ldots, L$,

$$\boldsymbol{\delta}_i'\boldsymbol{\tau} \notin A_i(\mathbf{y}) \quad \Leftrightarrow \quad \frac{|\boldsymbol{\delta}_i'\hat{\boldsymbol{\tau}} - \boldsymbol{\delta}_i'\boldsymbol{\tau}|}{(\boldsymbol{\delta}_i'\mathbf{C}\boldsymbol{\delta}_i)^{1/2}\hat{\sigma}} > \bar{t}_{\dot{\gamma}_i/2}(N - P_*)$$

and
$$\frac{\boldsymbol{\delta}_i'\hat{\boldsymbol{\tau}} - \boldsymbol{\delta}_i'\boldsymbol{\tau}}{(\boldsymbol{\delta}_i'\mathbf{C}\boldsymbol{\delta}_i)^{1/2}\hat{\sigma}} \sim St(N - P_*).$$

Thus, as an application of the result of Part (a) [that where the event E_i consists of the values of $\mathbf{y}$ such that $\boldsymbol{\delta}_i'\boldsymbol{\tau} \in A_i(\mathbf{y})$], we have that

$$\begin{aligned}\Pr[\boldsymbol{\delta}_i'\boldsymbol{\tau} \in A_i(\mathbf{y}) \ \ (i = 1, 2, \ldots, L)] &\geq 1 - \textstyle\sum_{i=1}^{L} \Pr[\boldsymbol{\delta}_i'\boldsymbol{\tau} \notin A_i(\mathbf{y})] \\ &= 1 - \textstyle\sum_{i=1}^{L} \Pr\left[\dfrac{|\boldsymbol{\delta}_i'\hat{\boldsymbol{\tau}} - \boldsymbol{\delta}_i'\boldsymbol{\tau}|}{(\boldsymbol{\delta}_i'\mathbf{C}\boldsymbol{\delta}_i)^{1/2}\hat{\sigma}} > \bar{t}_{\dot{\gamma}_i/2}(N - P_*)\right] \\ &= 1 - \textstyle\sum_{i=1}^{L} \dot{\gamma}_i \\ &= 1 - \dot{\gamma}.\end{aligned}$$

EXERCISE 18. Suppose that the data (of Section 4.2b) on the lethal dose of ouabain in cats are regarded as the observed values of the elements $y_1, y_2, \ldots, y_N$ of an $N(= 41)$-dimensional observable random vector $\mathbf{y}$ that follows a G–M model. Suppose further that (for $i = 1, 2, \ldots, 41$) $\mathrm{E}(y_i) = \delta(u_i)$, where $u_1, u_2, \ldots, u_{41}$ are the values of the rate u of injection and where $\delta(u)$ is the third-degree polynomial
$$\delta(u) = \beta_1 + \beta_2 u + \beta_3 u^2 + \beta_4 u^3.$$
And suppose that the distribution of the vector $\mathbf{e}$ of residual effects is $N(\mathbf{0}, \sigma^2\mathbf{I})$ (or is some other spherically symmetric distribution with mean vector $\mathbf{0}$ and variance-covariance matrix $\sigma^2\mathbf{I}$).

(a) Compute the values of the least squares estimators $\hat{\beta}_1, \hat{\beta}_2, \hat{\beta}_3$, and $\hat{\beta}_4$ of $\beta_1, \beta_2, \beta_3$, and β_4, respectively, and the value of the positive square root $\hat{\sigma}$ of the usual unbiased estimator of σ^2—it follows from the results of Section 5.3d that P_* (= rank $\mathbf{X}$) $= P = 4$, in which case $N - P_* = N - P = 37$, and that $\beta_1, \beta_2, \beta_3$, and β_4 are estimable.

(b) Find the values of $\bar{t}_{.05}(37)$ and $[4\bar{F}_{.10}(4, 37)]^{1/2}$, which would be needed if interval $I_{\mathbf{u}}^{(1)}(\mathbf{y})$ with end points (3.169) and interval $I_{\mathbf{u}}^{(2)}(\mathbf{y})$ with end points (3.170) (where in both cases $\dot{\gamma}$ is taken to be .10) were used to construct confidence bands for the response surface $\delta(u)$.

(c) By (for example) making use of the results in Liu's (2011) Appendix E, compute Monte Carlo approximations to the constants $c_{.10}$ and $c_{.10}^*$ that would be needed if interval $I_{\mathbf{u}}^{(3)}(\mathbf{y})$ with end points (3.171) and interval $I_{\mathbf{u}}^{(4)}(\mathbf{y})$ with end points (3.172) were used to construct confidence bands for $\delta(u)$; compute the approximations for the case where u is restricted to the interval $1 \leq u \leq 8$, and (for purposes of comparison) also compute $c_{.10}$ for the case where u is unrestricted.

(d) Plot (as a function of u) the value of the least squares estimator $\hat{\delta}(u) = \hat{\beta}_1 + \hat{\beta}_2 u + \hat{\beta}_3 u^2 + \hat{\beta}_4 u^3$ and (taking $\dot{\gamma} = .10$) the values of the end points (3.169) and (3.170) of intervals $I_{\mathbf{u}}^{(1)}(\mathbf{y})$ and $I_{\mathbf{u}}^{(2)}(\mathbf{y})$ and the values of the approximations to the end points (3.171) and (3.172) of intervals $I_{\mathbf{u}}^{(3)}(\mathbf{y})$ and $I_{\mathbf{u}}^{(4)}(\mathbf{y})$ obtained upon replacing $c_{.10}$ and $c_{.10}^*$ with their Monte Carlo approximations—assume (for purposes of creating the plot and for approximating $c_{.10}$ and $c_{.10}^*$) that u is restricted to the interval $1 \leq u \leq 8$.

Solution. (a) $\hat{\beta}_1 = 8.838$, $\hat{\beta}_2 = 8.602$, $\hat{\beta}_3 = 0.752$, $\hat{\beta}_4 = -0.109$, and $\hat{\sigma} = 12.357$.

(b) $\bar{t}_{.05}(37) = 1.687$ and $[4\bar{F}_{.10}(4, 37)]^{1/2} = 2.901$.

(c) Let $\mathbf{W}$ represent a 4×4 upper triangular matrix such that $(\mathbf{X}'\mathbf{X})^{-1} = \mathbf{W}'\mathbf{W}$ [in which case $(\mathbf{X}'\mathbf{X})^{-1} = \mathbf{W}'\mathbf{W}$ is the Cholesky decomposition of $(\mathbf{X}'\mathbf{X})^{-1}$], take $\mathbf{t}$ to be a 4×1 random vector whose distribution is $MVt(37, \mathbf{I}_4)$, and define (for $u \in \mathcal{R}^1$) $\mathbf{x}(u) = (1, u, u^2, u^3)'$ (so that $\delta(u) = [\mathbf{x}(u)]'\boldsymbol{\beta}$). Further, let (for $u \in \mathcal{R}^1$) $\mathbf{z}(u) = \mathbf{W}\mathbf{x}(u)$, and define (for $u \in \mathcal{R}^1$) functions $g(u)$ and $h(u)$ (of u) as follows:
$$g(u) = \frac{\mathbf{t}'\mathbf{z}(u)}{\{[\mathbf{z}(u)]'\mathbf{z}(u)\}^{1/2}} \quad \text{and} \quad h(u) = \mathbf{t}'\mathbf{z}(u).$$

By definition, $c_{.10}$ is the upper 10% point of the distribution of the random variable $\max_{\{u\,:\,1\le u\le 8\}}|g(u)|$ or the random variable $\max_{\{u\,:\,u\in\mathcal{R}^1\}}|g(u)|$ (depending on whether or not u is restricted to the interval $1\le u\le 8$), and $c^*_{.10}$ is the upper 10% point of the distribution of the random variable $\max_{\{u\,:\,1\le u\le 8\}}|h(u)|$.

To obtain a Monte Carlo approximation to $c_{.10}$, it is necessary to determine the value of $\max_{\{u\,:\,1\le u\le 8\}}|g(u)|$ or $\max_{\{u\,:\,u\in\mathcal{R}^1\}}|g(u)|$ for each of a large number of values of $\mathbf{t}$. Similarly, to obtain a Monte Carlo approximation to $c^*_{.10}$, it is necessary to determine the value of $\max_{\{u\,:\,1\le u\le 8\}}|h(u)|$ for each of a large number of values of $\mathbf{t}$.

Clearly,

$$\begin{aligned}\frac{d\,g(u)}{du} &= \frac{\{[\mathbf{z}(u)]'\mathbf{z}(u)\}^{1/2}\dfrac{d\,\mathbf{t}'\mathbf{z}(u)}{du} - [\mathbf{t}'\mathbf{z}(u)]\frac{1}{2}\{[\mathbf{z}(u)]'\mathbf{z}(u)\}^{-1/2}\dfrac{d\,[\mathbf{z}(u)]'\mathbf{z}(u)}{du}}{[\mathbf{z}(u)]'\mathbf{z}(u)}\\ &= \frac{\{[\mathbf{z}(u)]'\mathbf{z}(u)\}\dfrac{d\,\mathbf{t}'\mathbf{z}(u)}{du} - \frac{1}{2}[\mathbf{t}'\mathbf{z}(u)]\dfrac{d\,[\mathbf{z}(u)]'\mathbf{z}(u)}{du}}{\{[\mathbf{z}(u)]'\mathbf{z}(u)\}^{3/2}}. \end{aligned} \tag{S.18}$$

And

$$\frac{d\,h(u)}{du} = \frac{d\,\mathbf{t}'\mathbf{z}(u)}{du} = \mathbf{t}'\mathbf{W}\frac{d\,\mathbf{x}(u)}{du}, \tag{S.19}$$

and [making use of result (5.4.10) along with the chain rule]

$$\frac{d\,[\mathbf{z}(u)]'\mathbf{z}(u)}{du} = 2[\mathbf{x}(u)]'\mathbf{W}'\mathbf{W}\frac{d\,\mathbf{x}(u)}{du}. \tag{S.20}$$

Further,

$$\frac{d\,\mathbf{x}(u)}{du} = (0,\,1,\,2u,\,3u^2)'. \tag{S.21}$$

It follows from results (S.18), (S.19), and (S.20) that $\dfrac{d\,g(u)}{du} = 0$ (or, equivalently, that $\dfrac{d\,[-g(u)]}{du} = 0$) if and only if

$$[\mathbf{x}(u)]'\mathbf{W}'\mathbf{W}\mathbf{x}(u)\Big[\mathbf{t}'\mathbf{W}\frac{d\,\mathbf{x}(u)}{du}\Big] - \mathbf{t}'\mathbf{W}\mathbf{x}(u)\Big\{[\mathbf{x}(u)]'\mathbf{W}'\mathbf{W}\frac{d\,\mathbf{x}(u)}{du}\Big\} = 0. \tag{S.22}$$

As is evident from result (S.21), the left side of equality (S.22) is a polynomial (in u), so that the function polyroot (which is part of R) can be used to determine the values of u that satisfy equality (S.22). Moreover,

$$\max_{\{u\,:\,1\le u\le 8\}}|g(u)| = \max_{\{u\,:\,u\in\mathcal{U}\}}|g(u)| \text{ and } \max_{\{u\,:\,u\in\mathcal{R}^1\}}|g(u)| = \max_{\{u\,:\,u\in\mathcal{U}'\}}|g(u)|, \tag{S.23}$$

where $\mathcal{U}$ is the finite set whose elements consist of 1 and 8 and of any values of u in the interval $1\le u\le 8$ that satisfy equality (S.22) and where $\mathcal{U}'$ is the finite set whose elements consist of all of the values of u that satisfy equality (S.22).

Similarly, $\dfrac{d\,h(u)}{du} = 0$ (or, equivalently, $\dfrac{d\,[-h(u)]}{du} = 0$) if and only if

$$\mathbf{t}'\mathbf{W}\frac{d\,\mathbf{x}(u)}{du} = 0. \tag{S.24}$$

And the left side of equality (S.24) is a second-degree polynomial (in u), so that the values of u that satisfy equality (S.24) can be easily determined, and

$$\max_{\{u\,:\,1\le u\le 8\}}|h(u)| = \max_{\{u\,:\,u\in\mathcal{U}^*\}}|h(u)|, \tag{S.25}$$

where $\mathcal{U}^*$ is the finite set whose elements consist of 1 and 8 and of any values of u in the interval $1 \le u \le 8$ that satisfy equality (S.24).

Results (S.23) and (S.25) were used in obtaining Monte Carlo approximations to $c_{.10}$ and $c^*_{.10}$. Monte Carlo approximations were determined from 599999 draws with the following results:

$$c_{.10} \doteq 2.5183 \quad \text{or} \quad c_{.10} \doteq 2.5794,$$

depending on whether or not u is restricted to the interval $1 \le u \le 8$, and

$$c^*_{.10} \doteq 1.3516.$$

(d)

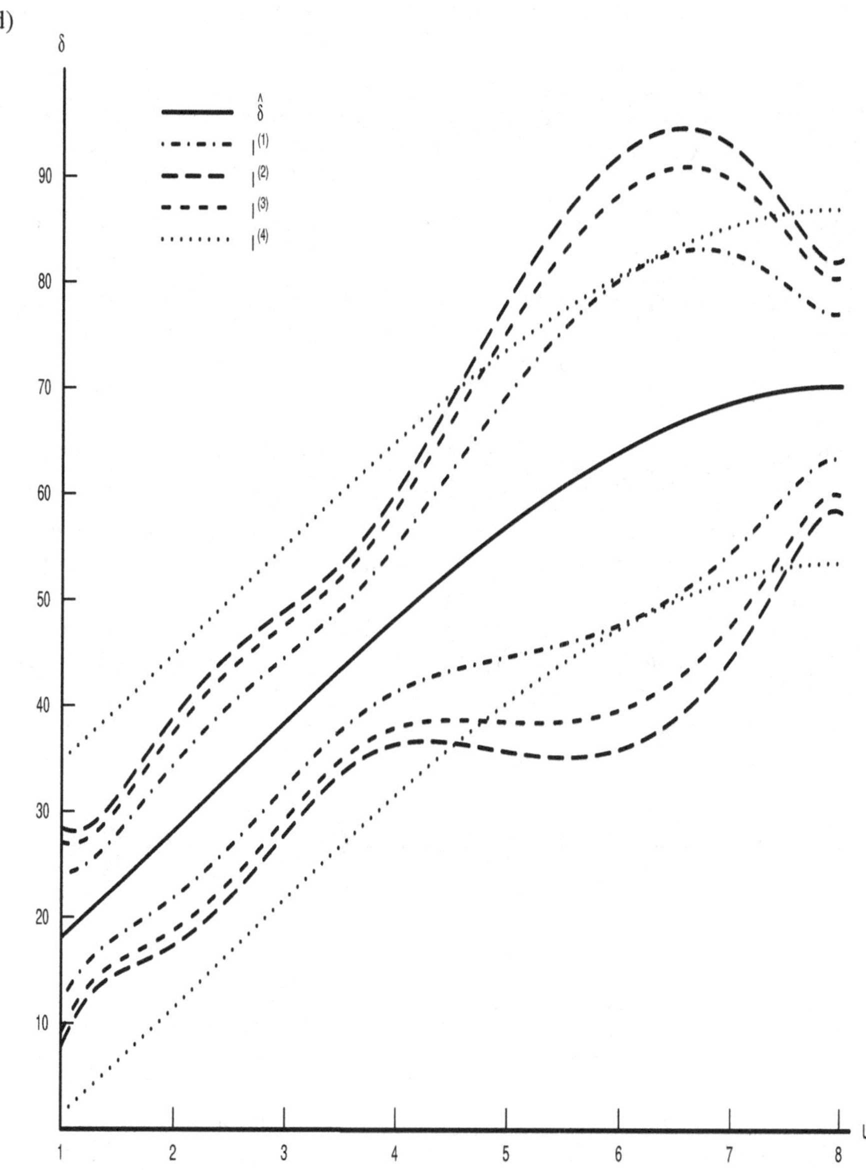

EXERCISE 19. Taking the setting to be that of the final four parts of Section 7.3b (and adopting the notation and terminology employed therein) and taking $\tilde{G}_2$ to be the group of transformations consisting of the totality of the groups $\tilde{G}_2(\boldsymbol{\alpha}^{(0)})$ $(\boldsymbol{\alpha}^{(0)} \in \mathcal{R}^{M_*})$ and $\tilde{G}_3$ the group consisting of

the totality of the groups $\tilde{G}_3(\boldsymbol{\alpha}^{(0)})$ $(\boldsymbol{\alpha}^{(0)} \in \mathbb{R}^{M_*})$, show that (1) if a confidence set $\tilde{A}(\mathbf{z})$ for $\boldsymbol{\alpha}$ is equivariant with respect to the groups $\tilde{G}_0$ and $\tilde{G}_2(\mathbf{0})$, then it is equivariant with respect to the group $\tilde{G}_2$ and (2) if a confidence set $\tilde{A}(\mathbf{z})$ for $\boldsymbol{\alpha}$ is equivariant with respect to the groups $\tilde{G}_0$ and $\tilde{G}_3(\mathbf{0})$, then it is equivariant with respect to the group $\tilde{G}_3$.

Solution. (1) Suppose tha $\tilde{A}(\mathbf{z})$ is equivariant with respect to the groups $\tilde{G}_0$ and $\tilde{G}_2(\mathbf{0})$. Then, for every strictly positive scalar k (and every value of $\mathbf{z}$),

$$\tilde{A}(k\mathbf{z}) = \tilde{A}[\tilde{T}_2(\mathbf{z};\mathbf{0})] = \{\bar{\boldsymbol{\alpha}} : \bar{\boldsymbol{\alpha}} = k\dot{\boldsymbol{\alpha}},\ \dot{\boldsymbol{\alpha}} \in \tilde{A}(\mathbf{z})\}. \tag{S.26}$$

And [in combination with the equivariance of $\tilde{A}(\mathbf{z})$ with respect to the group $\tilde{G}_0$] result (S.26) implies that for every $M_* \times 1$ vector $\mathbf{a}$ and every strictly positive scalar k,

$$\begin{aligned}\tilde{A}\left[\begin{pmatrix} k\mathbf{z}_1+\mathbf{a} \\ k\mathbf{z}_2 \\ k\mathbf{z}_3 \end{pmatrix}\right] &= \tilde{A}[\tilde{T}_0(k\mathbf{z})] = \{\ddot{\boldsymbol{\alpha}} : \ddot{\boldsymbol{\alpha}} = \mathbf{a}+\bar{\boldsymbol{\alpha}},\ \bar{\boldsymbol{\alpha}} \in \tilde{A}(k\mathbf{z})\} \\ &= \{\ddot{\boldsymbol{\alpha}} : \ddot{\boldsymbol{\alpha}} = \mathbf{a}+k\dot{\boldsymbol{\alpha}},\ \dot{\boldsymbol{\alpha}} \in \tilde{A}(\mathbf{z})\}. \end{aligned}\tag{S.27}$$

Moreover, upon setting $\mathbf{a} = \boldsymbol{\alpha}^{(0)} - k\boldsymbol{\alpha}^{(0)}$ in result (S.27), we find that (for every $M_* \times 1$ vector $\boldsymbol{\alpha}^{(0)}$ and every strictly positive scalar k)

$$\tilde{A}\left[\begin{pmatrix} \boldsymbol{\alpha}^{(0)} + k(\mathbf{z}_1 - \boldsymbol{\alpha}^{(0)}) \\ k\mathbf{z}_2 \\ k\mathbf{z}_3 \end{pmatrix}\right] = \{\ddot{\boldsymbol{\alpha}} : \ddot{\boldsymbol{\alpha}} = \boldsymbol{\alpha}^{(0)} + k(\dot{\boldsymbol{\alpha}} - \boldsymbol{\alpha}^{(0)}),\ \dot{\boldsymbol{\alpha}} \in \tilde{A}(\mathbf{z})\}.$$

Thus, $\tilde{A}(\mathbf{z})$ is equivariant with respect to the group $\tilde{G}_2$.

(2) That the equivariance of $\tilde{A}(\mathbf{z})$ with respect to the groups $\tilde{G}_0$ and $\tilde{G}_3(\mathbf{0})$ implies its equivariance with respect to the group $\tilde{G}_3$ can be established via an analogous argument.

EXERCISE 20. Taking the setting to be that of Section 7.4e (and adopting the assumption of normality and the notation and terminology employed therein), suppose that $M_* = 1$, and write $\hat{\alpha}$ for $\hat{\boldsymbol{\alpha}}$, α for $\boldsymbol{\alpha}$, and $\alpha^{(0)}$ for $\boldsymbol{\alpha}^{(0)}$. Further, let $\tilde{\phi}(\hat{\alpha}, \hat{\boldsymbol{\eta}}, \mathbf{d})$ represent the critical function of an arbitrary (possibly randomized) level-$\dot{\gamma}$ test of the null hypothesis $\tilde{H}_0 : \alpha = \alpha^{(0)}$ versus the alternative hypothesis $\tilde{H}_1 : \alpha \neq \alpha^{(0)}$, and let $\tilde{\gamma}(\alpha, \boldsymbol{\eta}, \sigma)$ represent its power function {so that $\tilde{\gamma}(\alpha, \boldsymbol{\eta}, \sigma) = \mathrm{E}[\tilde{\phi}(\hat{\alpha}, \hat{\boldsymbol{\eta}}, \mathbf{d})]$}. And define $s = (\hat{\alpha} - \alpha^{(0)})/[(\hat{\alpha} - \alpha^{(0)})^2 + \mathbf{d}'\mathbf{d}]^{1/2}$ and $w = (\hat{\alpha} - \alpha^{(0)})^2 + \mathbf{d}'\mathbf{d}$, denote by $\ddot{\phi}(s, w, \hat{\boldsymbol{\eta}})$ the critical function of a level-$\dot{\gamma}$ test (of $\tilde{H}_0$ versus $\tilde{H}_1$) that depends on $\hat{\alpha}$, $\hat{\boldsymbol{\eta}}$, and $\mathbf{d}$ only through the values of s, w, and $\hat{\boldsymbol{\eta}}$, and write E_0 for the expectation operator E in the special case where $\alpha = \alpha^{(0)}$.

(a) Show that if the level-$\dot{\gamma}$ test with critical function $\tilde{\phi}(\hat{\alpha}, \hat{\boldsymbol{\eta}}, \mathbf{d})$ is an unbiased test, then

$$\tilde{\gamma}(\alpha^{(0)}, \boldsymbol{\eta}, \sigma) = \dot{\gamma} \quad \text{for all } \boldsymbol{\eta} \text{ and } \sigma \tag{E.3}$$

and

$$\left.\frac{\partial \tilde{\gamma}(\alpha, \boldsymbol{\eta}, \sigma)}{\partial \alpha}\right|_{\alpha = \alpha^{(0)}} = 0 \quad \text{for all } \boldsymbol{\eta} \text{ and } \sigma. \tag{E.4}$$

(b) Show that

$$\frac{\partial \tilde{\gamma}(\alpha, \boldsymbol{\eta}, \sigma)}{\partial \alpha} = \mathrm{E}\left[\frac{\hat{\alpha} - \alpha}{\sigma^2}\tilde{\phi}(\hat{\alpha}, \hat{\boldsymbol{\eta}}, \mathbf{d})\right]. \tag{E.5}$$

(c) Show that corresponding to the level-$\dot{\gamma}$ test (of $\tilde{H}_0$) with critical function $\tilde{\phi}(\hat{\alpha}, \hat{\boldsymbol{\eta}}, \mathbf{d})$, there is a (level-$\dot{\gamma}$) test that depends on $\hat{\alpha}$, $\hat{\boldsymbol{\eta}}$, and $\mathbf{d}$ only through the values of s, w, and $\hat{\boldsymbol{\eta}}$ and that has the same power function as the test with critical function $\tilde{\phi}(\hat{\alpha}, \hat{\boldsymbol{\eta}}, \mathbf{d})$.

(d) Show that when $\alpha = \alpha^{(0)}$, (1) w and $\hat{\boldsymbol{\eta}}$ form a complete sufficient statistic and (2) s is statistically independent of w and $\hat{\boldsymbol{\eta}}$ and has an absolutely continuous distribution, the pdf of which is the pdf $h^*(\cdot)$ given by result (6.4.7).

(e) Show that when the critical function $\tilde{\phi}(\hat{\alpha}, \hat{\boldsymbol{\eta}}, \mathbf{d})$ (of the level-$\dot{\gamma}$ test) is of the form $\ddot{\phi}(s, w, \hat{\boldsymbol{\eta}})$, condition (E.3) is equivalent to the condition

$$\mathrm{E}_0[\ddot{\phi}(s, w, \hat{\boldsymbol{\eta}}) \mid w, \hat{\boldsymbol{\eta}}] = \dot{\gamma} \quad (\text{wp1}) \tag{E.6}$$

and condition (E.4) is equivalent to the condition

$$\mathrm{E}_0[s w^{1/2} \ddot{\phi}(s, w, \hat{\boldsymbol{\eta}}) \mid w, \hat{\boldsymbol{\eta}}] = 0 \quad (\text{wp1}). \tag{E.7}$$

(f) Using the generalized Neyman-Pearson lemma (Lehmann and Romano 2005b, sec. 3.6; Shao 2010, sec. 6.1.1), show that among critical functions of the form $\ddot{\phi}(s, w, \hat{\boldsymbol{\eta}})$ that satisfy (for any particular values of w and $\hat{\boldsymbol{\eta}}$) the conditions

$$\mathrm{E}_0[\ddot{\phi}(s, w, \hat{\boldsymbol{\eta}}) \mid w, \hat{\boldsymbol{\eta}}] = \dot{\gamma} \quad \text{and} \quad \mathrm{E}_0[s w^{1/2} \ddot{\phi}(s, w, \hat{\boldsymbol{\eta}}) \mid w, \hat{\boldsymbol{\eta}}] = 0, \tag{E.8}$$

the value of $\mathrm{E}[\ddot{\phi}(s, w, \hat{\boldsymbol{\eta}}) \mid w, \hat{\boldsymbol{\eta}}]$ (at those particular values of w and $\hat{\boldsymbol{\eta}}$) is maximized [for any particular value of α ($\neq \alpha^{(0)}$) and any particular values of $\boldsymbol{\eta}$ and σ] when the critical function is taken to be the critical function $\ddot{\phi}^*(s, w, \hat{\boldsymbol{\eta}})$ defined (for all s, w, and $\hat{\boldsymbol{\eta}}$) as follows:

$$\ddot{\phi}^*(s, w, \hat{\boldsymbol{\eta}}) = \begin{cases} 1, & \text{if } s < -c \text{ or } s > c, \\ 0, & \text{if } -c \le s \le c, \end{cases}$$

where c is the upper $100(\dot{\gamma}/2)\%$ point of the distribution with pdf $h^*(\cdot)$ [given by result (6.4.7)].

(g) Use the results of the preceding parts to conclude that among all level-$\dot{\gamma}$ tests of $\tilde{H}_0$ versus $\tilde{H}_1$ that are unbiased, the size-$\dot{\gamma}$ two-sided t test is a UMP test.

Solution. Let $\mathbf{z} = (\hat{\alpha}, \hat{\boldsymbol{\eta}}', \mathbf{d}')'$, and observe that (by definition)

$$\tilde{\gamma}(\alpha, \boldsymbol{\eta}, \sigma) = \int_{\mathbb{R}^N} \tilde{\phi}(\hat{\alpha}, \hat{\boldsymbol{\eta}}, \mathbf{d}) f_1(\hat{\alpha}; \alpha, \sigma) f_2(\hat{\boldsymbol{\eta}}; \boldsymbol{\eta}, \sigma) f_3(\mathbf{d}; \sigma)\, d\mathbf{z}, \tag{S.28}$$

where $f_1(\cdot\,; \alpha, \sigma)$ is the pdf of an $N(\alpha, \sigma^2)$ distribution, $f_2(\cdot\,; \boldsymbol{\eta}, \sigma)$ the pdf of an $N(\boldsymbol{\eta}, \sigma^2 \mathbf{I}_{P_*-1})$ distribution, and $f_3(\cdot\,; \sigma)$ the pdf of an $N(\mathbf{0}, \sigma^2 \mathbf{I}_{N-P_*})$ distribution. Observe also that

$$f_1(\hat{\alpha}; \alpha, \sigma) = (2\pi\sigma^2)^{-1/2} \exp[-(\hat{\alpha}-\alpha)^2/(2\sigma^2)]. \tag{S.29}$$

(a) To verify result (E.3), it suffices to observe that for any particular values of $\boldsymbol{\eta}$ and σ, $\tilde{\gamma}(\alpha, \boldsymbol{\eta}, \sigma)$ is a continuous function of α and hence if $\tilde{\gamma}(\alpha^{(0)}, \boldsymbol{\eta}, \sigma) < \dot{\gamma}$, then $\tilde{\gamma}(\alpha^{(0)}+\delta, \boldsymbol{\eta}, \sigma) < \dot{\gamma}$ for some sufficiently small nonzero scalar δ, in which case the test [with critical function $\tilde{\phi}(\hat{\alpha}, \hat{\boldsymbol{\eta}}, \mathbf{d})$] would not be unbiased. And to verify result (E.4), it suffices to observe that if (for any particular values of $\boldsymbol{\eta}$ and σ) the value of the partial derivative of $\tilde{\gamma}(\alpha, \boldsymbol{\eta}, \sigma)$ (with respect to α) at $\alpha = \alpha^{(0)}$ were either less than 0 or greater than 0, then [since $\tilde{\gamma}(\alpha^{(0)}, \boldsymbol{\eta}, \sigma) \le \dot{\gamma}$] $\tilde{\gamma}(\alpha^{(0)}+\delta, \boldsymbol{\eta}, \sigma) < \dot{\gamma}$ for some sufficiently small strictly positive or strictly negative scalar δ, in which case the test would not be unbiased.

(b) Result (E.5) follows from result (S.28) upon observing [in light of result (S.29)] that

$$\frac{\partial f_1(\hat{\alpha}; \alpha, \sigma)}{\partial \alpha} = (2\pi\sigma^2)^{-1/2} \exp[-(\hat{\alpha}-\alpha)^2/(2\sigma^2)] \frac{\partial[-(\hat{\alpha}-\alpha)^2/(2\sigma^2)]}{\partial \alpha} = \frac{\hat{\alpha}-\alpha}{\sigma^2} f_1(\hat{\alpha}; \alpha, \sigma).$$

(c) The observable random variables s and w and the observable random vector $\hat{\boldsymbol{\eta}}$ form a sufficient statistic as can be readily verified upon observing that [as α ranges over $\mathbb{R}^1$, $\boldsymbol{\eta}$ over $\mathbb{R}^{P_*-1}$, and σ over the interval $(0, \infty)$] the pdf of the joint distribution of $\hat{\alpha}$, $\hat{\boldsymbol{\eta}}$, and $\mathbf{d}$ forms an exponential family. Thus, the conditional expectation $\mathrm{E}[\tilde{\phi}(\hat{\alpha}, \hat{\boldsymbol{\eta}}, \mathbf{d}) \mid s, w, \hat{\boldsymbol{\eta}}]$ does not depend on α, $\boldsymbol{\eta}$, or σ and hence (when regarded as a function of s, w, and $\hat{\boldsymbol{\eta}}$) can serve as the critical function of a test of $\tilde{H}_0$. Moreover, this test is of level $\dot{\gamma}$ and in fact has the same power function as the test with critical function $\tilde{\phi}(\hat{\alpha}, \hat{\boldsymbol{\eta}}, \mathbf{d})$, as is evident upon observing that

$$\mathrm{E}\{\mathrm{E}[\tilde{\phi}(\hat{\alpha}, \hat{\boldsymbol{\eta}}, \mathbf{d}) \mid s, w, \hat{\boldsymbol{\eta}}]\} = \mathrm{E}[\tilde{\phi}(\hat{\alpha}, \hat{\boldsymbol{\eta}}, \mathbf{d})].$$

(d) Suppose that $\alpha = \alpha^{(0)}$. (1) That w and $\hat{\boldsymbol{\eta}}$ form a complete sufficient statistic is evident from the pdf of the joint distribution of $\hat{\alpha}$, $\hat{\eta}$, and $\mathbf{d}$ upon applying results [like those discussed by Lehmann and Romano (2005b, secs. 2.7 and 4.3)] on exponential families. (2) That s is statistically independent of w and $\hat{\eta}$ and has an absolutely continuous distribution with pdf $h^*(\cdot)$ is evident from the results of Section 6.4a (upon observing that $\hat{\alpha}$ and $\mathbf{d}$ are distributed independently of $\hat{\eta}$).

(e) That condition (E.3) is equivalent to condition (E.6) and condition (E.4) to condition (E.7) [when the critical function is of the form $\ddot{\phi}(s, w, \hat{\eta})$] is evident from the results of Parts (d) and (b) upon recalling the definition of completeness and upon observing that

$$E_0[\ddot{\phi}(s, w, \hat{\eta})] = E_0\{E_0[\ddot{\phi}(s, w, \hat{\eta}) \mid w, \hat{\eta}]\}, \tag{S.30}$$

that $\hat{\alpha} - \alpha^{(0)} = sw^{1/2}$, and that

$$E_0[sw^{1/2}\ddot{\phi}(s, w, \hat{\eta})] = E_0\{E_0[sw^{1/2}\ddot{\phi}(s, w, \hat{\eta}) \mid w, \hat{\eta}]\}. \tag{S.31}$$

(f) When $\alpha = \alpha^{(0)}$, the conditional distribution of s given w and $\hat{\eta}$ is the absolutely continuous distribution with pdf $h^*(\cdot)$ [as is evident from the result of Part (d)]. Moreover, $h^*(-s) = h^*(s)$ for "every" value of s. Thus,

$$E_0[\ddot{\phi}(s, w, \hat{\eta}) \mid w, \hat{\eta}] = \textstyle\int_c^1 h^*(s)\,ds + \int_{-1}^{-c} h^*(s)\,ds = 2\int_c^1 h^*(s)\,ds = \dot{\gamma}$$

and

$$\begin{aligned} E_0[sw^{1/2}\ddot{\phi}(s, w, \hat{\eta}) \mid w, \hat{\eta}] &= \textstyle\int_c^1 sw^{1/2}h^*(s)\,ds + \int_{-1}^{-c} sw^{1/2}h^*(s)\,ds \\ &= \textstyle\int_c^1 sw^{1/2}h^*(s)\,ds - \int_c^1 sw^{1/2}h^*(s)\,ds = 0. \end{aligned}$$

When $\alpha \neq \alpha^{(0)}$, the conditional distribution of s given w and $\hat{\eta}$ is an absolutely continuous distribution with a pdf $q^*(s \mid w)$ that is proportional (for any particular values of w and $\hat{\eta}$) to the pdf of the joint distribution of w and s; an expression for the latter pdf is obtainable from result (6.4.32). Together, results (6.4.7) and (6.4.32) imply that

$$\frac{q^*(s \mid w)}{h^*(s)} = k_0\, e^{w^{1/2}s(\alpha-\alpha^{(0)})}$$

for some "constant" k_0. Moreover, there exist "constants" k_1 and k_2 such that

$$k_0\, e^{w^{1/2}c(\alpha-\alpha^{(0)})} = k_1 + k_2 w^{1/2} c \quad \text{and} \quad k_0\, e^{w^{1/2}(-c)(\alpha-\alpha^{(0)})} = k_1 + k_2 w^{1/2}(-c).$$

And these constants are such that

$$\begin{aligned} k_0 e^{w^{1/2}s(\alpha-\alpha^{(0)})} > k_1 + k_2 w^{1/2} s \quad &\Leftrightarrow \quad s < -c \text{ or } s > c \\ &\Leftrightarrow \quad \ddot{\phi}^*(s, w, \hat{\eta}) = 1. \end{aligned}$$

Thus, it follows from the generalized Neyman-Pearson lemma that $E[\ddot{\phi}(s, w, \hat{\eta}) \mid w, \hat{\eta}]$ achieves its maximum value [with respect to the choice of the function $\ddot{\phi}(\cdot, \cdot, \cdot)$ from among those candidates that satisfy conditions (E.8)] when $\ddot{\phi}(\cdot, \cdot, \cdot)$ is taken to be the function $\ddot{\phi}^*(\cdot, \cdot, \cdot)$.

(g) The size-$\dot{\gamma}$ two-sided t test of $\tilde{H}_0$ versus $\tilde{H}_1$ is equivalent to the test with critical function $\ddot{\phi}^*(s, w, \hat{\eta})$, as can be readily verified by making use of relationship (6.4.31). Moreover, among tests with a critical function of the form $\ddot{\phi}(s, w, \hat{\eta})$ that satisfy conditions (E.6) and (E.7), the test with critical function $\ddot{\phi}^*(s, w, \hat{\eta})$ is a UMP test, as is evident from the result of Part (f). And in light of the results of Parts (c) and (e), it follows that the test with critical function $\ddot{\phi}^*(s, w, \hat{\eta})$ is a UMP test among all tests with a critical function $\tilde{\phi}(\hat{\alpha}, \hat{\eta}, \mathbf{d})$ that satisfies conditions (E.3) and (E.4). Thus, since the critical function $\tilde{\phi}(\hat{\alpha}, \hat{\eta}, \mathbf{d})$ of any level-$\dot{\gamma}$ unbiased test satisfies conditions (E.3) and (E.4) [as is evident from the result of Part (a)] and since the size-$\dot{\gamma}$ two-sided t test is an unbiased test, we conclude that among all level-$\dot{\gamma}$ unbiased tests, the size-$\dot{\gamma}$ two-sided t test is a UMP test.

EXERCISE 21. Taking the setting to be that of Section 7.4e and adopting the assumption of normality and the notation and terminology employed therein, let $\tilde{\gamma}(\boldsymbol{\alpha}, \boldsymbol{\eta}, \sigma)$ represent the power function of a size-$\dot{\gamma}$ similar test of $H_0 : \boldsymbol{\tau} = \boldsymbol{\tau}^{(0)}$ or $\tilde{H}_0 : \boldsymbol{\alpha} = \boldsymbol{\alpha}^{(0)}$ versus $H_1 : \boldsymbol{\tau} \neq \boldsymbol{\tau}^{(0)}$ or $\tilde{H}_1 : \boldsymbol{\alpha} \neq \boldsymbol{\alpha}^{(0)}$. Show that $\min_{\boldsymbol{\alpha} \in S(\boldsymbol{\alpha}^{(0)}, \rho)} \tilde{\gamma}(\boldsymbol{\alpha}, \boldsymbol{\eta}, \sigma)$ attains its maximum value when the size-$\dot{\gamma}$ similar test is taken to be the size-$\dot{\gamma}$ F test.

Solution. Define $\tilde{\gamma}^{\min}(\rho, \boldsymbol{\eta}, \sigma) = \min_{\boldsymbol{\alpha} \in S(\boldsymbol{\alpha}^{(0)}, \rho)} \tilde{\gamma}(\boldsymbol{\alpha}, \boldsymbol{\eta}, \sigma)$. Then, clearly,

$$\tilde{\gamma}^{\min}(\rho, \boldsymbol{\eta}, \sigma) \leq \bar{\gamma}(\rho, \boldsymbol{\eta}, \sigma). \tag{S.32}$$

Further, let $\tilde{\gamma}_F^{\min}(\rho, \boldsymbol{\eta}, \sigma)$ and $\bar{\gamma}_F(\rho, \boldsymbol{\eta}, \sigma)$ represent $\tilde{\gamma}^{\min}(\rho, \boldsymbol{\eta}, \sigma)$ and $\bar{\gamma}(\rho, \boldsymbol{\eta}, \sigma)$ in the special case where $\tilde{\gamma}(\boldsymbol{\alpha}, \boldsymbol{\eta}, \sigma)$ is the power function of the size-$\dot{\gamma}$ F test, and observe that

$$\tilde{\gamma}_F^{\min}(\rho, \boldsymbol{\eta}, \sigma) = \bar{\gamma}_F(\rho, \boldsymbol{\eta}, \sigma). \tag{S.33}$$

According to the main result of Section 7.4e,

$$\bar{\gamma}(\rho, \boldsymbol{\eta}, \sigma) \leq \bar{\gamma}_F(\rho, \boldsymbol{\eta}, \sigma). \tag{S.34}$$

Together, results (S.32), (S.33), and (S.34) imply that

$$\tilde{\gamma}^{\min}(\rho, \boldsymbol{\eta}, \sigma) \leq \tilde{\gamma}_F^{\min}(\rho, \boldsymbol{\eta}, \sigma).$$

Thus, $\tilde{\gamma}^{\min}(\rho, \boldsymbol{\eta}, \sigma)$ attains its maximum value when the size-$\dot{\gamma}$ similar test is taken to be the size-$\dot{\gamma}$ F test.

EXERCISE 22. Taking the setting to be that of Section 7.4e and adopting the assumption of normality and the notation and terminology employed therein, let $\tilde{\phi}(\hat{\boldsymbol{\alpha}}, \hat{\boldsymbol{\eta}}, \mathbf{d})$ represent the critical function of an arbitrary size-$\dot{\gamma}$ test of the null hypothesis $\tilde{H}_0 : \boldsymbol{\alpha} = \boldsymbol{\alpha}^{(0)}$ versus the alternative hypothesis $\tilde{H}_1 : \boldsymbol{\alpha} \neq \boldsymbol{\alpha}^{(0)}$. Further, let $\tilde{\gamma}(\cdot, \cdot, \cdot; \tilde{\phi})$ represent the power function of the test with critical function $\tilde{\phi}(\cdot, \cdot, \cdot)$, so that $\tilde{\gamma}(\boldsymbol{\alpha}, \boldsymbol{\eta}, \sigma; \tilde{\phi}) = \mathrm{E}[\tilde{\phi}(\hat{\boldsymbol{\alpha}}, \hat{\boldsymbol{\eta}}, \mathbf{d})]$. And take $\tilde{\gamma}^*(\cdot, \cdot, \cdot)$ to be the function defined as follows:

$$\tilde{\gamma}^*(\boldsymbol{\alpha}, \boldsymbol{\eta}, \sigma) = \sup_{\tilde{\phi}} \tilde{\gamma}(\boldsymbol{\alpha}, \boldsymbol{\eta}, \sigma; \tilde{\phi}).$$

This function is called the *envelope power function.*

(a) Show that $\tilde{\gamma}^*(\boldsymbol{\alpha}, \boldsymbol{\eta}, \sigma)$ depends on $\boldsymbol{\alpha}$ only through the value of $(\boldsymbol{\alpha} - \boldsymbol{\alpha}^{(0)})'(\boldsymbol{\alpha} - \boldsymbol{\alpha}^{(0)})$.

(b) Let $\tilde{\phi}_F(\hat{\boldsymbol{\alpha}}, \hat{\boldsymbol{\eta}}, \mathbf{d})$ represent the critical function of the size-$\dot{\gamma}$ F test of $\tilde{H}_0$ versus $\tilde{H}_1$. And as a basis for evaluating the test with critical function $\tilde{\phi}(\cdot, \cdot, \cdot)$, consider the use of the criterion

$$\max_{\boldsymbol{\alpha} \in S(\boldsymbol{\alpha}^{(0)}, \rho)} [\tilde{\gamma}^*(\boldsymbol{\alpha}, \boldsymbol{\eta}, \sigma) - \tilde{\gamma}(\boldsymbol{\alpha}, \boldsymbol{\eta}, \sigma; \tilde{\phi})], \tag{E.9}$$

which reflects [for $\boldsymbol{\alpha} \in S(\boldsymbol{\alpha}^{(0)}, \rho)$] the extent to which the power function of the test deviates from the envelope power function. Using the result of Exercise 21 (or otherwise), show that the size-$\dot{\gamma}$ F test is the "most stringent" size-$\dot{\gamma}$ similar test in the sense that (for "every" value of ρ) the value attained by the quantity (E.9) when $\tilde{\phi} = \tilde{\phi}_F$ is a minimum among those attained when $\tilde{\phi}$ is the critical function of some (size-$\dot{\gamma}$) similar test.

Solution. (a) Let $\boldsymbol{\alpha}_1$ and $\boldsymbol{\alpha}_2$ represent any two values of $\boldsymbol{\alpha}$ such that $(\boldsymbol{\alpha}_2 - \boldsymbol{\alpha}^{(0)})'(\boldsymbol{\alpha}_2 - \boldsymbol{\alpha}^{(0)}) = (\boldsymbol{\alpha}_1 - \boldsymbol{\alpha}^{(0)})'(\boldsymbol{\alpha}_1 - \boldsymbol{\alpha}^{(0)})$. It suffices to show that corresponding to any critical function $\tilde{\phi}_1(\cdot, \cdot, \cdot)$, there exists a second critical function $\tilde{\phi}_2(\cdot, \cdot, \cdot)$ such that $\tilde{\gamma}(\boldsymbol{\alpha}_2, \boldsymbol{\eta}, \sigma; \tilde{\phi}_2) = \tilde{\gamma}(\boldsymbol{\alpha}_1, \boldsymbol{\eta}, \sigma; \tilde{\phi}_1)$.

According to Lemma 5.9.9, there exists an orthogonal matrix $\mathbf{O}$ such that $\boldsymbol{\alpha}_2 - \boldsymbol{\alpha}^{(0)} = \mathbf{O}(\boldsymbol{\alpha}_1 - \boldsymbol{\alpha}^{(0)})$. Take $\tilde{\phi}_2(\cdot, \cdot, \cdot)$ to be a critical function defined [in terms of $\tilde{\phi}_1(\cdot, \cdot, \cdot)$] as follows:

$$\tilde{\phi}_2(\hat{\boldsymbol{\alpha}}, \hat{\boldsymbol{\eta}}, \mathbf{d}) = \tilde{\phi}_1[\boldsymbol{\alpha}^{(0)} + \mathbf{O}'(\hat{\boldsymbol{\alpha}} - \boldsymbol{\alpha}^{(0)}), \hat{\boldsymbol{\eta}}, \mathbf{d}].$$

And observe that

$$\begin{aligned}\tilde{\gamma}(\boldsymbol{\alpha}_2, \boldsymbol{\eta}, \sigma; \tilde{\phi}_2) &= \textstyle\int_{\mathbb{R}^N} \tilde{\phi}_2(\hat{\boldsymbol{\alpha}}, \hat{\boldsymbol{\eta}}, \mathbf{d}) f_1(\hat{\boldsymbol{\alpha}}\,; \boldsymbol{\alpha}_2, \sigma) f_2(\hat{\boldsymbol{\eta}}\,; \boldsymbol{\eta}, \sigma) f_3(\mathbf{d}\,; \sigma)\, d\mathbf{z} \\ &= \textstyle\int_{\mathbb{R}^N} \tilde{\phi}_1[\boldsymbol{\alpha}^{(0)} + \mathbf{O}'(\hat{\boldsymbol{\alpha}} - \boldsymbol{\alpha}^{(0)}), \hat{\boldsymbol{\eta}}, \mathbf{d}] \\ &\qquad f_1[\boldsymbol{\alpha}^{(0)} + \mathbf{O}'(\hat{\boldsymbol{\alpha}} - \boldsymbol{\alpha}^{(0)})\,; \boldsymbol{\alpha}_1, \sigma) f_2(\hat{\boldsymbol{\eta}}\,; \boldsymbol{\eta}, \sigma) f_3(\mathbf{d}\,; \sigma)\, d\mathbf{z}.\end{aligned}$$

Then, upon writing $\mathbf{z}_*$ for the vector $(\hat{\boldsymbol{\alpha}}_*', \hat{\boldsymbol{\eta}}', \mathbf{d}')'$, where $\hat{\boldsymbol{\alpha}}_* = \boldsymbol{\alpha}^{(0)} + \mathbf{O}'(\hat{\boldsymbol{\alpha}} - \boldsymbol{\alpha}^{(0)})$, and upon making a change of variables from $\mathbf{z}$ to $\mathbf{z}_*$, we find that

$$\tilde{\gamma}(\boldsymbol{\alpha}_2, \boldsymbol{\eta}, \sigma; \tilde{\phi}_2) = \textstyle\int_{\mathbb{R}^N} \tilde{\phi}_1(\hat{\boldsymbol{\alpha}}_*, \hat{\boldsymbol{\eta}}, \mathbf{d}) f_1(\hat{\boldsymbol{\alpha}}_*; \boldsymbol{\alpha}_1, \sigma) f_2(\hat{\boldsymbol{\eta}}\,; \boldsymbol{\eta}, \sigma) f_3(\mathbf{d}\,; \sigma)\, d\mathbf{z}_* = \tilde{\gamma}(\boldsymbol{\alpha}_1, \boldsymbol{\eta}, \sigma; \tilde{\phi}_1).$$

(b) According to Part (a), $\tilde{\gamma}^*(\boldsymbol{\alpha}, \boldsymbol{\eta}, \sigma)$ has the same value for all $\boldsymbol{\alpha} \in S(\boldsymbol{\alpha}^{(0)}, \rho)$. And as is evident from Exercise 21, $\min_{\boldsymbol{\alpha} \in S(\boldsymbol{\alpha}^{(0)}, \rho)} \tilde{\gamma}(\boldsymbol{\alpha}, \boldsymbol{\eta}, \sigma; \tilde{\phi})$ attains its maximum value and hence $-\min_{\boldsymbol{\alpha} \in S(\boldsymbol{\alpha}^{(0)}, \rho)} \tilde{\gamma}(\boldsymbol{\alpha}, \boldsymbol{\eta}, \sigma; \tilde{\phi})$ its minimum value [subject to the constraint that $\tilde{\phi}(\cdot, \cdot, \cdot)$ is the critical function of a (size-$\dot{\gamma}$) similar test] when $\tilde{\phi} = \tilde{\phi}_F$. Moreover,

$$\max_{\boldsymbol{\alpha} \in S(\boldsymbol{\alpha}^{(0)}, \rho)} [-\tilde{\gamma}(\boldsymbol{\alpha}, \boldsymbol{\eta}, \sigma; \tilde{\phi})] = -\min_{\boldsymbol{\alpha} \in S(\boldsymbol{\alpha}^{(0)}, \rho)} \tilde{\gamma}(\boldsymbol{\alpha}, \boldsymbol{\eta}, \sigma; \tilde{\phi}).$$

Thus, the quantity (E.9) attains its minimum value [subject to the constraint that $\tilde{\phi}(\cdot, \cdot, \cdot)$ is the critical function of a (size-$\dot{\gamma}$) similar test] when $\tilde{\phi} = \tilde{\phi}_F$.

EXERCISE 23. Take the setting to be that of Section 7.5a (and adopt the assumption of normality and the notation and terminology employed therein). Show that among all tests of the null hypothesis $H_0^+ : \tau \leq \tau^{(0)}$ or $\tilde{H}_0^+ : \alpha \leq \alpha^{(0)}$ (versus the alternative hypothesis $H_1^+ : \tau > \tau^{(0)}$ or $\tilde{H}_1^+ : \alpha > \alpha^{(0)}$) that are of level $\dot{\gamma}$ and that are unbiased, the size-$\dot{\gamma}$ one-sided t test is a UMP test. (*Hint*. Proceed stepwise as in Exercise 20.)

Solution. Let $\tilde{\phi}(\hat{\alpha}, \hat{\boldsymbol{\eta}}, \mathbf{d})$ represent the critical function of an arbitrary level-$\dot{\gamma}$ unbiased (possibly randomized) test of the null hypothesis $H_0^+ : \tau \leq \tau^{(0)}$ or $\tilde{H}_0^+ : \alpha \leq \alpha^{(0)}$ (versus the alternative hypothesis $H_1^+ : \tau > \tau^{(0)}$ or $\tilde{H}_1^+ : \alpha > \alpha^{(0)}$), and let $\tilde{\gamma}(\alpha, \boldsymbol{\eta}, \sigma)$ represent its power function {so that $\tilde{\gamma}(\alpha, \boldsymbol{\eta}, \sigma) = \mathrm{E}[\tilde{\phi}(\hat{\alpha}, \hat{\boldsymbol{\eta}}, \mathbf{d})]$}. Then,

$$\tilde{\gamma}(\alpha^{(0)}, \boldsymbol{\eta}, \sigma) = \dot{\gamma} \quad \text{for all } \boldsymbol{\eta} \text{ and } \sigma, \tag{S.35}$$

as is evident upon observing that for any particular values of $\boldsymbol{\eta}$ and σ, $\tilde{\gamma}(\alpha, \boldsymbol{\eta}, \sigma)$ is a continuous function of α, so that if $\tilde{\gamma}(\alpha^{(0)}, \boldsymbol{\eta}, \sigma) < \dot{\gamma}$ for some values of $\boldsymbol{\eta}$ and σ, then $\tilde{\gamma}(\alpha^{(0)}+\delta, \boldsymbol{\eta}, \sigma) < \dot{\gamma}$ for some sufficiently small strictly positive scalar δ (and for those values of $\boldsymbol{\eta}$ and σ), in which case the test [with critical function $\tilde{\phi}(\hat{\alpha}, \hat{\boldsymbol{\eta}}, \mathbf{d})$] would not be unbiased.

Now, let $s = (\hat{\alpha}-\alpha^{(0)})/[(\hat{\alpha}-\alpha^{(0)})^2 + \mathbf{d}'\mathbf{d}]^{1/2}$ and $w = (\hat{\alpha}-\alpha^{(0)})^2 + \mathbf{d}'\mathbf{d}$. And observe (in light of the discussion in the next-to-last part of Section 7.3a) that s, w, and $\hat{\boldsymbol{\eta}}$ form a sufficient statistic. Thus, the conditional expectation $\mathrm{E}[\tilde{\phi}(\hat{\alpha}, \hat{\boldsymbol{\eta}}, \mathbf{d}) \mid s, w, \hat{\boldsymbol{\eta}}]$ does not depend on α, $\boldsymbol{\eta}$, or σ and hence (when regarded as a function of s, w, and $\hat{\boldsymbol{\eta}}$) can serve as the critical function of a test of H_0^+ or $\tilde{H}_0^+$. Moreover,

$$\mathrm{E}\{\mathrm{E}[\tilde{\phi}(\hat{\alpha}, \hat{\boldsymbol{\eta}}, \mathbf{d}) \mid s, w, \hat{\boldsymbol{\eta}}]\} = \mathrm{E}[\tilde{\phi}(\hat{\alpha}, \hat{\boldsymbol{\eta}}, \mathbf{d})], \tag{S.36}$$

so that the test with critical function $\mathrm{E}[\tilde{\phi}(\hat{\alpha}, \hat{\boldsymbol{\eta}}, \mathbf{d}) \mid s, w, \hat{\boldsymbol{\eta}}]$ has the same power function as the test with critical function $\tilde{\phi}(\hat{\alpha}, \hat{\boldsymbol{\eta}}, \mathbf{d})$, implying in particular that [like the test with critical function $\tilde{\phi}(\hat{\alpha}, \hat{\boldsymbol{\eta}}, \mathbf{d})$] it is of level $\dot{\gamma}$ and is unbiased.

Let $\ddot{\phi}(s, w, \hat{\boldsymbol{\eta}})$ represent the critical function of a level-$\dot{\gamma}$ unbiased (possibly randomized) test of H_0^+ or $\tilde{H}_0^+$ (versus H_1^+ or $\tilde{H}_1^+$) that depends on $\hat{\alpha}$, $\hat{\boldsymbol{\eta}}$, and $\mathbf{d}$ only through the values of s, w, and $\hat{\boldsymbol{\eta}}$, and write E_0 for the expectation "operator" in the special case where $\alpha = \alpha^{(0)}$. In light of result (S.36), it suffices to show that the power of the size-$\dot{\gamma}$ one-sided t test of H_0^+ is uniformly (for every value of α that exceeds $\alpha^{(0)}$ and for all values of $\boldsymbol{\eta}$ and σ) greater than or equal to $\mathrm{E}[\ddot{\phi}(s, w, \hat{\boldsymbol{\eta}})]$ [for every choice of the critical function $\ddot{\phi}(\cdot, \cdot, \cdot)$]. And in light of result (S.35),

$$\mathrm{E}_0[\ddot{\phi}(s, w, \hat{\boldsymbol{\eta}})] = \dot{\gamma} \quad \text{for all } \boldsymbol{\eta} \text{ and } \sigma. \tag{S.37}$$

Moreover, when $\alpha = \alpha^{(0)}$, w and $\hat{\boldsymbol{\eta}}$ form a complete sufficient statistic [as is evident from the pdf of the joint distribution of $\hat{\alpha}$, $\hat{\boldsymbol{\eta}}$, and $\mathbf{d}$ upon applying results on exponential families like those discussed by Lehmann and Romano (2005b, secs. 2.7 and 4.3)], so that upon recalling the definition of completeness and upon observing that

$$\mathrm{E}_0[\ddot{\phi}(s, w, \hat{\boldsymbol{\eta}})] = \mathrm{E}_0\{\mathrm{E}_0[\ddot{\phi}(s, w, \hat{\boldsymbol{\eta}}) \mid w, \hat{\boldsymbol{\eta}}]\},$$

we find that condition (S.37) is equivalent to the condition

$$\mathrm{E}_0[\ddot{\phi}(s, w, \hat{\boldsymbol{\eta}}) \mid w, \hat{\boldsymbol{\eta}}] = \dot{\gamma} \quad \text{(wp1)}. \tag{S.38}$$

Clearly, s and w are distributed independently of $\hat{\eta}$. Moreover, the joint distribution of s and w is absolutely continuous with a pdf that is obtainable from result (6.4.32). When $\alpha = \alpha^{(0)}$, s is distributed independently of w as well as $\hat{\eta}$ with a pdf $h^*(\cdot)$ that is obtainable from result (6.4.7) simply by replacing N with $N-P_*$. When $\alpha \neq \alpha^{(0)}$, the conditional distribution of s given w (and $\hat{\eta}$) is absolutely continuous with a pdf $q^*(s \mid w)$ that is proportional (for any particular values of w and $\hat{\eta}$) to the pdf of the joint distribution of s and w. Together, results (6.4.7) and (6.4.32) imply that

$$\frac{q^*(s \mid w)}{h^*(s)} = k e^{w^{1/2} s(\alpha-\alpha^{(0)})/\sigma^2}$$

for some strictly positive scalar k that does not depend on s.

For $\alpha \geq \alpha^{(0)}$, the ratio $q^*(s \mid w)/h^*(s)$ is an increasing function of s, as is evident upon observing that

$$\frac{d\,[q^*(s \mid w)/h^*(s)]}{ds} = k w^{1/2}(\alpha-\alpha^{(0)})\sigma^{-2} e^{w^{1/2} s(\alpha-\alpha^{(0)})/\sigma^2}.$$

Thus, upon applying Theorem 7.4.1 (the Neyman-Pearson lemma) with $X = s$, $\theta = \alpha$, $\Theta = (-\infty, \infty)$, and $\theta^{(0)} = \alpha^{(0)}$ and upon observing that

$$\mathbf{z} \in \tilde{C}^+ \quad \Leftrightarrow \quad s \in \left\{ s : s \leq \frac{\bar{t}_{\dot{\gamma}}(N-P_*)}{\sqrt{N-P_* + [\bar{t}_{\dot{\gamma}}(N-P_*)]^2}} \right\}$$

[analogous to the equivalence (5.6)], we find that subject to the constraint $\mathrm{E}_0[\ddot{\phi}(s, w, \hat{\eta}) \mid w, \hat{\eta}] = \dot{\gamma}$, $\mathrm{E}[\ddot{\phi}(s, w, \hat{\eta}) \mid w, \hat{\eta}]$ attains its maximum value for any specified value of α that exceeds $\alpha^{(0)}$ and for any specified values of η and σ when the critical region of the test is taken to be the critical region $\tilde{C}^+$ or C^+ of the size-$\dot{\gamma}$ one-sided t test of H_0^+. And in light of the equivalence of conditions (S.37) and (S.38), it follows that subject to the constraint imposed on $\ddot{\phi}(s, w, \hat{\eta})$ by condition (S.37), $\mathrm{E}[\ddot{\phi}(s, w, \hat{\eta})]$ attains its maximum value for any specified value of α that exceeds $\alpha^{(0)}$ and for any specified values of η and σ when the critical region of the test is taken to be $\tilde{C}^+$ or C^+. Since there is no loss of generality in restricting attention to tests of H_0^+ or $\tilde{H}_0^+$ with critical functions of the form $\ddot{\phi}(s, w, \hat{\eta})$,since any level-$\dot{\gamma}$ unbiased test of H_0^+ or $\tilde{H}_0^+$ with a critical function of the form $\ddot{\phi}(s, w, \hat{\eta})$ satisfies condition (S.37), and since the size-$\dot{\gamma}$ one-sided t test of H_0^+ is an unbiased test, we conclude that among level-$\dot{\gamma}$ unbiased tests of H_0^+ or $\tilde{H}_0^+$ (versus H_1^+ or $\tilde{H}_1^+$), the size-$\dot{\gamma}$ one-sided t test of H_0^+ is a UMP test.

EXERCISE 24.

(a) Let (for an arbitrary positive integer M) $f_M(\cdot)$ represent the pdf of a $\chi^2(M)$ distribution. Show that (for $0 < x < \infty$)

$$x f_M(x) = M f_{M+2}(x).$$

(b) Verify [by using Part (a) or by other means] result (6.22).

Solution. (a) Upon recalling result (6.1.16) and that (for any positive scalar α) $\Gamma(\alpha+1) = \alpha\Gamma(\alpha)$, we find that (for $0 < x < \infty$)

$$\begin{aligned} x f_M(x) &= \frac{1}{\Gamma(M/2)2^{M/2}}\, x^{M/2} e^{-x/2} \\ &= \frac{M/2}{\Gamma[(M/2)+1]\,2^{M/2}}\, x^{M/2} e^{-x/2} \\ &= M \frac{1}{\Gamma[(M+2)/2)]\,2^{(M+2)/2}}\, x^{[(M+2)/2]-1} e^{-x/2} = M f_{M+2}(x). \end{aligned}$$

(b) It follows from Part (a) that (for $0 < u < \infty$)

$$\frac{u}{N-P_*} g(u) = g^*(u),$$

where $g^*(\cdot)$ is the pdf of the $\chi^2(N-P_*+2)$ distribution. Thus,

$$\begin{aligned}\int_{c_0}^{c_1}\left(\frac{u}{N-P_*}-1\right)g(u)\,du &= \int_{c_0}^{c_1}[g^*(u)-g(u)]\,du \\ &= \int_{0}^{c_1}[g^*(u)-g(u)]\,du - \int_{0}^{c_0}[g^*(u)-g(u)]\,du \\ &= G^*(c_1)-G(c_1)-[G^*(c_0)-G(c_0)].\end{aligned}$$

EXERCISE 25. This exercise is to be regarded as a continuation of Exercise 18. Suppose (as in Exercise 18) that the data (of Section 4.2 b) on the lethal dose of ouabain in cats are regarded as the observed values of the elements $y_1, y_2, \ldots, y_N$ of an $N(=41)$-dimensional observable random vector $\mathbf{y}$ that follows a G–M model. Suppose further that (for $i = 1, 2, \ldots, 41$) $\mathrm{E}(y_i) = \delta(u_i)$, where $u_1, u_2, \ldots, u_{41}$ are the values of the rate u of injection and where $\delta(u)$ is the third-degree polynomial

$$\delta(u) = \beta_1 + \beta_2 u + \beta_3 u^2 + \beta_4 u^3.$$

And suppose that the distribution of the vector $\mathbf{e}$ of residual effects is $N(\mathbf{0}, \sigma^2\mathbf{I})$.

(a) Determine for $\dot{\gamma} = 0.10$ and also for $\dot{\gamma} = 0.05$ (1) the value of the $100(1-\dot{\gamma})\%$ lower confidence bound for σ provided by the left end point of interval (6.2) and (2) the value of the $100(1-\dot{\gamma})\%$ upper confidence bound for σ provided by the right end point of interval (6.3).

(b) Obtain [via an implementation of interval (6.23)] a 90% two-sided strictly unbiased confidence interval for σ.

Solution. Let us adopt the notation employed in Section 7.6a. And observe that S (the residual sum of squares) equals 5650.109 and that $N-P_* = N-P = 37$ [so that the estimator $\hat{\sigma}$ of σ obtained by taking the square root of the usual (unbiased) estimator $S/(N-P_*)$ of σ^2 is equal to 12.357].

(a) When $\dot{\gamma} = 0.10$, $\bar{\chi}^2_{\dot{\gamma}} = 48.363$ and $\bar{\chi}^2_{1-\dot{\gamma}} = 26.492$; and when $\dot{\gamma} = 0.05$, $\bar{\chi}^2_{\dot{\gamma}} = 52.192$ and $\bar{\chi}^2_{1-\dot{\gamma}} = 24.075$. Thus, (1) the value of the 90% lower confidence bound for σ is 10.809, and the value of the 95% lower confidence bound is 10.405; and (2) the value of the 90% upper confidence bound for σ is 14.604, and the value of the 95% upper confidence bound is 15.320.

(b) When $\dot{\gamma} = 0.10$, the value $\dot{\gamma}_1^*$ of $\dot{\gamma}_1$ that is a solution to equation (6.19) is found to be 0.04205, and the corresponding value $\dot{\gamma}_2^*$ of $\dot{\gamma}_2$ is $\dot{\gamma}_2^* = \dot{\gamma}-\dot{\gamma}_1^* = 0.05795$. And $\bar{\chi}^2_{.04205} = 53.088$, and $\bar{\chi}^2_{1-.05795} = \bar{\chi}^2_{.94205} = 24.546$. Thus, upon setting $S = 5650.109$, $\dot{\gamma}_1^* = 0.04205$, and $\dot{\gamma}_2^* = 0.05795$ in the interval (6.23), we obtain as a 90% strictly unbiased confidence interval for σ the interval

$$10.316 \le \sigma \le 15.172.$$

EXERCISE 26. Take the setting to be that of the final part of Section 7.6c, and adopt the notation and terminology employed therein. In particular, take the canonical form of the G–M model to be that identified with the special case where $M_* = P_*$, so that $\boldsymbol{\alpha}$ and $\hat{\boldsymbol{\alpha}}$ are P_*-dimensional. Show that the (size-$\dot{\gamma}$) test of the null hypothesis $H_0^+ : \sigma \le \sigma_0$ (versus the alternative hypothesis $H_1^+ : \sigma > \sigma_0$) with critical region C^+ is UMP among all level-$\dot{\gamma}$ tests. Do so by carrying out the following steps.

(a) Let $\phi(T, \hat{\boldsymbol{\alpha}})$ represent the critical function of a level-$\dot{\gamma}$ test of H_0^+ versus H_1^+ [that depends on the vector $\mathbf{d}$ only through the value of T $(= \mathbf{d}'\mathbf{d}/\sigma_0^2)$]. And let $\gamma(\sigma, \boldsymbol{\alpha})$ represent the power function of the test with critical function $\phi(T, \hat{\boldsymbol{\alpha}})$. Further, let σ_* represent any particular value of σ greater than σ_0, let $\boldsymbol{\alpha}_*$ represent any particular value of $\boldsymbol{\alpha}$, and denote by $h(\cdot\,; \sigma)$ the pdf of the distribution of T, by $f(\cdot\,; \boldsymbol{\alpha}, \sigma)$ the pdf of the distribution of $\hat{\boldsymbol{\alpha}}$, and by $s(\cdot)$ the pdf of the $N(\boldsymbol{\alpha}_*, \sigma_*^2-\sigma_0^2)$ distribution. Show (1) that

$$\int_{\mathbb{R}^{P_*}} \gamma(\sigma_0, \boldsymbol{\alpha})\, s(\boldsymbol{\alpha})\, d\boldsymbol{\alpha} \le \dot{\gamma} \tag{E.10}$$

and (2) that

$$\int_{\mathbb{R}^{P_*}} \gamma(\sigma_0, \boldsymbol{\alpha}) s(\boldsymbol{\alpha})\, d\boldsymbol{\alpha} = \int_{\mathbb{R}^{P_*}} \int_0^\infty \phi(t, \hat{\boldsymbol{\alpha}}) h(t; \sigma_0) f(\hat{\boldsymbol{\alpha}}; \boldsymbol{\alpha}_*, \sigma_*)\, dt\, d\hat{\boldsymbol{\alpha}}. \tag{E.11}$$

(b) By, for example, using a version of the Neyman-Pearson lemma like that stated by Casella and Berger (2002) in the form of their Theorem 8.3.12, show that among those choices for the critical function $\phi(T, \hat{\boldsymbol{\alpha}})$ for which the power function $\gamma(\cdot, \cdot)$ satisfies condition (E.10), $\gamma(\sigma_*, \boldsymbol{\alpha}_*)$ can be maximized by taking $\phi(T, \hat{\boldsymbol{\alpha}})$ to be the critical function $\phi_*(T, \hat{\boldsymbol{\alpha}})$ defined as follows:

$$\phi_*(t, \hat{\boldsymbol{\alpha}}) = \begin{cases} 1, & \text{when } t > \bar{\chi}^2_{\dot{\gamma}}, \\ 0, & \text{when } t \le \bar{\chi}^2_{\dot{\gamma}}. \end{cases}$$

(c) Use the results of Parts (a) and (b) to reach the desired conclusion, that is, to show that the test of H_0^+ (versus H_1^+) with critical region C^+ is UMP among all level-$\dot{\gamma}$ tests.

Solution. (a)

(1) By definition, the test with critical function $\phi(T, \hat{\boldsymbol{\alpha}})$ is of level $\dot{\gamma}$, so that $\gamma(\sigma_0, \boldsymbol{\alpha}) \le \dot{\gamma}$ for all $\boldsymbol{\alpha}$. And it follows that

$$\int_{\mathbb{R}^{P_*}} \gamma(\sigma_0, \boldsymbol{\alpha}) s(\boldsymbol{\alpha})\, d\boldsymbol{\alpha} \le \dot{\gamma}.$$

(2) Clearly,

$$\int_{\mathbb{R}^{P_*}} \gamma(\sigma_0, \boldsymbol{\alpha}) s(\boldsymbol{\alpha})\, d\boldsymbol{\alpha} = \int_{\mathbb{R}^{P_*}} \int_0^\infty \phi(t, \hat{\boldsymbol{\alpha}}) h(t; \sigma_0) \int_{\mathbb{R}^{P_*}} f(\hat{\boldsymbol{\alpha}}; \boldsymbol{\alpha}, \sigma_0) s(\boldsymbol{\alpha})\, d\boldsymbol{\alpha}\, dt\, d\hat{\boldsymbol{\alpha}}.$$

Moreover,

$$\int_{\mathbb{R}^{P_*}} f(\hat{\boldsymbol{\alpha}}; \boldsymbol{\alpha}, \sigma_0) s(\boldsymbol{\alpha})\, d\boldsymbol{\alpha} = f(\hat{\boldsymbol{\alpha}}; \boldsymbol{\alpha}_*, \sigma_*),$$

as is evident upon regarding $\hat{\boldsymbol{\alpha}}$ and $\boldsymbol{\alpha}$ as random vectors whose joint distribution is $N\left\{\begin{pmatrix}\boldsymbol{\alpha}_* \\ \boldsymbol{\alpha}_*\end{pmatrix}, \begin{bmatrix}\sigma_*^2\mathbf{I} & (\sigma_*^2-\sigma_0^2)\mathbf{I} \\ (\sigma_*^2-\sigma_0^2)\mathbf{I} & (\sigma_*^2-\sigma_0^2)\mathbf{I}\end{bmatrix}\right\}$; when $\hat{\boldsymbol{\alpha}}$ and $\boldsymbol{\alpha}$ are so regarded, $f(\hat{\boldsymbol{\alpha}}; \boldsymbol{\alpha}, \sigma_0)$ is the pdf of the conditional distribution of $\hat{\boldsymbol{\alpha}}$ given $\boldsymbol{\alpha}$ and $s(\boldsymbol{\alpha})$ is the pdf of the marginal distribution of $\boldsymbol{\alpha}$, in which case $\int_{\mathbb{R}^{P_*}} f(\hat{\boldsymbol{\alpha}}; \boldsymbol{\alpha}, \sigma_0) s(\boldsymbol{\alpha})\, d\boldsymbol{\alpha}$ equals the pdf of the marginal distribution of $\hat{\boldsymbol{\alpha}}$.

(b) Clearly, the value $\gamma(\sigma_*, \boldsymbol{\alpha}_*)$ (when $\sigma = \sigma_*$ and $\boldsymbol{\alpha} = \boldsymbol{\alpha}_*$) of the power function $\gamma(\sigma, \boldsymbol{\alpha})$ of the test with critical function $\phi(T, \hat{\boldsymbol{\alpha}})$ is expressible as

$$\gamma(\sigma_*, \boldsymbol{\alpha}_*) = \int_{\mathbb{R}^{P_*}} \int_0^\infty \phi(t, \hat{\boldsymbol{\alpha}}) h(t; \sigma_*) f(\hat{\boldsymbol{\alpha}}; \boldsymbol{\alpha}_*, \sigma_*)\, dt\, d\hat{\boldsymbol{\alpha}}.$$

Moreover, the ratio $h(t; \sigma_*)/h(t; \sigma_0)$ is (for $t > 0$) a strictly increasing function of t, so that (for some strictly positive constant k) $\phi_*(t, \hat{\boldsymbol{\alpha}})$ is reexpressible in the form

$$\phi_*(t, \hat{\boldsymbol{\alpha}}) = \begin{cases} 1, & \text{when } h(t; \sigma_*) f(\hat{\boldsymbol{\alpha}}; \boldsymbol{\alpha}_*, \sigma_*) > k\, h(t; \sigma_0) f(\hat{\boldsymbol{\alpha}}; \boldsymbol{\alpha}_*, \sigma_*), \\ 0, & \text{when } h(t; \sigma_*) f(\hat{\boldsymbol{\alpha}}; \boldsymbol{\alpha}_*, \sigma_*) \le k\, h(t; \sigma_0) f(\hat{\boldsymbol{\alpha}}; \boldsymbol{\alpha}_*, \sigma_*). \end{cases}$$

And upon recalling result (E.11) and applying the Neyman-Pearson lemma, it follows that $\gamma(\sigma_*, \boldsymbol{\alpha}_*)$ can by maximized [subject to the constraint imposed on the choice of $\phi(T, \hat{\boldsymbol{\alpha}})$ by condition (E.10)] by taking $\phi(T, \hat{\boldsymbol{\alpha}})$ to be the critical function $\phi_*(T, \hat{\boldsymbol{\alpha}})$.

(c) Recall (from the final part of Section 7.6c) that corresponding to any (possibly randomized) test of H_0^+ (versus H_1^+) there is a test with a critical function of the form $\phi(T, \hat{\boldsymbol{\alpha}})$ that has the same power function. And observe that the test of H_0^+ with critical region C^+ is identical to the test with critical function $\phi_*(T, \hat{\boldsymbol{\alpha}})$. Accordingly, let us restrict attention to tests with a critical function of the form $\phi(T, \hat{\boldsymbol{\alpha}})$ and observe that (for purposes of showing that the test with critical region C^+ is UMP among all level-$\dot{\gamma}$ tests) there is no loss of generality in doing so.

Among tests with a critical function for which the power function $\gamma(\cdot, \cdot)$ satisfies condition (E.10), $\gamma(\sigma_*, \boldsymbol{\alpha}_*)$ is maximized by taking the critical function to be $\phi_*(T, \hat{\boldsymbol{\alpha}})$ [as is evident from Part b)]. Moreover, the set consisting of all tests that are of level $\dot{\gamma}$ is a subset of the set consisting of all tests with a power function that satisties condition (E.10). Thus, since the test with critical function $\phi_*(T, \hat{\boldsymbol{\alpha}})$ is of level $\dot{\gamma}$, we conclude that among all level-$\dot{\gamma}$ tests, $\gamma(\sigma_*, \boldsymbol{\alpha}_*)$ is maximized by taking the critical function of the test to be $\phi_*(T, \hat{\boldsymbol{\alpha}})$. It remains only to observe that this conclusion does not depend on the choice of σ_* or $\boldsymbol{\alpha}_*$.

EXERCISE 27. Take the context to be that of Section 7.7a, and adopt the notation employed therein. Using Markov's inequality (e.g., Casella and Berger 2002, lemma 3.8.3; Bickel and Doksum 2001, sec. A.15) or otherwise, verify inequality (7.12), that is, the inequality

$$\Pr(x_i > c \text{ for } k \text{ or more values of } i) \leq (1/k) \textstyle\sum_{i=1}^{M} \Pr(x_i > c).$$

Solution. Let $I(x)$ represent an indicator function defined (for $x \in \mathcal{R}^1$) as follows:

$$I(x) = \begin{cases} 1, & \text{if } x > c, \\ 0, & \text{otherwise.} \end{cases}$$

Then, upon making use of Markov's inequality, we find that

$$\begin{aligned} \Pr(x_i > c \text{ for } k \text{ or more values of } i) &= \Pr[I(x_i) = 1 \text{ for } k \text{ or more values of } i] \\ &= \Pr\big[\textstyle\sum_{i=1}^{M} I(x_i) \geq k\big] \\ &\leq (1/k)\, \mathrm{E}\big[\textstyle\sum_{i=1}^{M} I(x_i)\big] \\ &= (1/k) \textstyle\sum_{i=1}^{M} \mathrm{E}[I(x_i)] \\ &= (1/k) \textstyle\sum_{i=1}^{M} \Pr(x_i > c). \end{aligned}$$

EXERCISE 28.

(a) Letting X represent any random variable whose values are confined to the interval $[0, 1]$ and letting κ $(0 < \kappa < 1)$ represent a constant, show (1) that

$$\mathrm{E}(X) \leq \kappa \Pr(X \leq \kappa) + \Pr(X > \kappa) \tag{E.12}$$

and then use inequality (E.12) along with Markov's inequality (e.g., Casella and Berger 2002, sec. 3.8) to (2) show that

$$\frac{\mathrm{E}(X) - \kappa}{1 - \kappa} \leq \Pr(X > \kappa) \leq \frac{\mathrm{E}(X)}{\kappa}. \tag{E.13}$$

(b) Show that the requirement that the false discovery rate (FDR) satisfy condition (7.45) and the requirement that the false discovery proportion (FDP) satisfy condition (7.46) are related as follows:
(1) if FDR $\leq \delta$, then Pr(FDP $> \kappa) \leq \delta/\kappa$; and
(2) if Pr(FDP $> \kappa) \leq \epsilon$, then FDR $\leq \epsilon + \kappa(1 - \epsilon)$.

Solution. (a)
(1) Clearly,

$$\begin{aligned} \mathrm{E}(X) &= \mathrm{E}(X \mid X \leq \kappa) \Pr(X \leq \kappa) + \mathrm{E}(X \mid X > \kappa) \Pr(X > \kappa) \\ &\leq \kappa \Pr(X \leq \kappa) + \Pr(X > \kappa). \end{aligned}$$

(2) That $\dfrac{\mathrm{E}(X) - \kappa}{1 - \kappa} \leq \Pr(X > \kappa)$ is evident from inequality (E.12) upon observing that $\Pr(X \leq \kappa) = 1 - \Pr(X > \kappa)$. That $\Pr(X > \kappa) \leq \dfrac{\mathrm{E}(X)}{\kappa}$ follows from Markov's inequality.

(b)

(1) If FDR $\leq \delta$, then upon applying the inequality $\Pr(X > \kappa) \leq \dfrac{E(X)}{\kappa}$ with $X = \text{FDP}$, we find that $\Pr(\text{FDP} > \kappa) \leq \dfrac{\text{FDR}}{\kappa} \leq \dfrac{\delta}{\kappa}$.

(2) If $\Pr(\text{FDP} > \kappa) \leq \epsilon$, then upon applying the inequality $\dfrac{E(X) - \kappa}{1-\kappa} \leq \Pr(X > \kappa)$ (with $X = \text{FDP}$), we find that $\dfrac{\text{FDR} - \kappa}{1-\kappa} \leq \Pr(\text{FDP} > \kappa)$ and hence that

$$\text{FDR} \leq \kappa + (1-\kappa)\Pr(\text{FDP} > \kappa) \leq \kappa + (1-\kappa)\epsilon.$$

EXERCISE 29. Taking the setting to be that of Section 7.7 and adopting the terminology and notation employed therein, consider the use of a multiple-comparison procedure in testing (for every $i \in I = \{1, 2, \ldots, M\}$) the null hypothesis $H_i^{(0)} : \tau_i = \tau_i^{(0)}$ versus the alternative hypothesis $H_i^{(1)} : \tau_i \neq \tau_i^{(0)}$ (or $H_i^{(0)} : \tau_i \leq \tau_i^{(0)}$ versus $H_i^{(1)} : \tau_i > \tau_i^{(0)}$). And denote by T the set of values of $i \in I$ for which $H_i^{(0)}$ is true and by F the set for which $H_i^{(0)}$ is false. Further, denote by M_T the size of the set T and by X_T the number of values of $i \in T$ for which $H_i^{(0)}$ is rejected. Similarly, denote by M_F the size of the set F and by X_F the number of values of $i \in F$ for which $H_i^{(0)}$ is rejected. Show that

(a) in the special case where $M_T = 0$, FWER = FDR = 0;

(b) in the special case where $M_T = M$, FWER = FDR; and

(c) in the special case where $0 < M_T < M$, FWER $\geq$ FDR, with equality holding if and only if $\Pr(X_T > 0 \text{ and } X_F > 0) = 0$.

Solution. By definition,

$$\text{FDP} = \begin{cases} X_T/(X_T+X_F), & \text{if } X_T+X_F > 0, \\ 0, & \text{if } X_T+X_F = 0. \end{cases}$$

Thus,

$$\text{FDP} = 0 \;\Leftrightarrow\; X_T = 0; \tag{S.39}$$

$$\text{FDP} = 1 \;\Leftrightarrow\; X_T > 0 \text{ and } X_F = 0; \text{ and} \tag{S.40}$$

$$0 < \text{FDP} < 1 \;\Leftrightarrow\; X_T > 0 \text{ and } X_F > 0. \tag{S.41}$$

And in light of results (S.39), (S.40), and (S.41),

$$\text{FWER} = \Pr(X_T > 0) = \Pr(\text{FDP} > 0). \tag{S.42}$$

(a) Suppose that $M_T = 0$. Then, clearly, $X_T = 0$. And in light of result (S.39), FDP = 0. Thus,

$$\text{FWER} = \Pr(X_T > 0) = 0, \quad \text{and}$$
$$\text{FDR} = E(\text{FDP}) = 0.$$

(b) and (c) Suppose that $\Pr(X_T > 0 \text{ and } X_F > 0) = 0$. Then, in light of result (S.41),

$$\Pr(\text{FDP}=0 \text{ or } \text{FDP}=1) = 1.$$

Thus,

$$\text{FDR} = E(\text{FDP}) = 0\Pr(\text{FDP}=0) + 1\Pr(\text{FDP}=1) = \Pr(\text{FDP}=1) = \Pr(\text{FDP}>0).$$

And [in light of result (S.42)] it follows that

$$\text{FWER} = \text{FDR}, \tag{S.43}$$

which upon observing that

$$M_T = M \Rightarrow M_F = 0 \Rightarrow X_F = 0 \Rightarrow \Pr(X_T > 0 \text{ and } X_F > 0) = 0$$

establishes Part (b).

Now, suppose that $0 < M_T < M$ and that $\Pr(X_T > 0 \text{ and } X_F > 0) > 0$. And observe that if $X_F > 0$, then

$$\text{FDP} \le \frac{M_T}{M_T+1} < 1.$$

Then, upon recalling results (S.41) and (S.42), we find that

$$\begin{aligned}\text{FDR} &= \text{E(FDP)}\\ &= 0\Pr(\text{FDP}=0) + 1\Pr(\text{FDP}=1) + \text{E}(\text{FDP} \mid X_T > 0 \text{ and } X_F > 0)\Pr(0 < \text{FDP} < 1)\\ &< \Pr(\text{FDP}=1) + \Pr(0 < \text{FDP} < 1)\\ &= \Pr(\text{FDP} > 0)\\ &= \text{FWER},\end{aligned}$$

which [in combination with result (S.43)] establishes Part (c).

EXERCISE 30.

(a) Let $\hat{p}_1, \hat{p}_2, \ldots, \hat{p}_t$ represent p-values [so that $\Pr(\hat{p}_i \le u) \le u$ for $i = 1, 2, \ldots, t$ and for every $u \in (0, 1)$]. Further, let $\hat{p}_{(j)} = \hat{p}_{i_j}$ $(j = 1, 2, \ldots, t)$, where $i_1, i_2, \ldots, i_t$ is a permutation of the first t positive integers $1, 2, \ldots, t$ such that $\hat{p}_{i_1} \le \hat{p}_{i_2} \le \cdots \le \hat{p}_{i_t}$. And let s represent a positive integer such that $s \le t$, and let $c_0, c_1, \ldots, c_s$ represent constants such that $0 = c_0 \le c_1 \le \cdots \le c_s \le 1$. Show that

$$\Pr(\hat{p}_{(j)} \le c_j \text{ for 1 or more values of } j \in \{1, 2, \ldots, s\}) \le t\textstyle\sum_{j=1}^{s}(c_j - c_{j-1})/j. \tag{E.14}$$

(b) Take the setting to be that of Section 7.7d, and adopt the notation and terminology employed therein. And suppose that the α_j's of the step-down multiple-comparison procedure for testing the null hypotheses $H_1^{(0)}, H_2^{(0)}, \ldots, H_M^{(0)}$ are of the form

$$\alpha_j = \bar{t}_{([j\kappa]+1)\dot{\gamma}/\{2(M+[j\kappa]+1-j)\}}(N - P_*) \quad (j = 1, 2, \ldots, M) \tag{E.15}$$

[where $\dot{\gamma} \in (0, 1)$].

(1) Show that if

$$\Pr(|t_{u;T}| > \bar{t}_{u\dot{\gamma}/(2M_T)}(N - P_*) \text{ for 1 or more values of } u \in \{1, 2, \ldots, K\}) \le \epsilon, \tag{E.16}$$

then the step-down procedure [with α_j's of the form (E.15)] is such that $\Pr(\text{FDP} > \kappa) \le \epsilon$.

(2) Reexpress the left side of inequality (E.16) in terms of the left side of inequality (E.14).

(3) Use Part (a) to show that

$$\begin{aligned}&\Pr(|t_{u;T}| > \bar{t}_{u\dot{\gamma}/(2M_T)}(N - P_*)\\ &\qquad \text{for 1 or more values of } u \in \{1, 2, \ldots, K\}) \le \dot{\gamma}\textstyle\sum_{u=1}^{[M\kappa]+1} 1/u.\end{aligned} \tag{E.17}$$

(4) Show that the version of the step-down procedure [with α_j's of the form (E.15)] obtained upon setting $\dot{\gamma} = \epsilon/\sum_{u=1}^{[M\kappa]+1} 1/u$ is such that $\Pr(\text{FDP} > \kappa) \le \epsilon$.

Solution. (a) Let j' represent the smallest value of $j \in \{1, 2, \ldots, s\}$ for which $\hat{p}_{(j)} \le c_j$—if $\hat{p}_{(j)} > c_j$ for $j = 1, 2, \ldots, s$, set $j' = 0$. Then, upon regarding j' as a random variable and observing that $\{j' : j'=1\}, \{j' : j'=2\}, \ldots, \{j' : j'=s\}$, are disjoint events, we find that

$$\begin{aligned}&\Pr(\hat{p}_{(j)} \le c_j \text{ for 1 or more values of } j \in \{1, 2, \ldots, s\})\\ &\qquad = \Pr(j'=k \text{ for some integer } k \text{ between 1 and } s\text{, inclusive})\\ &\qquad = \textstyle\sum_{k=1}^{s} \Pr(j'=k).\end{aligned} \tag{S.44}$$

Now, for purposes of obtaining an upper bound on $\sum_{k=1}^{s}\Pr(j'=k)$, let N_j represent the number of p-values that are less than or equal to c_j, and observe that (for $j = 1, 2, \ldots, s$)

$$\textstyle\sum_{k=1}^{s} k\Pr(j'=k) \le \text{E}(N_j) \le \sum_{k=1}^{t}\Pr(\hat{p}_k \le c_j) \le \sum_{k=1}^{t} c_j = tc_j. \tag{S.45}$$

Then, upon multiplying both sides of inequality (S.45) by $1/[j(j+1)]$ for values of $j<s$ and by $1/s$ for $j=s$ (and upon summing over the s values of j), we obtain the inequality

$$\sum_{j=1}^{s-1} \frac{1}{j(j+1)} \sum_{k=1}^{j} k \Pr(j'=k) + \frac{1}{s} \sum_{k=1}^{s} k \Pr(j'=k) \leq \sum_{j=1}^{s-1} \frac{t c_j}{j(j+1)} + \frac{t c_s}{s}. \tag{S.46}$$

Moreover, upon interchanging the order of summation, the left side of inequality (S.46) is reexpressible as

$$\sum_{k=1}^{s-1} \left(\frac{1}{k} - \frac{1}{s}\right) k \Pr(j'=k) + \frac{1}{s} \sum_{k=1}^{s} k \Pr(j'=k) = \sum_{k=1}^{s} \Pr(j'=k). \tag{S.47}$$

And the right side of inequality (S.46) is reexpressible as

$$t \sum_{j=1}^{s-1} c_j \left(\frac{1}{j} - \frac{1}{j+1}\right) + t\, c_s \frac{1}{s} = t \sum_{j=1}^{s} \frac{c_j - c_{j-1}}{j}. \tag{S.48}$$

Thus, inequality (S.46) is reexpressible in the form

$$\textstyle\sum_{k=1}^{s} \Pr(j'=k) \leq t \sum_{j=1}^{s} (c_j - c_{j-1})/j$$

and hence [in light of result (S.44)] is reexpressible in the form of inequality (E.14).

(b)

(1) In light of inequality (7.66), we find that [for α_j's of the form (E.15)]

$$\begin{aligned}\Pr(|t_{u;T}| > \bar{t}_{u\dot{\gamma}/(2M_T)}(N-P_*) &\text{ for 1 or more values of } u \in \{1,2,\ldots,K\})\\ &\geq \Pr(|t_{u;T}| > \alpha_u^* \text{ for 1 or more values of } u \in \{1,2,\ldots,K\}).\end{aligned}$$

Thus, if condition (E.16) is satisfied, condition (7.62) is also satisfied, in which case the step-down procedure is such that $\Pr(\text{FDP} > \kappa) \leq \epsilon$.

(2) Set $t = M_T$. And letting $k_1, k_2, \ldots, k_{M_T}$ represent the elements of T, take (for $j = 1,2,\ldots,M_T$) $\hat{p}_j$ to be the value of $\dot{\gamma}$ for which $\bar{t}_{\dot{\gamma}/2}(N-P_*) = |t_{k_j}|$. Further, set $s = K$, and take (for $u = 1,2,\ldots,K$) $c_u = u\dot{\gamma}/M_T$. Then,

$$\begin{aligned}\Pr(|t_{u;T}| > \bar{t}_{u\dot{\gamma}/(2M_T)}(N-P_*) &\text{ for 1 or more values of } u \in \{1,2,\ldots,K\})\\ &= \Pr(\hat{p}_{(u)} \leq c_u \text{ for 1 or more values of } u \in \{1,2,\ldots,K\}).\end{aligned} \tag{S.49}$$

(3) Upon applying Part (a) to the right side of equality (S.49), we find that

$$\begin{aligned}\Pr(|t_{u;T}| > \bar{t}_{u\dot{\gamma}/(2M_T)}(N-P_*) &\text{ for 1 or more values of } u \in \{1,2,\ldots,K\})\\ &\leq M_T[c_1 + \textstyle\sum_{u=2}^{K} (c_u - c_{u-1})/u]\\ &= \dot{\gamma}\{1 + \textstyle\sum_{u=2}^{K} [u-(u-1)]/u\}\\ &= \dot{\gamma} \textstyle\sum_{u=1}^{K} 1/u\\ &\leq \dot{\gamma} \textstyle\sum_{u=1}^{[M\kappa]+1} 1/u.\end{aligned}$$

(4) When $\dot{\gamma} = \epsilon/\sum_{u=1}^{[M\kappa]+1} 1/u$, the right side of inequality (E.17) equals ϵ. Thus, when $\dot{\gamma} = \epsilon/\sum_{u=1}^{[M\kappa]+1} 1/u$, condition (E.16) is satisfied. And we conclude [on the basis of Part (1)] that upon setting $\dot{\gamma} = \epsilon/\sum_{u=1}^{[M\kappa]+1} 1/u$, the step-down procedure [with α_j's of the form (E.15)] is such that $\Pr(\text{FDP} > \kappa) \leq \epsilon$.

EXERCISE 31. Take the setting to be that of Section 7.7e, and adopt the notation and terminology employed therein. And take $\alpha_1^*, \alpha_2^*, \ldots, \alpha_M^*$ to be scalars defined implicitly (in terms of $\alpha_1, \alpha_2, \ldots, \alpha_M$) by the equalities

$$\alpha'_k = \textstyle\sum_{j=1}^{k} \alpha_j^* \qquad (k = 1, 2, \ldots, M) \tag{E.18}$$

or explicitly as

$$\alpha_j^* = \begin{cases} \Pr(\alpha_{j-1} \geq |t| > \alpha_j), & \text{for } j = 2, 3, \ldots, M, \\ \Pr(|t| > \alpha_j), & \text{for } j = 1, \end{cases}$$

where $t \sim St(N - P_*)$.

(a) Show that the step-up procedure for testing the null hypotheses $H_i^{(0)} : \tau_i = \tau_i^{(0)}$ $(i = 1, 2, \ldots, M)$ is such that (1) the FDR is less than or equal to $M_T \sum_{j=1}^{M} \alpha_j^*/j$; (2) when $M \sum_{j=1}^{M} \alpha_j^*/j < 1$, the FDR is controlled at level $M \sum_{j=1}^{M} \alpha_j^*/j$ (regardless of the identity of the set T); and (3) in the special case where (for $j = 1, 2, \ldots, M$) α'_j is of the form $\alpha'_j = j\dot{\gamma}/M$, the FDR is less than or equal to $\dot{\gamma}\,(M_T/M) \sum_{j=1}^{M} 1/j$ and can be controlled at level δ by taking $\dot{\gamma} = \delta \left(\sum_{j=1}^{M} 1/j\right)^{-1}$.

(b) The sum $\sum_{j=1}^{M} 1/j$ is "tightly" bounded from above by the quantity

$$\gamma + \log(M + 0.5) + [24(M + 0,5)^2]^{-1}, \tag{E.19}$$

where γ is the Euler-Mascheroni constant (e.g., Chen 2010)—to 10 significant digits, $\gamma = 0.5772156649$. Determine the value of $\sum_{j=1}^{M} 1/j$ and the amount by which this value is exceeded by the value of expression (E.19). Do so for each of the following values of M: 5, 10, 50, 100, 500, 1,000, 5,000, 10,000, 20,000, and 50,000.

(c) What modifications are needed to extend the results encapsulated in Part (a) to the step-up procedure for testing the null hypotheses $H_i^{(0)} : \tau_i \leq \tau_i^{(0)}$ $(i = 1, 2, \ldots, M)$?

Solution. (a)

(1) Starting with expression (7.91) and recalling that (for all i) the sets $A_{1;i}^*, A_{2;i}^*, \ldots, A_{M;i}^*$ are mutually disjoint and that their union equals $\mathcal{R}^{M-1}$, we find that

$$\begin{aligned}
\text{FDR} &= \textstyle\sum_{i\in T} \sum_{k=1}^{M} (\alpha'_k/k) \Pr\big(\mathbf{t}^{(-i)} \in A_{k;i}^* \mid |t_i| > \alpha_k\big) \\
&= \textstyle\sum_{i\in T} \sum_{k=1}^{M} \sum_{j=1}^{k} \alpha_j^* (1/k) \Pr\big(\mathbf{t}^{(-i)} \in A_{k;i}^* \mid |t_i| > \alpha_k\big) \\
&= \textstyle\sum_{i\in T} \sum_{j=1}^{M} \alpha_j^* \sum_{k=j}^{M} (1/k) \Pr\big(\mathbf{t}^{(-i)} \in A_{k;i}^* \mid |t_i| > \alpha_k\big) \\
&\leq \textstyle\sum_{i\in T} \sum_{j=1}^{M} \alpha_j^* \sum_{k=j}^{M} (1/j) \Pr\big(\mathbf{t}^{(-i)} \in A_{k;i}^* \mid |t_i| > \alpha_k\big) \\
&\leq \textstyle\sum_{i\in T} \sum_{j=1}^{M} \alpha_j^* \sum_{k=1}^{M} (1/j) \Pr\big(\mathbf{t}^{(-i)} \in A_{k;i}^* \mid |t_i| > \alpha_k\big) \\
&= \textstyle\sum_{i\in T} \sum_{j=1}^{M} (\alpha_j^*/j) \sum_{k=1}^{M} \Pr\big(\mathbf{t}^{(-i)} \in A_{k;i}^* \mid |t_i| > \alpha_k\big) \\
&= \textstyle\sum_{i\in T} \sum_{j=1}^{M} \alpha_j^*/j \\
&= M_T \textstyle\sum_{j=1}^{M} \alpha_j^*/j.
\end{aligned}$$

(2) Since $M_T \leq M$, it follows from Part (1) that (when $M \sum_{j=1}^{M} \alpha_j^*/j < 1$) the FDR is controlled at level $M \sum_{j=1}^{M} \alpha_j^*/j$ (regardless of the identity of the set T).

(3) Suppose that (for $j = 1, 2, \ldots, M$) α'_j is of the form $\alpha'_j = j\dot{\gamma}/M$. Then, $\alpha_1^* = \alpha'_1 = \dot{\gamma}/M$; and for $j = 2, 3, \ldots, M$,

$$\alpha_j^* = \alpha'_j - \alpha'_{j-1} = (j\dot{\gamma}/M) - (j-1)\dot{\gamma}/M = \dot{\gamma}/M.$$

And it follows from Part (2) that [when $\dot{\gamma} < \left(\sum_{j=1}^{M} 1/j\right)^{-1}$] the FDR is controlled at level $\dot{\gamma} \sum_{j=1}^{M} 1/j$. Moreover, when $\dot{\gamma} = \delta \left(\sum_{j=1}^{M} 1/j\right)^{-1}$, $\dot{\gamma} \sum_{j=1}^{M} 1/j = \delta$.

(b) Let $\text{Dev}(M)$ represent the function of M whose values are those of the difference between expression (E.19) and the sum $\sum_{j=1}^{M} 1/j$. Then, the values of $\sum_{j=1}^{M} 1/j$ and the values of $\text{Dev}(M)$ corresponding to the various values of M are as follows:

M	$\sum_{j=1}^{M} 1/j$	Dev(M)	M	$\sum_{j=1}^{M} 1/j$	Dev(M)
5	2.28	$< 10^{-5}$	1,000	7.49	$< 10^{-14}$
10	2.93	$< 10^{-6}$	5,000	9.09	$< 10^{-14}$
50	4.50	$< 10^{-8}$	10,000	9.79	$< 10^{-14}$
100	5.19	$< 10^{-10}$	20,000	10.48	$< 10^{-14}$
500	6.79	$< 10^{-12}$	50,000	11.40	$< 10^{-20}$

(c) For $j = 1, 2, \ldots, M$, take $\alpha_j' = \Pr(t > \alpha_j)$, rather than $\alpha_j' = \Pr(|t| > \alpha_j)$. And take $\alpha_1^*, \alpha_2^*, \ldots, \alpha_M^*$ to be the scalars redefined implicitly (via the redefined α_j''s) by the equalities (E.18) or explicitly as

$$\alpha_j^* = \begin{cases} \Pr(\alpha_{j-1} \geq t > \alpha_j), & \text{for } j = 2, 3, \ldots, M, \\ \Pr(t > \alpha_j), & \text{for } j = 1. \end{cases}$$

EXERCISE 32. Take the setting to be that of Section 7.7, and adopt the terminology and notation employed therein. Further, for $j = 1, 2, \ldots, M$, let

$$\dot{\alpha}_j = \bar{t}_{k_j \dot{\gamma}/[2(M+k_j-j)]}(N-P_*),$$

where [for some scalar κ $(0 < \kappa < 1)$] $k_j = [j\kappa]+1$, and let

$$\ddot{\alpha}_j = \bar{t}_{j\dot{\gamma}/(2M)}(N-P_*).$$

And consider two stepwise multiple-comparison procedures for testing the null hypotheses $H_1^{(0)}, H_2^{(0)}, \ldots, H_M^{(0)}$: a stepwise procedure for which α_j is taken to be of the form $\alpha_j = \dot{\alpha}_j$ [as in Section 7.7d in devising a step-down procedure for controlling Pr(FDP $> \kappa$)] and a stepwise procedure for which α_j is taken to be of the form $\alpha_j = \ddot{\alpha}_j$ (as in Section 7.7e in devising a step-up procedure for controlling the FDR). Show that (for $j = 1, 2, \ldots, M$) $\dot{\alpha}_j \geq \ddot{\alpha}_j$, with equality holding if and only if $j \leq 1/(1-\kappa)$ or $j = M$.

Solution. Clearly, it suffices to show that (for $j = 1, 2, \ldots, M$)

$$\frac{j}{M} - \frac{k_j}{M+k_j-j} \geq 0, \tag{S.50}$$

with equality holding if and only if $j \leq 1/(1-\kappa)$ or $j = M$. And as can be readily verified,

$$\frac{j}{M} - \frac{k_j}{M+k_j-j} = \frac{(M-j)(j-k_j)}{M[M-(j-k_j)]}. \tag{S.51}$$

Moreover, for $j = 2, 3, \ldots, M$,

$$[j\kappa] - 1 \leq j\kappa - 1 = (j-1)\kappa - (1-\kappa) < (j-1)\kappa,$$

implying (since $[j\kappa]-1$ is an integer) that

$$[j\kappa] - 1 \leq [(j-1)\kappa],$$

or equivalently that

$$[j\kappa] \leq [(j-1)\kappa] + 1 = k_{j-1},$$

so that

$$j - k_j = j - 1 - [j\kappa] \geq j - 1 - k_{j-1}; \tag{S.52}$$

and upon observing that

$$1 - k_1 = 0$$

and making repeated use of inequality (S.52), it follows that (for $j = 1, 2, \ldots, M$)

$$M - k_M \geq j - k_j \geq 0. \tag{S.53}$$

Together, results (S.51) and (S.53) validate inequality (S.50).

As is evident from equality (S.51), equality holds in inequality (S.50) if and only if $j = M$ or $j - k_j = 0$. Moreover,

$$j - k_j = 0 \iff [j\kappa] = j - 1,$$

and (since both $[j\kappa]$ and $j-1$ are integers and since $j\kappa < j$)

$$[j\kappa] = j-1 \;\Leftrightarrow\; j\kappa \geq j-1 \;\Leftrightarrow\; j \leq 1/(1-\kappa).$$

Thus, equality holds in inequality (S.50) if and only if $j \leq 1/(1-\kappa)$ or $j = M$.

EXERCISE 33. Take the setting to be that of Section 7.7f, and adopt the terminology and notation employed therein. And consider a multiple-compariaon procedure in which (for $i = 1, 2, \ldots, M$) the ith of the M null hypotheses $H_1^{(0)}, H_2^{(0)}, \ldots, H_M^{(0)}$ is rejected if $|t_i^{(0)}| > c$, where c is a strictly positive constant. Further, recall that T is the subset of the set $I = \{1, 2, \ldots, M\}$ such that $i \in T$ if $H_i^{(0)}$ is true, denote by R the subset of I such that $i \in R$ if $H_i^{(0)}$ is rejected, and (for $i = 1, 2, \ldots, M$) take X_i to be a random variable defined as follows:

$$X_i = \begin{cases} 1, & \text{if } |t_i^{(0)}| > c, \\ 0, & \text{if } |t_i^{(0)}| \leq c. \end{cases}$$

(a) Show that

$$\mathrm{E}[(1/M)\textstyle\sum_{i\in T} X_i] = (M_T/M)\Pr(|t| > c) \leq \Pr(|t| > c),$$

where $t \sim St(100)$.

(b) Based on the observation that [when $(1/M)\sum_{i=1}^{M} X_i > 0$]

$$\mathrm{FDP} = \frac{(1/M)\sum_{i\in T} X_i}{(1/M)\sum_{i=1}^{M} X_i},$$

on the reasoning that for large M the quantity $(1/M)\sum_{i=1}^{M} X_i$ can be regarded as a (strictly positive) constant, and on the result of Part (a), the quantity $M_T \Pr(|t| > c)/M_R$ can be regarded as an "estimator" of the FDR [$=$ E(FDP)] and $M \Pr(|t| > c)/M_R$ can be regarded as an estimator of $\max_T$ FDR (Efron 2010, chap. 2)—if $M_R = 0$, take the estimate of the FDR or of $\max_T$ FDR to be 0. Consider the application to the prostate data of the multiple-comparison procedure in the case where $c = c_{\dot\gamma}(k)$ and also in the case where $c = \bar t_{k\dot\gamma/(2M)}(100)$. Use the information provided by the entries in Table 7.5 to obtain an estimate of $\max_T$ FDR for each of these two cases. Do so for $\dot\gamma = .05, .10$, and $.20$ and for $k = 1, 5, 10$, and 20.

Solution. (a) For $i = 1, 2, \ldots, M$, $\mathrm{E}(X_i) = \Pr(|t_i^{(0)}| > c)$. And for $i \in T$, $t_i^{(0)} \sim St(100)$. Thus,

$$\begin{aligned} \mathrm{E}[(1/M)\textstyle\sum_{i\in T} X_i] &= (1/M)\textstyle\sum_{i\in T} \mathrm{E}(X_i) \\ &= (1/M)\textstyle\sum_{i\in T} \Pr(|t_i^{(0)}| > c)) \\ &= (1/M)\textstyle\sum_{i\in T} \Pr(|t| > c)) \\ &= (M_T/M)\Pr(|t| > c) \\ &\leq \Pr(|t| > c). \end{aligned}$$

(b) For the case where $c = c_{\dot\gamma}(k)$ and the case where $c = \bar t = \bar t_{k\dot\gamma/(2M)}(100)$, we obtain the following estimates of $\max_T$ FDR:

	$\dot\gamma = .05$		$\dot\gamma = .10$		$\dot\gamma = .20$	
k	$c = c_{\dot\gamma}(k)$	$c = \bar t$	$c = c_{\dot\gamma}(k)$	$c = \bar t$	$c = c_{\dot\gamma}(k)$	$c = \bar t$
1	.03	.03	.01	.01	.02	.02
5	.06	.02	.07	.04	.07	.05
10	.09	.04	.10	.05	.12	.06
20	.16	.05	.16	.06	.18	.08

For example, when $c = c_{\dot\gamma}(k), k = 10$, and $\dot\gamma = .05$, $M \Pr(|t| > c)/M_R = 6033 \Pr(|t| > 3.42)/58 = .09$.

EXERCISE 34. Take the setting to be that of Part 6 of Section 7.8a (pertaining to the testing of $H_0\colon \mathbf{w} \in S_0$ versus $H_1\colon \mathbf{w} \in S_1$), and adopt the notation and terminology employed therein.

(a) Write p_0 for the random variable $p_0(\mathbf{y})$, and denote by $G_0(\cdot)$ the cdf of the conditional distribution of p_0 given that $\mathbf{w} \in S_0$. Further, take k and c to be the constants that appear in the definition of the critical function $\phi^*(\cdot)$, take $k'' = [1 + (\pi_1/\pi_0)k]^{-1}$, and take $\phi^{**}(\cdot)$ to be a critical function defined as follows:

$$\phi^{**}(\underset{\sim}{y}) = \begin{cases} 1, & \text{when } p_0(\underset{\sim}{y}) < k'', \\ c, & \text{when } p_0(\underset{\sim}{y}) = k'', \\ 0, & \text{when } p_0(\underset{\sim}{y}) > k''. \end{cases}$$

Show that (1) $\phi^{**}(\underset{\sim}{y}) = \phi^*(\underset{\sim}{y})$ when $f(\underset{\sim}{y}) > 0$, (2) that k'' equals the smallest scalar $\underset{\sim}{p}_0$ for which $G_0(\underset{\sim}{p}_0) \geq \dot{\gamma}$, and (3) that

$$c = \frac{\dot{\gamma} - \Pr(p_0 < k'' \mid \mathbf{w} \in S_0)}{\Pr(p_0 = k'' \mid \mathbf{w} \in S_0)} \quad \text{when } \Pr(p_0 = k'' \mid \mathbf{w} \in S_0) > 0$$

—when $\Pr(p_0 = k'' \mid \mathbf{w} \in S_0) = 0$, c can be chosen arbitrarily.

(b) Show that if the joint distribution of $\mathbf{w}$ and $\mathbf{y}$ is MVN, then there exists a version of the critical function $\phi^*(\cdot)$ defined by equalities (8.25) and (8.26) for which $\phi^*(\mathbf{y})$ depends on the value of $\mathbf{y}$ only through the value of $\tilde{\mathbf{w}}(\mathbf{y}) = \boldsymbol{\tau} + \mathbf{V}'_{yw}\mathbf{V}_y^{-1}\mathbf{y}$ (where $\boldsymbol{\tau} = \boldsymbol{\mu}_w - \mathbf{V}'_{yw}\mathbf{V}_y^{-1}\boldsymbol{\mu}_y$).

(c) Suppose that $M = 1$ and that $S_0 = \{\underset{\sim}{w} : \ell \leq \underset{\sim}{w} \leq u\}$, where ℓ and u are (known) constants (with $\ell < u$). Suppose also that the joint distribution of w and $\mathbf{y}$ is MVN and that $\mathbf{v}_{yw} \neq \mathbf{0}$. And lettimg $\tilde{w} = \tilde{w}(\mathbf{y}) = \tau + \mathbf{v}'_{yw}\mathbf{V}_y^{-1}\mathbf{y}$ (with $\tau = \mu_w - \mathbf{v}'_{yw}\mathbf{V}_y^{-1}\boldsymbol{\mu}_y$) and $\tilde{v} = v_w - \mathbf{v}'_{yw}\mathbf{V}_y^{-1}\mathbf{v}_{yw}$, define

$$d = d(\mathbf{y}) = F\{[u - \tilde{w}(\mathbf{y})]/\tilde{v}^{1/2}\} - F\{[\ell - \tilde{w}(\mathbf{y})]/\tilde{v}^{1/2}\},$$

where $F(\cdot)$ is the cdf of the $N(0, 1)$ distribution. Further, let

$$C_0 = \{\underset{\sim}{y} \in \mathcal{R}^N : d(\underset{\sim}{y}) < \ddot{d}\},$$

where $\ddot{d}$ is the lower $100\dot{\gamma}\%$ point of the distribution of the random variable d. Show that among all $\dot{\gamma}$-level tests of the null hypothesis H_0, the nonrandomized $\dot{\gamma}$-level test with critical region C_0 achieves maximum power.

Solution. (a)

(1) Suppose that $f(\underset{\sim}{y}) > 0$. If $p_0(\underset{\sim}{y}) > 0$, then $f_0(\underset{\sim}{y}) > 0$, and it follows from result (8.27) that $\phi^{**}(\underset{\sim}{y}) = \phi^*(\underset{\sim}{y})$. If $p_0(\underset{\sim}{y}) = 0$ [in which case $p_1(\underset{\sim}{y}) = 1$], then $p_0(\underset{\sim}{y}) < k''$, $f_0(\underset{\sim}{y}) = 0$, and $f_1(\underset{\sim}{y}) > 0$, so that $\phi^{**}(\underset{\sim}{y}) = 1 = \phi^*(\underset{\sim}{y})$. Thus, in either case [i.e., whether $p_0(\underset{\sim}{y}) > 0$ or $p_0(\underset{\sim}{y}) = 0$], $\phi^{**}(\underset{\sim}{y}) = \phi^*(\underset{\sim}{y})$.

(2) and (3) Making use of result (8.27), we find that

$$\begin{aligned} \dot{\gamma} = \mathrm{E}[\phi^*(\mathbf{y}) \mid \mathbf{w} \in S_0] &= \textstyle\int_{\mathcal{R}^N} \phi^*(\underset{\sim}{y}) f_0(\underset{\sim}{y})\, d\underset{\sim}{y} \\ &= \textstyle\int_{\mathcal{R}^N} \phi^{**}(\underset{\sim}{y}) f_0(\underset{\sim}{y})\, d\underset{\sim}{y} \\ &= \Pr(p_0 < k'' \mid \mathbf{w} \in S_0) + c \Pr(p_0 = k'' \mid \mathbf{w} \in S_0). \end{aligned} \tag{S.54}$$

Thus,

$$G_0(k'') \geq \dot{\gamma} \geq \Pr(p_0 < k'' \mid \mathbf{w} \in S_0) = G_0(k'') - \Pr(p_0 = k'' \mid \mathbf{w} \in S_0),$$

which implies that k'' equals the smallest scalar $\underset{\sim}{p}_0$ for which $G_0(\underset{\sim}{p}_0) \geq \dot{\gamma}$. Moreover, it follows from equality (S.54) that

$$c = \frac{\dot{\gamma} - \Pr(p_0 < k'' \mid \mathbf{w} \in S_0)}{\Pr(p_0 = k'' \mid \mathbf{w} \in S_0)} \quad \text{when } \Pr(p_0 = k'' \mid \mathbf{w} \in S_0) > 0.$$

(b) Suppose that the joint distribution of $\mathbf{w}$ and $\mathbf{y}$ is MVN, in which case the $N(\boldsymbol{\mu}_y, \mathbf{V}_y)$ distribution is the marginal distribution of $\mathbf{y}$ and the $N[\tilde{\mathbf{w}}(\mathbf{y}), \mathbf{V}_w - \mathbf{V}'_{yw}\mathbf{V}_y^{-1}\mathbf{V}_{yw}]$ distribution is a conditional distribution of $\mathbf{w}$ given $\mathbf{y}$. Further, take the pdf $f(\cdot)$ of the marginal distribution of $\mathbf{y}$ to be such that $f(\mathbf{y}) > 0$ for every value of $\mathbf{y}$—clearly, this is consistent with the $N(\boldsymbol{\mu}_y, \mathbf{V}_y)$ distribution being the marginal distribution. And observe [in light of the result of Part (a)] that there exists a function, say

$\phi'(\cdot)$, such that $\phi^*(\mathbf{y}) = \phi'[p_0(\mathbf{y})]$ for every value of $\mathbf{y}$. Observe also that (for every value of $\mathbf{y}$) we can take

$$p_0(\mathbf{y}) = \int_{\mathcal{R}^M} g[\underline{\mathbf{w}}; \tilde{\mathbf{w}}(\mathbf{y})]\, d\underline{\mathbf{w}},$$

where $g[\cdot\,; \tilde{\mathbf{w}}(\mathbf{y})]$ is the pdf of the $N[\tilde{\mathbf{w}}(\mathbf{y}), \mathbf{V}_w - \mathbf{V}_{yw}'\mathbf{V}_y^{-1}\mathbf{V}_{yw})$ distribution, in which case $p_0(\mathbf{y})$ depends on $\mathbf{y}$ only through the value of $\tilde{\mathbf{w}}(\mathbf{y})$. It remains only to observe that if $p_0(\mathbf{y})$ depends on $\mathbf{y}$ only through the value of $\tilde{\mathbf{w}}(\mathbf{y})$, then (since $\phi^*(\mathbf{y}) = \phi'[p_0(\mathbf{y})]$) $\phi^*(\mathbf{y})$ depends on $\mathbf{y}$ only through the value of $\tilde{\mathbf{w}}(\mathbf{y})$.

(c) When the joint distribution of w and $\mathbf{y}$ is MVN, the $N[\tilde{w}(\mathbf{y}), \tilde{v}]$ distribution is a conditional distribution of w given $\mathbf{y}$, and the marginal distribution of $\mathbf{y}$ is absolutely continuous with a pdf $f(\cdot)$ for which $f(\underline{\mathbf{y}}) > 0$ for every $N \times 1$ vector $\underline{\mathbf{y}}$ [as was noted in the solution to Part (b)]. And upon applying the result of Part (a)-(1), we find that the critical function $\phi^*(\cdot)$ of a most-powerful $\dot{\gamma}$-level test is such that $\phi^*(\mathbf{y}) = \phi^{**}(\mathbf{y})$ for every value of $\mathbf{y}$ [where $\phi^{**}(\cdot)$ is as defined in Part (a)]. Moreover, when $M = 1$ and $S_0 = \{\underline{w} : \ell \le \underline{w} \le u\}$, we find that (for every $N \times 1$ vector $\underline{\mathbf{y}}$)

$$w \in S_0 \;\Leftrightarrow\; \ell \le w \le u \;\Leftrightarrow\; [\ell - \tilde{w}(\underline{\mathbf{y}})]/\tilde{v}^{1/2} \le [w - \tilde{w}(\underline{\mathbf{y}})]/\tilde{v}^{1/2} \le [u - \tilde{w}(\underline{\mathbf{y}})]/\tilde{v}^{1/2}$$

and hence that we can take

$$p_0(\underline{\mathbf{y}}) = d(\underline{\mathbf{y}}).$$

Now, supposing that $p_0(\mathbf{y}) = d(\mathbf{y})$, consider the constants k'' and c. When the joint distribution of w and $\mathbf{y}$ is MVN and when $\mathbf{v}_{yw} \neq \mathbf{0}$, the conditional distribution of $\tilde{w}$ given that $\ell \le w \le u$ is absolutely continuous. Thus, the conditional distribution of d given that $\ell \le w \le u$ is absolutely continuous and hence [in light of the results of Part (a)] is such that $k'' = \ddot{d}$ and is such that c can be taken to be 0. It remains only to observe that (when $c = 0$) the test with critical function $\phi^{**}(\cdot)$ is identical to the nonrandomized $\dot{\gamma}$-level test with critical region C_0.

8

Constrained Linear Models and Related Topics

EXERCISE 1. Using the technique described in Section 8.1a (or otherwise), extend the coverage of the following results in Section 5.3 [about estimability under an unconstrained G–M, Aitken, or general linear model (with model equation $\mathbf{y} = \mathbf{X}\boldsymbol{\beta} = \mathbf{e}$)] to estimability under a constrained G–M, Aitken, or general linear model (with model equation $\mathbf{y} = \mathbf{X}\boldsymbol{\beta} + \mathbf{e}$ and constraint $\mathbf{A}\boldsymbol{\beta} = \mathbf{d}$):

(a) the results pertaining to the issue of "how many essentially different estimable functions there are";

(b) the result that condition (5.3.6) is necessary and sufficient for the estimability of $\boldsymbol{\lambda}'\boldsymbol{\beta}$; and

(c) the result that condition (5.3.9) is necessary and sufficient for the estimability of $\boldsymbol{\lambda}'\boldsymbol{\beta}$.

Solution. Upon applying the technique described in Section 8.1a to the relevant results in Section 5.3, we obtain the following results on estimability under a constrained G–M, Aitken, or general linear model [with model equation $\mathbf{y} = \mathbf{X}\boldsymbol{\beta}+\mathbf{e}$ (where $\mathbf{X}$ is of dimensions $N\times P$) and with constraint $\mathbf{A}\boldsymbol{\beta} = \mathbf{d}$].

(a) Let $R = \text{rank}\begin{pmatrix}\mathbf{X}\\ \mathbf{A}\end{pmatrix}$. Then,

(1) there exists a set of R linearly independent estimable functions;

(2) no set of estimable functions contains more than R linearly independent estimable functions; and

(3) if $R < P$, then at least one and, in fact, at least $P-R$ of the individual parameters $\beta_1, \beta_2, \ldots, \beta_P$ (the P elements of $\boldsymbol{\beta}$) are nonestimable.

(b) For the linear combination $\boldsymbol{\lambda}'\boldsymbol{\beta}$ to be estimable, it is necessary and sufficient that

$$\boldsymbol{\lambda}'\begin{pmatrix}\mathbf{X}\\ \mathbf{A}\end{pmatrix}^{-}\begin{pmatrix}\mathbf{X}\\ \mathbf{A}\end{pmatrix} = \boldsymbol{\lambda}'.$$

(c) For the linear combination $\boldsymbol{\lambda}'\boldsymbol{\beta}$ to be estimable, it is necessary and sufficient that

$$\mathbf{k}'\boldsymbol{\lambda} = 0 \text{ for every } P\times 1 \text{ vector } \mathbf{k} \text{ in } \mathcal{N}\begin{pmatrix}\mathbf{X}\\ \mathbf{A}\end{pmatrix}.$$

EXERCISE 2. Suppose that $\mathbf{y}$ is an $N\times 1$ observable random vector that follows a constrained G–M (or Aitken or general linear) model, with model equation $\mathbf{y} = \mathbf{X}\boldsymbol{\beta}+\mathbf{e}$ and constraint $\mathbf{A}\boldsymbol{\beta} = \mathbf{d}$. Show that for any $P\times 1$ vector $\mathbf{r}$ (of constants), $\mathbf{r}'\mathbf{X}'\mathbf{y}$ is the constrained least squares estimator of its expected value (i.e., of $\mathbf{r}'\mathbf{X}'\mathbf{X}\boldsymbol{\beta}$) if and only if

$$\mathbf{X}\mathbf{r} \in \mathcal{C}[\mathbf{X}(\mathbf{I}-\mathbf{A}^{-}\mathbf{A})]. \tag{E.1}$$

Solution. The constrained least squares estimator of $\mathbf{r}'\mathbf{X}'\mathbf{X}\boldsymbol{\beta}$ is $\tilde{\mathbf{s}}'\mathbf{X}'\mathbf{y} + \tilde{\mathbf{t}}'\mathbf{d}$, where $\begin{pmatrix}\tilde{\mathbf{s}}\\ \tilde{\mathbf{t}}\end{pmatrix}$ is any solution to the constrained conjugate normal equations, that is, where $\tilde{\mathbf{s}}$ and $\tilde{\mathbf{t}}$ are any vectors for which

$$\begin{pmatrix}\mathbf{X}'\mathbf{X} & \mathbf{A}'\\ \mathbf{A} & \mathbf{0}\end{pmatrix}\begin{pmatrix}\tilde{\mathbf{s}}\\ \tilde{\mathbf{t}}\end{pmatrix} = \begin{pmatrix}\mathbf{X}'\mathbf{X}\mathbf{r}\\ \mathbf{0}\end{pmatrix}. \tag{S.1}$$

Thus, $\mathbf{r}'\mathbf{X}'\mathbf{y}$ is the constrained least squares estimator of its expected value if and only if

$$\mathbf{r}'\mathbf{X}'\mathbf{y} = \tilde{\mathbf{s}}'\mathbf{X}'\mathbf{y} + \tilde{\mathbf{t}}'\mathbf{d} \qquad \text{for every value of } \mathbf{y},$$

or equivalently if and only if

$$\tilde{\mathbf{t}}'\mathbf{d} = 0 \quad \text{and} \quad \mathbf{X}\mathbf{r} = \mathbf{X}\tilde{\mathbf{s}},$$

and hence (since $\mathbf{X}\mathbf{r} = \mathbf{X}\tilde{\mathbf{s}} \Rightarrow \mathbf{A}'\tilde{\mathbf{t}} = \mathbf{0}$ and since $\mathbf{d} = \mathbf{A}\mathbf{k}$ for some vector $\mathbf{k}$) if and only if

$$\mathbf{X}\mathbf{r} = \mathbf{X}\tilde{\mathbf{s}}. \tag{S.2}$$

Moreover, since (by definition) $\mathbf{A}\tilde{\mathbf{s}} = \mathbf{0}$, it follows from Theorem 2.11.3 that $\tilde{\mathbf{s}} = (\mathbf{I}-\mathbf{A}^-\mathbf{A})\mathbf{u}$ for some vector $\mathbf{u}$, so that

$$\mathbf{X}\tilde{\mathbf{s}} = \mathbf{X}(\mathbf{I}-\mathbf{A}^-\mathbf{A})\mathbf{u} \in \mathcal{C}[\mathbf{X}(\mathbf{I}-\mathbf{A}^-\mathbf{A})]. \tag{S.3}$$

It is clear from result (S.3) that if condition (S.2) is satisfied, then so is condition (E.1). It remains to verify the converse, that is, that condition (E.1) implies condition (S.2). Accordingly, suppose that condition (E.1) is satisfied, in which case

$$\mathbf{X}\mathbf{r} = \mathbf{X}(\mathbf{I}-\mathbf{A}^-\mathbf{A})\mathbf{u}$$

for some vector $\mathbf{u}$. Then, among the vectors $\tilde{\mathbf{s}}$ and $\tilde{\mathbf{t}}$ that satisfy equality (S.1) are those obtained by taking $\tilde{\mathbf{s}} = (\mathbf{I}-\mathbf{A}^-\mathbf{A})\mathbf{u}$ and $\tilde{\mathbf{t}} = \mathbf{0}$. And when $\tilde{\mathbf{s}} = (\mathbf{I}-\mathbf{A}^-\mathbf{A})\mathbf{u}$ and $\tilde{\mathbf{t}} = \mathbf{0}$, $\mathbf{X}'\mathbf{X}\tilde{\mathbf{s}} = \mathbf{X}'\mathbf{X}\mathbf{r}$, implying (in light of Corollary 2.3.4) that condition (S.2) is satisfied.

EXERCISE 3. Let $\mathbf{B}$ represent a partitioned matrix of the form $\mathbf{B} = \begin{pmatrix} \mathbf{X}'\mathbf{X} & \mathbf{A}' \\ \mathbf{A} & \mathbf{0} \end{pmatrix}$, and let $\mathbf{G}$ represent a generalized inverse (of $\mathbf{B}$) that has been partitioned as $\mathbf{G} = \begin{pmatrix} \mathbf{G}_{11} & \mathbf{G}_{12} \\ \mathbf{G}_{21} & \mathbf{G}_{22} \end{pmatrix}$, conformally to the partitioning of $\mathbf{B}$ (so that the dimensions of $\mathbf{G}_{11}$ are the same as those of $\mathbf{X}'\mathbf{X}$). Add to the results of Corollary 8.1.4 by using Lemma 8.1.3 (or other means) to show (1) that

$$\mathbf{A}\mathbf{G}_{11}\mathbf{A}' = \mathbf{0}, \quad \mathbf{X}'\mathbf{X}\mathbf{G}_{11}\mathbf{A}' = \mathbf{0}, \text{ and } \mathbf{A}\mathbf{G}_{11}\mathbf{X}'\mathbf{X} = \mathbf{0},$$

(2) that

$$\mathbf{X}'\mathbf{X}\mathbf{G}_{12}\mathbf{A} = \mathbf{A}'\mathbf{G}_{21}\mathbf{X}'\mathbf{X} = -\mathbf{A}'\mathbf{G}_{22}\mathbf{A},$$

and (3) that

$$\mathbf{X}'\mathbf{X} = \mathbf{X}'\mathbf{X}\mathbf{G}_{11}\mathbf{X}'\mathbf{X} - \mathbf{A}'\mathbf{G}_{22}\mathbf{A}.$$

Solution. In light of Lemma 8.1.1, we find [upon applying Lemma 8.1.3 with $\mathbf{A}_1 = \begin{pmatrix} \mathbf{X}'\mathbf{X} \\ \mathbf{A} \end{pmatrix}$ and $\mathbf{A}_2 = \begin{pmatrix} \mathbf{A}' \\ \mathbf{0} \end{pmatrix}$] that

$$\begin{pmatrix} \mathbf{X}'\mathbf{X} \\ \mathbf{A} \end{pmatrix} (\mathbf{G}_{11}, \mathbf{G}_{12}) \begin{pmatrix} \mathbf{X}'\mathbf{X} \\ \mathbf{A} \end{pmatrix} = \begin{pmatrix} \mathbf{X}'\mathbf{X} \\ \mathbf{A} \end{pmatrix}, \tag{S.4}$$

$$\begin{pmatrix} \mathbf{A}' \\ \mathbf{0} \end{pmatrix} (\mathbf{G}_{21}, \mathbf{G}_{22}) \begin{pmatrix} \mathbf{X}'\mathbf{X} \\ \mathbf{A} \end{pmatrix} = \mathbf{0}, \tag{S.5}$$

and

$$\begin{pmatrix} \mathbf{X}'\mathbf{X} \\ \mathbf{A} \end{pmatrix} (\mathbf{G}_{11}, \mathbf{G}_{12}) \begin{pmatrix} \mathbf{A}' \\ \mathbf{0} \end{pmatrix} = \mathbf{0}. \tag{S.6}$$

Moreover, equality (S.4) implies that

$$\mathbf{X}'\mathbf{X} = \mathbf{X}'\mathbf{X}\mathbf{G}_{11}\mathbf{X}'\mathbf{X} + \mathbf{X}'\mathbf{X}\mathbf{G}_{12}\mathbf{A} \tag{S.7}$$

and

$$\mathbf{A}\mathbf{G}_{11}\mathbf{X}'\mathbf{X} = \mathbf{A} - \mathbf{A}\mathbf{G}_{12}\mathbf{A}, \tag{S.8}$$

equality (S.5) implies that

$$\mathbf{A}'\mathbf{G}_{21}\mathbf{X}'\mathbf{X} = -\mathbf{A}'\mathbf{G}_{22}\mathbf{A}, \tag{S.9}$$

and equality (S.6) implies that

$$\mathbf{X'XG}_{11}\mathbf{A'} = \mathbf{0} \quad \text{and} \quad \mathbf{AG}_{11}\mathbf{A'} = \mathbf{0}. \tag{S.10}$$

And in light of Corollary 8.1.4, equality (S.8) implies that

$$\mathbf{AG}_{11}\mathbf{X'X} = \mathbf{0}. \tag{S.11}$$

Together, results (S.10) and (S.11) verify Part (1) of the exercise. Now, consider Parts (2) and (3). Upon observing that $\mathbf{G'}$, like $\mathbf{G}$ itself, is a generalized inverse of the (symmetric) matrix $\mathbf{B}$ and that $\mathbf{G'} = \begin{pmatrix} \mathbf{G}'_{11} & \mathbf{G}'_{21} \\ \mathbf{G}'_{12} & \mathbf{G}'_{22} \end{pmatrix}$ and upon applying result (S.9) with $\mathbf{G}'_{12}$ in place of $\mathbf{G}_{21}$ and $\mathbf{G}'_{22}$ in place of $\mathbf{G}_{22}$, we find that

$$\mathbf{X'XG}_{12}\mathbf{A} = (\mathbf{A'G}'_{12}\mathbf{X'X})' = (-\mathbf{A'G}'_{22}\mathbf{A})' = -\mathbf{A'G}_{22}\mathbf{A},$$

which in combination with result (S.9) verifies Part (2). Moreover, upon substituting $-\mathbf{A'G}_{22}\mathbf{A}$ for $\mathbf{X'XG}_{12}\mathbf{A}$ in expression (S.7), we find that

$$\mathbf{X'X} = \mathbf{X'XG}_{11}\mathbf{X'X} - \mathbf{A'G}_{22}\mathbf{A},$$

which verifies Part (3).

EXERCISE 4. Suppose that $\mathbf{y}$ is an $N \times 1$ observable random vector that follows a constrained G–M model, with model equation $\mathbf{y} = \mathbf{X}\boldsymbol{\beta} + \mathbf{e}$ and constraint $\mathbf{A}\boldsymbol{\beta} = \mathbf{d}$. Further, let $\hat{\boldsymbol{\alpha}}$ represent a $P \times 1$ vector whose elements are the constrained least squares estimators of the corresponding elements of the vector $\mathbf{X'X}\boldsymbol{\beta}$, and let $\tilde{\boldsymbol{\alpha}}$ represent a $P \times 1$ vector whose elements are the unconstrained least squares estimators. And take $\mathbf{G}$ to be a generalized inverse of the partitioned matrix $\begin{pmatrix} \mathbf{X'X} & \mathbf{A'} \\ \mathbf{A} & \mathbf{0} \end{pmatrix}$, and partition $\mathbf{G}$ conformally as $\mathbf{G} = \begin{pmatrix} \mathbf{G}_{11} & \mathbf{G}_{12} \\ \mathbf{G}_{21} & \mathbf{G}_{22} \end{pmatrix}$.

(a) Making use of the results of Exercise 3 (or otherwise), show that

$$\text{var}(\tilde{\boldsymbol{\alpha}}) = \text{var}(\hat{\boldsymbol{\alpha}}) + \sigma^2(-\mathbf{A'G}_{22}\mathbf{A}).$$

(b) Show that the result of Part (a) implies that the matrix $-\mathbf{A'G}_{22}\mathbf{A}$ is symmetric and nonnegative definite.

Solution. (a) It follows from results (1.38) and (1.39) that

$$\text{var}(\hat{\boldsymbol{\alpha}}) = \sigma^2\mathbf{X'XG}_{11}\mathbf{X'X}$$

and from the results of the final part of Section 5.4c that

$$\text{var}(\tilde{\boldsymbol{\alpha}}) = \sigma^2\mathbf{X'X}(\mathbf{X'X})^-\mathbf{X'X} = \sigma^2\mathbf{X'X}.$$

Thus, making use of Part (3) of Exercise 3, we find that

$$\text{var}(\tilde{\boldsymbol{\alpha}}) = \sigma^2(\mathbf{X'XG}_{11}\mathbf{X'X} - \mathbf{A'G}_{22}\mathbf{A}) = \text{var}(\hat{\boldsymbol{\alpha}}) + \sigma^2(-\mathbf{A'G}_{22}\mathbf{A}).$$

(b) Let $\boldsymbol{\ell}$ represent a $P \times 1$ vector of constants. Then, $\boldsymbol{\ell}'\mathbf{X'X}\boldsymbol{\beta}$ is estimated unbiasedly by both the unconstrained least squares estimator $\boldsymbol{\ell}'\tilde{\boldsymbol{\alpha}}$ and the constrained least squares estimator $\boldsymbol{\ell}'\hat{\boldsymbol{\alpha}}$. Moreover, $\boldsymbol{\ell}'\tilde{\boldsymbol{\alpha}}$ and $\boldsymbol{\ell}'\hat{\boldsymbol{\alpha}}$ are both linear estimators, and it follows from the results of Section 8.1g that $\boldsymbol{\ell}'\hat{\boldsymbol{\alpha}}$ has minimum variance among all linear unbiased estimators. Thus,

$$\boldsymbol{\ell}'[\text{var}(\tilde{\boldsymbol{\alpha}}) - \text{var}(\hat{\boldsymbol{\alpha}})]\boldsymbol{\ell} = \text{var}(\boldsymbol{\ell}'\tilde{\boldsymbol{\alpha}}) - \text{var}(\boldsymbol{\ell}'\hat{\boldsymbol{\alpha}}) \geq 0,$$

which implies that $\text{var}(\tilde{\boldsymbol{\alpha}}) - \text{var}(\hat{\boldsymbol{\alpha}})$ is a nonnegative definite matrix and hence [in light of the result of Part (a)] that $\sigma^2(-\mathbf{A'G}_{22}\mathbf{A})$ is a nonnegative definite matrix, leading (since $\sigma^2 > 0$) to the conclusion that the matrix $-\mathbf{A'G}_{22}\mathbf{A}$ is nonnegative definite. That $-\mathbf{A'G}_{22}\mathbf{A}$ is symmetric follows immediately from the result of Part (a) upon observing that variance-covariance matrices are inherently symmetric.

EXERCISE 5. Suppose that $\mathbf{y}$ is an $N \times 1$ observable random vector that follows a G–M model with model equation $\mathbf{y} = \mathbf{X}\boldsymbol{\beta} + \mathbf{e}$ (where $\boldsymbol{\beta}$ is a $P \times 1$ vector of unconstrained parameters), assume that the distribution of $\mathbf{e}$ is MVN (with mean vector $\mathbf{0}$ and variance-covariance matrix $\sigma^2\mathbf{I}$), let $\boldsymbol{\tau} = \boldsymbol{\Lambda}'\boldsymbol{\beta}$ represent an $M \times 1$ vector of linearly independent estimable linear combinations of the elements of $\boldsymbol{\beta}$, and take F to be the F statistic for testing the null hypothesis $\boldsymbol{\tau} = \boldsymbol{\tau}^{(0)}$ (versus the alternative $\boldsymbol{\tau} \neq \boldsymbol{\tau}^{(0)}$). Further, let $\boldsymbol{\tau}_1 = \boldsymbol{\Lambda}'_1\boldsymbol{\beta}$ represent an M_1-dimensional subvector of $\boldsymbol{\tau}$, and take F_1 to be the test statistic that would be applicable for testing the null hypothesis $\boldsymbol{\tau}_1 = \boldsymbol{\tau}_1^{(0)}$ (versus $\boldsymbol{\tau}_1 \neq \boldsymbol{\tau}_1^{(0)}$) if $\boldsymbol{\beta}$ were subject to the constraint $\mathbf{A}\boldsymbol{\beta} = \mathbf{d}$, where $\boldsymbol{\Lambda}' = \begin{pmatrix} \boldsymbol{\Lambda}'_1 \\ \mathbf{A} \end{pmatrix}$ (so that the elements of $\boldsymbol{\tau}_1$ are the first M_1 elements of $\boldsymbol{\tau}$) and where $\boldsymbol{\tau}^{(0)} = \begin{pmatrix} \boldsymbol{\tau}_1^{(0)} \\ \mathbf{d} \end{pmatrix}$. Recalling that F and F_1 have noncentral F distributions, letting $\boldsymbol{\delta} = (1/\sigma)(\boldsymbol{\tau} - \boldsymbol{\tau}^{(0)})$ and $\boldsymbol{\delta}_1 = (1/\sigma)(\boldsymbol{\tau}_1 - \boldsymbol{\tau}_1^{(0)})$, and observing that the noncentrality parameter of the noncentral F distribution is expressible as a function, say $\lambda(\boldsymbol{\delta})$, of $\boldsymbol{\delta}$ in the case of the distribution of F and as a function, say $\lambda_1(\boldsymbol{\delta}_1)$, of $\boldsymbol{\delta}_1$ in the case of the distribution of F_1, (1) compare the numerator and denominator degrees of freedom of the distribution of F_1 with those of the distribution of F, and (2) show that when $\boldsymbol{\delta} = \begin{pmatrix} \boldsymbol{\delta}_1 \\ \mathbf{0} \end{pmatrix}$, $\lambda(\boldsymbol{\delta}) = \lambda_1(\boldsymbol{\delta}_1)$.

Solution. (1) In light of the linear independence of the elements of $\boldsymbol{\tau}$, the numerator degrees of freedom of the distribution of F equals M and the denominator degrees of freedom equals $N - P_*$ (where $P_* = \operatorname{rank} \mathbf{X}$). And by way of comparison, the numerator degrees of freedom of the distribution of F_1 equals M_1 and the denominator degrees of freedom equals $N - P_* + (M - M_1)$, so that the distribution of F_1 has $M - M_1$ fewer numerator degrees of freedom than the distribution of F but $M - M_1$ more denominator degrees of freedom.

(2) As is evident from the results of Section 7.3b,

$$\lambda(\boldsymbol{\delta}) = \boldsymbol{\delta}'\mathbf{C}^{-1}\boldsymbol{\delta},$$

where $\operatorname{var}(\hat{\boldsymbol{\tau}}) = \sigma^2\mathbf{C}$ and $\hat{\boldsymbol{\tau}}$ is the least squares estimator of $\boldsymbol{\tau}$. And as is evident from the results of Section 8.1e,

$$\lambda_1(\boldsymbol{\delta}_1) = \boldsymbol{\delta}'_1\mathbf{B}^{-1}\boldsymbol{\delta}_1,$$

where $\operatorname{var}(\tilde{\boldsymbol{\tau}}_1) = \sigma^2\mathbf{B}$ and $\tilde{\boldsymbol{\tau}}_1$ is the constrained least squares estimator of $\boldsymbol{\tau}_1$.

Upon partitioning the $M \times M$ matrix $\mathbf{C}$ as

$$\mathbf{C} = \begin{pmatrix} \mathbf{C}_{11} & \mathbf{C}_{12} \\ \mathbf{C}_{21} & \mathbf{C}_{22} \end{pmatrix}$$

(where $\mathbf{C}_{11}$ is of dimensions $M_1 \times M_1$) and upon making use of Theorem 2.6.6, we find that when $\boldsymbol{\delta} = \begin{pmatrix} \boldsymbol{\delta}_1 \\ \mathbf{0} \end{pmatrix}$

$$\lambda(\boldsymbol{\delta}) = \begin{pmatrix} \boldsymbol{\delta}_1 \\ \mathbf{0} \end{pmatrix}' \mathbf{C}^{-1} \begin{pmatrix} \boldsymbol{\delta}_1 \\ \mathbf{0} \end{pmatrix} = \boldsymbol{\delta}'_1\mathbf{Q}^{-1}\boldsymbol{\delta}_1$$

where $\mathbf{Q} = \mathbf{C}_{11} - \mathbf{C}_{12}\mathbf{C}_{22}^{-1}\mathbf{C}_{21}$. Moreover, $\mathbf{C}$ is expressible in the form

$$\mathbf{C} = \mathbf{K}'\mathbf{X}'\mathbf{X}\mathbf{K}$$

where $\mathbf{K}$ is any matrix that satisfies the condition $\mathbf{X}'\mathbf{X}\mathbf{K} = \boldsymbol{\Lambda}$, in which case the submatrices of $\mathbf{C}$ are [upon partitioning $\mathbf{K}$ as $\mathbf{K} = (\mathbf{K}_1, \mathbf{K}_2)$ (where $\mathbf{K}_1$ has M_1 columns)] expressible as

$$\mathbf{C}_{11} = \mathbf{K}'_1\mathbf{X}'\mathbf{X}\mathbf{K}_1, \mathbf{C}_{12} = \mathbf{K}'_1\mathbf{X}'\mathbf{X}\mathbf{K}_2, \mathbf{C}_{21} = \mathbf{K}'_2\mathbf{X}'\mathbf{X}\mathbf{K}_1, \text{ and } \mathbf{C}_{22} = \mathbf{K}'_2\mathbf{X}'\mathbf{X}\mathbf{K}_2.$$

Thus, to complete the solution to Part (2), it suffices to show that

$$\mathbf{B} = \mathbf{K}'_1\mathbf{X}'\mathbf{X}\mathbf{K}_1 - \mathbf{K}'_1\mathbf{X}'\mathbf{X}\mathbf{K}_2(\mathbf{K}'_1\mathbf{X}'\mathbf{X}\mathbf{K}_1)^{-1}\mathbf{K}'_2\mathbf{X}'\mathbf{X}\mathbf{K}_1 \qquad \text{(S.12)}$$

(in which case $\mathbf{B} = \mathbf{Q}$).

For purposes of establishing the validity of expression (S.12), observe that

$$\tilde{\boldsymbol{\tau}}_1 = \boldsymbol{\Lambda}'_1\mathbf{b}$$

where $\mathbf{b}$ is the first (M_1-dimensional) part of any solution to the constrained normal equations

$$\begin{pmatrix} \mathbf{X}'\mathbf{X} & \mathbf{A}' \\ \mathbf{A} & \mathbf{0} \end{pmatrix} \begin{pmatrix} \mathbf{b} \\ \mathbf{r} \end{pmatrix} = \begin{pmatrix} \mathbf{X}'\mathbf{y} \\ \mathbf{d} \end{pmatrix},$$

observe that a solution to these equations is obtainable by taking

$$\mathbf{r} = (\mathbf{K}_2'\mathbf{X}'\mathbf{X}\mathbf{K}_2)^{-1}(\mathbf{K}_2'\mathbf{X}'\mathbf{y} - \mathbf{d}) \quad \text{and} \quad \mathbf{b} = \boldsymbol{\ell} - \mathbf{K}_2\mathbf{r}$$

where $\boldsymbol{\ell}$ is any solution to the (ordinary) normal equations $\mathbf{X}'\mathbf{X}\boldsymbol{\ell} = \mathbf{X}'\mathbf{y}$, in which case

$$\tilde{\boldsymbol{\tau}}_1 = \mathbf{K}_1'\mathbf{X}'\mathbf{X}\mathbf{b} = \mathbf{K}_1'\mathbf{X}'\mathbf{X}(\boldsymbol{\ell} - \mathbf{K}_2\mathbf{r}) = \mathbf{K}_1'\mathbf{X}'\mathbf{y} - \mathbf{K}_1'\mathbf{X}'\mathbf{X}\mathbf{K}_2(\mathbf{K}_2'\mathbf{X}'\mathbf{X}\mathbf{K}_2)^{-1}(\mathbf{K}_2'\mathbf{X}'\mathbf{y} - \mathbf{d}), \quad \text{(S.13)}$$

and finally observe that the variance of $\tilde{\boldsymbol{\tau}}_1$ as determined from expression (S.13) is that given by expression (S.12).

EXERCISE 6. Take the setting to be that of the final part of Section 8.1f, and adopt the notation and terminology employed therein. Suppose that the transformed vector $\mathbf{z}$ were taken to be $\mathbf{z} = \mathbf{y} - \mathbf{X}\mathbf{O}_1\underline{\mathbf{h}}$, where $\underline{\mathbf{h}}$ is the unique solution to the linear system $\mathbf{U}_1'\mathbf{h} = \mathbf{d}$ (in $\mathbf{h}$), rather than $\mathbf{z} = \mathbf{y} - \mathbf{X}\tilde{\mathbf{b}}$. Show that if (under the constrained G–M model with model equation $\mathbf{y} = \mathbf{X}\boldsymbol{\beta} + \mathbf{e}$ and constraint $\mathbf{A}\boldsymbol{\beta} = \mathbf{d}$) $\boldsymbol{\lambda}'\boldsymbol{\beta}$ is estimable, then for any solution $\begin{pmatrix} \underline{\mathbf{b}} \\ \underline{\mathbf{r}} \end{pmatrix}$ to the constrained normal equations $\begin{pmatrix} \mathbf{X}'\mathbf{X} & \mathbf{A}' \\ \mathbf{A} & \mathbf{0} \end{pmatrix} \begin{pmatrix} \mathbf{b} \\ \mathbf{r} \end{pmatrix} = \begin{pmatrix} \mathbf{X}'\mathbf{y} \\ \mathbf{d} \end{pmatrix}$ and any solution $\underline{\mathbf{t}}$ to the unconstrained normal equations $\mathbf{W}'\mathbf{W}\mathbf{t} = \mathbf{W}'\mathbf{z}$ (where $\mathbf{W} = \mathbf{X}\mathbf{O}_2$),

$$\boldsymbol{\lambda}'\underline{\mathbf{b}} = (\mathbf{O}_1'\boldsymbol{\lambda})'\underline{\mathbf{h}} + (\mathbf{O}_2'\boldsymbol{\lambda})'\underline{\mathbf{t}}. \quad \text{(E.2)}$$

Solution. Assuming that $\boldsymbol{\lambda}'\boldsymbol{\beta}$ is estimable and letting $\tilde{\mathbf{t}}$ represent an arbitrary solution to what would have been the unconstrained normal equations $\mathbf{W}'\mathbf{W}\mathbf{t} = \mathbf{W}'\mathbf{z}$ if $\mathbf{z}$ had been taken to be $\mathbf{y} - \mathbf{X}\tilde{\mathbf{b}}$, observe [in light of result (1.115)] that

$$\boldsymbol{\lambda}'\underline{\mathbf{b}} = \boldsymbol{\lambda}'\tilde{\mathbf{b}} + \boldsymbol{\lambda}'\mathbf{O}_2\tilde{\mathbf{t}}. \quad \text{(S.14)}$$

Observe also that

$$\mathbf{O}_1'\tilde{\mathbf{b}} = \underline{\mathbf{h}}.$$

Moreover,

$$\mathbf{W}'\mathbf{W}(\underline{\mathbf{t}} - \mathbf{O}_2'\tilde{\mathbf{b}}) = \mathbf{W}'(\mathbf{y} - \mathbf{X}\mathbf{O}_1\underline{\mathbf{h}} - \mathbf{W}\mathbf{O}_2'\tilde{\mathbf{b}})$$

and

$$\mathbf{W}\mathbf{O}_2'\tilde{\mathbf{b}} = \mathbf{X}\mathbf{O}_2\mathbf{O}_2'\tilde{\mathbf{b}} = \mathbf{X}\tilde{\mathbf{b}} - \mathbf{X}\mathbf{O}_1\mathbf{O}_1'\tilde{\mathbf{b}} = \mathbf{X}\tilde{\mathbf{b}} - \mathbf{X}\mathbf{O}_1\underline{\mathbf{h}},$$

implying that

$$\mathbf{W}'\mathbf{W}(\underline{\mathbf{t}} - \mathbf{O}_2'\tilde{\mathbf{b}}) = \mathbf{W}'(\mathbf{y} - \mathbf{X}\tilde{\mathbf{b}})$$

and hence that

$$\mathbf{W}\tilde{\mathbf{t}} = \mathbf{W}(\underline{\mathbf{t}} - \mathbf{O}_2'\tilde{\mathbf{b}}).$$

And since $\boldsymbol{\lambda}'\boldsymbol{\beta}$ is estimable, it follows from the results in the next-to-last part of Section 8.1f that $\boldsymbol{\lambda}'\mathbf{O}_2\boldsymbol{\tau}$ is estimable, in which case $\boldsymbol{\lambda}'\mathbf{O}_2 = \mathbf{a}'\mathbf{W}$ for some vector $\mathbf{a}$, and it follows that

$$\boldsymbol{\lambda}'\mathbf{O}_2\tilde{\mathbf{t}} = \boldsymbol{\lambda}'\mathbf{O}_2(\underline{\mathbf{t}} - \mathbf{O}_2'\tilde{\mathbf{b}}). \quad \text{(S.15)}$$

Finally, by making use of result (S.15), expression (S.14) can be reexpressed as

$$\begin{aligned} \boldsymbol{\lambda}'\underline{\mathbf{b}} &= \boldsymbol{\lambda}'\tilde{\mathbf{b}} + \boldsymbol{\lambda}'\mathbf{O}_2(\underline{\mathbf{t}} - \mathbf{O}_2'\tilde{\mathbf{b}}) \\ &= \boldsymbol{\lambda}'(\mathbf{I} - \mathbf{O}_2\mathbf{O}_2')\tilde{\mathbf{b}} + \boldsymbol{\lambda}'\mathbf{O}_2\underline{\mathbf{t}} \\ &= \boldsymbol{\lambda}'\mathbf{O}_1\mathbf{O}_1'\tilde{\mathbf{b}} + \boldsymbol{\lambda}'\mathbf{O}_2\underline{\mathbf{t}} \\ &= (\mathbf{O}_1'\boldsymbol{\lambda})'\underline{\mathbf{h}} + (\mathbf{O}_2'\boldsymbol{\lambda})'\underline{\mathbf{t}}, \end{aligned}$$

thereby verifying expression (E.2).

EXERCISE 7. Let $\mathbf{y}$ represent an $N \times 1$ observable random vector and $\mathbf{d}$ a $Q \times 1$ observable random vector, and suppose that the combined $(N+Q)$-dimensional vector $\begin{pmatrix} \mathbf{y} \\ \mathbf{d} \end{pmatrix}$ follows a G–M model with model equation $\begin{pmatrix} \mathbf{y} \\ \mathbf{d} \end{pmatrix} = \begin{pmatrix} \mathbf{X} \\ \mathbf{A} \end{pmatrix} \boldsymbol{\beta} + \mathbf{e}$, where $\boldsymbol{\beta}$ is a $P \times 1$ vector of unknown (and unconstrained) parameters (and where $\mathbf{X}$ is of dimensions $N \times P$ and the residual vector $\mathbf{e}$ is of dimension $N+Q$). Further, take $\boldsymbol{\lambda}$ to be a $P \times 1$ vector of constants such that $\boldsymbol{\lambda}' \in \mathcal{R}(\mathbf{X})$ (so that $\boldsymbol{\lambda}'\boldsymbol{\beta}$ is estimable from $\mathbf{y}$ alone), and suppose that $\mathcal{R}(\mathbf{X}) \cap \mathcal{R}(\mathbf{A}) = \{\mathbf{0}\}$. Show that the least squares estimator of $\boldsymbol{\lambda}'\boldsymbol{\beta}$ is the same (when both $\mathbf{y}$ and $\mathbf{d}$ are observable) as when only $\mathbf{y}$ is observable. And show that for $\mathbf{d} = \mathbf{0}$ or more generally for $\mathbf{d} \in \mathcal{C}(\mathbf{A})$, the residual sum of squares is also the same.

Solution. Let $\underline{\mathbf{y}}$ represent the observed value of $\mathbf{y}$ and $\underline{\mathbf{d}}$ the observed value of $\mathbf{d}$. Then, the normal equations (when both $\mathbf{y}$ and $\mathbf{d}$ are observable) are expressible as follows:

$$(\mathbf{X}'\mathbf{X} + \mathbf{A}'\mathbf{A})\mathbf{b} = \mathbf{X}'\underline{\mathbf{y}} + \mathbf{A}'\underline{\mathbf{d}}. \tag{S.16}$$

And when only $\mathbf{y}$ is observable, the normal equations are

$$\mathbf{X}'\mathbf{X}\mathbf{b} = \mathbf{X}'\underline{\mathbf{y}}. \tag{S.17}$$

Let $\underline{\mathbf{b}}$ represent any value of $\mathbf{b}$ that satisfies equation (S.16). Then, clearly,

$$\mathbf{X}'(\underline{\mathbf{y}} - \mathbf{X}\underline{\mathbf{b}}) = \mathbf{A}'(\mathbf{A}\underline{\mathbf{b}} - \underline{\mathbf{d}}).$$

And upon observing that $\mathcal{R}(\mathbf{X}) \cap \mathcal{R}(\mathbf{A}) = \{\mathbf{0}\}$ implies that $\mathcal{C}(\mathbf{X}') \cap \mathcal{C}(\mathbf{A}') = \{\mathbf{0}\}$, it follows that

$$\mathbf{X}'(\underline{\mathbf{y}} - \mathbf{X}\underline{\mathbf{b}}) = \mathbf{0} \quad \text{and} \quad \mathbf{A}'(\mathbf{A}\underline{\mathbf{b}} - \underline{\mathbf{d}}) = \mathbf{0}, \tag{S.18}$$

implying in particular that

$$\mathbf{X}'\mathbf{X}\underline{\mathbf{b}} = \mathbf{X}'\underline{\mathbf{y}}. \tag{S.19}$$

Thus, any solution to equation (S.16) is also a solution to equation (S.17), leading to the conclusion that the least squares estimator of $\boldsymbol{\lambda}'\boldsymbol{\beta}$ is the same (when both $\mathbf{y}$ and $\mathbf{d}$ are observable) as when only $\mathbf{y}$ is observable.

The residual sum of squares (when both $\mathbf{y}$ and $\mathbf{d}$ are observable) is

$$\left[\begin{pmatrix} \underline{\mathbf{y}} \\ \underline{\mathbf{d}} \end{pmatrix} - \begin{pmatrix} \mathbf{X} \\ \mathbf{A} \end{pmatrix} \underline{\mathbf{b}}\right]' \left[\begin{pmatrix} \underline{\mathbf{y}} \\ \underline{\mathbf{d}} \end{pmatrix} - \begin{pmatrix} \mathbf{X} \\ \mathbf{A} \end{pmatrix} \underline{\mathbf{b}}\right]. \tag{S.20}$$

And when only $\mathbf{y}$ is observable, the residual sum of squares is exprssible (in terms of $\underline{\mathbf{b}}$) as

$$(\underline{\mathbf{y}} - \mathbf{X}\underline{\mathbf{b}})'(\underline{\mathbf{y}} - \mathbf{X}\underline{\mathbf{b}}), \tag{S.21}$$

as is evident from result (S.19).

Now, suppose that $\underline{\mathbf{d}} \in \mathcal{C}(\mathbf{A})$. Then, $\underline{\mathbf{d}} = \mathbf{A}\mathbf{k}$ for some $P \times 1$ vector $\mathbf{k}$. And upon making use of result (S.18), we find that

$$\mathbf{A}'\mathbf{A}(\underline{\mathbf{b}} - \mathbf{k}) = \mathbf{0},$$

which implies that

$$\mathbf{A}(\underline{\mathbf{b}} - \mathbf{k}) = \mathbf{0}$$

and hence that

$$\mathbf{A}\underline{\mathbf{b}} = \mathbf{A}\mathbf{k} = \underline{\mathbf{d}}.$$

Moreover, when $\mathbf{A}\underline{\mathbf{b}} = \underline{\mathbf{d}}$, expression (S.20) simplifies to expression (S.21). Thus, for $\mathbf{d} \in \mathcal{C}(\mathbf{A})$, the residual sum of squares (when both $\mathbf{y}$ and $\mathbf{d}$ are observable) is the same as when only $\mathbf{y}$ is observable.

EXERCISE 8. Suppose that $\mathbf{y}$ is an $N \times 1$ observable random vector that follows a G–M model with model equation $\mathbf{y} = \mathbf{X}\boldsymbol{\beta} + \mathbf{e}$, where $\boldsymbol{\beta}$ is a $P \times 1$ vector of unknown parameters and where $\mathrm{E}(\mathbf{e}) = \mathbf{0}$ and $\mathrm{var}(\mathbf{e}) = \sigma^2\mathbf{I}$. Suppose further—refer to Section 5.9d—that $\mathbf{e} \sim \sigma\mathbf{u}$, where $\mathbf{u}$ is a random vector that has an absolutely continuous spherical distribution with variance-covariance matrix $\mathbf{I}$ and with a pdf $h(\cdot)$ that is expressible in the form $h(\mathbf{u}) = c^{-1}g(\mathbf{u}'\mathbf{u})$ for some strictly positive constant c and some nonnegative strictly decreasing function $g(\cdot)$. And let $L(\boldsymbol{\beta}, \sigma)$ represent the likelihood function.

(a) Show that (regardless of the value of σ) $L(\boldsymbol{\beta}, \sigma)$ attains its maximum value with respect to $\boldsymbol{\beta}$ at any solution to the normal equations $\mathbf{X}'\mathbf{X}\mathbf{b} = \mathbf{X}'\mathbf{y}$ and that if $\boldsymbol{\beta}$ were subject to the constraint $\mathbf{A}\boldsymbol{\beta} = \mathbf{d}$ [where $\mathbf{d} \in \mathcal{C}(\mathbf{A})$] $L(\boldsymbol{\beta}, \sigma)$ would attain its maximum value with respect to $\boldsymbol{\beta}$ at the first ($\mathbf{b}$-part) of any solution to the constrained normal equations $\begin{pmatrix} \mathbf{X}'\mathbf{X} & \mathbf{A}' \\ \mathbf{A} & \mathbf{0} \end{pmatrix} \begin{pmatrix} \mathbf{b} \\ \mathbf{r} \end{pmatrix} = \begin{pmatrix} \mathbf{X}'\mathbf{y} \\ \mathbf{d} \end{pmatrix}$.

(b) Let $k(s)$ represent the function of s defined (for $s>0$) as follows: $k(s) = \dfrac{d \log g(s)}{ds}$. Further, let $\underline{\boldsymbol{\beta}}$ represent any particular value of $\boldsymbol{\beta}$ for which $q(\boldsymbol{\beta})>0$. And take a to be the strictly positive scalar defined implicitly (in terms of σ) by the equality $\sigma^2 = aq(\underline{\boldsymbol{\beta}})/N$. Show that $L(\underline{\boldsymbol{\beta}}, \sigma)$ attains its maximum value (with respect to σ) when the value of a is such that $a = -2k(N/a)$.

(c) (1) Show that among the possibilities for the random vector $\mathbf{u}$ (having an absolutely continuous spherical distribution with variance-covariance matrix $\mathbf{I}$) is that obtained by taking

$$\mathbf{u} \sim \sqrt{(M-2)/M}\,\mathbf{t}, \tag{E.3}$$

where (for $M>2$) $\mathbf{t} \sim MVt(M, \mathbf{I}_N)$; and (2) show that in the special case (E.3), the solution for the strictly positive scalar a in Part (b) is $a = M/(M-2)$.

(d) Let $\boldsymbol{\tau} = \boldsymbol{\Lambda}'\boldsymbol{\beta}$ represent an $M\times 1$ vector of linear combinations of the elements of $\boldsymbol{\beta}$, and consider the test of the null hypothesis $H_0 : \boldsymbol{\tau} = \boldsymbol{\tau}^{(0)}$ [where $\boldsymbol{\tau}^{(0)} \in \mathcal{C}(\boldsymbol{\Lambda}')$] versus the alternative hypothesis $H_1 : \boldsymbol{\tau} \neq \boldsymbol{\tau}^{(0)}$. Show that the $\dot{\gamma}$-level F test is the equivalent of a $\dot{\gamma}$-level likelihood ratio test.

Solution. (a) As is evident from expression (5.9.74),

$$L(\boldsymbol{\beta}, \sigma) = c^{-1}\sigma^{-N} g[q(\boldsymbol{\beta})/\sigma^2],$$

where $q(\boldsymbol{\beta})$ is the function $q(\boldsymbol{\beta}) = (\mathbf{y}-\mathbf{X}\boldsymbol{\beta})'(\mathbf{y}-\mathbf{X}\boldsymbol{\beta})$.

Let $\bar{\boldsymbol{\beta}}$ represent any solution to the (unconstrained) normal equations. Then, $q(\boldsymbol{\beta})$ attains its minimum value at $\boldsymbol{\beta} = \bar{\boldsymbol{\beta}}$ and hence [since $g(\cdot)$ is a strictly decreasing function] $g[q(\boldsymbol{\beta})/\sigma^2]$ attains its maximum value at $\boldsymbol{\beta} = \bar{\boldsymbol{\beta}}$ (for any particular value of σ). And it follows that $L(\boldsymbol{\beta}, \sigma)$ attains its maxiumum value with respect to $\boldsymbol{\beta}$ (for any particular value of σ) at $\boldsymbol{\beta} = \bar{\boldsymbol{\beta}}$.

Now, let $\tilde{\boldsymbol{\beta}}$ represent the $\mathbf{b}$-part of any solution to the constrained normal equations. Then, subject to the constraint $\mathbf{A}\boldsymbol{\beta} = \mathbf{d}$, $q(\boldsymbol{\beta})$ attains its minimum value at $\boldsymbol{\beta} = \tilde{\boldsymbol{\beta}}$. And it follows from the same line of reasoning as in the unconstrained case that $L(\boldsymbol{\beta}, \sigma)$ attains its maximum value (with respect to $\boldsymbol{\beta}$) over the set of $\boldsymbol{\beta}$-values for which $\mathbf{A}\boldsymbol{\beta} = \mathbf{d}$ (and for any particular value of σ) at $\boldsymbol{\beta} = \tilde{\boldsymbol{\beta}}$.

(b) Clearly,

$$\begin{aligned} \frac{dL(\underline{\boldsymbol{\beta}}, \sigma)}{d\sigma} &= -(N/\sigma) + k[q(\underline{\boldsymbol{\beta}})/\sigma^2]q(\underline{\boldsymbol{\beta}})(-2/\sigma^3) \\ &= -(N/\sigma)[1 + 2k(N/a)(1/a)] \\ &= -[N/(a\sigma)][a + 2k(N/a)]. \end{aligned}$$

Thus, $\dfrac{dL(\underline{\boldsymbol{\beta}}, \sigma)}{d\sigma}$ equals 0 if $a = -2k(N/a)$, is greater than 0 if $a < -2k(N/a)$, and is less than 0 if $a > -2k(N/a)$. And it follows that $L(\underline{\boldsymbol{\beta}}, \sigma)$ attains its maximum value (with respect to σ) when $a = -2k(N/a)$.

(c) (1) Let $\mathbf{x} = \sqrt{(M-2)/M}\,\mathbf{t}$. Then, $\text{var}(\mathbf{x}) = \mathbf{I}$, as is evident from result (6.4.64). And in light of result (6.4.55), the pdf of the distribution of $\mathbf{x}$, say $h(\cdot)$, is expressible in the form

$$h(\mathbf{x}) = c^{-1}\Big(1 + \frac{\mathbf{x}'\mathbf{x}}{M-2}\Big)^{-(M+N)/2},$$

where c is a strictly positive constant. Then, $\mathbf{x}$ has an absolutely continuous spherical distribution with variance-covariance matrix $\mathbf{I}$.

(2) Suppose that $g(s) = \Big(1 + \dfrac{s}{M-2}\Big)^{-(M+N)/2}$. Then,

$$k(s) = \frac{-(M+N)}{2}\Big(1 + \frac{s}{M-2}\Big)^{-1}\frac{1}{M-2} = \frac{-(M+N)}{2}\,\frac{1}{M-2+s}.$$

Further,

$$-2k(N/a) = \frac{(M+N)a}{(M-2)a + N}.$$

And upon equating $-2k(N/a)$ to a, we obtain the equation

$$\frac{(M+N)a}{(M-2)a+N} = a.$$

Clearly, this equation has as its solution $a = M/(M-2)$.

(d) Suppose that $\mathbf{y} \notin \mathcal{C}(\mathbf{X})$ (as is the case with probability 1). And take $\bar{\boldsymbol{\beta}}$ to be any value of $\boldsymbol{\beta}$ and $\bar{\sigma}$ any value of σ that maximize $L(\boldsymbol{\beta}, \sigma)$ (in the absence of any constraints on $\boldsymbol{\beta}$). Further, take $\tilde{\boldsymbol{\beta}}$ to be any value of $\boldsymbol{\beta}$ and $\tilde{\sigma}$ any value of σ that maximize $L(\boldsymbol{\beta}, \sigma)$ when $\boldsymbol{\beta}$ is subject to the restriction $\boldsymbol{\Lambda}'\boldsymbol{\beta} = \boldsymbol{\tau}^{(0)}$. Then, as in the special case where the distribution of $\mathbf{e}$ is MVN [and as is evident from Part (b)], the likelihood ratio test statistic is expressible as $[q(\bar{\boldsymbol{\beta}})/q(\tilde{\boldsymbol{\beta}})]^{N/2}$.

Now, recall (from Section 7.3) that the F statistic has the same distribution (under H_0) as in the special case where the distribution of $\mathbf{e}$ is MVN. Then, upon employing the same reasoning as in result (3.10), we conclude that (as in the case where the distribution of $\mathbf{e}$ is MVN) the $\dot{\gamma}$-level F test is the equivalent of a $\dot{\gamma}$-level likelihood ratio test.

EXERCISE 9. Suppose that $\mathbf{y}$ is an $N \times 1$ observable random vector that follows a constrained G–M model with model equation $\mathbf{y} = \mathbf{X}\boldsymbol{\beta} + \mathbf{e}$, where $\boldsymbol{\beta}$ is a $P \times 1$ vector of unknown parameters that [for some $Q \times P$ matrix $\mathbf{A}$ and $Q \times 1$ vector $\mathbf{d} \in \mathcal{C}(\mathbf{A})$] is confined to the set $\{\boldsymbol{\beta} : \mathbf{A}\boldsymbol{\beta} = \mathbf{d}\}$. Suppose further that $\mathbf{k}'\mathbf{d} \neq 0$ for some $Q \times 1$ vector $\mathbf{k}$ for which $\mathbf{k}'\mathbf{A} \in \mathcal{R}(\mathbf{X})$. Show that when $\mathbf{y} = \mathbf{0}$, $ResSS > 0$ (thereby verifying an assertion made in the introductory part of Section 8.4).

Solution. Suppose that $\mathbf{y} = \mathbf{0}$. And suppose (for purposes of establishing a contradiction) that $ResSS$ is not greater than 0 and hence (since $ResSS \geq 0$) that $ResSS = 0$. Then, for any solution, say that with $\mathbf{b} = \tilde{\mathbf{b}}$ and $\mathbf{r} = \tilde{\mathbf{r}}$, to the constrained normal equations $\begin{pmatrix} \mathbf{X}'\mathbf{X} & \mathbf{A}' \\ \mathbf{A} & \mathbf{0} \end{pmatrix} \begin{pmatrix} \mathbf{b} \\ \mathbf{r} \end{pmatrix} = \begin{pmatrix} \mathbf{X}'\mathbf{y} \\ \mathbf{d} \end{pmatrix}$,

$$(\mathbf{X}\tilde{\mathbf{b}})'\mathbf{X}\tilde{\mathbf{b}} = (\mathbf{0} - \mathbf{X}\tilde{\mathbf{b}})'(\mathbf{0} - \mathbf{X}\tilde{\mathbf{b}}) = 0,$$

as is evident from result (4.11). And it follows that

$$\mathbf{X}\tilde{\mathbf{b}} = \mathbf{0}.$$

Now, since $\mathbf{k}'\mathbf{A} \in \mathcal{R}(\mathbf{X})$, $\mathbf{k}'\mathbf{A} = \boldsymbol{\ell}'\mathbf{X}$ for some column vector $\boldsymbol{\ell}$. Thus,

$$\mathbf{k}'\mathbf{A}\tilde{\mathbf{b}} = \boldsymbol{\ell}'\mathbf{X}\tilde{\mathbf{b}} = 0.$$

However, $\mathbf{A}\tilde{\mathbf{b}} = \mathbf{d}$, which (since $\mathbf{k}'\mathbf{d} \neq 0$) leads to the contradictory conclusion that $\mathbf{k}'\mathbf{A}\tilde{\mathbf{b}} \neq 0$. It is now clear that $ResSS > 0$.

EXERCISE 10. Suppose that $\mathbf{y}$ is an $N \times 1$ observable random vector that follows a G–M model with model equation $\mathbf{y} = \mathbf{X}\boldsymbol{\beta} + \mathbf{e}$, where $\boldsymbol{\beta}$ is a $P \times 1$ vector of unknown parameters. Further, denote the coefficient of determination by R^2 and adopt the general definition (5.12). And let $ResSS = \mathbf{y}'(\mathbf{I} - \mathbf{P_X})\mathbf{y}$ represent the residual SS; and extend the definition of the regression SS to this setting by taking $RegSS = \mathbf{y}'\mathbf{P_X}\mathbf{y} - N\bar{y}^2$ (where $\bar{y} = N^{-1}\mathbf{1}'\mathbf{y}$) or $RegSS = \mathbf{y}'\mathbf{P_X}\mathbf{y}$ depending on whether or not $\mathbf{1} \in \mathcal{C}(\mathbf{X})$. Show that

$$\frac{RegSS}{ResSS} = \frac{R^2}{1-R^2}.$$

Solution. In the special case where $\mathbf{1} \in \mathcal{C}(\mathbf{X})$,

$$\frac{RegSS}{ResSS} = \frac{RegSS}{ResSS + RegSS - RegSS} = \frac{RegSS}{\mathbf{y}'\mathbf{y} - N\bar{y}^2 - RegSS} = \frac{RegSS}{(\mathbf{y}'\mathbf{y} - N\bar{y}^2)(1-R^2)} = \frac{R^2}{1-R^2}.$$

Similarly, in the special case where $\mathbf{1} \notin \mathcal{C}(\mathbf{X})$,

$$\frac{RegSS}{ResSS} = \frac{RegSS}{ResSS + RegSS - RegSS} = \frac{RegSS}{\mathbf{y}'\mathbf{y} - RegSS} = \frac{RegSS}{\mathbf{y}'\mathbf{y}\,(1-R^2)} = \frac{R^2}{1-R^2}.$$

EXERCISE 11. Let y represent a random variable and $\mathbf{x}$ a $K \times 1$ random vector. Further, let $\sigma_y = \sqrt{\text{var}(y)}$, $\boldsymbol{\sigma}_{\mathbf{x}y} = \text{cov}(\mathbf{x}, y)$, and $\boldsymbol{\Sigma}_{\mathbf{x}} = \text{var}(\mathbf{x})$, so that $\text{var}\begin{pmatrix} y \\ \mathbf{x} \end{pmatrix} = \begin{pmatrix} \sigma_y^2 & \boldsymbol{\sigma}_{\mathbf{x}y}' \\ \boldsymbol{\sigma}_{\mathbf{x}y} & \boldsymbol{\Sigma}_{\mathbf{x}} \end{pmatrix}$ and there exists a matrix $\boldsymbol{\Gamma} = (\boldsymbol{\gamma}_y, \boldsymbol{\Gamma}_{\mathbf{x}})$ (where $\boldsymbol{\gamma}_y$ is a column vector) such that $\text{var}\begin{pmatrix} y \\ \mathbf{x} \end{pmatrix} = \boldsymbol{\Gamma}'\boldsymbol{\Gamma}$ and hence such that $\sigma_y^2 = \boldsymbol{\gamma}_y'\boldsymbol{\gamma}_y$, $\boldsymbol{\sigma}_{\mathbf{x}y} = \boldsymbol{\Gamma}_{\mathbf{x}}'\boldsymbol{\gamma}_y$, and $\boldsymbol{\Sigma}_{\mathbf{x}} = \boldsymbol{\Gamma}_{\mathbf{x}}'\boldsymbol{\Gamma}_{\mathbf{x}}$. And let $\boldsymbol{\beta}$ represent any $K \times 1$ vector such that $\boldsymbol{\Sigma}_{\mathbf{x}}\boldsymbol{\beta} = \boldsymbol{\sigma}_{\mathbf{x}y}$.

(a) Show that for any constant a and any $K \times 1$ vector of constants $\mathbf{b}$,
$$\text{var}(a + \mathbf{b}'\mathbf{x}) = \mathbf{b}'\boldsymbol{\Sigma}_{\mathbf{x}}\mathbf{b} \quad \text{and} \quad \text{cov}(y,\ a + \mathbf{b}'\mathbf{x}) = \boldsymbol{\beta}'\boldsymbol{\Sigma}_{\mathbf{x}}\mathbf{b}. \tag{E.4}$$

(b) Using the Cauchy-Schwarz inequality (or other means), show that
$$|\boldsymbol{\beta}'\boldsymbol{\Sigma}_{\mathbf{x}}\mathbf{b}| \le (\boldsymbol{\beta}'\boldsymbol{\Sigma}_{\mathbf{x}}\boldsymbol{\beta})^{1/2}(\mathbf{b}'\boldsymbol{\Sigma}_{\mathbf{x}}\mathbf{b})^{1/2}. \tag{E.5}$$

(c) Show that (for any constant a and any $K \times 1$ vector of constants $\mathbf{b}$)
$$|\text{corr}(y,\ a + \mathbf{b}'\mathbf{x})| \le (\boldsymbol{\beta}'\boldsymbol{\Sigma}_{\mathbf{x}}\boldsymbol{\beta})^{1/2}/\sigma_y = \text{corr}(y,\ a + \boldsymbol{\beta}'\mathbf{x}),$$
so that the maximum value (with respect to $\mathbf{b}$) of $|\text{corr}(y,\ a + \mathbf{b}'\mathbf{x})|$ is $(\boldsymbol{\beta}'\boldsymbol{\Sigma}_{\mathbf{x}}\boldsymbol{\beta})^{1/2}/\sigma_y$ and that value is attained at $\mathbf{b} = \boldsymbol{\beta}$ [a value of $\mathbf{b}$ for which $\text{corr}(y,\ a + \mathbf{b}'\mathbf{x})$ is nonnegative].

(d) Let $\alpha = \mu_y - \boldsymbol{\beta}'\boldsymbol{\mu}_{\mathbf{x}}$, where $\mu_y = \text{E}(y)$ and $\boldsymbol{\mu}_{\mathbf{x}} = \text{E}(\mathbf{x})$. And take $\eta(\mathbf{x})$ to be a function of $\mathbf{x}$ defined as follows: $\eta(\mathbf{x}) = \alpha + \boldsymbol{\beta}'\mathbf{x}$. If α and $\boldsymbol{\beta}$ were known, $\eta(\mathbf{x})$ could serve as a point predictor (for predicting the value of y from that of $\mathbf{x}$). What can be said about the MSE (mean squared error) of that predictor (relative to the MSE of other point predictors)? Provide justification for your answer.

(e) The correlation $\text{corr}(y,\ \alpha + \boldsymbol{\beta}'\mathbf{x}) = (\boldsymbol{\beta}'\boldsymbol{\Sigma}_{\mathbf{x}}\boldsymbol{\beta})^{1/2}/\sigma_y$ between y and $\alpha + \boldsymbol{\beta}'\mathbf{x}$ [where α is as defined in Part (d)] is referred to as the multiple correlation coefficient. Let $\mathbf{x}_1, \mathbf{x}_2, \ldots, \mathbf{x}_N$ represent N values of $\mathbf{x}$ and $y_1, y_2, \ldots, y_N$ the corresponding values of y. Show that in the special case where the joint distribution of y and $\mathbf{x}$ is the empirical distribution that assigns probability $1/N$ to each of the vectors $\begin{pmatrix} y_1 \\ \mathbf{x}_1 \end{pmatrix}, \begin{pmatrix} y_2 \\ \mathbf{x}_2 \end{pmatrix}, \ldots, \begin{pmatrix} y_N \\ \mathbf{x}_N \end{pmatrix}$,
$$\frac{\boldsymbol{\beta}'\boldsymbol{\Sigma}_{\mathbf{x}}\boldsymbol{\beta}}{\sigma_y^2} = \frac{\mathbf{y}'\mathbf{P}_{\mathbf{X}}\mathbf{y} - N\bar{y}^2}{\mathbf{y}'\mathbf{y} - N\bar{y}^2},$$
where $\mathbf{X} = (\mathbf{1},\ \mathbf{X}_*)$ with $\mathbf{X}_* = (\mathbf{x}_1, \mathbf{x}_2, \ldots, \mathbf{x}_N)'$, $\mathbf{y} = (y_1, y_2, \ldots, y_N)'$, and $\bar{y} = N^{-1}\sum_i y_i$, so that (in this special case) the square of the multiple correlation coefficient equals the coefficient of determination [that for an observable random vector $\mathbf{y}$ that follows a G–M model with a model matrix $\mathbf{X} = (\mathbf{1},\ \mathbf{X}_*)$ for which $\mathbf{1} \in \mathcal{C}(\mathbf{X})$ and whose observed value is $\mathbf{y} = (y_1, y_2, \ldots, y_N)'$].

Solution. (a) That $\text{var}(a + \mathbf{b}'\mathbf{x}) = \mathbf{b}'\boldsymbol{\Sigma}_{\mathbf{x}}\mathbf{b}$ follows immediately from result (3.2.43). And upon applying formula (3.2.42), we find that
$$\text{cov}(y,\ a + \mathbf{b}'\mathbf{x}) = \text{cov}(y, \mathbf{x})\,\mathbf{b} = \boldsymbol{\sigma}_{\mathbf{x}y}'\mathbf{b} = \boldsymbol{\beta}'\boldsymbol{\Sigma}_{\mathbf{x}}\mathbf{b}.$$

(b) Making use of the Cauchy-Schwarz inequality [in the form of inequality (2.4.10)], we find that
$$\begin{aligned} |\boldsymbol{\beta}'\boldsymbol{\Sigma}_{\mathbf{x}}\mathbf{b}| &= |\boldsymbol{\beta}'\boldsymbol{\Gamma}_{\mathbf{x}}'\boldsymbol{\Gamma}_{\mathbf{x}}\mathbf{b}| = |(\boldsymbol{\Gamma}_{\mathbf{x}}\boldsymbol{\beta})'\boldsymbol{\Gamma}_{\mathbf{x}}\mathbf{b}| \\ &\le [(\boldsymbol{\Gamma}_{\mathbf{x}}\boldsymbol{\beta})'\boldsymbol{\Gamma}_{\mathbf{x}}\boldsymbol{\beta}]^{1/2}[(\boldsymbol{\Gamma}_{\mathbf{x}}\mathbf{b})'\boldsymbol{\Gamma}_{\mathbf{x}}\mathbf{b}]^{1/2} = (\boldsymbol{\beta}'\boldsymbol{\Sigma}_{\mathbf{x}}\boldsymbol{\beta})^{1/2}(\mathbf{b}'\boldsymbol{\Sigma}_{\mathbf{x}}\mathbf{b})^{1/2}. \end{aligned}$$

(c) Making use of results (E.4) and (E.5), we find that
$$\begin{aligned} |\text{corr}(y,\ a + \mathbf{b}'\mathbf{x})| = \frac{|\boldsymbol{\beta}'\boldsymbol{\Sigma}_{\mathbf{x}}\mathbf{b}|}{(\mathbf{b}'\boldsymbol{\Sigma}_{\mathbf{x}}\mathbf{b})^{1/2}\sigma_y} &\le \frac{(\boldsymbol{\beta}'\boldsymbol{\Sigma}_{\mathbf{x}}\boldsymbol{\beta})^{1/2}(\mathbf{b}'\boldsymbol{\Sigma}_{\mathbf{x}}\mathbf{b})^{1/2}}{(\mathbf{b}'\boldsymbol{\Sigma}_{\mathbf{x}}\mathbf{b})^{1/2}\sigma_y} \\ &= (\boldsymbol{\beta}'\boldsymbol{\Sigma}_{\mathbf{x}}\boldsymbol{\beta})^{1/2}/\sigma_y = \frac{\boldsymbol{\beta}'\boldsymbol{\Sigma}_{\mathbf{x}}\boldsymbol{\beta}}{(\boldsymbol{\beta}'\boldsymbol{\Sigma}_{\mathbf{x}}\boldsymbol{\beta})^{1/2}\sigma_y} = \text{corr}(y,\ a + \boldsymbol{\beta}'\mathbf{x}). \end{aligned}$$

(d) If the joint distribution of y and $\mathbf{x}$ is such that $\mathrm{E}(y \mid \mathbf{x}) = \eta(\mathbf{x})$ (with probability 1) as is the case when the joint distribution is MVN (refer to Theorem 3.5.10), then $\eta(\mathbf{x})$ has minimum MSE (conditionally as well as unconditionally) among all predictors (not just among predictors of the form $a + \mathbf{b}'\mathbf{x}$)—that this is so is evident from the results of Section 5.10a. Moreover, regardless of the form of the joint distribution, $\eta(\mathbf{x})$ has minimum (unconditional) MSE among all predictors of the form $a+\mathbf{b}'\mathbf{x}$. For verification, it suffices to observe that the (unconditional) MSE of any predictor of the form $a+\mathbf{b}'\mathbf{x}$ depends on the joint distribution of y and $\mathbf{x}$ only through its first- and second-order moments. Thus, the (unconditional) MSE of any predictor of the form $a + \mathbf{b}'\mathbf{x}$ is the same whether the joint distribution is MVN or whether it is of some other form. And if we were to assume the existence of a predictor of the form $a+\mathbf{b}'\mathbf{x}$ with a smaller MSE than $\eta(\mathbf{x})$ for some joint distribution of y and $\mathbf{x}$, we would [since $\eta(\mathbf{x})$ is of the form $a + \mathbf{b}'\mathbf{x}$] be led to the contradictory conclusion that there is a predictor that has a smaller MSE than $\eta(\mathbf{x})$ when the joint distribution is MVN.

(e) Suppose that the joint distribution of y and $\mathbf{x}$ is the empirical distribution that assigns probability $1/N$ to each of the vectors $\begin{pmatrix} y_1 \\ \mathbf{x}_1 \end{pmatrix}, \begin{pmatrix} y_2 \\ \mathbf{x}_2 \end{pmatrix}, \ldots, \begin{pmatrix} y_N \\ \mathbf{x}_N \end{pmatrix}$. Then,

$$\mathrm{E}(y) = N^{-1}\mathbf{1}'\mathbf{y} = \bar{y} \quad \text{and} \quad \mathrm{E}(\mathbf{x}) = N^{-1}\mathbf{X}_*'\mathbf{1},$$

in which case

$$\alpha = \bar{y} - N^{-!}\boldsymbol{\beta}'\mathbf{X}_*'\mathbf{1}.$$

Further,

$$\sigma_y^2 = N^{-1}\mathbf{y}'\mathbf{y} - \bar{y}^2,$$

$$\boldsymbol{\sigma}_{\mathbf{x}y} = N^{-1}\mathbf{X}_*'\mathbf{y} - N^{-1}\mathbf{X}_*'\mathbf{1}\bar{y} = N^{-1}\mathbf{X}_*'(\mathbf{y} - \mathbf{1}\bar{y}), \quad \text{and}$$

$$\boldsymbol{\Sigma}_x = N^{-1}\mathbf{X}_*'\mathbf{X}_* - N^{-1}\mathbf{X}_*'\mathbf{1}(N^{-1}\mathbf{X}_*'\mathbf{1})' = N^{-1}\mathbf{X}_*'(\mathbf{I} - N^{-1}\mathbf{1}\mathbf{1}')\mathbf{X}_*.$$

Now, let us reexpress $\mathbf{y}'\mathbf{P}_{\mathbf{X}}\mathbf{y} - N\bar{y}^2$ in terms of the solution (for the scalar b_1 and $K \times 1$ vector $\mathbf{b}_*$) to the equations

$$\begin{pmatrix} N & \mathbf{1}'\mathbf{X}_* \\ \mathbf{X}_*'\mathbf{1} & \mathbf{X}_*'\mathbf{X}_* \end{pmatrix} \begin{pmatrix} b_1 \\ \mathbf{b}_* \end{pmatrix} = \begin{pmatrix} N\bar{y} \\ \mathbf{X}_*'\mathbf{y} \end{pmatrix}$$

—these are the relevant normal equations. A solution to these equations is obtainable by taking $b_1 = \alpha$ and $\mathbf{b}_* = \boldsymbol{\beta}$, as can be verified via a relatively straightforward exercise. Thus,

$$\mathbf{y}'\mathbf{P}_{\mathbf{X}}\mathbf{y} - N\bar{y}^2 = \begin{pmatrix} \alpha \\ \boldsymbol{\beta} \end{pmatrix}' \begin{pmatrix} N\bar{y} \\ \mathbf{X}_*'\mathbf{y} \end{pmatrix} - N\bar{y}^2 = N\bar{y}^2 - \boldsymbol{\beta}'\mathbf{X}_*'\mathbf{1}\bar{y} + \boldsymbol{\beta}'\mathbf{X}_*'\mathbf{y} - N\bar{y}^2 = N\boldsymbol{\beta}'\boldsymbol{\sigma}_{\mathbf{x}y} = N\boldsymbol{\beta}'\boldsymbol{\Sigma}_{\mathbf{x}}\boldsymbol{\beta}.$$

And it follows that

$$\frac{\mathbf{y}'\mathbf{P}_{\mathbf{X}}\mathbf{y} - N\bar{y}^2}{\mathbf{y}'\mathbf{y} - N\bar{y}^2} = \frac{\boldsymbol{\beta}'\boldsymbol{\Sigma}_{\mathbf{x}}\boldsymbol{\beta}}{\sigma_y^2}.$$

EXERCISE 12. Take the setting to be that in Part 2 of Section 8.5e, and adopt the terminology and notation employed therein. Further, for $j = 1, 2, \ldots, M$, denote by $\bar{\boldsymbol{\tau}}_j$ the ordinary (unconstrained) least squares estimator of $\boldsymbol{\tau}_j$. Show that each of the following two conditions is equivalent to the condition that [as in Condition (5.30)]

$$\boldsymbol{\Lambda}_j = \mathbf{X}'\mathbf{X}[\mathbf{I} - (\tilde{\boldsymbol{\Lambda}}_{j+1}')^{-}\tilde{\boldsymbol{\Lambda}}_{j+1}]\mathbf{U}_j$$

for some matrix $\mathbf{U}_j$ (where $M-1 \geq j \geq 1$):

(1) $\boldsymbol{\Lambda}_j = \mathbf{X}'\mathbf{X}\mathbf{T}_j$ for some matrix $\mathbf{T}_j$ such that $\tilde{\boldsymbol{\Lambda}}_{j+1}'\mathbf{T}_j = \mathbf{0}$ [i.e., some matrix $\mathbf{T}_j$ whose columns are contained in the null space of $\tilde{\boldsymbol{\Lambda}}_{j+1}'$];

(2) $\mathrm{cov}(\bar{\boldsymbol{\tau}}_j, \bar{\boldsymbol{\tau}}_k) = \mathbf{0}$ for $k = j+1, j+2, \ldots, M$.

Solution. (1) The equivalence of condition (1) is evident upon observing (in light of Theorem 2.11.3) that the matrix $\mathbf{T}_j$ is a solution to the linear system $\mathbf{X}'\mathbf{X}\mathbf{T} = \mathbf{0}$ (in $\mathbf{T}$) if and only if $\mathbf{T}_j = [\mathbf{I} - (\tilde{\boldsymbol{\Lambda}}_{j+1}')^{-}\tilde{\boldsymbol{\Lambda}}_{j+1}]\mathbf{U}_j$ for some matrix $\mathbf{U}_j$.

(2) Clearly, it suffices to show that condition (2) is equivalent to condition (1). Accordingly, for $j = 1, 2, \ldots, M$, observe that (since $\boldsymbol{\tau}_j$ is estimable) there exists a matrix $\mathbf{T}_j$ such that $\mathbf{X}'\mathbf{X}\mathbf{T}_j = \boldsymbol{\Lambda}_j$ and subsequently regarding $\mathbf{T}_j$ as any such matrix, observe that $\bar{\boldsymbol{\tau}}_j = \mathbf{T}_j'\mathbf{X}'\mathbf{y}$. Then, for $j = 1, 2, \ldots, M-1$ and $k = j+1, j+2, \ldots, M$,

$$\operatorname{cov}(\bar{\boldsymbol{\tau}}_j, \bar{\boldsymbol{\tau}}_k) = \sigma^2 \mathbf{T}_j'\mathbf{X}'\mathbf{X}\mathbf{T}_k = \sigma^2 \mathbf{T}_j'\boldsymbol{\Lambda}_k.$$

Thus, for $M-1 \geq j \geq 1$, $\operatorname{cov}(\bar{\boldsymbol{\tau}}_j, \bar{\boldsymbol{\tau}}_k) = \mathbf{0}$ for $k = j+1, j+2, \ldots, M$ if and only if $\mathbf{T}_j'\tilde{\boldsymbol{\Lambda}}_{j+1} = \mathbf{0}$ or equivalenty if and only if $\tilde{\boldsymbol{\Lambda}}_{j+1}'\mathbf{T}_j = \mathbf{0}$.

It is now clear that if condition (2) is satisfied, then $\tilde{\boldsymbol{\Lambda}}_{j+1}'\mathbf{T}_j = \mathbf{0}$ for any matrix $\mathbf{T}_j$ for which $\mathbf{X}'\mathbf{X}\mathbf{T}_j = \boldsymbol{\Lambda}_j$, which (since the estimability of $\boldsymbol{\tau}_j$ implies the existence of such a matrix) implies that condition (1) is satisfied. Conversely, if condition (1) is satisfied, there is an implication that $\operatorname{cov}(\bar{\boldsymbol{\tau}}_j, \bar{\boldsymbol{\tau}}_k) = \mathbf{0}$ for $k = j+1, j+2, \ldots, M$, that is, an implication that condition (2) is satisfied.

EXERCISE 13. Consider further the application to the ouabain data of the results of Section 8.5e [on K or $(K-1)$-line ANOVA tables]. Suppose that instead of taking the "independent variable" in this application to be the rate of injection (as in the approach to the application taken in Part 5 of Section 8.5e), it were taken to be the log to the base 2 of the rate. Determine the effects of this change.

Solution. Continue to regard the 41 lethal doses as the realizations of the elements $y_1, y_2, \ldots, y_{41}$ of an observable random vector $\mathbf{y}$ that follows a G–M model and to denote by $u_1, u_2, \ldots, u_{41}$ the corresponding rates of injection. And (for $i = 1, 2, \ldots, 41$) suppose that

$$\mathrm{E}(y_i) = \beta_1 + \beta_2 w_i + \beta_3 w_i^2 + \beta_4 w_i^3,$$

where $w_i = \log_2 u_i$.

When there are $L = 4$ segments (consisting of one column each), the ANOVA associated with the partitioning of the model matrix $\mathbf{X}$ into L segments (as in Part 1 of Section 8.5e) is the ANOVA with degrees of freedom and sums of squares as follows:

Source	df	SS
β_1	1	61971.61
β_2 after β_1	1	15683.55
β_3 after β_1 and β_2	1	376.37
β_4 after β_1, β_2, and β_3	1	34.36
Residual	37	5650.11
Total	41	83716

An ANOVA with the same degrees of freedom and sums of squares can be generated from the $M = 4$ null hypotheses $H_j^{(0)} : \beta_j = 0$ $(j = 1, 2, 3, 4)$ via an implementation of the approach described in Part 2 of Section 8.5e.

It remains to consider the approach to the generation of an ANOVA described in Part 4 of Section 8.5e. As a unit upper triangular matrix $\mathbf{Q}$ and a diagonal matrix $\mathbf{D}$ for which

$$\mathbf{Q}'\mathbf{X}'\mathbf{X}\mathbf{Q} = \mathbf{D},$$

we obtain

$$\mathbf{Q} = \begin{pmatrix} 1 & -1.37 & 0.89 & -0.25 \\ 0 & 1 & -2.93 & 4.50 \\ 0 & 0 & 1 & -4.43 \\ 0 & 0 & 0 & 1 \end{pmatrix} \quad \text{and} \quad \mathbf{D} = \begin{pmatrix} 41 & 0 & 0 & 0 \\ 0 & 51.51 & 0 & 0 \\ 0 & 0 & 40.76 & 0 \\ 0 & 0 & 0 & 17.89 \end{pmatrix}.$$

By making use of the matrix $\mathbf{Q}$, we find that (for $i = 1, 2, \ldots, 41$)

$$\begin{aligned} \mathrm{E}(y_i) &= (1, w_i, w_i^2, w_i^3)\mathbf{Q}\mathbf{Q}^{-1}(\beta_1, \beta_2, \beta_3, \beta_4)' \\ &= \gamma_1 + \gamma_2 s_1(w_i) + \gamma_3 s_2(w_i) + \gamma_4 s_3(w_i), \end{aligned}$$

where

$$
\begin{aligned}
s_1(w_i) &= w_i - 1.37, \\
s_2(w_i) &= w_i^2 - 2.93w_i + 0.89, \\
s_3(w_i) &= w_i^3 - 4.43w_i^2 + 4.50w_i - 0.25, \\
\gamma_1 &= \beta_1 + 1.37\beta_2 + 3.12\beta_3 + 7.95\beta_4, \\
\gamma_2 &= \beta_2 + 2.93\beta_3 + 8.52\beta_4, \\
\gamma_3 &= \beta_3 + 4.43\beta_4, \quad \text{and} \\
\gamma_4 &= \beta_4.
\end{aligned}
$$

Here, $s_1(\cdot)$, $s_2(\cdot)$, and $s_3(\cdot)$ are polynomials of degrees one, two, and three, respectively.

The nature of the polynomials $s_1(\cdot)$, $s_2(\cdot)$, and $s_3(\cdot)$ is such that when the model is "parameterized" in terms of γ_1, γ_2, γ_3, and γ_4 rather than in terms of β_1, β_2, β_3, and β_4 and an ANOVA is generated from the $M = 4$ null hypotheses $H_j^{(0)} : \gamma_j = 0$ ($j = 1, 2, 3, 4$) via an additional implementation of the approach described in Part 2 of Section 8.5e, the resultant degrees of freedom and sums of squares are the same as before.

EXERCISE 14. Take the setting to be that of Section 8.6, and adopt the notation and terminology employed therein.

(a) Devise a $(K+1)(d+1)$-dimensional column vector $\tilde{\mathbf{x}}(u)$ (that is functionally dependent on u) for which
$$\tilde{s}(u) = \tilde{\mathbf{x}}'(u)\boldsymbol{\beta} \quad \text{(for all } u\text{)},$$
where $\tilde{\mathbf{x}}'(u) = [\tilde{\mathbf{x}}(u)]'$.

(b) Let $\tilde{s}^{(1)}(\cdot), \tilde{s}^{(2)}(\cdot), \ldots, \tilde{s}^{(M)}(\cdot)$ represent any splines of the same degree d and with the same knots $\lambda_1, \lambda_2, \ldots, \lambda_K$, so that they are completely characterized by their $(K+1)(d+1) \times 1$ vectors of coefficients, say $\boldsymbol{\beta}^{(1)}, \boldsymbol{\beta}^{(2)}, \ldots, \boldsymbol{\beta}^{(M)}$. Making use of your solution to Part (a) (or other means), show that for any constants $a_1, a_2, \ldots, a_M$, the linear combination $\sum_{i=1}^M a_i \tilde{s}^{(i)}(\cdot)$ is a spline function of degree d with knots $\lambda_1, \lambda_2, \ldots, \lambda_K$ and with vector of coefficients $\sum_{i=1}^M a_i \boldsymbol{\beta}^{(i)}$. Note that there is an implication that the set consisting of spline functions of the same degree and with the same number and placement of knots constitutes what is known as a vector space (or linear space).

(c) Taking the matrices $\mathbf{X}$ and $\mathbf{A}$ to be as defined in Subsection a of Section 8.6 and taking the functionally dependent row vector $\tilde{\mathbf{x}}'(\cdot)$ to be the same as in Part (a), letting $R = \dim[\mathfrak{N}(\mathbf{A})]$, letting $\boldsymbol{\beta}^{(1)}, \boldsymbol{\beta}^{(2)}, \ldots, \boldsymbol{\beta}^{(M)}$ represent any $(K+1)(d+1)$-dimensional column vectors that span $\mathfrak{N}(\mathbf{A})$, letting $\mathbf{L} = (\boldsymbol{\beta}^{(1)}, \boldsymbol{\beta}^{(2)}, \ldots, \boldsymbol{\beta}^{(M)})$, and letting $\mathbf{W} = \mathbf{XL}$, show (1) that the ith row of $\mathbf{X}$ equals $\tilde{\mathbf{x}}'(u_i)$ ($i = 1, 2, \ldots, N$), (2) that
$$R = K + d + 1,$$
(3) that the constrained least squares estimator of the value of $\tilde{s}(u)$ corresponding to any value of u for which $\tilde{s}(u)$ is estimable equals $\tilde{\mathbf{x}}'(u)\mathbf{Lt}$, where $\mathbf{t}$ is any solution to the linear system $\mathbf{W}'\mathbf{Wt} = \mathbf{W}'\mathbf{y}$, and that the constrained least squares estimator is reexpressible in the form $\sum_{j=1}^M t_j \tilde{s}(u; j)$, where (for $j = 1, 2, \ldots, M$) t_j is the jth element of $\mathbf{t}$ and $\tilde{s}(u; j)$ equals $\tilde{\mathbf{x}}'(u)\beta^{(j)}$, (4) that if the values of $\tilde{s}(u)$ corresponding to any $d+1$ distinct values of u contained in the kth of the intervals defined by the knots $\lambda_1, \lambda_2, \ldots, \lambda_K$ are estimable, then $\beta_{k1}, \beta_{k2}, \ldots, \beta_{k,d+1}$ are all estimable [which implies that the value of $\tilde{s}(u)$ corresponding to every value of u in the kth of these intervals is estimable], (5) that
$$\text{rank}(\mathbf{W}) \leq K + d + 1,$$
with equality holding if and only if all $(K+1)(d+1)$ of the parameters β_{kj} ($k = 1, 2, \ldots, K+$

1; $j = 1, 2, \ldots, d+1$) are estimable, and (6) that if $M = R$ [so that $\boldsymbol{\beta}^{(1)}, \boldsymbol{\beta}^{(2)}, \ldots, \boldsymbol{\beta}^{(M)}$ form a basis for $\mathcal{N}(\mathbf{A})$] and if all $(K+1)(d+1)$ of the parameters β_{kj} ($k = 1, 2, \ldots, K+1$; $j = 1, 2, \ldots, d+1$) are estimable, then the linear system $\mathbf{W}'\mathbf{W}\mathbf{t} = \mathbf{W}'\mathbf{y}$ has a unique solution.

(d) What modifications need to be made to the results of Part (c) to make them applicable to natural splines [for which $d = 3$ and for which the constraint is $\mathbf{A}_{+}\boldsymbol{\beta} = \mathbf{0}$ rather than $\mathbf{A}\boldsymbol{\beta} = \mathbf{0}$].

(e) Let us add to the development in Part (d) pertaining to natural splines by taking $M = K$ and considering the effect of imposing on $\mathbf{L}$ [and hence on $\boldsymbol{\beta}^{(1)}, \boldsymbol{\beta}^{(2)}, \ldots, \boldsymbol{\beta}^{(K)}$] the additional constraint $\mathbf{UL} = \mathbf{I}$,where $\mathbf{U}$ is the $K \times 4(K+1)$ matrix with kth row $\tilde{\mathbf{x}}'(\boldsymbol{\lambda}_k)$. Show that there is a unique matrix $\mathbf{L}$ that satisfies this constraint as well as the constraint $\mathbf{A}_{+}\mathbf{L} = \mathbf{0}$ and that (for this choice of $\mathbf{L}$) the kth element of the solution $\mathbf{t}$ to the linear system $\mathbf{W}'\mathbf{W}\mathbf{t} = \mathbf{W}'\mathbf{y}$ is the least squares estimator of $\tilde{\mathbf{x}}'(\boldsymbol{\lambda}_k)\boldsymbol{\beta}$—the K splines $\tilde{s}(u; k) = \tilde{\mathbf{x}}'(u)\boldsymbol{\beta}^{(k)}$ ($k = 1, 2, \ldots, K$) are known as cardinal splines.

Solution. (a) For $k = 1, 2, \ldots, K+1$, let $\tilde{\mathbf{x}}_k(u)$ represent a $(d+1)$-dimensional column vector with transpose

$$\tilde{\mathbf{x}}_k'(u) = \begin{cases} (1, u, u^2, \ldots, u^d), & \text{if } u \in I_k, \\ \mathbf{0}, & \text{otherwise,} \end{cases}$$

where $I_1 = (-\infty, \lambda_1]$, $I_k = (\lambda_{k-1}, \lambda_k]$ ($k = 2, 3, \ldots, K$), and $I_{K+1} = (\lambda_K, \infty)$. And take

$$\tilde{\mathbf{x}}'(u) = [\tilde{\mathbf{x}}_1'(u), \tilde{\mathbf{x}}_2'(u), \ldots, \tilde{\mathbf{x}}_{K+1}'(u)].$$

(b) Making use of the solution presented herein to Part (a), we find that (for all u)

$$\textstyle\sum_{i=1}^M a_i \tilde{s}^{(i)}(u) = \sum_{i=1}^M a_i \tilde{\mathbf{x}}'(u)\boldsymbol{\beta}^{(i)} = \tilde{\mathbf{x}}'(u)\sum_{i=1}^M a_i \boldsymbol{\beta}^{(i)}.$$

(c) (1) As is evident from the expression for $\tilde{\mathbf{x}}(u)$ presented herein as the solution to Part (a),

$$\tilde{\mathbf{x}}'(u_i) = [\tilde{\mathbf{x}}_1'(u_i), \tilde{\mathbf{x}}_2'(u_i), \ldots, \tilde{\mathbf{x}}_{K+1}'(u_i)] = (\mathbf{x}_1', \mathbf{x}_2', \ldots, \mathbf{x}_{K+1}').$$

(2) Making use of Lemma 2.11.5, we find that

$$\dim[\mathcal{N}(\mathbf{A})] = (K+1)(d+1) - \text{rank}(\mathbf{A}) = (K+1)(d+1) - Kd = K + d + 1.$$

(3) Upon [in light of Part (a)] expressing $\tilde{s}(u)$ in the form $\tilde{s}(u) = \tilde{\mathbf{x}}'(u)\boldsymbol{\beta}$, it follows from result (1.115) that the constrained least squares estimator of the value of $\tilde{s}(u)$ [corresponding to any value of u for which $\tilde{s}(u)$ is estimable] equals $\tilde{\mathbf{x}}'(u)\mathbf{Lt}$. Moreover,

$$\tilde{\mathbf{x}}'(u)\mathbf{Lt} = \tilde{\mathbf{x}}'(u)\textstyle\sum_{j=1}^M t_j \boldsymbol{\beta}^{(j)} = \sum_{j=1}^M t_j \tilde{\mathbf{x}}'(u)\boldsymbol{\beta}^{(j)} = \sum_{j=1}^M t_j \tilde{s}(u; j).$$

(4) For $u \in \mathbf{I}_k$ (where $\mathbf{I}_k$ is the kth of the intervals defined by the knots $\lambda_1, \lambda_2, \ldots, \lambda_K$),

$$\tilde{s}(u) = \beta_{k1} + \beta_{k2}u + \cdots + \beta_{k,d+1}u^d.$$

Thus, it follows from the results of Section 5.3d (on the Vandermonde matrix) that if the values of $\tilde{s}(u)$ corresponding to any $d+1$ distinct values of $u \in \mathbf{I}_k$ are estimable, then all $d+1$ of the parameters $\beta_{k1}, \beta_{k2}, \ldots, \beta_{k,d+1}$ are estimable.

(5) According to result (1.121),

$$\text{rank}(\mathbf{W}) = \text{rank}\begin{pmatrix}\mathbf{X}\\ \mathbf{A}\end{pmatrix} - \text{rank}(\mathbf{A}).$$

Moreover, $\text{rank}\begin{pmatrix}\mathbf{X}\\ \mathbf{A}\end{pmatrix} \le (K+1)(d+1)$ with equality holding if and only if β_{kj} ($k = 1, 2, \ldots, K+1$; $j = 1, 2, \ldots, d+1$) are all estimable. And $\text{rank}(\mathbf{A}) = Kd$. Thus, $\text{rank}(\mathbf{W}) \le (K+1)(d+1) - Kd$, with equality holding if and only if β_{kj} ($k = 1, 2, \ldots, K+1$; $j = 1, 2, \ldots, d+1$) are all estimable. Or equivalently $\text{rank}(\mathbf{W}) \le K+d+1$, with equality holding if and only if β_{kj} $(k = 1,2,\ldots,K{+}1;\ j = 1,2,\ldots,d+1)$ are all estimable.

(6) By definition, the matrix $\mathbf{W}$ has M columns. Thus, when $M = R$, $\mathbf{W}$ has [in light of the result of Part (2)] $K{+}d{+}1$ columns. And when in addition β_{kj} $(k = 1,2,\ldots,K{+}1;\ j = 1,2,\ldots,d{+}1)$ are all estimable, we conclude [in light of the result of Part (5)] that $\mathbf{W}$ is of full column rank and hence that $\mathbf{W}'\mathbf{W}$ is nonsingular, in which case the linear system $\mathbf{W}'\mathbf{W}\mathbf{t} = \mathbf{W}'\mathbf{y}$ has a unique solution.

(d) Instead of taking $R = \dim[\mathcal{N}(\mathbf{A})]$ and taking $\boldsymbol{\beta}^{(1)}, \boldsymbol{\beta}^{(2)}, \ldots, \boldsymbol{\beta}^{(M)}$ to be vectors that span $\mathcal{N}(\mathbf{A})$, take $R = \dim[\mathcal{N}(\mathbf{A}_+)]$ and take $\boldsymbol{\beta}^{(1)}, \boldsymbol{\beta}^{(2)}, \ldots, \boldsymbol{\beta}^{(M)}$ to be [$4(K+1)$-dimensional] vectors that span $\mathcal{N}(\mathbf{A}_+)$. Then,

$$R = 4(K+1) - (3K+4) = K.$$

And when $\tilde{s}(\cdot)$ is a natural spline, whether or not the value of $\tilde{s}(u)$ corresponding to any particular value of u is estimable is determined by whether or not $\tilde{\mathbf{x}}'(u) \in \mathcal{R}\begin{pmatrix}\mathbf{X}\\ \mathbf{A}_+\end{pmatrix}$ rather than by whether or not $\tilde{\mathbf{x}}'(u) \in \mathcal{R}\begin{pmatrix}\mathbf{X}\\ \mathbf{A}\end{pmatrix}$. Moreover,

$$\text{rank}(\mathbf{W}) = \text{rank}\begin{pmatrix}\mathbf{X}\\ \mathbf{A}_+\end{pmatrix} - \text{rank}(\mathbf{A}_+) \le 4(K+1) - (3K+4) = K,$$

with equality holding in the inequality if and only if all $4(K+1)$ of the parameters β_{kj} $(k = 1,2,\ldots,K+1;\ j = 1,2,3,4)$ are estimable; if in addition $M = K$, then the solution to the linear system $\mathbf{W}'\mathbf{W}\mathbf{t} = \mathbf{W}'\mathbf{y}$ is unique..

(e) The rows of the $(4K+4) \times (4K+4)$ partitioned matrix $\begin{pmatrix}\mathbf{U}\\ \mathbf{A}_+\end{pmatrix}$ are linearly independent, as can be readily verified. Thus, $\begin{pmatrix}\mathbf{U}\\ \mathbf{A}_+\end{pmatrix}$ is nonsingular and the two constraints $\mathbf{U}\mathbf{L} = \mathbf{I}$ and $\mathbf{A}_+\mathbf{L} = \mathbf{0}$ can be simultaneously (and uniquely) satisfied by taking $\mathbf{L} = \begin{pmatrix}\mathbf{U}\\ \mathbf{A}_+\end{pmatrix}^{-1}\begin{pmatrix}\mathbf{I}\\ \mathbf{0}\end{pmatrix}$, so the least squares estimator of $\tilde{\mathbf{x}}'(\lambda_k)\boldsymbol{\beta}$ equals $\tilde{\mathbf{x}}'(\lambda_k)\mathbf{L}\mathbf{t}$, which is the kth element of $\begin{pmatrix}\mathbf{U}\\ \mathbf{A}_+\end{pmatrix}\mathbf{L}\mathbf{t} = \begin{pmatrix}\mathbf{I}\\ \mathbf{0}\end{pmatrix}\mathbf{t}$ and hence the kth element of $\mathbf{t}$.

EXERCISE 15. Use the tree-height data of Section 4.4h to further illustrate the results of Section 8.5e [on K- or $(K-1)$-line ANOVA tables]. For purposes of doing so, let $y = \log(\text{tree height})$ and $u = \log(\text{tree diameter})$. Further, regard the 101 values of y as the observed values of random variables $y_1, y_2, \ldots, y_{101}$ that follow a G–M model with

$$\mathrm{E}(y_t) = \beta_1 + \beta_2 u_t + \beta_3 u_t^2 + \beta_4 u_t^3 + \beta_5 u_t^4 + \beta_6 u_t^5 \qquad (t = 1,2,\ldots,101),$$

where $u_1, u_2, \ldots, u_{101}$ are the corresponding values of u—assume (for purposes of this exercise) that (as in the use of these data in Section 8.6b) the information on the relative locations of the trees is to be disregarded. Subject these data to the same kinds of procedures that were applied to the ouabain data in Part 5 of Section 8.5e (and in Exercise 13).

Solution. When there are $L = 6$ segments (consisting of one column each), the ANOVA associated with the partitioning of the model matrix $\mathbf{X}$ into L segments (as in Part 1 of Section 8.5e) is the ANOVA with degrees of freedom and sums of squares as follows:

Source	df	SS
β_1	1	877.6203
β_2 after β_1	1	15.9139
β_3 after β_1 and β_2	1	0.3836
β_4 after β_1, β_2, and β_3	1	0.0874
β_5 after β_1, β_2, β_3, and β_4	1	0.0205
β_6 after β_1, β_2, β_3, β_4, and β_5	1	0.0680
Residual	95	2.1783
Total	101	896.2721

An ANOVA with the same degrees of freedom and sums of squares can be generated from the $M = 6$ null hypotheses $H_j^{(0)} : \beta_j = 0$ $(j = 1, 2, \ldots, 6)$ via an implementation of the approach described in Part 2 of Section 8.5e.

It remains to consider the approach to the generation of an ANOVA described in Part 4 of Section 8.5e. As a unit upper triangular matrix $\mathbf{Q}$ and a diagonal matrix $\mathbf{D}$ for which

$$\mathbf{Q}'\mathbf{X}'\mathbf{X}\mathbf{Q} = \mathbf{D},$$

we obtain

$$\mathbf{Q} = \begin{pmatrix} 1 & -3.43 & 11.08 & -40.12 & 139.56 & -493.18 \\ 0 & 1 & -6.79 & 36.12 & -167.55 & 737.53 \\ 0 & 0 & 1 & -10.54 & 73.85 & -433.73 \\ 0 & 0 & 0 & 1 & -14.17 & 125.43 \\ 0 & 0 & 0 & 0 & 1 & -17.85 \\ 0 & 0 & 0 & 0 & 0 & 1 \end{pmatrix}$$

and

$$\mathbf{D} = \begin{pmatrix} 101 & 0 & 0 & 0 & 0 & 0 \\ 0 & 45.05 & 0 & 0 & 0 & 0 \\ 0 & 0 & 18.33 & 0 & 0 & 0 \\ 0 & 0 & 0 & 10.31 & 0 & 0 \\ 0 & 0 & 0 & 0 & 5.05 & 0 \\ 0 & 0 & 0 & 0 & 0 & 1.62 \end{pmatrix}.$$

By making use of the matrix $\mathbf{Q}$, we find that (for $t = 1, 2, \ldots, 101$)

$$\begin{aligned} \mathrm{E}(y_t) &= (1, u_t, u_t^2, u_t^3, u_t^4, u_t^5)\mathbf{Q}\mathbf{Q}^{-1}(\beta_1, \beta_2, \beta_3, \beta_4, \beta_5, \beta_6)' \\ &= \gamma_1 + \gamma_2 s_1(u_t) + \gamma_3 s_2(u_t) + \gamma_4 s_3(u_t) + \gamma_5 s_4(u_t) + \gamma_6 s_5(u_t), \end{aligned}$$

where

$$\begin{aligned} s_1(u_t) &= u_t - 3.43, \\ s_2(u_t) &= u_t^2 - 6.79u_t + 11.08, \\ s_3(u_t) &= u_t^3 - 10.54u_t^2 + 36.12u_t - 40.12, \\ s_4(u_t) &= u_t^4 - 14.17u_t^3 + 73.85u_t^2 - 167.55u_t + 139.56, \\ s_5(u_t) &= u_t^5 - 17.85u_t^4 + 125.43u_t^3 - 433.73u_t^2 + 737.53u_t - 493.18, \\ \gamma_1 &= \beta_1 + 3.43\beta_2 + 12.22\beta_3 + 44.95\beta_4 + 170.02\beta_5 + 658.38\beta_6, \\ \gamma_2 &= \beta_2 + 6.79\beta_3 + 35.43\beta_4 + 168.21\beta_5 + 765.67\beta_6, \\ \gamma_3 &= \beta_3 + 10.54\beta_4 + 75.49\beta_5 + 459.14\beta_6, \\ \gamma_4 &= \beta_4 + 14.17\beta_5 + 127.47\beta_6, \\ \gamma_5 &= \beta_5 + 17.85\beta_6, \quad \text{and} \\ \gamma_6 &= \beta_6. \end{aligned}$$

Here, $s_1(\cdot)$, $s_2(\cdot)$, $s_3(\cdot)$, $s_4(\cdot)$, and $s_5(\cdot)$are polynomials of degrees one, two, three, four, and five, respectively, and are orthogonal with respect to the inner product defined by the quadratic form with matrix $\mathbf{X}'\mathbf{X}$.

The nature of the polynomials $s_1(\cdot)$, $s_2(\cdot)$, $s_3(\cdot)$, $s_4(\cdot)$, and $s_5(\cdot)$ is such that when the model is "parameterized" in terms of $\gamma_1, \gamma_2, \gamma_3, \gamma_4, \gamma_5$, and γ_6 rather than in terms of $\beta_1, \beta_2, \beta_3, \beta_4, \beta_5$, and β_6 and an ANOVA is generated from the $M = 6$ null hypotheses $H_j^{(0)} : \gamma_j = 0$ $(j = 1, 2, \ldots, 6)$ via an additional implementation of the approach described in Part 2 of Section 8.5e, the resultant degrees of freedom and sums of squares are the same as before.

EXERCISE 16. Consider the two "multiple regression" models defined implicitly by expressions (4.2.10) and (4.2.11). Show that these two models can be regarded as reparameterizations of each other. Do so by regarding one of these models as the original model with model matrix $\mathbf{X}$ and the other as the alternative model with model matrix $\mathbf{W}$ and by finding matrices $\mathbf{F}$ and $\mathbf{G}$ that satisfy condition (7.22).

Solution. Adopting the notation and terminology employed in Section 8.7a, regard the model defined by expression (4.2.10) as the original model (with model matrix $\mathbf{X}$) and the model defined by expression (4.2.11) as the alternative model (with model matrix $\mathbf{W}$). Then, letting (for $j = 1, 2, \ldots, C$) $\mathbf{u}_j$ represent an N-dimensional column vector whose elements are the N values of the jth explanatory variable u_j (and letting $\mathbf{1}$ represent an N-dimensional column vector of 1's), the model matrix $\mathbf{X}$ is expressible as

$$\mathbf{X} = (\mathbf{1},\ \mathbf{u}_1 - a_1\mathbf{1},\ \mathbf{u}_2 - a_2\mathbf{1}, \ldots,\ \mathbf{u}_C - a_C\mathbf{1})$$

and the model matrix $\mathbf{W}$ as

$$\mathbf{W} = (\mathbf{1},\ \mathbf{u}_1,\ \mathbf{u}_2, \ldots,\ \mathbf{u}_C).$$

And upon taking

$$\mathbf{F} = \begin{pmatrix} 1 & a_1 & a_2 & \ldots & a_C \\ 0 & 1 & 0 & \ldots & 0 \\ 0 & 0 & 1 & & 0 \\ \vdots & \vdots & & & \\ 0 & 0 & 0 & & 1 \end{pmatrix} \quad \text{and} \quad \mathbf{G} = \begin{pmatrix} 1 & -a_1 & -a_2 & \ldots & -a_C \\ 0 & 1 & 0 & \ldots & 0 \\ 0 & 0 & 1 & & 0 \\ \vdots & \vdots & & & \\ 0 & 0 & 0 & & 1 \end{pmatrix},$$

we find that

$$\mathbf{W} = \mathbf{XF} \quad \text{and} \quad \mathbf{X} = \mathbf{WG}.$$

EXERCISE 17. Take the setting to be that of Section 8.7a, and adopt the notation and terminology employed therein. Suppose that the alternative model (with model equation $\mathbf{y} = \mathbf{W}\boldsymbol{\tau} + \mathbf{e}$ and $N \times R$ model matrix $\mathbf{W}$) is a reparameterization of the original model (with model equation $\mathbf{y} = \mathbf{X}\boldsymbol{\beta} + \mathbf{e}$ and $N \times P$ model matrix $\mathbf{X}$). And take the matrix $\mathbf{G}$ to be as defined by condition (7.22), that is, to be an $R \times P$ matrix for which $\mathbf{X} = \mathbf{WG}$.

(a) Show that $\mathcal{R}(\mathbf{X}) \subset \mathcal{R}(\mathbf{G})$.

(b) Show that in the special case where the alternative model is a full-rank reparameterization of the original model, (1) $\mathcal{R}(\mathbf{X}) = \mathcal{R}(\mathbf{G})$ and (2) a linear combination $\boldsymbol{\lambda}'\boldsymbol{\beta}$ of the elements of $\boldsymbol{\beta}$ is estimable if and only if $\boldsymbol{\lambda}' \in \mathcal{R}(\mathbf{G})$.

Solution. (a) That $\mathcal{R}(\mathbf{X}) \subset \mathcal{R}(\mathbf{G})$ is an immediate consequence of Lemma 2.4.3.

(b) Suppose that the alternative model is a full-rank reparameterization of the original model. Then, (1) $\operatorname{rank}\mathbf{X} = \operatorname{rank}\mathbf{G}$, as is evident upon observing that

$$\operatorname{rank}\mathbf{X} = \operatorname{rank}(\mathbf{WG}) \le \operatorname{rank}\mathbf{G} \le R = \operatorname{rank}\mathbf{X},$$

leading [in light of Theorem 2.4.16 and the result of Part (a)] to the conclusion that $\mathcal{R}(\mathbf{X}) = \mathcal{R}(\mathbf{G})$. And (2) upon recalling that $\boldsymbol{\lambda}'\boldsymbol{\beta}$ is estimable if and only if $\boldsymbol{\lambda}' \in \mathcal{R}(\mathbf{X})$, Part (2) follows from Part (1).

EXERCISE 18. Take the setting to be that of Section 8.7a, and adopt the notation and terminology employed therein. Further, let $\boldsymbol{\Lambda}'\boldsymbol{\beta}$ represent a vector of linearly independent linear combinations of the elements of the vector $\boldsymbol{\beta}$ that are estimable and that are equal in number to $\text{rank}(\mathbf{X})$. And suppose that the model matrix $\mathbf{W}$ of the alternative model is that for which $\mathbf{W} = \mathbf{X}\boldsymbol{\Lambda}(\boldsymbol{\Lambda}'\boldsymbol{\Lambda})^{-1}$.

(a) Show that the alternative model is a full-rank reparameterization of the original model and find matrices $\mathbf{F}$ and $\mathbf{G}$ that satisfy condition (7.22).

(b) Show that the least squares estimator of $\boldsymbol{\tau}$ equals the least squares estimator of $\boldsymbol{\Lambda}'\boldsymbol{\beta}$.

Solution. (a) Since the elements of $\boldsymbol{\Lambda}'\boldsymbol{\beta}$ are estimable,

$$\boldsymbol{\Lambda}' = \mathbf{A}'\mathbf{X} \text{ for some matrix } \mathbf{A};$$

and since the rows of $\boldsymbol{\Lambda}'$ are linearly independent and equal in number to $\text{rank}(\mathbf{X})$,

$$\text{rank}(\boldsymbol{\Lambda}') = \text{rank}(\mathbf{X}).$$

And upon recalling Corollary 2.4.17, it follows that

$$\mathcal{R}(\boldsymbol{\Lambda}') = \mathcal{R}(\mathbf{X})$$

and hence that

$$\mathbf{X} = \mathbf{K}\boldsymbol{\Lambda}' \text{ for some matrix } \mathbf{K},$$

so that

$$\mathbf{X} = \mathbf{K}\boldsymbol{\Lambda}' = \mathbf{K}\boldsymbol{\Lambda}'\boldsymbol{\Lambda}(\boldsymbol{\Lambda}'\boldsymbol{\Lambda})^{-1}\boldsymbol{\Lambda}' = \mathbf{X}\boldsymbol{\Lambda}(\boldsymbol{\Lambda}'\boldsymbol{\Lambda})^{-1}\boldsymbol{\Lambda}' = \mathbf{W}\boldsymbol{\Lambda}'.$$

Thus, condition (7.22) can be satisfied by taking

$$\mathbf{F} = \boldsymbol{\Lambda}(\boldsymbol{\Lambda}'\boldsymbol{\Lambda})^{-1} \quad \text{and} \quad \mathbf{G} = \boldsymbol{\Lambda}'.$$

It remains only to observe that when condition (7.22) is satisfied, the alternative model is a reparameterization of the original model, and that when in addition the number R of columns of the model matrix $\mathbf{W}$ of the alternative model equals $\text{rank}(\mathbf{X})$, the alternative model is a full-rank reparameterization.

(b) Let $\tilde{\mathbf{t}}$ represent the solution to the normal equations (7.27) for the alternative model—since $\mathbf{W}$ is of full column rank, the solution is unique. Then, upon applying result (7.28) and making use of the result of Part (a), we find that a solution $\tilde{\mathbf{b}}$ to the normal equations (7.26) for the original model is obtainable by taking

$$\tilde{\mathbf{b}} = \mathbf{F}\tilde{\mathbf{t}} = \boldsymbol{\Lambda}(\boldsymbol{\Lambda}'\boldsymbol{\Lambda})^{-1}\tilde{\mathbf{t}};$$

and it follows that the least squares estimator of $\boldsymbol{\Lambda}'\boldsymbol{\beta}$ equals

$$\boldsymbol{\Lambda}'\tilde{\mathbf{b}} = \boldsymbol{\Lambda}'\boldsymbol{\Lambda}(\boldsymbol{\Lambda}'\boldsymbol{\Lambda})^{-1}\tilde{\mathbf{t}} = \tilde{\mathbf{t}},$$

which is the least squares estimator of $\boldsymbol{\tau}$.

EXERCISE 19. Suppose that $\mathbf{y}$ is an $N \times 1$ observable random vector. And suppose that $\mathbf{y}$ follows a G–M model for which the $N \times P$ model matrix $\mathbf{X}$ is expressible in the form $\mathbf{X} = (\mathbf{1}, \mathbf{X}_1, \mathbf{X}_2)$, where for some number a $(0 < a < 1)$ and some integer C $(2 \le C \le P-2)$ $\mathbf{X}_1$ has C columns that sum to $a\mathbf{1}$ and $\mathbf{X}_2$ $P-C-1$ columns that sum to $(1-a)\mathbf{1}$. Suppose also that $\text{rank}(\mathbf{X}) = P-2$. A G–M model with these characteristics might be encountered in the analysis of mixture data of the kind considered by Cornell (2002) in his Section 4.14.

(a) Show that for a linear combination $\sum_{j=1}^{P} \lambda_j \beta_j$ of the elements $\beta_1, \beta_2, \ldots, \beta_P$ of $\boldsymbol{\beta}$ to be estimable, it is necessary and sufficient that (1) $\lambda_1 = \sum_{j=2}^{P} \lambda_j$ and (2) $a \sum_{j=C+2}^{P} \lambda_j = (1-a) \sum_{j=2}^{C+1} \lambda_j$.

(b) Show that each of the two quantities, β_1 and $(a/C) \sum_{j=2}^{C+1} \beta_j + [(1-a)/(P-C-1)] \sum_{j=C+2}^{P} \beta_j$, is nonestimable but that the two of them are not jointly nonestimable.

(c) Show that the two quantities, $\sum_{j=2}^{C+1} \beta_j$ and $\sum_{j=C+2}^{P} \beta_j$, are jointly nonestimable.

(d) Let $\mathbf{B}_1$ represent a $(C-1)\times C$ matrix whose rows are the last $C-1$ rows of a $C\times C$ orthogonal matrix (such as a $C\times C$ Helmert matrix) whose first row is $C^{-1/2}\mathbf{1}'_C$ and $\mathbf{B}_2$ a $(P-C-2)\times(P-C-1)$ matrix whose rows are the last $P-C-2$ rows of a $(P-C-1)\times(P-C-1)$ orthogonal matrix whose first row is $(P-C-1)^{-1/2}\mathbf{1}'_{P-C-1}$. And let $\mathbf{U} = \mathbf{X}\mathbf{B}$, where $\mathbf{B} = \begin{pmatrix} 1 & \mathbf{0} & \mathbf{0} \\ \mathbf{0} & \mathbf{B}'_1 & \mathbf{0} \\ \mathbf{0} & \mathbf{0} & \mathbf{B}'_2 \end{pmatrix}$. Show that the unconstrained G–M model

$$\mathbf{y} = \mathbf{U}\boldsymbol{\alpha} + \mathbf{e}$$

[with model matrix $\mathbf{U}$ and in which $\boldsymbol{\alpha}$ is an (unconstrained) parameter vector] is a reparameterization of the constrained G–M model with model matrix $\mathbf{X}$ and with constraints $\sum_{j=2}^{C+1} \beta_j = 0$ and $\sum_{j=C+2}^{P} \beta_j = 0$ and is also a reparameterization of the unconstrained G–M model with model matrix $\mathbf{X}$.

(e) Letting $\mathbf{b}_j$ represent the jth column of the $P\times(P-2)$ matrix $\mathbf{B}$ [defined in Part (d)], show that for $j = 2, 3, \ldots, P-2$ or for $j = 1, 2, \ldots, P-2$ (depending on whether the model is the unconstrained G–M model with model matrix $\mathbf{X}$ or the constrained G–M model with model matrix $\mathbf{X}$ and constraints $\sum_{j=2}^{C+1} \beta_j = 0$ and $\sum_{j=C+2}^{P} \beta_j = 0$), $\mathbf{b}'_j\boldsymbol{\beta}$ is estimable and its unconstrained or constrained least squares estimator equals the (unconstrained) least squares estimator of α_j (as determined under the unconstrained G–M model $\mathbf{y} = \mathbf{U}\boldsymbol{\alpha} + \mathbf{e}$), where α_j is the jth element of $\boldsymbol{\alpha}$.

Solution. (a) Let $\mathbf{k}_1 = \begin{pmatrix} 1 \\ -\mathbf{1}_C \\ -\mathbf{1}_{P-C-1} \end{pmatrix}$ and $\mathbf{k}_2 = \begin{pmatrix} 0 \\ (1-a)\mathbf{1}_C \\ -a\mathbf{1}_{P-C-1} \end{pmatrix}$. Clearly, $\mathbf{k}_1$ and $\mathbf{k}_2$ are contained in $\mathcal{N}(\mathbf{X})$ and are linearly independent. Moreover, in light of Lemma 2.11.5,

$$\dim[\mathcal{N}(\mathbf{X})] = P - \text{rank}(\mathbf{X}) = 2.$$

Thus, for $\sum_{j=1}^{P} \lambda_j\beta_j$ to be estimable, it follows from the results of Section 5.3b that it is necessary and sufficient that the vector $\boldsymbol{\lambda} = (\lambda_1, \lambda_2, \ldots, \lambda_P)'$ satisfy the two conditions $\mathbf{k}'_1\boldsymbol{\lambda} = 0$ and $\mathbf{k}'_2\boldsymbol{\lambda} = 0$. Clearly, these two conditions are equivalent to the two conditions $\lambda_1 = \sum_{j=2}^{P} \lambda_j$ and $a\sum_{j=C+2}^{P} \lambda_j = (1-a)\sum_{j=2}^{C+1} \lambda_j$.

(b) Clearly, the first of these two quantities is expressible as $\sum_{j=1}^{P} \lambda_j\beta_j$, where $\lambda_1 = 1$ and $\lambda_2 = \lambda_3 = \cdots = \lambda_P = 0$, and the second is expressible as $\sum_{j=1}^{P} \lambda_j\beta_j$, where $\lambda_1 = 0$, $\lambda_2 = \lambda_3 = \cdots = \lambda_{C+1} = a/C$, and $\lambda_{C+2} = \lambda_{C+3} = \cdots = \lambda_P = (1-a)/(P-C-1)$. And we find that each of these two quantities satisfies condition (2) of Part (a), but does not satisfy condition (1), so that each of them is nonestimable. However, the sum of the two quantities satisties both condition (1) and condition (2), as can be readily verified. Thus, the sum is estimable, which establishes the existence of a nontrivial linear combination of the two quantities that is estimable and leads to the conclusion that the two quantities are not jointly nonestimable.

(c) Clearly, the first of these two quantities is expressible as $\sum_{j=1}^{P} \lambda_j\beta_j$, where $\lambda_1 = 0$, $\lambda_2 = \lambda_3 = \cdots = \lambda_{C+1} = 1$, and $\lambda_{C+2} = \lambda_{C+3} = \cdots = \lambda_P = 0$, and the second is expressible as $\sum_{j=1}^{P} \lambda_j\beta_j$, where $\lambda_1 = \lambda_2 = \cdots = \lambda_{C+1} = 0$ and $\lambda_{C+2} = \lambda_{C+3} = \cdots = \lambda_P = 1$. And a linear combination of these two quantities, say $\ell_1\sum_{j=2}^{C+1} \beta_j + \ell_2\sum_{j=C+2}^{P} \beta_j$, is expressible as $\sum_{j=1}^{P} \lambda_j\beta_j$, where $\lambda_1 = 0$, $\lambda_2 = \lambda_3 = \cdots = \lambda_{C+1} = \ell_1$, and $\lambda_{C+2} = \lambda_{C+3} = \cdots = \lambda_P = \ell_2$.

Neither $\sum_{j=2}^{C+1} \beta_j$ nor $\sum_{j=C+2}^{P} \beta_j$ satisfies condition (1) of Part (a) [or condition (2)], so that each of these quantities is nonestimable. The linear combination $\ell_1\sum_{j=2}^{C+1} \beta_j + \ell_2\sum_{j=C+2}^{P} \beta_j$ satisfies condition (1) if and only if $(P-C-1)\ell_2 = -C\ell_1$ and satisfies condition (2) if and only if $a(P-C-1)\ell_2 = (1-a)C\ell_1$. Thus, $\ell_1\sum_{j=2}^{C+1} \beta_j + \ell_2\sum_{j=C+2}^{P} \beta_j$ satisfies both of these conditions if and only if $\ell_2 = \ell_1 = 0$ and hence is estimable if and only if $\ell_2 = \ell_1 = 0$, leading to the conclusion that $\sum_{j=2}^{C+1} \beta_j$ and $\sum_{j=C+2}^{P} \beta_j$ are jointly nonestimable.

(d) Clearly, the constraints $\sum_{j=2}^{C+1} \beta_j = 0$ and $\sum_{j=C+2}^{P} \beta_j = 0$ in the constrained G–M model are reexpressible in the form

$$\mathbf{A}\boldsymbol{\beta} = \mathbf{0},$$

where

$$\mathbf{A} = \begin{pmatrix} 0 & \mathbf{1}'_C & \mathbf{0} \\ 0 & \mathbf{0} & \mathbf{1}'_{P-C-1} \end{pmatrix}.$$

Moreover,

$$\mathbf{AB} = \mathbf{0},$$

so that all P–2 columns of the matrix $\mathbf{B}$ are contained in $\mathcal{N}(\mathbf{A})$. And upon observing that the columns of $\mathbf{B}$ are orthonormal and hence linearly independent and upon observing (in light of Lemma 2.11.5) that

$$\dim[\mathcal{N}(\mathbf{A})] = P - \text{rank}(\mathbf{A}) = P - 2,$$

it is evident that the columns of $\mathbf{B}$ span $\mathcal{N}(\mathbf{A})$. Thus, it follows from the results of Section 8.7b that the unconstrained G–M model $\mathbf{y} = \mathbf{U}\boldsymbol{\alpha} + \mathbf{e}$ is a reparameterization of the constrained G–M model with model matrix $\mathbf{X}$ and with constraints $\sum_{j=2}^{C+1} \beta_j = 0$ and $\sum_{j=C+2}^{P} \beta_j = 0$.

That $\sum_{j=2}^{C+1} \beta_j$ and $\sum_{j=C+2}^{P} \beta_j$ are jointly nonestimable [as was established in Part (c)] implies (in light of the results of Section 8.2a) that $\mathcal{R}(\mathbf{X})$ and $\mathcal{R}(\mathbf{A})$ are essentially disjoint. Thus, it follows from result (7.16) that $\mathcal{C}(\mathbf{U}) = \mathcal{C}(\mathbf{X})$, leading to the conclusion that the unconstrained G–M model $\mathbf{y} = \mathbf{U}\boldsymbol{\alpha} + \mathbf{e}$ is a reparameterization of the unconstrained G–M model with model matrix $\mathbf{X}$.

(e) Suppose that the G–M model (with model matrix $\mathbf{X}$) were subject to the constraints $\sum_{j=2}^{C+1} \beta_j = 0$ and $\sum_{j=C+2}^{P} \beta_j = 0$. Then, every linear combination $\boldsymbol{\lambda}'\boldsymbol{\beta}$ of the elements of $\boldsymbol{\beta}$ is estimable, as is evident upon observing that $\boldsymbol{\lambda}'\boldsymbol{\beta}$ is estimable if $\boldsymbol{\lambda}' \in \mathcal{R}\begin{pmatrix} \mathbf{X} \\ \mathbf{A} \end{pmatrix}$ [where $\mathbf{A}$ is as defined in the solution to Part (d)] and upon observing [in light of result (2.2)] that

$$\text{rank}\begin{pmatrix} \mathbf{X} \\ \mathbf{A} \end{pmatrix} = \text{rank}(\mathbf{X}) + \text{rank}(\mathbf{A}) = P - 2 + 2 = P.$$

Thus, $\mathbf{b}'_j\boldsymbol{\beta}$ is estimable for $j = 1, 2, \ldots, P$–2. And since [as established in Part (d)] the unconstrained G–M model $\mathbf{y} = \mathbf{U}\boldsymbol{\alpha} + \mathbf{e}$ is a reparameterization of the constrained G–M model, it follows from the results of Section 8.7b that (for $j = 1, 2, \ldots, P-2$) the constrained least squares estimator of $\mathbf{b}'_j\boldsymbol{\beta}$ equals the (unconstrained) least squares estimator of $\mathbf{b}'_j\mathbf{B}\boldsymbol{\alpha}$. Moreover, $\mathbf{b}'_j\mathbf{B}\boldsymbol{\alpha} = \alpha_j$.

Alternatively, suppose that the model is the unconstrained G–M model with model matrix $\mathbf{X}$. And observe that $\mathbf{b}'_1\boldsymbol{\beta} = \beta_1$, which [as noted in Part (b)] is nonestimable. Observe also that while $\mathbf{b}'_j\boldsymbol{\beta}$ is not estimable for $j = 1$, it is estimable for $j = 2, 3, \ldots, P-2$, as can be verified by letting b_{ij} represent the ith element of $\mathbf{b}_j$ and by noting that $b_{1j} = 0$, $\sum_{i=2}^{C+1} b_{ij} = 0$, and $\sum_{i=C+2}^{P} b_{ij} = 0$—for $j = 2, 3, \ldots, C$, the vector $(b_{2j}, b_{3j}, \ldots, b_{C+1,j})'$ is orthogonal to the vector $\mathbf{1}_C$ and for $j = C+1, C+2, \ldots, P-2$, the vector $(b_{C+2,j}, b_{C+3,j}, \ldots, b_{Pj})'$ is orthogonal to the vector $\mathbf{1}_{P-C-1}$—and by then making use of conditions (1) and (2) of Part (a). And since [as established in Part (d)] the unconstrained G–M model $\mathbf{y} = \mathbf{U}\boldsymbol{\alpha} + \mathbf{e}$ is a reparameterization of the unconstrained G–M model with model matrix $\mathbf{X}$ (and since $\mathbf{U} = \mathbf{XB}$), it follows from the results of Section 8.7a that (for $j = 2, 3, \ldots, P-2$) the unconstrained least squares estimator of $\mathbf{b}'_j\boldsymbol{\beta}$ equals the (unconstrained) least squares estimator of $\mathbf{b}'_j\mathbf{B}\boldsymbol{\alpha}$ $(= \alpha_j)$.

Additional References

Escobar, L. A., and Skarpness, B. (1986), "The Bias of the Least Squares Estimator Over Interval Constraints," *Economics Letters*, 20, 331–335.

Griffiths, W. E., and Hill, R. C. (2022), "On the Power of the F-Test for Hypotheses in a Linear Model," *The American Statistician*, 76, 78–84.

Harrell, F. E., Jr. (2015), *Regression Modeling Strategies: With Applications to Linear Models, Logistic and Ordinal Regression, and Survival Analysis (2nd ed.)*, New York: Springer.

Hocking, R. R. (1985), *The Analysis of Linear Models*, Monterey, CA: Brooks/Cole.

Magnus, J. R., and Neudecker, H. (1988), *Matrix Differential Calculus With Applications in Statistics and Econometrics*, New York: Wiley.

Ohtani, K. (1987), "The MSE of the Least Squares Estimator Over an Interval Constraint," *Economics Letters*, 25, 351–354.

Perperoglou, A., Sauerbrei, W., Abrahamowicz, M., and Schmid, M. (2019), "A Review of Spline Function Procedures in R," *BMC Medical Research Methodology*, 19:46.

Printed in the United States
by Baker & Taylor Publisher Services